*energies*

BioEnergy and BioChemicals Production from Biomass and Residual Resources

Edited by
Dimitar Karakashev and Yifeng Zhang

Printed Edition of the Special Issue Published in *Energies*

www.mdpi.com/journal/energies

 MDPI

BioEnergy and BioChemicals Production from Biomass and Residual Resources

BioEnergy and BioChemicals Production from Biomass and Residual Resources

Special Issue Editors

Dimitar Karakashev
Yifeng Zhang

MDPI • Basel • Beijing • Wuhan • Barcelona • Belgrade

MDPI

Special Issue Editors
Dimitar Karakashev
Danish Technological Institute
Denmark

Yifeng Zhang
Technical University of Denmark
Denmark

Editorial Office
MDPI
St. Alban-Anlage 66
Basel, Switzerland

This is a reprint of articles from the Special Issue published online in the open access journal *Energies* (ISSN 1996-1073) in 2018 (available at: http://www.mdpi.com/journal/energies/special_issues/ BioEnergy_BioChemicals)

For citation purposes, cite each article independently as indicated on the article page online and as indicated below:

LastName, A.A.; LastName, B.B.; LastName, C.C. Article Title. *Journal Name* **Year**, *Article Number,* Page Range.

ISBN 978-3-03897-214-3 (Pbk)
ISBN 978-3-03897-215-0 (PDF)

Contents

About the Special Issue Editors

Dimitar Karakashev, PhD, is an academic officer at Aalborg University, Denmark, and is responsible for scientific and administrative quality of project funding proposals. He was employed as a Senior Project Manager at Danish Technological Institute, and as a Senior Researcher at Technical University of Denmark. His research is in the area of applied microbiology and biotechnology for biofuels, biochemicals and biorefinery. He has 46 ISI publications and an H-index of 24.

Yifeng Zhang is an Associate Professor at the Department of Environmental Engineering, Technical University of Denmark. He is now leading the Microbial Environmental Electrochemistry group in this department. His research regarding microbial electrochemistry has been innovative and has led to a number of demonstrations of cutting-edge technologies including microbial electrosynthesis, smart wastewater treatment, energy-gained resources recovery, and novel biosensors for environmental monitoring. So far, he has about 50 ISI scientific publications (as first/corresponding author) and two patents, with an H-index of 21. He has also been invited to speak at several international conferences.

Editorial

BioEnergy and BioChemicals Production from Biomass and Residual Resources

Dimitar Karakashev [1,*,†] **and Yifeng Zhang** [2,*]

1   Danish Technological Institute, Biomass and Biorefinery, 2630 Taastrup, Denmark
2   Department of Environmental Engineering, Technical University of Denmark, 2800 Kgs. Lyngby, Denmark
*   Correspondence: mitac_bg@yahoo.com (D.K.); yifz@env.dtu.dk (Y.Z.)
†   Current address: Snogegårdsvej 139 D, 2860 Søborg, Denmark.

Received: 17 July 2018; Accepted: 13 August 2018; Published: 15 August 2018

1. Introduction

Research and technology developments in bioenergy and biochemical production systems are of the utmost important for the development of next generation, highly efficient biomass conversion concepts maximizing the total energy and chemical output. The utilization of non-conventional biomasses and unexploited residual resources (e.g., agriculture and agroindustry wastes), innovative solutions for online monitoring and process control, novel biochemical pathways, microbial platforms and reactor technologies are key issues to be addressed. Though conventional technologies are constantly developing and novel processes are continually emerging, major challenges have still to be solved, such as the design of high performance and cost-effective technologies for the production of bioenergy (gaseous, liquid, sold biofuels, heat, renewable electricity) and biochemicals from residual resources in a biorefinery concept, where the potential of the biomass and residual waste streams is fully valorized. In this context, evaluation of the environmental, technological, economical, and social sustainability of the concepts developed are of extreme importance. The main objective of this Special Issue is, hence, to provide cost-effective and technologically sound solutions for next generation bioenergy and biochemical production systems.

The particular topics of interest in the original call for papers included, but were not limited to:

- Novel and unexploited residual resources for next generation biorefineries
- New emerging bioenergy and biochemicals production technologies
- Biochemicals pathways involved in biofuels and biochemicals production
- Microbial ecology of the biomass conversion processes
- Bioreactors for bioenergy and biochemicals production
- Novel approaches for biosystems sustainability evaluation

This book contains the successful invited submissions [1–21] to a Special Issue of *Energies* on the subject of "BioEnergy and BioChemicals Production from Biomass and Residual Resources".

2. Statistics of the Special Issue

The response to our call had the following statistics:

- Submissions (33);
- Publications (21);
- Rejections (12);
- Article types: research article (19); review article (2).

The authors' geographical distribution (published papers) is:

- China (3);
- USA (2);
- Italy (2);
- Germany (2);
- Poland (2);
- Mexico (2);
- Austria (1);
- Denmark (1);
- Iceland (1);
- Hungary (1);
- Portugal (1);
- Costa Rica (1);
- Thailand (1);
- United Arab Emirates (1).

Published submissions are related to the most important techniques and analysis applied to the bioeconomy of biofuels and biochemicals derived from varied residual biomasses.

We found the edition and selections of papers for this book very inspiring and rewarding. We also thank the editorial staff and reviewers for their efforts and help during the process.

3. Brief Overview of the Contributions to This Special Issue

The twenty-one published papers cover a variety of biomass or waste residuals that have been converted into different types of energy, biofuels or biochemicals including heat [1,11,16], methane [2,4,7–9,17,19–21], electricity [2,11,16], short chain fatty acids [2,19], ethanol [3,12,19], syngas [5,19], nutrient pellets [6], hydroxymethylfurfura [15], and hydrogen [2,10,14,18,19]. The key information including biomass or residuals, products, and technology for the production and type of research are summarized in Table 1. Among the published papers, fourteen papers were experiment-based research, while five papers were based on sustainability and tech–economic analysis. For more specific information, a brief description of each paper is provided in the following.

The first contribution by Fuller et al. [1] assessed the impact of co-combustion of an herbaceous biomass with low-quality Greek lignite on the quality of the fly ash. The authors compared the results with those of fly ash samples from an industrial facility using the same fuel qualities. Their work offers insights into ash management, a circular economy of herbaceous biomass, and sustainable power plant operations.

Bastidas-Oyanedel and Schmidt [2] conducted a techno-economic analysis of a food-waste-based biorefinery process. It was found in their study that dark fermentation with separation and purification of acetic and butyric acids could gain the highest profit among other scenarios.

Safarian and Unnthorsson [3] made a comprehensive assessment of the sustainability of producing lignocellulosic bioethanol from municipal organic wastes in Iceland, including timber, wood, paper and paperboard, and garden waste. They also evaluated the potential of the total wastes and bioethanol production in Iceland.

Small-volume bioreactors could serve as a test platform to provide an indication of potential challenges that are important for the development of large-scale bioreactors. Kasprzycka and Kuna [4] investigated the fermentation process in small-volume fermenters with a volume of 100–120 mL, which was an effective methodology to facilitate the preliminary selection before large-scale operation.

Syngas is a promising feedstock for the chemical and energy industries. Biomass residues from agriculture and agroindustry with less water content are ideal sources of syngas. Cerone and Zimbardi [5] investigated the effect of air and steam supply on the overall process performance and syngas composition in a pilot updraft gasifier.

Nagy et al. [6] conducted an economic analysis of a process that utilized digestate and waste heat from a biogas plant as the raw materials to produce a pellet. The pellet can be subsequently used as a nutrient in soil or heat energy. The authors reported that the pellets can be produced at a cost of 88–90 EUR/ton with a dry matter substrate content around 6 to 10%.

Seneesrisakul et al. [7] discussed the effect of temperature on the activity of methanogens in relation to the availability of micronutrients. The authors conducted the experiment using anaerobic sequencing batch reactors treating ethanol wastewater at 37 °C (mesophilic) and 55 °C (thermophilic) temperatures.

Beside temperature, organic loading is also a key factor for successful anaerobic digestion. Gao et al. [8] investigated the effect of the organic loading rate on the microbial responses in both mesophilic and thermophilic anaerobic digestion reactors treating municipal solid waste.

Microorganisms are key players in the anaerobic digestion process for biogas production. Jin et al. [9] explored the feasibility of using rumen microorganisms as an inoculum to produce biogas from corn straw and livestock manure in a modified pilot-scale anaerobic digestion system under different solid contents and mixture ratios.

While the conversion of biomass and waste into syngas in gasification systems is technically mature, the production of hydrogen as the primary product is rarely reported, which still requires development and optimization to address various technical and economic challenges. Smoliński et al. [10] conducted experimental work on hydrogen production through the gasification of energy crop biomass and sewage sludge.

The European Union (EU) has ambitious targets to increase the use of renewable energy in the near future. Bioenergy is expected to meet these targets for 2020 and beyond. This situation has created the need to evaluate potential bioenergy technologies and related feedstocks. Madsen and Bentsen [11] analyzed the carbon debt and payback time of using forest residues to replace coal for heat and power generation. Unlike most other studies, the authors conducted the analysis based on empirical data from a retrofit of a heat and power generation plant in northern Europe.

Coffee is one of the most popular food commodities all over the world. During the coffee manufacture process, mucilage, husk, skin (pericarp), parchment, silver-skin, and pulp can be produced as by-products. Using these residual by-products such as mucilage for bioenergy production has rarely been studied. Orrego et al. [12] investigated for the first time the production of bioethanol from mucilage.

The thermal treatment of biomass, for example through torrefaction, could increase the energy density of the biomass and reduce its hygroscopicity. Gaitán-Álvarez et al. [13] studied the thermogravimetric and devolatilization rates of hemicellulose and cellulose from five tropical woody species using three torrefaction temperatures and three torrefaction times. The calorimetric behavior of the torrefied biomass was demonstrated.

Hamedani et al. [14] conducted a life cycle assessment of an agro-industrial residue-based gasification process for hydrogen production. The authors found that an improvement in the hydrogen production efficiency does not necessarily lead to a decrease in environmental impacts.

Steinbach et al. [15] reported a novel process to synthesize hydroxymethylfurfural using highly available disaccharide sucrose as raw material. Hydroxymethylfurfural, as high-value polymer, has unique merits among the bio-based platform chemicals used as precursors and fuel additives. The authors demonstrated that sucrose is the ideal feedstock for the large-scale production of hydroxymethylfurfural.

Ramos et al. [16] provided a case study for the environmental analysis and life cycle assessment of the waste-to-energy process in Portugal. Nine environmental impact categories were grouped from the raw data for the energy recovery plant. The authors also compared the results to two European average situations (incineration plant and sanitary landfill without pre-treatment). This paper showed that these facilities in Portugal were better or at the same level as the average European situation.

Tapia-Tussell et al. [17] investigated the potential for energetic macroalgae conversion to biomethane using biological pretreatment. Their results revealed that an increase in methane yield of 20% was achieved by using fungus for pretreatment. The authors concluded that macroalgae with high concentrations of ash and alkali metal biological processes is more suitable for the biomass utilization for than combustion or pyrolysis.

Rodríguez-Félix et al. [18] for the first time identify and quantify the volatile compounds in tequila vinasses from two different processes of tequila production. The authors also determined the correlation between the volatile compounds and the type of production process (cooked or uncooked stems). The results could serve as a basis for the further development of bioenergy processes using vinasses as the feedstock.

Sharara and Sadaka [19] provided a comprehensive review of the opportunities and barriers for swine manure conversion technologies. This work shed light on the gaps for further investigation and improvement of the technical applicability. The global growth of swine production brings challenges to manure management. The review discussed various technologies that were developed for the production of energy, fuels and bioproducts from swine manure. The authors found that full-scale research in the area of the thermochemical conversion of swine manure could be a core research interest in the future.

Co-digestion is a more cost-effective strategy, compared to dilution with water, to increase the biodegradability of waste streams. Duan et al. [20] investigated the feasibility of the co-digestion of chicken manure and algal digestate water. The kinetic parameters and mass flow involved in such processes were compared and discussed.

Lignocellulosic materials are regarded as an inexhaustible, ubiquitous natural resource, and using them as a basis for the second-generation of biofuel and biogas contributes significantly to social and environmental sustainability. Wagner et al. [21] reviewed the existing pretreatment techniques that have been developed to overcome the structural impediments of lignocellulosic materials. Special attention has been paid to biological pretreatment for the downstream processing of lignocellulosic biomass in anaerobic digestion.

Table 1. Brief summary of the key information from the submissions to this Special Issue.

Biomass	Products	Technology	Experimental	Assessment	Reference
Herbaceous biomass	Heat	Combustion	√ [a]		[1]
Food waste	Methane, Power, lactic acid, polylactic acid, hydrogen, acetic acid & butyric acid	Biorefinery		√	[2]
Lignocellulosic municipal organic wastes	Ethanol	Pretreatment & Fermentation		√	[3]
Chopped maize silage	Biogas	Anaerobic digestion	√		[4]
Agroresidues	Syngas	Gasification	√		[5]
Digestate	Pellet	Anaerobic digestion		√	[6]
Ethanol wastewater	Biogas	Anaerobic digestion	√		[7]
Organic fraction of municipal solid waste	Biogas	Anaerobic digestion	√		[8]
Corn straw & livestock manure	Biogas	Anaerobic digestion	√		[9]
Crops biomass & sewage sludge	Hydrogen	Gasification	√		[10]

Table 1. *Cont.*

Biomass	Products	Technology	Type of Research		Reference
			Experimental	Assessment	
Forest residues	Heat and power generation	Combustion	✓		[11]
Coffee Mucilage	Ethanol	Fermentation	✓		[12]
Tropical woody species	Torrefied biomass	Torrefaction	✓		[13]
Agro-industrial residue	Hydrogen	Gasification		✓	[14]
Disaccharide sucrose	Hydroxymethylfurfura	Acid-catalyzed conversion	✓		[15]
Organic residues	Heat/electricity	Incineration & landfill		✓	[16]
Macroalgae	Biogas	Anaerobic digestion	✓		[17]
Tequila vinasses	Hydrogen	Dark fermentation	✓		[18]
Swine Manure	Multiple products	Multiple products		✓	[19]
Chicken manure & algal digestate	Biogas	Anaerobic digestion	✓		[20]
Lignocellulosic Resources	Biogas	Anaerobic digestion		✓	[21]

[a] The type of research conducted in the selected study.

Fuding: This research was supported financially by The Danish Council for Independent Research (DFF-1335-00142) and Novo Nordisk Foundation (NNF16OC0021568).

Conflicts of Interest: The author declares no conflict of interest.

References

1. Fuller, A.; Maier, J.; Karampinis, E.; Kalivodova, J.; Grammelis, P.; Kakaras, E.; Scheffknecht, G. Fly Ash Formation and Characteristics from (co-)Combustion of an Herbaceous Biomass and a Greek Lignite (Low-Rank Coal) in a Pulverized Fuel Pilot-Scale Test Facility. *Energies* **2018**, *11*, 1581. [CrossRef]
2. Bastidas-Oyanedel, J.-R.; Schmidt, J.E. Increasing Profits in Food Waste Biorefinery—A Techno-Economic Analysis. *Energies* **2018**, *11*, 1551. [CrossRef]
3. Safarian, S.; Unnthorsson, R. An Assessment of the Sustainability of Lignocellulosic Bioethanol Production from Wastes in Iceland. *Energies* **2018**, *11*, 1493. [CrossRef]
4. Kasprzycka, A.; Kuna, J. Methodical Aspects of Biogas Production in Small-Volume Bioreactors in Laboratory Investigations. *Energies* **2018**, *11*, 1378. [CrossRef]
5. Cerone, N.; Zimbardi, F. Gasification of Agroresidues for Syngas Production. *Energies* **2018**, *11*, 1280. [CrossRef]
6. Nagy, D.; Balogh, P.; Gabnai, Z.; Popp, J.; Oláh, J.; Bai, A. Economic Analysis of Pellet Production in Co-Digestion Biogas Plants. *Energies* **2018**, *11*, 1135. [CrossRef]
7. Seneesrisakul, K.; Sutabutr, T.; Chavadej, S. The Effect of Temperature on the Methanogenic Activity in Relation to Micronutrient Availability. *Energies* **2018**, *11*, 1057. [CrossRef]
8. Gao, Y.; Kong, X.; Xing, T.; Sun, Y.; Zhang, Y.; Luo, X.; Sun, Y. Digestion Performance and Microbial Metabolic Mechanism in Thermophilic and Mesophilic Anaerobic Digesters Exposed to Elevated Loadings of Organic Fraction of Municipal Solid Waste. *Energies* **2018**, *11*, 952. [CrossRef]
9. Jin, W.; Xu, X.; Yang, F. Application of Rumen Microorganisms for Enhancing Biogas Production of Corn Straw and Livestock Manure in a Pilot-Scale Anaerobic Digestion System: Performance and Microbial Community Analysis. *Energies* **2018**, *11*, 920. [CrossRef]
10. Smoliński, A.; Howaniec, N.; Bąk, A. Utilization of Energy Crops and Sewage Sludge in the Process of Co-Gasification for Sustainable Hydrogen Production. *Energies* **2018**, *11*, 809. [CrossRef]

11. Madsen, K.; Bentsen, N.S. Carbon Debt Payback Time for a Biomass Fired CHP Plant—A Case Study from Northern Europe. *Energies* **2018**, *11*, 807. [CrossRef]

12. Orrego, D.; Zapata-Zapata, A.D.; Kim, D. Optimization and Scale-Up of Coffee Mucilage Fermentation for Ethanol Production. *Energies* **2018**, *11*, 786. [CrossRef]

13. Gaitán-Álvarez, J.; Moya, R.; Puente-Urbina, A.; Rodriguez-Zúñiga, A. Thermogravimetric, Devolatilization Rate, and Differential Scanning Calorimetry Analyses of Biomass of Tropical Plantation Species of Costa Rica Torrefied at Different Temperatures and Times. *Energies* **2018**, *11*, 696. [CrossRef]

14. Rajabi Hamedani, S.; Villarini, M.; Colantoni, A.; Moretti, M.; Bocci, E. Life Cycle Performance of Hydrogen Production via Agro-Industrial Residue Gasification—A Small Scale Power Plant Study. *Energies* **2018**, *11*, 675. [CrossRef]

15. Steinbach, D.; Kruse, A.; Sauer, J.; Vetter, P. Sucrose Is a Promising Feedstock for the Synthesis of the Platform Chemical Hydroxymethylfurfural. *Energies* **2018**, *11*, 645. [CrossRef]

16. Ramos, A.; Afonso Teixeira, C.; Rouboa, A. Environmental Analysis of Waste-to-Energy—A Portuguese Case Study. *Energies* **2018**, *11*, 548. [CrossRef]

17. Tapia-Tussell, R.; Avila-Arias, J.; Domínguez Maldonado, J.; Valero, D.; Olguin-Maciel, E.; Pérez-Brito, D.; Alzate-Gaviria, L. Biological Pretreatment of Mexican Caribbean Macroalgae Consortiums Using Bm-2 Strain (*Trametes hirsuta*) and Its Enzymatic Broth to Improve Biomethane Potential. *Energies* **2018**, *11*, 494. [CrossRef]

18. Rodríguez-Félix, E.; Contreras-Ramos, S.M.; Davila-Vazquez, G.; Rodríguez-Campos, J.; Marino-Marmolejo, E.N. Identification and Quantification of Volatile Compounds Found in Vinasses from Two Different Processes of Tequila Production. *Energies* **2018**, *11*, 490. [CrossRef]

19. Sharara, M.A.; Sadaka, S.S. Opportunities and Barriers to Bioenergy Conversion Techniques and Their Potential Implementation on Swine Manure. *Energies* **2018**, *11*, 957. [CrossRef]

20. Duan, N.; Ran, X.; Li, R.; Kougias, P.G.; Zhang, Y.; Lin, C.; Liu, H. Performance Evaluation of Mesophilic Anaerobic Digestion of Chicken Manure with Algal Digestate. *Energies* **2018**, *11*, 1829. [CrossRef]

21. Wagner, A.O.; Lackner, N.; Mutschlechner, M.; Prem, E.M.; Markt, R.; Illmer, P. Biological Pretreatment Strategies for Second-Generation Lignocellulosic Resources to Enhance Biogas Production. *Energies* **2018**, *11*, 1797. [CrossRef]

energies

MDPI

Article

Identification and Quantification of Volatile Compounds Found in Vinasses from Two Different Processes of Tequila Production

**Elizabeth Rodríguez-Félix [1], Silvia Maribel Contreras-Ramos [1], Gustavo Davila-Vazquez [1,†],
Jacobo Rodríguez-Campos [2,*] and Erika Nahomy Marino-Marmolejo [3,*]**

[1] Environmental Technology Unit, Center for Research and Assistance in Technology and Design of the State of Jalisco (CIATEJ), Guadalajara 44270, Jalisco, Mexico; elrodriguez_al@ciatej.edu.mx (E.R.-F.); smcontreras@ciatej.mx (S.M.C.-R); gdv@ciatej.mx (G.D.-V)

[2] Analytical and Metrological Unit, Center for Research and Assistance in Technology and Design of the State of Jalisco (CIATEJ), Guadalajara 44270, Jalisco, Mexico

[3] Medical and Pharmaceutical Unit, Center for Research and Assistance in Technology and Design of the State of Jalisco (CIATEJ), Guadalajara 44270, Jalisco, Mexico

* Correspondence: jarodriguez@ciatej.mx (J.R.-C.); emarino@ciatej.mx (E.N.M.-M.); Tel.: +52-333-345-5200 (ext. 1671) (E.N.M.-M.)

† Deceased, 13 March 2017.

Received: 18 January 2018; Accepted: 11 February 2018; Published: 26 February 2018

Abstract: Vinasses are the main byproducts of ethanol distillation and distilled beverages worldwide and are generated in substantial volumes. Tequila vinasses (TVs) could be used as a feedstock for biohydrogen production through a dark fermentative (DF) process due to their high content of organic matter. However, TV components have not been previously assayed in order to evaluate if they may dark ferment. This work aimed to identify and quantify volatile compounds (VC) in TV and determine if the VC profile depends upon the type of production process (whether the stems were initially cooked or not). TVs were sampled from 3 agave stems with a not-cooking (NC) process, and 3 agave stems with a cooking (C) process, and volatile compounds were determined by gas chromatography coupled with mass spectrometry (GC–MS). A total of 111 volatile compounds were identified, the TV from the cooking process (C) showed the higher presence of furanic compounds (furfural and 5-(hydroxymethyl) furfural) and organic acids (acetic acid and butyric acid), which have been reported as potential inhibitors for DF. To our knowledge, this is the first description of the VC composition from TVs. This study could serve as a base for further investigations related to vinasses from diverse sources.

Keywords: stillage; volatile compounds; tequila; inhibitors; dark fermentative

1. Introduction

Vinasses are produced as byproducts of the fermentation and distillation of ethanol [1,2] from sugarcane (*Saccharum officinarum* L.) in South America [3], beet molasses (*Beta vulgaris ssp. vulgaris var. altissima Döll*) in Europe [4,5], or from the distillation of fermented beverages, such as mezcal, bacanora, and tequila in Mexico [6–8]. Large quantities of vinasses are produced worldwide; on average, 12–15 L vinasses are obtained for each liter of ethanol produced [9]. According to the Renewable Fuel Association (http://www.ethanolrfa.org), global ethanol production was 25.7 billions of gallons in 2016, leading to approximately 2.56×10^{12} L of vinasses, which are released without any treatment into agricultural soils or water bodies [9,10]. Regardless of the ethanol production process or the sugar source, these may have similar characteristics. They are complex wastewaters and have high biological oxygen demand (BOD) and chemical oxygen demand (COD), 16–45 g/L and 26–91 g/L, respectively, pH ranges from 3 to 5, high concentrations of suspended solids (2–8.4 g/L), volatile solids (1.1–9 g/L), and phenols (0.04–0.08 g/L). Occasionally they may contain heavy metals, and display a dark brown color [2,4,11–13]. The toxic and recalcitrant nature of this effluent can produce negative environmental impacts like anoxia, eutrophication, and the death of aquatic microorganism and wildlife [13,14].

Tequila distilleries are one of the most important agro-industries [15] producing a popular alcoholic beverage called tequila, which is obtained from the fermentation of hydrolyzable sugars from the stems of *Agave tequilana* Weber var. Azul [10,14]. The agave juice or "mieles de agave" (as locals call it) extraction process requires fructans to be transformed by thermic treatments into fermentable sugars (fructose and glucose), so they can be subsequently fermented by yeast [16,17]. Currently, two processes are used to accomplish sugar hydrolysis. Some distilleries employ traditional methods, i.e., (cooking (C) of agave stems, whereby the mature stems (with no leaves) of agave, so-called "piñas", are cooked in ovens or autoclaves with steam injection (0.5–1.4 kg/cm^2) at 95–120 °C for approximately 48 h or 8–12 h. Once the agave stems are cooked, they are transferred on conveyor belts to a mill with rotatory knives where they are shredded and washed under pressure with potable water to dissolve the sugars. Finally, the stems are placed in a press to extract the agave juice, containing hydrolyzable sugars, soluble, and insoluble compounds which are generated during the cooking process [6,18–20]. However, most producers use a "not-cooking" (NC) process, i.e., [20,21], they used equipment called a "diffuser" to obtain the raw agave juice with hot water (80 °C) directly from previously shredded raw agave stems. Afterwards, the raw agave juice is hydrolyzed through heat (80–85 °C) in acidic conditions (pH 1.8–3) and thermal conditions [18,20,22]. The agave juice obtained either from cooking or not-cooking are submitted to a fermentation process, where hydrolizable sugars are biotransformed to ethanol, carbon dioxide, and other organoleptic compounds such as volatile compounds (esters, aldehydes, ketones, and furans, among others) [23]. Once fermentation is completed, the fermented juice, "must", is transferred to a distiller, where two tandem distillations are performed to obtain tequila. After the final distillation, vinasses are generated as the residual liquid [22,23].

Both processes (cooking or not-cooking the agave stems) might influence the composition of distillation wastewater. Like other vinasses, tequila vinasses (TVs) can also be used as a feedstock for a dark fermentative process (DF) to produce hydrogen, due to its high organic matter content. However, the vinasses' organic matter is not entirely used for biohydrogen production during DF [7,24–26]. Nevertheless, there is not sufficient information about the presence of volatile or complex organic compounds in vinasses, and even less in tequila vinasses.

Some work has tried to determine the composition of molasses and sugarcane vinasses to predict their toxicity. Fagier et al. [27] identified phenolic compounds and some carboxylic acids as volatile compounds in sugarcane vinasses, some of which are highly toxic for *Daphnia magna* (IC50 = 0.9 mg/L). Additionally, Lima et al. [28] identified some fatty acids, alcohols, and esters in hydrolyzed sugarcane vinasses.

As has been previously observed in sugarcane vinasses, volatile compounds, like furans (furfural, 5-(hydroxymethyl) furfural and, methyl-2-furoate), organic acids (acetic acid, butyric acid, propanoic acid, and pentanoic acid) and phenols (2,5-di-tert-butylphenol, eugenol, guaiacol and, 4-ethyl-3-methoxy phenol) have been found also in tequila and cooked agave juice [27,29,30]. This suggests that those or similar volatile compounds might be present in tequila vinasses. Moreover, some of these volatile compounds are toxic and might inhibit subsequent biological treatments of vinasses, like the DF process to produce hydrogen [31–34]. More detailed knowledge on vinasse composition would help scientists to develop more efficient uses of vinasses, and reduce their adverse effects.

There is still a gap in the knowledge of vinasse composition from various sources regarding the presence of volatile compounds, which could inhibit downstream processes or uses, such as biohydrogen production. Furthermore, the extent of the effect of the initial process for tequila production (cooking or not-cooking of agave stems) on the final vinasse composition have not yet been described. Thus, the aim of this work was to identify and quantify the profile of volatile compounds in vinasses, obtained from two different tequila-production processes: (i) cooking (C) and not-cooking (NC) agave stems.

2. Results

2.1. Identification and Concentration of Volatile Compounds

The vinasse samples have different profiles and volatile compound concentration. It was possible to identify 104 compounds belonging to different chemical families (Table 1). The most frequent chemical families were alcohols (20), acids (16), and furans (11); also, alkanes, aldehydes, esters, ketones, phenols, and pyrans were identified.

2.1.1. Organic Acids

The organic acids detected in tequila vinasses were: acetic (1), isobutyric (2), butyric (3), valeric (4), caproic (6), oenanthic (7), caprylic (8), lauric (12), palmitic (15) acids, and others (Table 1).

Acetic acid was found in all studied vinasses, its concentration (13.20–181.25 mg/L) showed significant differences ($p < 0.05$), between C and NC processes, where the vinasse C2 showed the highest acetic acid concentration (181.25 mg/L). Butyric acid was identified in all samples too, with significant differences ($p < 0.05$) among vinasse samples, ranging from 7.84 to 38.86 mg/L, NC2 vinasse showed the highest concentration than other vinasses (Figure 1).

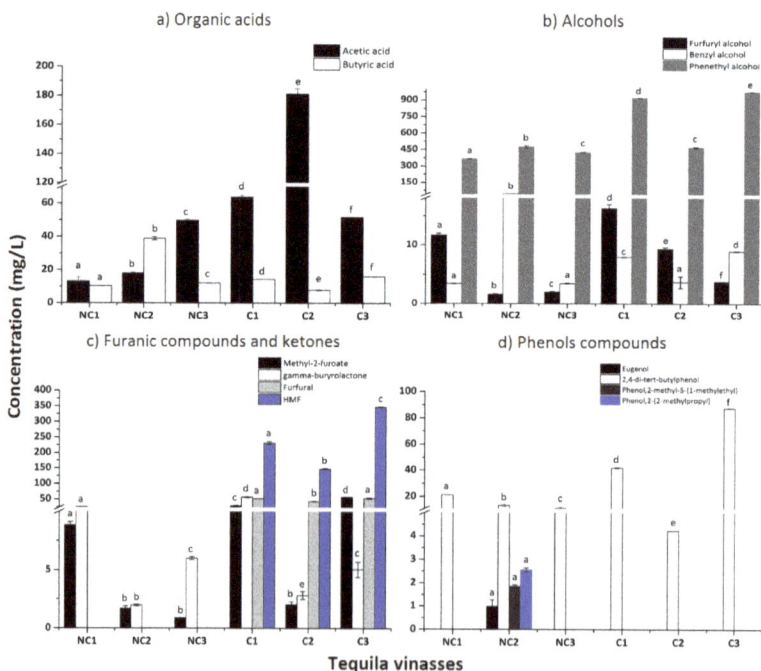

Figure 1. The total concentration of volatile compounds found in tequila vinasses from both processes (not-cooking (NC) and cooking (C)) by functional group. Bars represents mean ± standard deviation. Different lowercase letters indicate significant differences for each compound among the analyzed vinasses ($p < 0.05$, an analysis of variance (ANOVA)).

Table 1. Volatile compounds identified in tequila vinasses, using gas chromatography coupled with mass spectrometry (GC–MS).

Number	Compound	Retention Time (min)	Odor Quality[a]	Kovats Retention Index[b]	Sample Found	Identification[c]	Reference
					Acids		
1	Acetic acid	11.89	sour, astringent, viniegra	1680[b]	All samples	MS, IK, STD	Wanakhachornkrai & Lertsiri (2003)
2	Isobutyric acid	14.48	rancid, butter, cheese, hammy	>1500[b]	NC1, NC2, NC3, C1, C3	MS, IK	Chuenchomrat et al. (2008)
3	Butyric acid	16.06	-	1720[b]	All samples	MS, IK, STD	Wanakhachornkrai & Lertsiri (2003)
4	Isovaleric acid	16.43	sweat, acid, rancid	1864[b]	All samples	MS, IK, STD	Jerković et al. (2012)
5	Valeric acid	19.4	sweet	-	NC2	MS, IK, STD	Chung et al. (1993)
6	Capric acid	22.5	cheese, oil, pungent, sour, rancid, sickening	-	All samples	MS, IK	Wanakhachornkrai & Lertsiri (2003)
7	Oenanthic acid	25.82	rancid, sour, sweat	-	NC2	MS	-
8	Caprylic acid	29.09	cheese, fat, grass, oil	-	NC1, NC2, NC3, C1, C3	-	-
9	Benzoic acid, hexahydro-	29.79	fruit	-	NC2	MS	-
10	9-Decenoic acid	37.15	-	-	NC2	MS	-
11	Hendecanoic acid	41.25	oil	-	NC1, NC3	MS	-
12	Lauric acid	41.5	metal	-	NC2	MS	-
13	3-Methyl-benzoic acid	42.56	-	-	All samples	MS	-
14	Benzenepropanoic acid	44.48	-	-	NC2, C2	MS	-
15	Palmitic acid	51.775	-	-	All samples	MS	-
16	Myristic acid	57.6	-	-	All samples	MS	-
					Esters		
17	Ethyl orthoformate	7.07	-	1465[b]	NC2, C2	MS	-
18	Ethyl butanoate	8.88	-	>1493[b]	C2	MS, IK	Wanakhachornkrai & Lertsiri (2003)
19	1,1-Dimethylpropyl ester, pentanoic acid	9.38	sweet	>1500[b]	NC1, NC3	MS	-
20	Ethyl lactate	9.88	pungent, rancid, soy	>1500[b]	NC2, C1, C2, C3	MS, IK	Wanakhachornkrai & Lertsiri (2003)
21	Methylthiohexanoate	17.11	-	-	NC2	MS	-
22	Diethyl succinate	17.4	-	-	NC2	MS	-
23	Ethyl acetate	19.37	aromatic, brandy, grape	-	C2	MS, IK	-
24	Methyl salicylate	20.43	-	-	NC2	MS	-
25	Allyl phenylacetate	21.62	floral	-	C2	MS	-
26	Ethyl palmitate	34.65	-	-	NC1, NC2, NC3, C1, C3	MS	-
27	2-Phenylethyl acetate	35.4	flower, honey, rose	-	NC1, NC3	MS	-
28	Monoethyl succinate	38.62	-	-	NC1, NC2, NC3, C1, C3	MS	-

Table 1. *Cont.*

Number	Compound	Retention Time (min)	Odor Quality [a]	Kovats Retention Index [b]	Sample Found	Identification [c]	Reference
				Alcohols			
29	2-Methyl-1-propanol	6.36	apple, bitter, cocoa, wine	1107 [b]	NC1, NC3	MS, IK	Chuenchomrat et al. (2008)
30	3,7-Dimethyl-1-octanol	6.57	-	-	NC1, NC3	MS	-
31	1-Butanol	7.21	fruit	1156 [b]	NC2	MS, IK	Chuenchomrat et al. (2008)
32	3-Penten-2-ol	7.56	-	-	All samples	MS	-
33	2-Butyl-1-Octanol	9.26	-	-	NC1, NC3, C1, C2, C3	MS	-
34	3-Ethoxy-1-propanol	10.37	-	-	C2	MS	-
35	2-Ethyl-1-hexanol	12.52	-	>1361 [b]	C2	MS, IK	Chung et al. (1993)
36	1-Decanol, 2-hexyl-	13.50	-	-	NC2	MS	-
37	2,3-Butanediol	14.69	-	>1493 [b]	All samples	MS, IK	Chuenchomrat et al. (2008)
38	1,2-Propanediol	15.1	-	-	NC2, C1, C2	MS	-
39	Furfuryl alcohol	17.13	burnt, caramel, cooked	>1493 [b]	All samples	MS, IK, STD	Chuenchomrat et al. (2008)
40	3-(Methylthio)-1-propanol	18.65	-	-	C1, C2	MS	-
41	1,3-Propanediol	20.67	-	-	C2	MS	-
42	Benzyl alcohol	23.54	sweet, flowery, boiled cherries, moss, roasted bread, rose	1943 [b]	All samples	MS, IK	Chuenchomrat et al. (2008)
43	Phenylethyl Alcohol	24.58	honey, spice, rose, flowery, caramel	1997 [b]	All samples	MS, IK	Jerkovic et al. (2012)
44	Benzene propanol	28.66	-	-	All samples	MS	-
45	p-Menthane-1,8-diol	30.17	-	-	NC1, NC3	MS	-
47	1-Phenyl-1-decanol	58.78	-	-	NC1, NC3	MS	-
48	3-(p-Hydroxyphenyl)-1-propanol	58.99	-	-	NC2	MS	-
				Aldehydes			
49	Benzaldehyde	13.63	bitter almond, burnt sugar, cherry, malt, roasted pepper	>1500 [b]	C1, C3	MS, IK	Wanakhachornkrai & Lertsiri (2003)
50	4-Methyl-benzaldehyde	16.8	-	>1500 [b]	NC1, NC2, NC3	MS, IK	Wanakhachornkrai & Lertsiri (2003)
51	3-Methyl-benzaldehyde	37.35	-	>1500 [b]	NC1, NC3, C2	MS, IK	Wanakhachornkrai & Lertsiri (2003)
				Alkanes			
52	Dodecane	7.823	-	-	All samples	MS	-
53	4,6-Dimethyl-dodecane	7.90	-	1200 [b]	C1, C3, NC1, NC3	MS, IK	Chung et al. (1993)
54	2,3,6,7-Tetramethyl-octane	8.064	-	1300 [b]	C1, C3	MS, IK	Chung et al. (1993)
55	5-Methyl-tridecane	9.67	-	-	NC1, NC3, C1, C3	MS	-

Table 1. *Cont.*

Number	Compound	Retention Time (min)	Odor Quality [a]	Kovats Retention Index [b]	Sample Found	Identification [c]	Reference
					Alkanes		
56	Tetradecane	10.54	-	1400[b]	All samples	MS, IK	Jerković et al. (2012)
57	3,3-Dimethyl-heptane	10.61	-	-	NC1, NC3	MS	-
58	Farnesan	11.13	-	-	NC1, NC3, C2	MS	-
59	3-Ethyl-3-methylheptane	15.30	-	-	All samples	MS	-
60	Nonadecane	17.80	-	1900[b]	All samples	MS, IK	Jerković et al. (2012)
61	Eicosane	12.48	-	2000[b]	All samples	MS, IK	Chung et al. (1993)
					Furanic		
62	2,2,3,3,4,4-Hexamethyltetrahydrofuran	8.905	-	-	C1, C3	MS	-
63	Furfural	12.30	almond, baked potatoes, bread, burnt, spice	1493[b]	C1, C2, C3	MS, IK, STD	Chuenchomrat et al. (2008)
64	5-Methylfurfural	12.38	-	>1493[b]	C1, C2, C3	MS, IK	Chuenchomrat et al. (2008)
65	Acetylfuran	13.20	balsamic, cocoa, coffee	-	C1, C3	MS	-
66	2-Furoate-methyl	14.88	fruit	-	All samples	MS, STD	-
67	Furan, 2-(1,2-diethoxyethyl)-	21.07	-	-	All samples	MS, IK	-
68	2-(Hydroxyacetyl)furan	27.83	-	-	NC1, NC3	MS	-
69	2,5-Dimethyl-2-(2-tetrahydrofuryl) tetrahydrofuran	30.37	-	-	All samples	MS	-
70	HMF	41.99	almond, baked potatoes, bread, burnt, spice	>1493[b]	C1, C2, C3	MS, IK, STD	Chuenchomrat et al. (2008)
71	Furan, 2-(1,2-dimethoxyethyl)-	32.16	-	-	NC2	MS	-
72	Furan, 2-ethoxy-4-ethyl-2,3-dihydro	36.04	-	-	C1, C3	MS	-
					Ketones		
73	3(2H)-Furanone, dihydro-2-methyl	8.8	nuts	1207[b]	C1, C2, C3	MS, IK	Wanakhachornkrai & Lertsiri (2003)
74	Acetoin	9.057	butter, creamy, green pepper	1306[b]	NC1, NC2, C1, C2, C3	MS, IK	Chuenchomrat et al. (2008)
75	2-Cyclopenten-1-one	10.26	-	-	C1, C3	MS	-
76	2-Butanone, 3,4-epoxy-3-ethyl	13.18	-	-	NC2	MS	-
77	2(3H)-Furanone, dihydro-5-methyl-	15.87	-	>1500[b]	NC2, C1, C2, C3	MS, IK	Wanakhachornkrai & Lertsiri (2003)
78	γ-Butyrolactone	16.394	caramel, cheese, roasted nut	1610[b]	NC1, NC2, NC3, C1, C2,C3	MS, IK	Marquez et al. (2010)
79	2-Cyclopenten-1-one, 2-hydroxy-3-methyl	22.04	-	-	C1, C2, C3	MS	-
80	α, β-Angelica lactone	17.75	floral	-	C1, C2	MS, IK	-

Table 1. *Cont.*

Number	Compound	Retention Time (min)	Odor Quality [a]	Kovats Retention Index [b]	Sample Found	Identification [c]	Reference
	Ketones						
81	2,5-Dimethyl-4-hydroxy-3(2H)-furanone	22.96	burnt, caramel, cotton candy, honey	2024 [b]	C1, C3	MS, IK	Fuhrmann & Grosch (2002)
82	Furyl hydroxymethyl ketone	27.51	-	-	NC3, C1, C2, C3	MS	-
83	2(3H)-Furanone, dihydro-3-hydroxy-4,4-dimethyl	28.34	-	-	C1, C3	MS	-
84	Ethanone, 1-(2,6-dihydroxy-4-methoxyphenyl)	39.39	-	-	NC2	MS	-
85	Ethanone, 1-(4-hydroxy-3,5-dimethoxyphenyl)	54.24	-	-	NC2	MS	-
	Phenols						
86	p-Guaiacol	23.06	burnt, phenol, wood	>1500 [b]	NC2	MS, IK	Wanakhachornkrai & Lertsiri (2003)
87	p-Methylguaiacol	26.06	phenol	>1500 [b]	NC2	MS, IK	Wanakhachornkrai & Lertsiri (2003)
88	3-Methyl phenol	29.93	-	-	NC2	MS	-
89	Phenol, 2-methyl-5-(1-methylethyl)	32.31	caraway, spice, thyme	2189 [b]	NC2	MS, IK	Jerković et al. (2012)
90	2,4-Di-tert-butylphenol	36.67	-	-	All samples	MS	-
91	Phenol, 2-(5-isoxazolyl)	45.90	-	-	NC2	MS	-
92	Eugenol	45.31	burnt, clove, spice	-	NC2	MS, STD	-
93	Phenol, 2-(2-methylpropyl)	51.53	-	-	NC2	MS	-
94	4-(2-Hydroxyethyl) phenol	54.59	-	-	C1, C3	MS	-
	Pyrans						
95	2H-Pyran-2,6(3H)-dione	27.26	-	-	NC1, NC3	MS	-
96	2-Pentoxy-tetrahydropyran	28.10	-	-	NC1, NC3	MS	-
97	4H-Pyran-4-one, 2,3-dihydro-3,5-dihydroxy-6-methyl	35.27	-	-	NC1, NC2, NC3	MS	-
98	Pyrrolo[1,2-a] pyrazine-1,4-dione, hexahydro-3-(2-methylpropyl)	60.62	-	-	C1, C2, C3	MS	-
	Others						
99	m-Dioxane, 2-methyl	5.96	-	-	C2	MS	-
100	p-Dichlorobenzene	11.78	-	-	NC1, NC3	-	-
101	m-Di-tert-butylbenzene	11.28	-	-	C1, C2, C3	MS	-
102	6,7-Dihydro-7-hydroxylinalool	26.69	-	-	NC1, NC3	MS	-
103	4-Acetoxy-3-methoxystyrene	33.29	-	-	NC1, NC3	MS	-
104	Ethyl linoleate	60.85	-	-	All samples	MS	-

[a] Flavor notes reported (Flavor and Extract Manufacturers Association (FEMA)); [b] obtained from literature; [c] identification method: MS = mass spectrometry, IK = Kovats index, STD = pure compound (standard).

2.1.2. Esters

Esters ethyl orthoformate (17), ethyl butanoate (18), ethyl lactate (20), ethyl palmitate (26), among others esters were found in most tequila vinasses samples (NC and C).

2.1.3. Alcohols

Alcohols 2-methyl-1-propanol (29), 1-butanol (31), 2,3-butanediol (37), furfuryl alcohol (39), benzyl alcohol (42), phenylethyl alcohol (43), benzenepropanol (44), and others (Table 1) were found in the present work. Their concentrations ranged from 1.6–977.95 mg/L (NC and C processes).

In this research, furfuryl alcohol (39), benzyl alcohol (42) and phenylethyl alcohol (43) were identified and quantified in all vinasse samples (Figure 1). Furfuryl alcohol in vinasses from the C process presented significantly higher concentrations (3.76–16.25 mg/L) than those from the NC process ($p < 0.05$). The vinasse C1 showed the highest concentration (16.26 mg/L). Benzyl alcohol was present in all the samples, showing significant differences ($p < 0.05$) between vinasses from the NC and C processes. The concentrations ranged from 3.42–46.6 mg/L. Among them, the NC2 vinasse had the highest concentration of benzyl alcohol. Phenylethyl alcohol was found in both processes ranging from 366.41–470.92 mg/L, with significant differences ($p < 0.05$) between both kinds of vinasse evaluated. The highest concentration was found in the C3 vinasse (Figure 1).

2.1.4. Aldehydes

Only three aldehydes were detected in tequila vinasses: benzaldehyde (49), benzaldehyde, 4-methyl (50), and benzaldehyde, 3-methyl (51) (Table 1), regardless of the process (NC or C).

2.1.5. Alkanes

Similarly, 10 alkanes were identified in the vinasses evaluated regardless of the process involved: dodecane (52), tetradecane (56), nonadecane (60), and eicosane (61) (Table 1).

2.1.6. Furanic Compounds

Eleven different furanic compounds were identified in tequila vinasses, 10 in the C vinasses and five in the NC vinasses. Furfural (63), 5-methylfurfural (64), acetylfuran (65), HMF (70) and furan,2-ethoxy-4-ethyl-2,3-dihydro (72) were found only in the C vinasses, while 2-furoate-methyle (66), furan, 2-(1,2-diethoxyethyl)- (67) and 2,5-dimethyl-2(2-tetrahydrofuryl) tetrahydrofuran (69) were found in all the vinasses (Table 1).

Furfural was found in C1 and C3 (52.11 and 53.57 mg/L, respectively); HMF presented the highest level in the C3 vinasse (347.61 mg/L) (Figure 1).

2.1.7. Ketones

Thirteen ketones were identified in the vinasses. Ten were mostly found in the C vinasses, while only seven were found in some NC. γ-butyrolactone (78) was identified and quantified in all TVs, showing significantly higher concentration in the C vinasses ($p < 0.05$) than in the NC vinasses. The concentration range was 1.5–56.78 mg/L, where vinasse C1 presented the highest amount (Figure 1).

2.1.8. Phenols

Nine phenols were identified in the vinasses. p-Guaiacol (86), p-methylguaiacol (87), 3-methylphenol (88), eugenol (92), among others, were detected in the NC vinasses (Table 1). The 2,4-di-tert-butylphenol (90) was found in all the vinasses. The NC vinasses had a significantly lower concentration (11.20–21.35 mg/L) than the C vinasses (4.23–87.44 mg/L). The highest presence of phenolic compounds (eugenol, p-guaiacol, 4-(2-hydroxyethyl) phenol) was observed in NC2 vinasse (Figure 1).

2.1.9. Pyrans

Four pyrans were detected in the evaluated vinasses: (2H-pyran-2,6(3H)-dione (95), 2-pentoxy-te trahydropyran (96), 4H-pyran-4-one,2,3-dihydro-3,5-dihydroxy-6-methyl (97), and pyrrolo[1,2-a] pyrazine-1,4-dione (98) (Table 1). The latter was found only in the C vinasses while the others were found in the NC vinasses.

2.2. Principal Component Analysis (PCA)

Principal component analysis (PCA) was used to determine the main sources of variability of the data sets and establish the relationship between tequila vinasses (objects) and volatile compounds (variables) [35]. PCA facilitated the interpretation in this study. The analysis allowed us to identify what volatile compounds were correlated with the kind of vinasse (NC or C). The two principal components (PCs) were enough to explain 60% of total variability from the data set, 32.29% and 28.09% by PC1 and PC2, respectively (Figure 2).

The PCs showed four distinct groups (Figure 2). The PC1 separated the volatile compounds according to the kind of vinasse. Vinasses from the cooking process were found in the positive side of the PC1 axis. Also, on this side, volatile compounds that correlate strongly with those vinasses are found, e.g., acetic acid (1), furfuryl alcohol (39), and furan compounds such as furfural (63), acetylfuran (65), 2-furoate-methyl (66), and HMF (70), among others. However, in the negative side of PC1, two groups were detected: the first includes NC1 and NC3 related to butyric acid (3), caproic acid (6), hendecanoic acid (11), among others; the second includes, NC2, which was correlated with phenolic compounds like p-guaiacol (86), p-methylguaiacol (87), 3-methyl phenol (88) eugenol (92), among others.

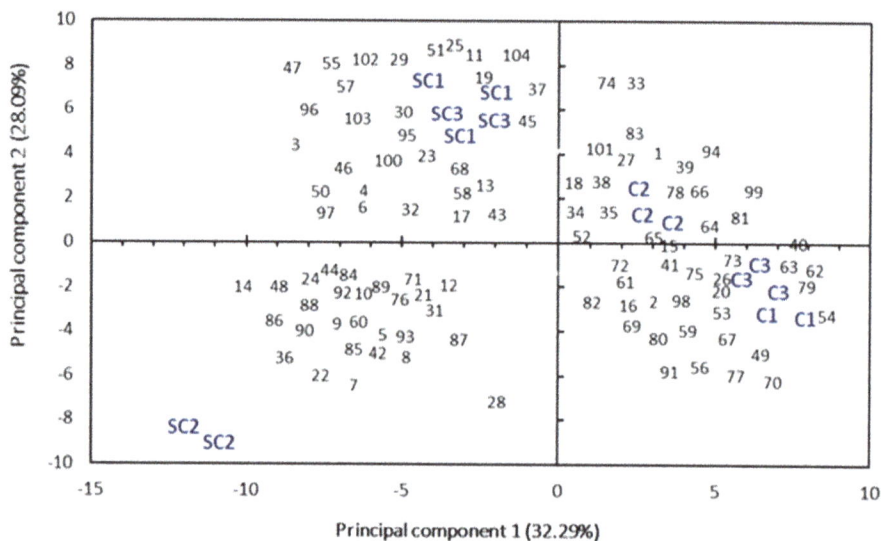

Figure 2. Principal component analysis (PCA), loadings (volatile compounds), and score plot (NC and C processes) of PC1 and PC2, from all volatile compounds in treatments.

3. Discussion

The identification of approximately 37% of the compounds in the present work is in agreement with previous work published with similar matrices, such as cooked agave juice, tequila beverage, and sugarcane vinasses [27–30]. This work focuses on the volatile compounds found in higher concentrations or with inhibitory potential for the biological process in the vinasse treatment.

3.1. Organic Acids

Some organic acids found in tequila vinasses have been previously reported in similar matrices (i.e., tequila, cooked agave juice, and sugarcane vinasses) by other studies. For instance, acetic, butyric, isobutyric, valeric, palmitic, and myristic acids have been reported in tequila by Prado-Jaramillo et al. [30]. Likewise, Mancilla-Margalli and López [29] published the existence of acetic, butyric, valeric, caproic, oenanthic, caprylic and lauric acids in cooked agave juice. Also, Fagier et al. [27] reported the presence of palmitic acid in sugarcane vinasses. Lima et al. [28] identified lauric, myristic, and palmitic acids in hydrolyzed sugarcane vinasses.

The presence of organic acids in high concentrations confer toxicity to the vinasses, hindering their treatment by biological process or their exploitation via biohydrogen production. Acetic and butyric acids have negatives effects in the DF process of acetic acid-decreased Hmax (maximum potential of H2 production) and Rmax (maximum H2 production rate) in batch production [36]. In addition, Wang, Wan, and Wang [37] reported inhibitory effects for Hmax and Rmax in batch conditions, using 6 g/L and 8 g/L of acetic and butyric acids, respectively. In this work, these acids were found in both processes studied, NC (146.14 mg/L) and C (512.47 mg/L), at lower concentration than those reported in the literature [34,36,37]. Nonetheless, a synergistic effect might occur, as both are present in the vinasses, which can boost their potential individual inhibitory effects. Franden et al. (2013) demonstrated that the concomitant presence of acetic and formic acids (IC25 = 50.3 mg/L) inhibited the growth of *Zymomonas mobilis* [38].

Organic acids may be either protonated or unprotonated, both species inhibiting biohydrogen production [34]. They can uncouple hydrogen-producing bacteria (HPB) growth in two ways: (i) on the one hand, the nonpolar un-dissociated form can penetrate cell membrane HPB whereby they dissociate due to higher intracellular pH, releasing protons in the cell cytoplasm; as a result, this creates a pH imbalance and decreases intracellular pH, producing a reduction in the available energy used in HPB growth; (ii) on the other hand, if the polar-dissociated part of organic acids is present in the fermentative hydrogen production system at a high concentration, the ionic strength in the solution will be increased, causing HPB growth inhibition and cell lysis [34,36,37,39].

3.2. Esters

Prado-Jaramillo et al. [30] identified ethyl butyrate, ethyl caprate, ethyl caproate, ethyl ethanoate, ethyl furoate, among other esters, in tequila. Likewise, Fagier et al. [27] found palmitic acid methyl ester in vinasses from sugarcane distillation. Similarly, in hydrolyzed sugarcane vinasses, Lima et al. [28] found ethyl myristate, ethyl palmitoleate, ethyl palmitate, 2-phenylethyl laurate, ethyl oleate, and ethyl stearate.

3.3. Alcohols

In a previous study performed by Prado-Jaramillo et al. [30] in tequila, the alcohols reported were 1-decanol, 1-dodecanol, 1-heptanol, butanol, isobutanol, phenylethyl alcohol, n-propanol, among others. Similarly, Mancilla-Margalli and López [29] found 1-butanol, 3-methyl-1-butanol, 1-hexanol, 1-octen-3-ol, 1,2-butanediol, 1,3-butanediol, 1,2-ethanediol, benzyl alcohol, phenylethyl alcohol in cooked agave juice. Also, they quantified benzyl alcohol and phenylethyl alcohol concentration, which were 3.78 mg/L and 4.58 mg/L, respectively. Additionaly, Fagier et al. [27] detected the presence of 2-phenyl ethanol in sugarcane vinasses. In other works, the presence of 2-phenylethanol, pentadecan-1-ol, and hexadecane-1-ol was identified in hydrolyzed sugarcane vinasse [28].

It is known that furfuryl alcohol might cause a significant membrane leakage in some bacteria (i.e., *Escherichia coli*) and exhibits synergism when present along with other inhibitors of microbial growth [40]. Currently, there is no available information about the inhibitory nature of the identified alcohols on the DF process.

3.4. Aldehydes

Prado-Jaramillo et al. [30] reported the presence of aldehydes in tequila, such as benzaldehyde, benzene acetaldehyde, hexanaldehyde, myristaldehyde, nonaldehyde, among others. Likewise, Mancilla-Margalli and López [29] detected benzaldehyde, phenylacetaldehyde, and 4-hydroxybenzaldehyde in cooked agave juice. Furthermore, pentadecanal was found in hydrolyzed sugarcane vinasse by Lima et al. [28].

It is known that the presence of some aldehydes (i.e., 4-hydroxybenzaldehyde) has been reported to impair the growth of an ethanologenic *Escherichia coli* B (LY01) at concentrations of 600 mg/L, where a 50% growth inhibition was observed [41]. Aldehydes inhibit some pathways such as glycolysis and fermentation [41–43]. Therefore, their presence in tequila vinasses could be considered as a potential inhibitor for DF in hydrogen production.

3.5. Alkanes

Several hydrocarbons have been detected in tequila by Prado-Jaramillo et al. [30], who reported the presence of dodecane, eicosane, heptadecane, hexadecane, nonadecane, tetradecane, tricosane, and others. Furthermore, Fagier et al. [27] reported the presence of one cycloalkane (1,1,4,4-tetramethyl-2,5-dimethylenecyclohexane) in sugarcane vinasses. In addition, Lima et al. [28] detected heptacosane in hydrolyzed sugarcane vinasse.

The chemical group of alkanes has been not reported as potential inhibitors of the DF process. However, the presence of some hydrocarbons (i.e., alkanes, branched alkanes, among others) in soils can alter soil enzymatic activities and microbial biomass carbon [44,45]. Also, they may inhibit some bacterial populations [46].

3.6. Furanic Compounds

The presence of some furanic compounds has been previously reported in tequila and cooked agave juice. Prado-Jaramillo et al. [30] identified 10 furanic compounds in tequila from the cooking process, such as 1-furan-2-yl ethenone, 2-(1,2-diethoxyethyl) furan, furfural, 5-(hydroxymethyl) furfural (HMF), 5-methylfurfural, among others. Similarly, Mancilla-Margalli and López [29] identified eight furanic compounds in cooked agave juice, among which are furfural, HMF, methyl-2-furoate, 2-furanmethanol, and tetrahydro-2-methylfuran. Recently, furfural has also been identified in hydrolyzed sugarcane vinasses by Lima et al. [28].

Furanic compounds such as furfural and HMF have been reported as inhibitors during fermentative hydrogen production [32,34,36,47]. The presence of furfural and HMF can inhibit several enzymes (i.e., alcohol dehydrogenase and pyruvate dehydrogenase) in *Saccharomyces cerevisiae* at concentrations of 192–480 mg/L for furfural and 252–630 mg/L for HMF [48]. Even furfural is known to be more toxic than HMF for industrial catalysts, in *E. coli* and *S. cerevisiae*. Furthermore, ethanol production by *E. coli* B LYO1 is inhibited at concentrations ≥ 2.6 g/L of furfural. This may be due to the fact that this compound has a direct effect on glycolytic and fermentative enzymes [41].

In agreement with previous studies performed in batch conditions for biohydrogen production, concentrations higher than 250 and 100 mg/L of furfural and HMF, respectively, directly affected Hmax and Rmax [32,36]. In this study, furfural concentration in vinasses was lower than the inhibitory concentration previously reported, but HMF concentration in tequila vinasses (C) was higher than that reported [32,34,36]. Also, both compounds in vinasses could increase the inhibitory effects.

Furfural and HMF are known to suppress cell growth, induce DNA damage, and inhibit several enzymes of the glycolytic and fermentative pathways (i.e., hexokinase, phosphofructokinase, triosephosphate, dehydrogenase, aldolase, among others) [32,34,48,49]. Furfural may decrease the activities of NADH-dependent enzymes involved in fermentation reactions; moreover, HMF and furfural reduction to furfuryl alcohol requires NADH in yeasts, which decrease H_2 production. Additionally, NADH may reduce furan derivatives as it is oxidized to NAD_+, leading to lower levels of NADH and lowering H_2 production. Furthermore, HMF bears a hydroxyl group, which can bind to DNA nitrogen bases (specifically, adenine and thymine), causing DNA damage and mutations in many HPB [34,36,42,43]. In addition, it is known that furfural can inhibit fermentation and growth synergistically with phenols in *E. coli* and, when applied with acetic acid, can inhibit growth in *S. cerevisiae* [38].

3.7. Ketones

Some ketones have been found by Prado-Jaramillo et al. [30] in tequila, who reported acetoin, α-angelica lactone, β-angelica lactone, 1-(2-hydroxyphenyl) ethenone, 2-furyl acetone, acetovanillone, among others. Furthermore, Mancilla-Margalli and López [29] reported the existence of five ketones in cooked agave juice: 2-hydroxy-2-butanone, 1-hydroxy-2-propanone, butyrolactone, cyclotene and β-damascenone. Moreover, the presence of one ketone (E-1-(3-hydroxy-2,6,6-trimethylcyclohex-1-enyl) but-2-en-1-one) was reported by Fagier et al. [27] in sugarcane vinasse.

There are no reports about ketones with potential inhibition for the DF process. In fact, Mills, Sandoval and Gill [40] state that ketones can be generated during the acidogenic stage, but are not considered as potential inhibitors because they are produced at low concentrations (<0.05 g/L). However, there is still a lack of knowledge about the effect of ketones on the DF process.

3.8. Phenol Compounds

Some phenols have been previously reported by Prado-Jaramillo et al. [30] in tequila; like eugenol, carvacrol, phenol, thymol, and butylated hydroxyanisole. Furthermore, Fagier et al. [27] reported the presence of four phenols in sugarcane vinasses: 4-ethyl-3-methoxy phenol, 2,6-dimethoxy phenol, 3,4,5- trimethoxy phenol and 4-allyl-2,6-dimethoxy phenol.

Several studies have shown the inhibitory effects on the DF process caused by other phenolic compounds such as hydroxybenzoic acid, syringaldehyde and vanillin [32,34,36,49,50]. However, those phenols were not identified in tequila vinasses, and other phenols were found that may have other adverse effects. High concentrations (0.12–50 mg/L) of phenolic compounds can inhibit fungi- or bacteria-fermenting carbohydrates into ethanol, by deactivation of lignocellulose-hydrolyzing enzymes [51]. Phenolic compounds are more inhibitory than furan compounds and organic acids, as phenols can lead to precipitation and irreversible inhibition of lignocellulose-hydrolyzing enzymes [51,52]. Also, since phenols are the most abundant components in soils [53], these compounds may affect the pools and fluxes of soil nutrients by disturbing the communities of microbial decomposers. Generally, vinasses are disposed into soils without any treatment [10], which could cause a disturbance in soils by altering the natural concentration of phenols, thereby affecting soil microbiota and plants. Phenolics, depending on their concentration, have been found to either stimulate or inhibit spore germination. For instance, hyphal growth of saprophytic fungi and hyphal branching are altered by quercetin (30.22 mg/L), p-coumaric acid (16.4 mg/L), ferulic acid (19.41 mg/L), hydroxybenzoic acid (13.81 mg/L) [54–56]. Arbuscular mycorrhizal fungi colonization of asparagus roots is decreased by ferulic acid (>50 µg/g) [55]. Also, other phenolic compounds (trans-Cinnamic acid and p-coumaric acid) inhibit growth of the etiolated seedling and seeding growth of lettuce [57]. Moreover, phenolic compounds present antimicrobial activity against *Escherichia coli* [58], but also against other Gram-negative and positive bacteria such as *Salmonella typhimurium*, *Shigella dysenteriae*, *Oenococcus oen* and *Lactobacillus hilgardii* by changing the cell's membrane permeability DNA binding [59,60], and as a consequence, affecting their ability to serve as selective barriers causing leakage and intracellular damage, leaving cells vulnerable to extracellular toxic compounds [32,34,36,50]. In the tequila vinasses phenolic compounds were found at 108.13–206.32 g/moL, which suggests that p-guaiacol, p-methylguaiacol, phenol, 3-methyl and 4-2(hydroxyethyl)phenol may negatively affect HPB

3.9. Pyrans

Prado-Jaramillo et al. [30] detected the presence of one pyran (3,4-dihydro-2H-pyran) in tequila. Mancilla-Margalli and López [29] reported the presence of three pyrans, such as 2,3-dihydro-4(H)-pyran-4-one maltol, 2,3-dihydroxy-3,5-dihydro-6-methyl-4(H)-pyran-4-one, and 2,3-dihydro-2-methyl-4(H)-pyran-4-one in cooked agave juice. These suggest that pyranic compounds could be present in the vinasses independent of the cooking or not-cooking process.

There is no report about the impact of pyrans on the DF process. Although, 15 mg/L, 5,6 dihydro-2(H)- pyran-2-one has been shown to inhibit growth, and biofilm formation of the Gram-negative bacteria *Ralstonia solanacearum* [61,62]. All this suggest that the biological processes in all matrices where vinasses are disposed (soil and water) are negatively influenced.

3.10. Relationship between PCA and Not-Cooking (NC) and Cooking (C) Processes

According to PCA, it can be inferred from C1 and C3 that thermal hydrolysis carried out by cooking the agave stems with autoclave show a significant effect on furan compounds formation, (i.e., furfural and HMF), which may inhibit subsequently the fermentative process. Fructans are insoluble in water at room temperature (25 °C) but are soluble at >50 °C [63]. Therefore, hydrolysis is required to obtain fermentable sugars. When the cooking process is carried out, sugars are subjected to a series of complex reactions, mainly caramelization, Maillard and oxidation–dehydration reactions [22]. These reactions are influenced by several factors, such as temperature and period of the cooking process. Furfural is produced by xylose dehydration at elevated temperatures and pressures [34,64]. HMF is generated during the cooking process because of the dehydration of fructose in the initial stages of the Maillard reaction [16,65]. The generation of furfural and HMF is continuously increased throughout the cooking process of agave juice [29]. In our research, C3 vinasse showed a higher presence of acetic acid, which is known to be a degradation product of the Maillard reaction of glucose and fructose [66].

During the not-cooking agave process, acidic hydrolysis is carried out at 80 °C. NC2 vinasses showed a higher presence of phenolic compounds. They might be already present in the feedstock or generated during the initial stages of acid-thermal degradation (NC) of [29,30]. Also, the stems of *Agave tequilana* Weber var. Azul presented high contents of α-cellulose (65%), hemicellulose (5.5%) and lignin (16.8%) [66]; and when fructans hydrolysis is performed under acidic conditions, the splitting of β-O-4 ether and other acid-labile linkages in lignin forms phenolic compounds [67]. Also, vinasses NC1 and NC3 presented a higher presence of organic acids than might be formed during hemicellulose acid-hydrolysis [68], contributing to the formation of phenolic compounds.

4. Materials and Methods

4.1. Chemicals and Standards Used

All experiments were performed with compounds standard grade and solvents high-performance liquid chromatography (HPLC) grade. 2,4-di-tert-butylphenol, 5-(hydroxymethyl) furfural (HMF), acetic acid, butyric acid, eugenol, formic acid, furfural, furfuryl alcohol, isobutyric acid, methyl-2-furoate, propionic acid, stearic acid and valeric acid standards were purchased from Sigma-Aldrich (Sigma-Aldrich, St. Louis, MO, USA). The HPLC-grade dichloromethane and methanol were supplied by Sigma-Aldrich. The anhydrous sodium sulfate was purchased from Fermont (Fermont, Monterrey, Mexico).

4.2. Samples

Vinasses were obtained from six tequila distilleries located in Tequila, Jalisco, Mexico. The tequila factories were selected from two types of agave juice extraction process, three of them were manufactured by using the traditional process (cooking agave stems) and were coded from C1 to C3; the other three come from a not cooking process and were labeled as NC1 to NC3. Twenty liters of vinasses from each industry were sampled after the distillation process was accomplished. Vinasses were kept at 4 °C until their analysis.

4.3. Volatile Compound Extractions

Volatile compounds were extracted by the liquid–liquid extraction method proposed by Prado-Jaramillo et al. [30] using dichloromethane. Vinasses were placed in a centrifuge tube (40-mL) and 10 mL of CH_2Cl_2 were added. Subsequently, the samples were shaken for 5 min and centrifuged for 10 min at 5000 rpm and 10 °C. The organic layer was separated and dried with anhydrous Na_2SO_4. All extracts were concentrated to a final volume of 1.5 mL with a rotatory evaporator IKA® RV 10 basic (IKA, Wilmington, NC, USA). The extracts were settled in suitable amber vials and preserved at −19 °C, until chromatographic analysis. Each vinasse sample was extracted and analyzed in triplicate.

4.4. Separation and Identification of Volatile Compounds

The volatile compounds were analyzed by a gas chromatography (GC) 6890N Network system (Agilent Technology, Santa Clara, CA, USA), coupled to a selective mass spectrometer detector 5975 (Agilent Technology, USA). Compound separations were performed using an HP-FFAP capillary column of 25 m × 0.32 mm (i.d), coated with a 0.52 μm film (Agilent Technologies, Santa Clara, CA, USA). Helium was used as carrier gas, using a 2 mL/min of flow with an initial temperature of 40 °C for 5 min. Followed by a temperature program of 20 °C/min to 100 °C for 1 min, and by a second rate of 3 °C/min to a final temperature of 230 °C kept for 40 min. The injector temperature was 220 °C, and the injection mode was splitless. The selective mass detector operated an electronic impact ionization system at 70 eV, and at 260 °C. The identification of compounds was based on three criteria: (1) by comparing the mass spectra with the National Institute of Standards and Technology (NIST) library mass spectra; (2) by comparing the retention index with literature data; and (3) whenever possible, the identification was confirmed by using pure standards. Chromatograms obtained from the GC–MS analysis were integrated, and the peak areas were recorded for each identified compound. Quantification was performed by standard curves, obtained by preparing a solution with the different standard compounds in concentrations of 0.5, 1, 5, 10, 15 and 20 mg/L. Interpolation within a linear regression equation (R^2 from 0.995 to 0.999) was used to calculate each compound concentration. When a commercial standard was not available, the quantification was achieved using a slope obtained for a standard of an analog compound that was structurally similar, but slightly different, as has been described elsewhere [69].

4.5. Statistical Analysis

All the statistical analyses were achieved with Statgraphics Centurion XVI (Statgraphics, The Plains, VA, USA). An analysis of variance (ANOVA) as well as Fisher's multiple-range tests of the minimal significant differences (LSD) were performed to find significant differences between C and NC processes in the volatile profile. A PCA was used with the whole data.

5. Conclusions

The profile of volatile compounds was similar in cooking vinasses (C) and not-cooking steam vinasses, but some differences were found in the concentration of volatile compounds, suggesting that the cooking process has an influence on the profile obtained in the vinasse composition. The cooking process increased the content of furanic compounds and organic acids, which are described as inhibitors of DF biohydrogen production from wastewater. The tequila vinasses obtained from a not-cooking process showed a higher presence of phenolic compounds. Although, the identified phenolic compounds have not been reported as inhibitors of the DF process, they may inhibit other biological processes in other matrices such as soil. Detailed description of the vinasses' composition could help in the development of any process to eliminate compounds that may interfere with biological processes to treat or use those or similar waste products around the world.

Acknowledgments: The investigation presented in this paper was supported by the National Council for Science and Technology (CONACYT-BASIC SCIENCE) through project No. CB 2013-222677. In addition, Elizabeth Rodríguez-Félix thanks CONACYT for the masters scholarship No. 421959. The authors want to dedicate this manuscript to the memory of Gustavo Davila-Vazquez (R.I.P.), outstanding scientist, friend, academic tutor, co-author and co-worker in this research whose ideas triggered the basis for this and several investigations in this field, who stood out for his dedication and enthusiasm at every moment.

Author Contributions: Gustavo Davila-Vazquez (R.I.P) was the lead researcher in this project who conceived and designed the experiments. Elizabeth Rodríguez-Félix conducted the experiments and wrote the manuscript. Jacobo Rodríguez-Campos contributed with analytical methods (gas chromatography) and statistical analysis capacitation. Silva Maribel Contreras-Ramos and Erika Nahomy Marino-Marmolejo contributed with reagents/material/analytical tools, provided insightful suggestions on the work, and revised the paper.

Conflicts of Interest: The authors declare no conflict of interest.

References

1. Morin Couallier, E.; Salgado Ruiz, B.; Lameloise, M.-L.; Decloux, M. Usefulness of reverse osmosis in the treatment of condensates arising from the concentration of distillery vinasses. *Desalination* **2006**, *196*, 306–317. [CrossRef]

2. Zhang, P.J.; Zhao, Z.G.; Yu, S.J.; Guan, Y.G.; Li, D.; He, X. Using strong acid-cation exchange resin to reduce potassium level in molasses vinasses. *Desalination* **2012**, *286*, 210–216. [CrossRef]

3. Alves, P.R.L.; Natal-da-Luz, T.; Sousa, J.P.; Cardoso, E.J.B.N. Ecotoxicological characterization of sugarcane vinasses when applied to tropical soils. *Sci. Total Environ.* **2015**, *526*, 222–232. [CrossRef] [PubMed]

4. Moran-Salazar, R.G.; Sanchez-Lizarraga, A.L.; Rodriguez-Campos, J.; Davila-Vazquez, G.; Marino-Marmolejo, E.N.; Dendooven, L.; Contreras-Ramos, S.M. Utilization of vinasses as soil amendment: Consequences and perspectives. *SpringerPlus* **2016**, *5*, 1007. [CrossRef] [PubMed]

5. Díaz, B.; Gomes, A.; Freitas, M.; Fernandes, E.; Nogueira, D.R.; González, J.; Moure, A.; Levoso, A.; Vinardell, M.P.; Mitjans, M.; et al. Valuable Polyphenolic Antioxidants from Wine Vinasses. *Food Bioprocess Technol.* **2012**, *5*, 2708–2716. [CrossRef]

6. Cedeño, M.C. Tequila Production. *Crit. Rev. Biotechnol.* **1995**, *15*, 1–11. [CrossRef] [PubMed]

7. Méndez-Acosta, H.O.; Snell-Castro, R.; Alcaraz-González, V.; González-Álvarez, V.; Pelayo-Ortiz, C. Anaerobic treatment of Tequila vinasses in a CSTR-type digester. *Biodegradation* **2010**, *21*, 357–363. [CrossRef] [PubMed]

8. Robles-González, V.; Galíndez-Mayer, J.; Rinderknecht-Seijas, N.; Poggi-Varaldo, H.M. Treatment of mezcal vinasses: A review. *J. Biotechnol.* **2012**, *157*, 524–546. [CrossRef] [PubMed]

9. Rajagopal, V.; Paramjit, S.M.; Suresh, K.P.; Yogeswar, S.; Nageshwar, R.D.V.K.; Avinash, N. Significance of vinasses waste management in agriculture and environmental quality-Review. *Afr. J. Agric. Res.* **2014**, *9*, 2862–2873. [CrossRef]

10. López-López, A.; Davila-Vazquez, G.; León-Becerril, E.; Villegas-García, E.; Gallardo-Valdez, J. Tequila vinasses: Generation and full scale treatment processes. *Rev. Environ. Sci. Bio/Technol.* **2010**, *9*, 109–116. [CrossRef]

11. Decloux, M.; Bories, A.; Lewandowski, R.; Fargues, C.; Mersad, A.; Lameloise, M.L.; Bonnet, F.; Dherbecourt, B.; Osuna, L.N. Interest of electrodialysis to reduce potassium level in vinasses. Preliminary experiments. *Desalination* **2002**, *146*, 393–398. [CrossRef]

12. España-Gamboa, E.; Mijangos-Cortes, J.; Barahona-Perez, L.; Dominguez-Maldonado, J.; Hernández-Zarate, G.; Alzate-Gaviria, L. Vinasses: Characterization and treatments. *Waste Manag. Res.* **2011**, *29*, 1235–1250. [CrossRef] [PubMed]

13. Cruz-Salomón, A.; Meza-Gordillo, R.; Rosales-Quintero, A.; Ventura-Canseco, C.; Lagunas-Rivera, S.; Carrasco-Cervantes, J. Biogas production from a native beverage vinasse using a modified UASB bioreactor. *Fuel* **2016**, *198*, 170–174. [CrossRef]

14. Da Silva, M.A.; Griebeler, N.P.; Borges, L.C. Uso de vinhaça e impactos nas propriedades do solo e lençol freático (Use of stillage and its impact on soil properties and groundwater). *Rev. Bras. Eng. Agríc. Ambient* **2007**, *11*, 108–114. [CrossRef]

15. Espinoza-Escalante, F.M.; Pelayo-Ortíz, C.; Navarro-Corona, J.; González-García, Y.; Bories, A.; Gutiérrez-Pulido, H. Anaerobic digestion of the vinasses from the fermentation of Agave tequilana Weber to tequila: The effect of pH, temperature and hydraulic retention time on the production of hydrogen and methane. *Biomass Bioenergy* **2009**, *33*, 14–20. [CrossRef]

16. Waleckx, E.; Gschaedler, A.; Colonna-Ceccaldi, B.; Monsan, P. Hydrolysis of fructans from Agave tequilana Weber var. azul during the cooking step in a traditional tequila elaboration process. *Food Chem.* **2008**, *108*, 40–48. [CrossRef]

17. Valenzuela, A.G. *The Tequila Industry in Jalisco, Mexico*; University of Arizona: Tucson, AZ, USA, 2016.

18. Martínez, M.D.C.; Pérez, E.L. Análisis del Mercado Potencial del Tequila 100% Agave. PhD. Thesis, Instituto Politécnico Nacional, Mexico City, Mexico, 2008.

19. Vazquez-Landaverde, P.A.; Rodriguez-Olvera, M.G. Tequila processing and flavor. *Am. Chem. Soc.* **2012**, *1104*, 237–276. [CrossRef]

20. Salgado, O. *Perspectivas Económicas de la Producción de Tequila vs. Tequila 100% Agave*; Universidad Autónoma de Querétaro: Queretaro, Mexico, 2012.

21. Cedeño Cruz, M.; Alvarez Jacobs, J. Production of tequila from agave: Historical influences and contemporary processes. In *The Alcohol Textbook: A Reference for the Beverage, Fuel and Industrial Alcohol Industries*; Nottingham University Press; Alltech Inc.: Nottingham, UK, 1999; pp. 225–241.

22. Villanueva-Rodríguez, S.; Rodríguez-Garay, B.; Prado-Ramírez, R.; Gschaedler, A. Tequila: Tequila Raw Material, Classification, Process, and Quality Parameters. In *Encyclopedia of Food and Health (FOHE)*; Academic Press: Waltham, MA, USA, 2015; pp. 1–7.

23. Villanueva-Rodriguez, S.; Escalona-Buendia, H. *Tequila and mezcal: Sensory Attributes and Sensory Evaluation*; Woodhead Publishing Limited: Cambridge, UK, 2011; ISBN 9780857090515.

24. Buitrón, G.; Carvajal, C. Biohydrogen production from Tequila vinasses in an anaerobic sequencing batch reactor: Effect of initial substrate concentration, temperature and hydraulic retention time. *Bioresour. Technol.* **2010**, *101*, 9071–9077. [CrossRef] [PubMed]

25. González, C.; Durán, J.E. Producción de hidrógeno a partir del tratamiento anaerobio de vinazas en un reactor UASB. *Tecnol. Marcha* **2014**, *27*, 3–12. [CrossRef]

26. Buitrón, G.; Prato-Garcia, D.; Zhang, A. Biohydrogen production from tequila vinasses using a fixed bed reactor. *Water Sci. Technol.* **2014**, *70*, 1919–1925. [CrossRef] [PubMed]

27. Fagier, M.A.; Elmugdad, A.A.; Aziz, M.E.A.; Gabra, N.M. Identification of some Organic Compounds in Sugarcane vinasse by Gas Chromatography-Mass Spectrometry and Prediction of their Toxicity Using TEST Method. *J. Chem. Pharm. Sci.* **2015**, *8*, 899–905.

28. Lima, A.V.A.; Barbosa, M.A.S.; Cunha, L.C.S.; De Morais, S.A.L.; De Aquino, F.J.T.; Chang, R.; Do Nascimento, E.A. Volatile compounds obtained by the hydrodistillation of sugarcane vinasse, a residue from ethanol production. *Rev. Virtual Quim.* **2017**, *9*, 764–773. [CrossRef]

29. Mancilla-Margalli, N.A.; López, M.G. Generation of Maillard compounds from inulin during the thermal processing of Agave tequilana Weber var. azul. *J. Agric. Food Chem.* **2002**, *50*, 806–812. [CrossRef] [PubMed]

30. Prado-Jaramillo, N.; Estarrón-Espinosa, M.; Escalona-Buendía, H.; Cosío-Ramírez, R.; Martín-del-Campo, S.T. Volatile compounds generation during different stages of the tequila production process: A preliminary study. *LWT-Food Sci. Technol.* **2015**, *61*, 471–483. [CrossRef]

31. Chen, Y.; Cheng, J.J.; Creamer, K.S. Inhibition of anaerobic digestion process: A review. *Bioresour. Technol.* **2008**, *99*, 4044–4064. [CrossRef] [PubMed]

32. Quéménur, M.; Hamelin, J.; Barakat, A.; Steyer, J.; Carrere, H.; Trably, E. Inhibition of fermentative hydrogen production by lignocellulose-derived compounds in mixed cultures. *Int. J. Hydrogen Energy* **2012**, *7*, 3150–3159. [CrossRef]

33. De Sá, L.R.V.; De Oliveira Moutta, R.; Da Silva Bon, E.P.; Cammarota, M.C.; Ferreira-Leitão, V.S. Fermentative biohydrogen production using hemicellulose fractions: Analytical validation for C5 and C6-sugars, acids and inhibitors by HPLC. *Int. J. Hydrogen Energy* **2015**, *40*, 13888–13900. [CrossRef]

34. Bundhoo, M.A.Z.; Mohee, R. Inhibition of dark fermentative bio-hydrogen production: A review. *Int. J. Hydrogen Energy* **2016**, *41*, 6713–6733. [CrossRef]

35. Jolliffe, I.T. Principal Component Analysis. *J. Technometrics* **1988**, *30*, 487. [CrossRef]

36. Siqueira, M.R.; Reginatto, V. Inhibition of fermentative H_2 production by hydrolysis byproducts of lignocellulosic substrates. *Renew. Energy* **2015**, *80*, 109–116. [CrossRef]

37. Wang, B.; Wan, W.; Wang, J. Inhibitory effect of ethanol, acetic acid, propionic acid and butyric acid on fermentative hydrogen production. *Int. J. Hydrog. Energy* **2008**, *33*, 7013–7019. [CrossRef]

38. Franden, M.A.; Pilath, H.M.; Mohagheghi, A.; Pienkos, P.T.; Zhang, M. Inhibition of growth of Zymomonas mobilis by model compounds found in lignocellulosic hydrolysates. *Biotechnol. Biofuels* **2013**, *6*, 99. [CrossRef] [PubMed]

39. Ciranna, A.; Ferrari, R.; Santala, V.; Karp, M. Inhibitory effects of substrate and soluble end products on biohydrogen production of the alkalithermophile Caloramator celer: Kinetic, metabolic and transcription analyses. *Int. J. Hydrogen Energy* **2014**, *39*, 6391–6401. [CrossRef]

40. Mills, T.Y.; Sandoval, N.R.; Gill, R.T. Cellulosic hydrolysate toxicity and tolerance mechanisms in Escherichia coli. *Biotechnol. Biofuels* **2009**, *2*, 26. [CrossRef] [PubMed]

41. Zaldivar, J.; Martinez, A.; Ingram, L.O. Effect of selected aldehydes on the growth and fermentation of ethanologenic Escherichia coli. *Biotechnol. Bioeng.* **1999**, *65*, 24–33. [CrossRef]

42. Banerjee, N.; Bhatnagar, R.; Viswanathan, L. Inhibition of glycolysis by furfural in Saccharomyces cerevisiae. *Eur. J. Appl. Microbiol. Biotechnol.* **1981**, *11*, 226–228. [CrossRef]

43. Palmqvist, E.; Hahn-Hagerald, B. Fermentation of lignocellulosic hydrolysates. II: Inhibitors and mechanisms of inhibition mechanisms of inhibition. *Bioresour. Technol.* **2000**, *74*, 25–33. [CrossRef]

44. Abha, S.; Singh, C.S. Hydrocarbon Pollution: Effects on Living Organisms, Remediation of Contaminated Environments, and Effects of Heavy Metals Co-Contamination on Bioremediation. In *Introduction to Enhanced Oil Recovery (EOR) Processes and Bioremediation of Oil-Contaminated Sites*; INTECH: Rijeka, Croatia, 2012; pp. 185–206.

45. Alrumman, S.A.; Standing, D.B.; Paton, G.I. Effects of hydrocarbon contamination on soil microbial community and enzyme activity. *J. King Saud Univ. Sci.* **2015**, *27*, 31–41. [CrossRef]

46. Hamamura, N.; Ward, D.M.; Inskeep, W.P. Effects of petroleum mixture types on soil bacterial population dynamics associated with the biodegradation of hydrocarbons in soil environments. *FEMS Microbiol. Ecol.* **2013**, *85*, 168–178. [CrossRef] [PubMed]

47. Haroun, B.M.; Nakhla, G.; Hafez, H.; Nasr, F.A. Impact of furfural on biohydrogen production from glucose and xylose in continuous-flow systems. *Renew. Energy* **2016**, *93*, 302–311. [CrossRef]

48. Modig, T.; Lidén, G.; Taherzadeh, M.J. Inhibition effects of furfural on alcohol dehydrogenase, aldehyde dehydrogenase and pyruvate dehydrogenase. *Biochem. J.* **2002**, *363*, 769–776. [CrossRef] [PubMed]

49. Monlau, F.; Sambusiti, C.; Barakat, A.; Quéméneur, M.; Trably, E.; Steyer, J.P.; Carrère, H. Do furanic and phenolic compounds of lignocellulosic and algae biomass hydrolyzate inhibit anaerobic mixed cultures? A comprehensive review. *Biotechnol. Adv.* **2014**, *32*, 934–951. [CrossRef] [PubMed]

50. Lin, R.; Cheng, J.; Ding, L.; Song, W.; Zhou, J.; Cen, K. Inhibitory effects of furan derivatives and phenolic compounds on dark hydrogen fermentation. *Bioresour. Technol.* **2015**, *196*, 250–255. [CrossRef] [PubMed]

51. Tejirian, A.; Xu, F. Inhibition of enzymatic cellulolysis by phenolic compounds. *Enzyme Microb. Technol.* **2011**, *48*, 239–247. [CrossRef] [PubMed]

52. Qin, L.; Li, W.-C.; Liu, L.; Zhu, J.-Q.; Li, X.; Li, B.-Z.; Yuan, Y.-J. Inhibition of lignin-derived phenolic compounds to cellulase. *Biotechnol. Biofuels* **2016**, *9*, 70. [CrossRef] [PubMed]

53. Min, K.; Freeman, C.; Kang, H.; Choi, S.-U.; Min, K.; Freeman, C.; Kang, H.; Choi, S.-U. The Regulation by Phenolic Compounds of Soil Organic Matter Dynamics under a Changing Environment. *Biomed Res. Int.* **2015**, *2015*, 825098. [CrossRef] [PubMed]

54. Wacker, T.L.; Safir, G.R.; Stephens, C.T. Effects of ferulic acid on *Glomus fasciculatum* and associated effects on phosphorus uptake and growth of asparagus (*Asparagus officinalis* L.). *J. Chem. Ecol.* **1990**, *16*, 901–909. [CrossRef] [PubMed]

55. Fries, L.L.M.; Pacovsky, R.S.; Safir, G.R.; Siqueira, J.O. Plant growth and arbuscular mycorrhizal fungal colonization affected by exogenously applied phenolic compounds. *J. Chem. Ecol.* **1997**, *23*, 1755–1767. [CrossRef]

56. Lattanzio, V.; Lattanzio, V.M.T.; Cardinali, A.; Amendola, V. Role of phenolics in the resistance mechanisms of plants against fungal pathogens and insects. *Phytochem. Adv. Res.* **2006**, *661*, 23–67.

57. Li, H.H.; Inoue, M.; Nishimura, H. Interactions oftrans-cinnamic acid, its related phenolic allelochemicals, and abscisic acid in seedling growth and seed germination of lettuce. *J. Chem. Ecol.* **1993**, *19*, 1775–1787. [CrossRef] [PubMed]

58. Heipieper, H.J.; Keweloh, H.; Rehm, H.J. Influence of phenols on growth and membrane permeability of free and immobilized Escherichia coli. *Appl. Environ. Microbiol.* **1991**, *57*, 1213–1217. [PubMed]

59. Alves, M.J.; Ferreira, I.C.F.R.; Froufe, H.J.C.; Abreu, R.M.V.; Martins, A.; Pintado, M. Antimicrobial activity of phenolic compounds identified in wild mushrooms, SAR analysis and docking studies. *J. Appl. Microbiol.* **2013**, *115*, 346–357. [CrossRef] [PubMed]

60. Campos, F.M.; Couto, J.A.; Hogg, T.A. Influence of phenolic acid on growth and inactivation of Oenococcus oeni and Lactobacillus hilgardii. *J. Appl. Microbiol.* **2003**, *94*, 167–174. [CrossRef] [PubMed]

61. Priou, S.; Aley, P.; Chujoy, E.; Lemaga, B.; French, E. *Control Integrado de la Marchitez Bacteriana de la Papa;* Food and Agriculture Orgnization of the United Nations: Rome, Italy, 1999.

62. Gallego, A.; Giordano, W.; Rosero, E.; Echeverri, F. Effect of furans and a Pyran on Several Quorum Sensing Factors in Ralstonia solanacearum. *J. Microb. Biochem. Technol.* **2016**, *8*, 1–5. [CrossRef]

63. Salazar-Leyva, J.A.; Osuna-Ruiz, I.; Rodríguez-Tirado, V.A.; Zazueta-Patrón, I.E.; Brito-Rojas, H.D. Optimization study of fructans extraction from Agave tequilana Weber azul variety. *Food Sci. Technol.* **2016**, *36*, 631–637. [CrossRef]

64. Rong, C.; Ding, X.; Zhu, Y.; Li, Y.; Wang, L.; Qu, Y.; Ma, X.; Wang, Z. Production of furfural from xylose at atmospheric pressure by dilute sulfuric acid and inorganic salts. *Carbohydr. Res.* **2012**, *350*, 77–80. [CrossRef] [PubMed]

65. Gomes, F.N.D.C.; Pereira, L.R.; Ribeiro, N.F.P.; Souza, M.M.V.M. Production of 5-hydroxymethylfurfural (HMF) via fructose dehydration: Effect of solvent and salting-out. *Braz. J. Chem. Eng.* **2015**, *32*, 119–126. [CrossRef]

66. Iñiguez-Covarrubias, G.; Díaz-Teres, R.; Sanjuan-Dueñas, R.; Anzaldo-Hernández, J.; Rowell, R. Utilization of by-products from the tequila industry. Part 2: Potential value of Agave tequilana Weber azul leaves. *Bioresour. Technol.* **2001**, *77*, 101–108. [CrossRef]

67. Jönsson, L.J.; Martín, C. Pretreatment of lignocellulose: Formation of inhibitory by-products and strategies for minimizing their effects. *Bioresour. Technol.* **2016**, *199*, 103–112. [CrossRef] [PubMed]

68. Verardi, A.; De Bari, I.; Ricca, E.; Calabrò, V. Hydrolysis of Lignocellulosic Biomass: Current Status of Processes and Technologies and Future Perspectives. In *Bioethanol*; INTECH: Rijeka, Croatia, 2012; p. 290.

69. Rodriguez-Campos, J.; Escalona-Buendía, H.B.; Contreras-Ramos, S.M.; Orozco-Avila, I.; Jaramillo-Flores, E.; Lugo-Cervantes, E. Effect of fermentation time and drying temperature on volatile compounds in cocoa. *Food Chem.* **2012**, *132*, 277–288. [CrossRef] [PubMed]

energies

MDPI

Article

Biological Pretreatment of Mexican Caribbean Macroalgae Consortiums Using Bm-2 Strain (*Trametes hirsuta*) and Its Enzymatic Broth to Improve Biomethane Potential

Raúl Tapia-Tussell [1], Julio Avila-Arias [1], Jorge Domínguez Maldonado [1], David Valero [1], Edgar Olguin-Maciel [1], Daisy Pérez-Brito [2] and Liliana Alzate-Gaviria [1,*]

[1]   Unidad de Energía Renovable, Centro de Investigación Científica de Yucatán AC, Carretera Sierra Papacal-Chuburná Puerto, Km 5, Sierra Papacal, Mérida, Yucatán CP 97302, Mexico; rtapia@cicy.mx (R.T.-T.); ja.avila.arias@gmail.com (J.A.-A.); joe2@cicy.mx (J.D.M.); davidvalero1986@gmail.com (D.V.); edgar.olguin@cicy.mx (E.O.-M.)

[2]   Laboratorio GeMBio, Centro de Investigación Científica de Yucatán AC, Calle No. 130 43 x 32 y 34 Col. Chuburná de Hidalgo, Mérida, Yucatán CP 97205, Mexico; daisypb@cicy.mx

*   Correspondence: lag@cicy.mx; Tel.: +52-999-942-8330

Received: 30 January 2018; Accepted: 19 February 2018; Published: 27 February 2018

Abstract: The macroalgae consortium biomass in the Mexican Caribbean represents an emerging and promising biofuel feedstock. Its biological pretreatment and potential for energetic conversion to biomethane were investigated, since some macroalgae have hard cell walls that present an obstacle to efficient methane production when those substrates are used. It has been revealed by anaerobic digestion assays that pretreatment with a Bm-2 strain (*Trametes hirsuta*) isolated from decaying wood in Yucatan, Mexico was 104 L $CH_4 \cdot kg$ VS^{-1}; In fact, the fungal pretreatment produced a 20% increase in methane yield, with important amounts of alkali metals Ca, K, Mg, Na of 78 g/L, ash 35.5% and lignin 15.6%. It is unlikely that high concentrations of ash and alkali metals will produce an ideal feedstock for combustion or pyrolysis, but they can be recommended for a biological process.

Keywords: biological pretreatment; macroalgae consortium; biomethane potential; *Trametes hirsuta*

1. Introduction

Although there are a large number of species of macroalgae, the majority of shoreline inundation incidents are caused by two genera: *Ulva* is a green macroalga, which causes green tides, and *Sargassum*, a golden/brown floating or pelagic macroalga responsible for golden tides, especially in the Caribbean and West Africa [1].

The genus *Sargassum* contains over 350 species. Two of these, *S. natans* and *S. fluitans*, are holopelagic and the major contributory species in golden tides. Both of these have a vegetative reproduction process, and do not physically connect to the ocean floor during their lifecycle [1].

Pelagic *Sargassum* (gulfweed) has been called a floating jungle or floating golden rainforest, as it provides a habitat for a wide range of invertebrates, fishes, sea turtles, birds, and mammals: over 100 species of fish and 145 species of invertebrate have been found linked to it. *Sargassum*'s importance lies not only in its ecological role, but also in the part it plays worldwide in oceanic carbon sequestration: the *Sargassum* of the Sargasso Sea is a net sink of CO_2, and represents 7% of the planet's net "carbon pump". It is also the largest accumulation of macroalgae in the world, containing a total biomass of 10 million tons. Approximately one million tons of *Sargassum* passes through the Florida Straits from the Gulf of Mexico towards that sector of the Atlantic Ocean known as the Sargasso Sea each year [2].

During the 2015 Caribbean inundation, 10,000 wet tons of macroalgae were being cast up onto Caribbean island beaches every day. The precise cause of the *Sargassum* inundations in the last few years is not completely understood, although it is believed that both climate change and coastal sea eutrophication are involved. Nor is it clear whether the large golden tides of 2011–2015 will constitute a phenomenon that will be repeated in following years [1].

Sargassum inundations have a part to play in maintaining the stability of beaches, and they also provide a food source for beach and dune vegetation. *Sargassum* cast up on a beach is not thought to present any risk to human health, although beach users may dislike the presence of large quantities. It can also have unwelcome environmental effects, suffocating animals such as turtle hatchlings. Tourism provided more than 80% of the Caribbean region's GDP in 2014, bringing $29.2 billion in local spending. On the other hand, removing *Sargassum* from beaches or stopping its arrival might prove extremely expensive. Taking as a basis the figure of $5 million spent on cleaning Mexican beaches, it is possible to calculate that at least $120 million would be required to remove *Sargassum* inundations throughout the Caribbean. Clean, renewable biomass from the ocean might provide an outstanding addition to our present range of alternative energy sources [3,4]. At the present time, the application and viability of anaerobic digestion technology on macroalgae in the long term have not been established, and it is worth researching both mono- and codigestion methods. The currently available literature on monodigestion of seaweeds in continuous processes is limited [5]. Specifically, it is possible to convert macroalgae into biogas (60% methane) through anaerobic digestion. Algal biogas could potentially reduce greenhouse gas emissions by more than half and fossil depletion by nearly 70%, in comparison with natural gas [6].

White rot fungi have a well-known ability to decompose a wide spectrum of environmentally persistent xenobiotics and organopollutants. Similarly, recent tests have been made on biological methods such as using enzymes. They are considered relatively cheap, environmentally friendly pretreatments for improving the anaerobic biodegradability of macroalgal and microalgal biomass. Choice of enzymes is determined by the main compounds forming the macroalgal and microalgal cell wall—specifically, cellulose, hemicelluloses, pectin, glycoproteins, and even lignin. Depending on the culture conditions, Ligninolytic fungi generate nonnspecific intracellular and extracellular enzymes. The presence of various inducers leads to the production of laccases, and how the latter affect metabolic activity and cell growth depends on environmental conditions and specific regulatory mechanisms. The white-rot fungus *Trametes* sp. is one of the best-known laccase-producing fungi [7,8].

Thus, the objective of this work was to evaluate the potential of biogas production from macroalgal biomass of the Mexican Caribbean in tests of biochemical methane potential (BMP), using a strain Bm-2 (*Trametes hirsuta*) isolated in Yucatan, Mexico, and its enzymatic broth as biological pretreatments, as well as structural changes occurred during the pretreatment of this biomass.

2. Materials and Methods

2.1. Macroalgae Consortium Biomass and Characterization

Samples of a mixture of macroalgae biomass was manually collected on the shore in Progreso (Yucatan, Mexico) in January 2017, in order to reproduce the conditions of harvesting readily available biomass on the beach. The samples were washed with tap water several times to remove impurities like salts and sands, and then dried at 105 °C (APHA 2005). The biomass was stored in a cold room. The characteristics of macroalgae consortiums are summarized in Table 1.

Table 1. Main characteristics of macroalgae biomass.

Component Name	Average (%)
Lignin	15.6 ± 0.51
Cellulose	31.2 ± 1.2
Hemicellulose	10.5 ± 0.83
Phenols	18.7 ± 1.5
Extractable in solvent	3.1 ± 0.04
Ash	35.5 ± 1.9

2.2. Fungal Strain and Culture Conditions

Trametes hirsuta strain Bm-2 (GQ280373) was isolated from decaying wood in Yucatan, Mexico [9], and used throughout this work. The strain was maintained by periodical subculturing on plates with 2% malt extract and 2% bacteriological agar. A mycelia suspension of *T. hirsuta* was obtained by inoculating four 1-cm diameter plugs in a 250 mL Erlenmeyer flask containing Kirk's liquid basal medium, pH 6, which consisted of 10 g glucose, 5 g ammonium tartrate, 0.2 g $MgSO_4 \cdot 7H_2O$, 2 g KH_2PO_4, 0.1 g $CaCl_2 \cdot 2H_2O$, 1 mg Thiamine, and 10 mL trace compound solution, without Tween 20 [10,11]. The flask was incubated at 35 °C, 150 rpm for 3 days. The biomass obtained was homogenized with the T18 digital ULTRA-TURRAX® by IKA®. The resulting suspension was used as inoculum in this work.

2.3. Fungal Broth Production (Fb)

T. hirsuta broth was produced in the YMPG medium whose composition ($g \cdot L^{-1}$) was as follows: Yeast extract 2 g, Malt extract 10 g, Peptone 2 g, Glucose 10 g, KH_2PO_4 2 g, $MgSO_4 \cdot 7H_2O$ 50 g, Thiamine hydrochloride 1 mg, Asparagine 1 g and wheat bran 2%. The pH was adjusted to 4.5 with HCl. 50 mL of the medium was transferred to 250 mL flasks and sterilized at 121 °C for 15 min in an autoclave. One mL of fungus inoculum was added in each flask and incubated at 35 °C, 150 rpm for 8 days. The liquid cultures obtained after fungal biomass separation were spectrophotometrically analyzed for enzyme activity at 25 °C. Laccase activity was determined using the substrate 2,2′-azino-bis(3-ethylbenzothiazoline-6-sulfonic acid, Sigma-Aldrich) (ABTS) and was measured at 420 nm (ε max = 36,000 $L \cdot mol^{-1} \cdot cm^{-1}$) [12]. One unit of laccase activity was defined as the amount necessary to catalyze the formation of 1 mmol of oxidized ABTS min^{-1}.

2.4. Biologic Pretreatment

Two biological pretreatments were carried out. In both pretreatments, 10 g of the dry macroalgae consortium (Mc) were suspended in 100 mL of water (10% w/v). The pH of the Mc suspension was adjusted to 5 with an acid solution of HCl (1 N).

In the first case suspended Mc were inoculated with 5 mL of a mycelial suspension of *T. hirsuta* (McF), obtained as described in Section 2.3 and incubated at 35 °C, 150 rpm in an orbital shaker for 6 days prior to BMP tests. In the second case, the Mc suspension was inoculated with 10 mL fungal broth (Fb) with laccase activity of 7000 $U \cdot mL^{-1}$, obtained as described in Section 2.4. The resultant suspension (McFb) was incubated at 40 °C, 150 rpm for 24 h in an orbital shaker. Both pretreatments were assayed in triplicate.

2.5. Fourier Transforms Infrared (FT-IR) Analysis

Changes in the functional groups of untreated macroalgae and after biological and enzymatic pretreatment were observed using a Bruker spectrophotometer FT-IR Tensor II model (Milton, ON, Canada). Samples were analyzed in Platinum ATR (Attenuated Total Reflection) and spectra were obtained in the region of 4000–500 cm^{-1}, at a resolution of 4 cm^{-1} with 32 scans.

2.6. Scanning Electron Microscopy (SEM)

To study the microscopic structure of the macroalgae and the effect of pretreatment, a scanning electron microscope (SEM, model JSM-6360LV, JEOL, Tokyo, Japan) was used. Flour samples were mounted on a metallic stub using double-sided adhesive tape coated with a 15-nm gold layer and observed at 20 kV.

2.7. Biochemical Methane Potential (BMP) Tests

The BMP tests followed protocols and calculation in accordance with Valero et al. [13]. After biological pretreatment of macroalgae, BMP tests were carried out in 115 mL triplicate serum bottles capped with rubber septum sleeve stoppers, with a useful volume of 55 mL and a headspace volume of 60 mL. The inoculum consisted of a native mixed microbial consortium containing 30 g/L of deep soil, 300 g/L of cattle manure, 150 g/L of pig manure, 1.5 g/L of commercial Na_2CO, 5 g/L of sugar, and 1000 mL of tap water as described [14]. Previous to the BMP test, the inoculum was degassed for five days at 38 °C. The inoculum/substrate ratio was 2.05 g VS inoculum/g VS substrate for the three trials set-up: macroalgae consortium (Mc), macroalgae consortium + Fungal broth production (McFb), macroalgae consortium + *T. hirsuta* (McF).

Furthermore, the bottles were filled with distilled water up to 55 mL. Nitrogen was flushed to remove the air in the headspace and they were placed in an incubator at 38 °C for a period of 29 days. All the reactors were manually agitated once a day. Biogas production was measured by the pressure increase in the headspace volume. After each measurement, the reactors were vented until atmospheric pressure in the reactor was reached. Three blanks with 55 g of inoculum were also tested to measure the methane potential of the inoculum.

2.8. Analytical Methods

A digital pressure transducer with silicon measuring cell (ifm, type PN78, up to 2000 mbar, Essen, Germany)—was used to measure headspace pressure of the reactors through the septum with a connected syringe to the transducer. Biogas characterization was determined on a Molesieve column (30 m long, 0.53 mm internal diameter and 0.25 µm film thickness) in a gas chromatograph (Clarus 500-Perkin Elmer, Waltham, MA, USA) with the thermal conductivity detector (TCD), nitrogen was used as gas carrier and temperatures of 75, 30 and 200 °C for the injector, oven and detector respectively. Conductivity, temperature, pH and total solids (TS), volatile solids (VS) were analyzed following standard methods [15], colorimetric methods (Hach Company DR-890, Loveland, CO, USA) was used to determinate chemical oxygen demand (COD).

2.9. Statistical Analysis

All measurements on the physicochemical parameters were performed in triplicate. Results were expressed as means ± standard deviation (SD) subtracting methane production from the inoculum using Statistica 9 software.

2.10. Numerical Calculations

The accumulated volumes of methane were calculated by the cumulative summation of methane volumes in accordance with Valero et al. [13]

3. Results and Discussion

3.1. Biological Pretreatment

The fungal broth, obtained from the growth of *T. hirsuta* in a 2% wheat-bran-enriched Kirk medium, showed an extracellular laccase activity of 7000 U·mL^{-1} much higher than the value (170 U·mL^{-1} at four days) reported by Hom-Diaz et al. [8], for *T. versicolor* grown in Kirk medium.

Our results demonstrate that wheat bran (lignocellulosic substrate) is an efficient inducer for the production of this enzyme at an extracellular level. Previous works have shown that the Bm-2 strain has a battery of laccase genes, which are differentially expressed when different phenolic compounds are present in the lignin structure [7]. Zapata-Castillo et al. [16] reported that *T. hirsuta* Bm-2 produces three laccase isoenzymes (Lac I, II, III) on wheat bran. These three share some properties, but are different in other characteristics. They are very tolerant to temperature and organic solvents, which underlines their significant role in a variety of processes.

The fungal broth produced by *T. hirsuta* culture in Kirk's medium with wheat bran 2% is mostly rich in laccase enzymes among other enzymes. Sing et al., 2013 [17] observed the secretion of other enzymes such as cellulases and hemicellulases after three days cultivation of *T. versicolor*. This is very important for macroalgae consortium cell wall degradation. These results offer a new direction for biotechnological processes using mixtures or combinations of oxidizing enzymes for biological pretreatment in order to improve biomethane production.

3.2. FT-IR and SEM Analysis Biology Pretreatment

Figure 1 shows the FT-IR spectral of the macroalgae consortium (Mc), both untreated and after biological pretreatment with fungi (McF) from strain *T. hirsuta* and fungal broth (McFb), which presented typical bands of major lignocellulosic biomass components.

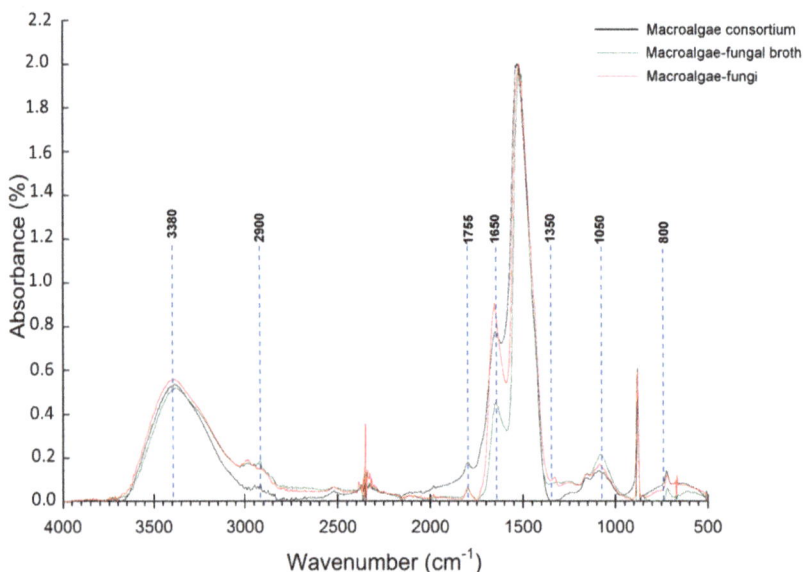

Figure 1. FT-IR spectral of untreated and pretreated macroalgae consortium sample. Black line: macroalgae consortium (Mc), green line: macroalgae–fungal broth (McFb) and red line: macroalgae fungi (McF).

The FT-IR spectral data display a strong, broad absorption band around 3380 cm^{-1}, which arises from the stretching of –OH groups. This band may be associated with the hydroxyl groups of cellulose, hemicellulose and lignin. These findings coincide with previously reported results for sugarcane bagasse submitted to a pretreatment process with steam explosion and sodium hydroxide [18–20].

An increase in signals for C–H stretching at approximately 2900 cm^{-1} and 1350 cm^{-1} was observed in the treatments with the fungi (McF) and fungal broth (McFb), which can be attributed to exposure of the amorphous and crystalline cellulose after the lignin has been solubilized. It coincides with the proposal of Nelson and O'Conner [21] on a method for calculating the "Total Crystalline Index" (TCI) as the ratio of absorption intensities from 1372 cm^{-1} to 2900 cm^{-1}. C–O–C stretching that is typical of xylans of the hemicellulose is believed to be responsible for the sharp band at 1050 cm^{-1} [19]. This confirms the delignification of the macroalgae consortium in the biologic treatment.

The lower band intensity at 1755 cm^{-1}, which can be seen in the spectrums of McF and McFb, is due to the fact that delignification of the biomass promotes a breakdown of the carbonyl ester bonds between hemicellulose and lignin. Laccase enzymes catalyze this delignification, which has been the subject of much research showing it to be effective both for the removal and/or the modification of the lignin polymer, and for the reduction of the phenolic content of pretreated lignocellulosic materials [22]. In general, the absorption spectral bands at 3380, 2900, 1350, 1050, and 800 cm^{-1} are associated with cellulose and hemicellulose core structures, as described by Ju et al. [19] in the chemical treatment of sugarcane bagasse.

3.3. SEM Analysis of Untreated and Pretreated Macroalgae Consortium

Generally, an efficient pretreatment strategy includes the removal of the cross-linked matrix of lignin and hemicelluloses, and an increase in the porosity and surface area of cellulose, for later enzymatic hydrolysis. In the scanning electron micrographs presented in Figure 2, the compact structure of lignocellulosic biomass of the macroalgae consortium before pretreatment can be observed (Figure 2a′), as can the structural alterations that took place on the surface of fibers subsequent to biological pretreatment with fungal broth (McFb; Figure 2b′) and fungi (McF; Figure 2c′) from the *T. hirsuta* strain.

In Figure 2a′, a smooth and compact surface of the epidermis of untreated cells can be seen, whereas in the treatment with the enzymatic extract (Figure 2b′) it is possible to observe a degradation of the fiber at the superficial level, which may be linked to the fact that the action of the laccases removed the lignin only partially [22].

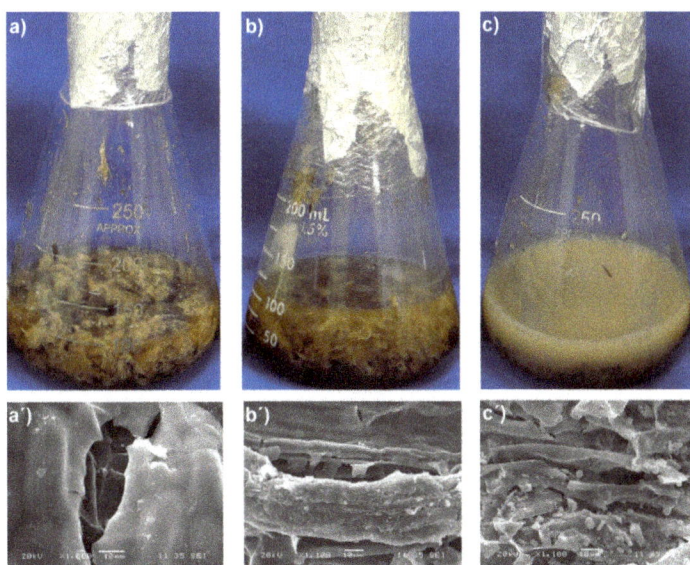

Figure 2. Scanning electron micrographs of macroalgae consortium samples before and after pretreatment. (**a,a′**): macroalgae consortium (Mc), (**b,b′**): macroalgae + fungal broth (McFb), (**c,c′**): macroalgae + fungi (McF).

Treatment with Bm-2 strain (*T. hirsuta*) (Figure 2c) promoted the formation of grooves on the surface of the material. These were probably a result of the lignocellulose fiber being completely destroyed, and the hemicellulose partially removed. This in turn is probably the result of the presence of other enzymes and other mediators generated by *T. hirsuta* during its culture, and these could also play a part in the solubilization of macroalgae consortium biomass.

3.4. Production of Biogas in BMP Test

Macroalgae Consortiums and BMP Tests

For the purposes of evaluating the increase in anaerobic biodegradability in BMP tests, the experiment was extended over 29 days, until accumulated production of methane leveled off as shown in Figure 3.

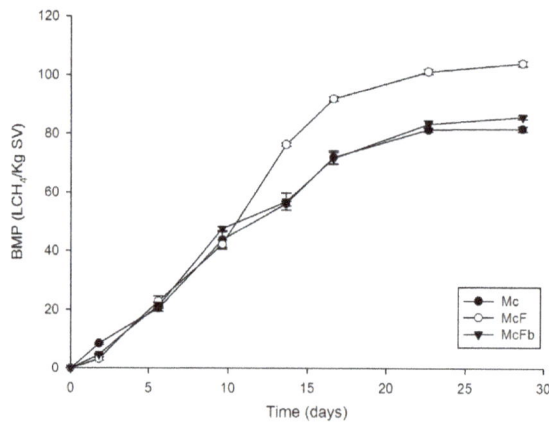

Figure 3. Accumulated production of methane for the anaerobic digestion of macroalgae consortium employing two pretreatments and a control. Macroalgae consortium (Mc); macroalgae consortium + fungal broth (McFb); macroalgae consortium + fungi (McF).

As can be seen from the results (Table 2), the macroalgae consortium + fungi trial produced the highest methane yield as compared to non-pretreated macroalgae consortium and macroalgae consortium + enzymatic Broth.

Table 2. Methane content and production for the various BMP test trials.

Trial	Methane Content (%)	BMP L $CH_4 \cdot kg \, VS^{-1}$
Macroalgae consortium	40 ± 1.0	81 ± 1.2
Macroalgae consortium + enzymatic broth	46 ± 0.5	86 ± 0.7
Macroalgae consortium + fungi	52 ± 1.0	104 ± 1.4

The macroalgae consortium used in this work was characterized by a $15.6 \pm 0.51\%$ of lignin, in accordance with Carrere et al. [23] lignin, cellulose, hemicelluloses, total uronic acids, proteins and soluble sugars content, as well as crystallinity of lignocellulosic substrates, were used to identify the most influential parameters linked to the biomethane potential values, the most important parameter being the lignin content (above 12%), which was negatively correlated to the BMP. Likewise, the large amount of ash in the macroalgae is linked to salt. Table 1 shows as that the ash content was high of 35.5% ± 1.9 with a biodegradability of 0.35 ± 0.02. Furthermore, this was observed in SEM analysis (Figure 4). Thus the buildup of salt in the digester over time may be a further factor in achieving long term anaerobic digestion, biodegradability and biomethane yields [5].

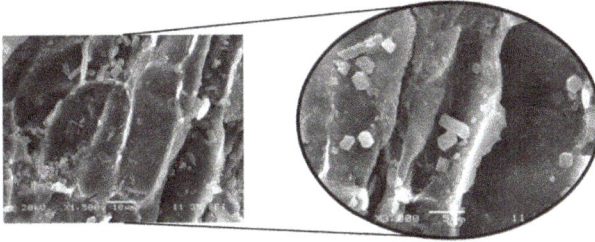

Figure 4. Content of salts, mainly sodium chloride and potassium chloride, in the macroalgae consortium.

Figure 2 (Figure 2a,b), shows how the feedstock had a lignocellulosic appearance, and it was evident, as in Figure 2c) that biological pretreatment with macroalgae consortium + fungi, which obtained 104 L $CH_4 \cdot kg \ VS^{-1}$, was higher to macroalgae consortium + enzymatic broth and macroalgae consortium, whose values were 85 and 82 L $CH_4 \cdot kg \ VS^{-1}$ respectively.

Furthermore, the C:N ratio of macroalgae consortium harvested in Progreso was 20:5 (in January), a value which is within the ideal range for anaerobic digestion (20–30:1). Adams et al. [24] and Oliveira et al. [25] found that at the beginning of the year, the macroalgae contained the lowest quantities of carbohydrate and the highest of ash and metals, whereas samples collected in July had the highest carbohydrate content, and the lowest quantity of metals and ash. Ross et al. [26] asserted that as the macroalgae progresses through its growth cycle, the biochemical composition alters. In March, the macroalgae normally have a high level of protein and alginic acid, and a low level of carbohydrate. During the spring the sugar content (e.g., mannitol) rises as photosynthesis increases, and there is a correlative drop in ash, proteins and alginic acid. These results also coincide with those of Tabassum et al. [5], who mentioned that *Fucus vesiculosus* and *Fucus serratus* harvested in March and subjected to biologic pretreatment had values of 126 L $CH_4 \cdot kg \ VS^{-1}$ and 101 L $CH_4 \cdot kg \ VS^{-1}$ respectively. Likewise the carbohydrate values were on average 31%, whereas in the macroalgae consortium in this study they were 22%, The carbohydrate content (alginic acid) in the seaweed or macroalgae is a significant parameter in the BMP; higher amounts produce higher BMPs [5,27].

The content of phenols in the macroalgae consortium was 19%, and these results are in agreement with Allen et al. [27], who stated that high levels of phenols (up to 14%) are natural inhibitors of the anaerobic digestion process, and a BMP yield of 166.3 L $CH_4 \cdot kg \ VS^{-1}$ and 110 L $CH_4 \cdot kg \ VS^{-1}$ was reported in Ireland and Norway [27,28], respectively, where a high level of phenol gives a low index of biodegradability and low kinetic decay values. Therefore, macroalgae consortiums are very good candidates for pretreatment processes and evaluation of variation depending on the season, because several shown that the level of polyphenols in seaweed or macroalgae varies during the year [5,29,30]. And as Tabassum et al. [5] mentions in his article, further investigation is necessary to assess the optimum month for harvesting macroalgae consortiums and a suitable pretreatment to improve their potential for biomethane.

The metals contents can be seen in Table 3, of the eight elements which were found, those showing the highest concentrations were Mg, K and Na. These results agree with Adams et al. [24]. Elements existing at comparatively elevated concentrations in macroalgae can also be poisonous or problematic when released as volatiles. Likewise, in this study the value of Zn was 60.1 and Adams et al. [24] found that Zn content in *Fucus vesiculosis* of 358 ppm contrasts to the 30 ppm detected in *L. digitata*. The size of the difference makes it improbable that it indicates simply a difference in the ability of the species to accumulate Zn.

Table 3. Methane content and production for the different Biochemical Methane Potential (BMP) test trials.

Component Name	Average (ppm)
Fe	584.9 ± 3
Mn	12.9 ± 2
Cu	7.5 ± 0.2
Zn	60.1 ± 0.1
Na	$45,210 \pm 25$
K	$28,500 \pm 55$
Cd	<0.1
Mg	4806.1 ± 20

In this study, the macroalgae consortium contains macrominerals such as Na, K, and Mg, and trace elements like Fe, Cu, Zn and Mn. These metals occurring in biomass species are significant in conversion processes because they have an effect on slagging, fouling and other problems linked to ash [26]. The sum of the alkali metals Ca, K, Mg, Na was 78 g·L^{-1}. This value was similar to that obtained by Ross et al. [26] in the first months of the year (98 ± 5 g·L^{-1}). Knowing the potassium and sodium content permits predicting serious problems in the possible use of this fuel in combustion or gas production. It likewise provides an estimate of the fouling potential of the ash [26].

4. Conclusions

The availability of macroalgae consortium waste used in this study, is promising for biomethane production, and it is a readily available resource that in the future, could meet goals for advanced biofuels in the Mexican Caribbean zone. Likewise, the effect of Bm-2 strain on macroalgae consortiums as a pretreatment step before proceeding to anaerobic digestion, is promising for the future scaling in continuous reactors. The enhancement of biogas production that reached 104 L CH$_4$·kg VS^{-1}, was statistically higher than values obtained with enzymatic pretreatments. Thus, the process we propose in this study can play a role in providing a solution to a problem currently faced by Mexican beaches, and contribute to providing clean, sustainable energy.

Acknowledgments: The authors would like to acknowledge the financial support received from CONACYT (National Council of Science and Technology Mexico) for postdoctoral fellowships (Grant No. 2138).

Author Contributions: All the authors contributed to this work. Raul Tapia-Tussell and Liliana Alzate-Gaviria conceived, designed and wrote the paper; Julio Avila-Arias, David Valero and Edgar Olguin-Maciel performed the experiments and analyzed the data; Daisy Perez-Brito and Jorge Dominguez-Maldonado participated in the data analysis and writing of the paper.

Conflicts of Interest: The authors declare no conflicts of interest.

References

1. Smetacek, V.; Zingone, A. Green and golden seaweed tides on the rise. *Nature* **2013**, *504*, 84–88. [CrossRef] [PubMed]
2. Krause-Jensen, D.; Duarte, C.M. Substantial role of Macroalgae in marine carbon sequestration. *Nat. Geosci.* **2016**, *9*, 737–742. [CrossRef]
3. Barbot, Y.N.; Al-Ghaili, H.; Benz, R. A review on the valorization of macroalgal wastes for biomethane production. *Mar. Drugs* **2016**, *14*, 120. [CrossRef] [PubMed]
4. Milledge, J.J.; Harvey, P.J. Golden tides: Problem or golden opportunity? The valorisation of sargassum from beach inundations. *J. Mar. Sci. Eng.* **2016**, *4*, 60. [CrossRef]
5. Tabassum, M.R.; Xia, A.; Murphy, J.D. Potential of seaweed as a feedstock for renewable gaseous fuel production in Ireland. *Renew. Sustain. Energy Rev.* **2017**, *68*, 136–146. [CrossRef]

6. Montingelli, M.E.; Benyounis, K.Y.; Quilty, B.; Stokes, J.; Olabi, A.G. Optimisation of biogas production from the Macroalgae *Laminaria* sp. at different periods of harvesting in Ireland. *Appl. Energy* **2016**, *177*, 671–682. [CrossRef]

7. Tapia-Tussell, R.; Pérez-Brito, D.; Torres-Calzada, C.; Cortés-Velázquez, A.; Alzate-Gaviria, L.; Chablé-Villacís, R.; Solís-Pereira, S. Laccase Gene Expression and Vinasse Biodegradation by *Trametes hirsuta* Strain Bm-2. *Molecules* **2015**, *20*, 15147–15157. [CrossRef] [PubMed]

8. Hom-Diaz, A.; Passos, F.; Ferrer, I.; Vicent, T.; Blánquez, P. Enzymatic pretreatment of microalgae using fungal broth from *Trametes versicolor* and commercial laccase for improved biogas production. *Algal Res.* **2016**, *19*, 184–188. [CrossRef]

9. Tapia-Tussell, R.; Pérez-Brito, D.; Rojas-Herrera, R.; Cortes-Velazquez, A.; Rivera-Muñoz, G.; Solis-Pereira, S. New laccase-producing fungi isolates with biotechnological potential in dye decolorization. *Afr. J. Biotechnol.* **2011**, *10*, 10134–10142.

10. Tien, M.; Kirk, T.K. Lignin-degrading enzyme from *Phanerochaete chrysosporium*: Purification, characterization, and catalytic properties of a unique H_2O_2-requiring oxygenase. *Proc. Natl. Acad. Sci. USA* **1984**, *81*, 2280–2284. [CrossRef] [PubMed]

11. Kirk, T.K.; Croan, S.; Tien, M.; Murtagh, K.E.; Farrell, R.L. Production of multiple ligninases by *Phanerochaete chrysosporium*: Effect of selected growth conditions and use of a mutant strain. *Enzym. Microb. Technol.* **1986**, *8*, 27–32. [CrossRef]

12. Wolfenden, B.S.; Willson, R.L. Radical-cations as reference chromogens in kinetic studies of ono-electron transfer reactions: Pulse radiolysis studies of 2,2'-azinobis-(3-ethylbenzthiazoline-6-sulphonate). *J. Chem. Soc. Perkin Trans. 2* **1982**, *0*, 805–812. [CrossRef]

13. Valero, D.; Montes, J.A.; Rico, J.L.; Rico, C. Influence of headspace pressure on methane production in Biochemical Methane Potential (BMP) tests. *Waste Manag.* **2016**, *48*, 193–198. [CrossRef] [PubMed]

14. Poggi-Varaldo, H.; Valdés, L.; Esparza-Garcia, F.; Fernández-Villagómez, G. Solid substrate anaerobic co-digestion of paper mill sludge, biosolids, and municipal solid waste. *Water Sci. Technol.* **1997**, *35*, 197–204.

15. Association, A.P.H.; Association, A.W.W.; Federation, W.P.C.; Federation, W.E. *Standard Methods for the Examination of Water and Wastewater*; American Public Health Association: Washington, DC, USA, 1915; Volume 2.

16. Zapata-Castillo, P.; Villalonga-Santana, L.; Islas-Flores, I.; Rivera-Muñoz, G.; Ancona-Escalante, W.; Solís-Pereira, S. Synergistic action of laccases from *Trametes hirsuta* Bm2 improves decolourization of indigo carmine. *Lett. Appl. Microbiol.* **2015**, *61*, 252–258. [CrossRef] [PubMed]

17. Singh, P.; Sulaiman, O.; Hashim, R.; Peng, L.C.; Singh, R.P. Evaluating biopulping as an alternative application on oil palm trunk using the white-rot fungus *Trametes versicolor*. *Int. Biodeterior. Biodegrad.* **2013**, *82*, 96–103. [CrossRef]

18. Yang, H.; Wang, K.; Song, X.; Xu, F.; Sun, R.-C. Enhanced enzymatic hydrolysis of triploid poplar following stepwise acidic pretreatment and alkaline fractionation. *Process Biochem.* **2012**, *47*, 619–625. [CrossRef]

19. Ju, Y.-H.; Huynh, L.-H.; Kasim, N.S.; Guo, T.-J.; Wang, J.-H.; Fazary, A.E. Analysis of soluble and insoluble fractions of alkali and subcritical water treated sugarcane bagasse. *Carbohydr. Polym.* **2011**, *83*, 591–599. [CrossRef]

20. Silva, T.A.L.; Zamora, H.D.Z.; Varão, L.H.R.; Prado, N.S.; Baffi, M.A.; Pasquini, D. Effect of Steam Explosion Pretreatment Catalysed by Organic Acid and Alkali on Chemical and Structural Properties and Enzymatic Hydrolysis of Sugarcane Bagasse. *Waste Biomass Valoriz.* **2017**, 1–11. [CrossRef]

21. Nelson, M.L.; O'Connor, R.T. Relation of certain infrared bands to cellulose crystallinity and crystal lattice type. Part II. A new infrared ratio for estimation of crystallinity in celluloses I and II. *J. Appl. Polym. Sci.* **1964**, *8*, 1325–1341.

22. Fillat, Ú.; Ibarra, D.; Eugenio, M.E.; Moreno, A.D.; Tomás-Pejó, E.; Martín-Sampedro, R. Laccases as a Potential Tool for the Efficient Conversion of Lignocellulosic Biomass: A Review. *Fermentation* **2017**, *3*, 17. [CrossRef]

23. Carrere, H.; Passos, F.; Antonopoulou, G.; Rouches, E.; Affes, R.; Battimelli, A.; Ferrer, I.; Steyer, J.; Lyberatos, G. Enhancement of Anaerobic Digestion Performance: Which Pretreatment for which Waste? In Proceedings of the 5th International Conference on Engineering for Waste and Biomass Valorisation (WasteEng14), Rio de Janeiro, Brazil, 25–28 August 2014.

24. Adams, J.; Ross, A.Á.; Anastasakis, K.; Hodgson, E.; Gallagher, J.; Jones, J.; Donnison, I. Seasonal variation in the chemical composition of the bioenergy feedstock Laminaria digitata for thermochemical conversion. *Bioresour. Technol.* **2011**, *102*, 226–234. [CrossRef] [PubMed]

25. Oliveira, J.; Alves, M.; Costa, J. Optimization of biogas production from *Sargassum* sp. using a design of experiments to assess the co-digestion with glycerol and waste frying oil. *Bioresour. Technol.* **2015**, *175*, 480–485. [PubMed]

26. Ross, A.; Jones, J.; Kubacki, M.; Bridgeman, T. Classification of Macroalgae as fuel and its thermochemical behaviour. *Bioresour. Technol.* **2008**, *99*, 6494–6504. [CrossRef] [PubMed]

27. Allen, E.; Wall, D.M.; Herrmann, C.; Xia, A.; Murphy, J.D. What is the gross energy yield of third generation gaseous biofuel sourced from seaweed? *Energy* **2015**, *81*, 352–360. [CrossRef]

28. Hanssen, J.F.; Indergaard, M.; Østgaard, K.; Bævre, O.A.; Pedersen, T.A.; Jensen, A. Anaerobic digestion of *Laminaria* spp. and *Ascophyllum nodosum* and application of end products. *Biomass* **1987**, *14*, 1–13. [CrossRef]

29. Apostolidis, E.; Karayannakidis, P.D.; Kwon, Y.-I.; Lee, C.M.; Seeram, N.P. Seasonal variation of phenolic antioxidant-mediated α-glucosidase inhibition of *Ascophyllum nodosum*. *Plant Foods Hum. Nutr.* **2011**, *66*, 313–319. [CrossRef] [PubMed]

30. Tabassum, M.R.; Xia, A.; Murphy, J.D. Seasonal variation of chemical composition and biomethane production from the brown seaweed *Ascophyllum nodosum*. *Bioresour. Technol.* **2016**, *216*, 219–226. [CrossRef] [PubMed]

energies

MDPI

Article

Environmental Analysis of Waste-to-Energy— A Portuguese Case Study

Ana Ramos [1], Carlos Afonso Teixeira [2] and Abel Rouboa [2,3,*]

[1] INEGI-FEUP, Institute of Science and Innovation in Mechanical and Industrial Engineering, Faculty of Engineering of the University of Porto, 4200-465 Porto, Portugal; aramos@inegi.up.pt

[2] CITAB, Centre for the Research and Technology of Agro-Environmental and Biological Sciences, University of Trás-os-Montes and Alto Douro, 5001-801 Vila Real, Portugal; cafonso@utad.pt

[3] CIENER-INEGI, Centre for Renewable Energy Research, Institute of Science and Innovation in Mechanical and Industrial Engineering, Faculty of Engineering of the University of Porto, 4200-465 Porto, Portugal

* Correspondence: rouboa@utad.pt; Tel.: +351-259-350-317; Fax: +351-259-350-356

Received: 26 January 2018; Accepted: 28 February 2018; Published: 4 March 2018

Abstract: Environmental evaluation of the waste treatment processes for the area of Greater Porto (Portugal) is presented for the year 2015. The raw data for the energy recovery plant (ERP) provided by the waste management entity were modelled into nine environmental impact categories, resorting to a life cycle assessment dedicated software (GaBi) for the treatment of 1 tonne of residues. Also, a sensitivity analysis was conducted for five scenarios in order to verify the assessment quality. Results were compared to two European average situations (typical incineration plant and sanitary landfill with no waste pre-treatment), which showed that these facilities perform better or at the same level as the average European situation, mostly due to the high efficiency observed at the ERP and to the electricity production in the incineration process. A detailed analysis concluded that these helped to mitigate the environmental impacts caused by some of the processes involved in the waste-to-energy technology (landfill showing the harder impacts), by saving material resources as well as avoiding emissions to fresh water and air. The overall performance of the energy recovery plant was relevant, 1 tonne of waste saving up to 1.3 million kg of resources and materials. Regarding the environmental indicators, enhanced results were achieved especially for the global warming potential (-171 $kg_{CO_2\text{-eq.}}$), eutrophication potential (-39×10^{-3} $kg_{PO_4\text{-eq.}}$) and terrestrial ecotoxicity potential (-59×10^{-3} $kg_{DCB\text{-eq.}}$) categories. This work was the first to characterize this Portuguese incineration plant according to the used methodology, supporting the necessary follow-up required by legal frameworks proposed by European Union (EU), once this facility serves a wide populational zone and therefore is representative of the current waste management tendency in the country. LCA (life cycle assessment) was confirmed as a suitable and reliable approach to evaluate the environmental impacts of the waste management scenarios, acting as a functional tool that helps decision-makers to proceed accordingly.

Keywords: waste to energy; LCA; sustainability; waste; incineration

---

1. Introduction

Waste management is currently a global concern in view of an exponential population growth accompanied by lifestyle improvements and their consequences, such as higher demand for plastic products and packaging, evidenced by the steadily growing number of field-related published literature. Chen et al. [1] performed a bibliometric analysis of the research concerning municipal solid wastes from 1997 to 2014 and concluded that this type of publications has progressively increased, especially at the beginning of the 21st century. Recently, Eriksson [2] has also published a special issue on energy and waste management, compiling more than 20 works which cover the technical aspects as well as some future perspectives on the energy systems. Zhang et al. [3] reported on the key challenges and opportunities on the waste-to-energy (WtE) in China, referring some hints of the economic and social benefits related to the implementation of standardized and regulated waste management processes. Environmental regulations and directives seek sustainable solutions to this problem, regarding the implementation of new technologies as well as using the existing ones, to assure environmental quality and aiding to meet the set goals [4–8]. The European Union has established well-defined waste management policies, preconizing preventive measures and promoting reducing ones, with the aim to take control over the progressively increasing amounts of solid residues produced nowadays [9].

Cucchiella et al. [10] studied the sustainability of Italian waste-to-energy (WtE) plants according to environmental, financial, economic and social interpretations. A remarkable conclusion of their work was that WtE processes are effective in combating climate change arising from global warming potential causes, once it is possible to generate renewable energy, reducing carbon emissions. This is corroborated by a myriad of studies compiled some years ago by Cherubini and Stromman [11]. Also, as waste is combusted instead of disposed of, these techniques reduce the amount of methane released by landfills. The authors found an interesting solution to balance the need to manage waste with a safe and controlled release of pollutant emissions through the use of mixed waste strategies, therefore promoting sustainability as well as complying with waste legislations. For a deeper understanding of waste management evolution, a thorough review on this topic was published by Brunner and Rechberger [12], where incineration is highlighted as a featured WtE technique. This technology was also pointed as more environment-friendly when compared to others such as sanitary landfill and mechanical-biological treatment [13] or even recycling in specific cases [14]. There are even published works on the waste management balance between some techniques, showing that as landfilling is reduced and other options such as incineration raise, more easily attainable are the EU goals, while high efficiency rates are reached [4]. While the first incinerators were built only for hygienic and waste volume reduction purposes, with no interest in energy recovery, nowadays besides environmental protection, modern WtE plants show significant contributions to the so-claimed resource conservation once some of their by-products may substitute primary resources [3,4]. In countries where waste streams are already seen as important assets for energy production, WtE outcomes are intensively scrutinized in order to determine the overall amount of biogenic CO_2 emissions, as in the case of Austria [15], and also to interpret the effect of changing waste fractions by adding different types of residues, recalling a recent work published for Norway [16].

Portugal also struggles to reach the so desired environmental sustainability, hence progresses have been made in the last years. Back in 2006, Magrinho et al. [17] published a review on the municipal waste disposal, reporting on the waste management practices at that time. Main findings were that since 1998 separate collection of residues was growing as the most common way of disposal, until 2002 when WtE plants became the most important disposal means. In 2009, Ferreira et al. [18] conducted an overview of the bioenergy production highlighting that, although by that time the country was the fourth-largest share of renewable electricity generation in Europe, bioenergy production was not at the desired level. The authors suggested that the energy from animal origin had high potential but was still not well developed. Regarding biomass, it was and still is a highly available resource, enabling the use of several technologies for power production. More recently, Margallo et al. [19] assessed incineration in Iberian Peninsula, so that the overall process was better known and discussed, in order to understand the influence of some critical factors such as waste composition, moisture and heating value on the environmental burdens associated with each fraction. The trigger for this conscious behaviour towards environment and public health protection as well as materials and energy return was given by the settlement of PERSU I (strategic plan dealing with municipal solid waste management between 1996 and 2006, establishing major goals such as ending up waste discharges in Nature, creating waste recovery plants and sanitary landfills, among others), followed by PERSU II (proceeding with municipal solid waste management between 2007 and 2016 and rectifying possible flaws from the previous plan) [17]. Nowadays the prevailing plan is PERSU 2020, which constitutes an improvement of PERSU II for the period between 2014 and 2020 aiming at specific targets like reducing waste deposition from 63% to 35% of the reference values for 1995, raising the reuse and recycling rates from 24% to 50% and also ensuring levels of selective waste collection of 47 kg/inhabitant/year [20].

The waste management entity for the area of Greater Porto (the most densely populated district on the north of the country), LIPOR, holds responsibility for the management, recovery and treatment of municipal wastes from eight associated municipalities, produced by 1 million inhabitants at a 500,000 t/year rate. Its integrated waste management system (IWMS) includes separated units for waste valorisation, incineration, recovery, composting of the organic residue and landfilling of a small pre-treated fraction (Table 1). Despite recyclables may be seen as treasured resources due to their origin and heating value (enhancing the capacity of the plant to produce renewable energy), the company makes the effort of instilling the idea that these items can be transformed in better assets through the sorting plant than sending them to energy valorisation process, according to the waste management hierarchy.

Table 1. Waste streams separation considered for this study by final destination, for 2015.

Waste Final Destination	Weight %
Sorting Plant	9.12
Composting Plant	9.79
Energy Recovery Plant	81.08
Landfill	0.01

A life cycle assessment (LCA) approach is a very useful tool in the evaluation of the contribution of each of the processes to the overall efficiency picture of the disposal options. Arena et al. [21] and Tarantini et al. [22] compared the performance of alternative solid waste management in Italy quantifying the relative advantages and disadvantages for several options, while Liu et al. [23] evaluated the urban solid waste handling options in China and Menikpura et al. [24] assessed the sustainability of an integrated waste management system in Thailand, all of them using LCA as a decision-support tool. Although this is a very powerful mean, aspects such as the lack of transparency or wrong methodology assumptions may lead to difficult comparisons or even deficient interpretation of the results as reviewed for municipal solid waste by Cleary [25]. A summary of the methodology for correctly applying LCA was reported by Clift et al. [26], special attention being paid to the system definition and to the environmental credits achieved from materials or energy recovery. A detailed discussion on the importance of a complete life cycle inventory may be accomplished elsewhere [27]. System boundaries are also a crucial element to be clearly defined, once they have a direct effect on the magnitude of the inputs, accounting for totally different outputs and consequently distinct LCA features [25,28], along with other technical issues [26,29]. LCA may also be seen as a tool that provides decision makers with key information that can help them plan and opt between different waste management scenarios [6,23,24,30,31]. Astrup et al. [32] published a recent review including major recommendations to perform a correct LCA study for WtE technologies. Parkes et al. [33] assessed three different scenarios of waste generation (mixed residential/commercial, mainly residential or mainly commercial/industrial), thus generating diverse streams. The authors found that advanced thermal treatments depicted lower global warming potential than landfill process. Toniolo et al. [34] conducted an environmental assessment on the design phase and on the operational phase of a municipal solid waste (MSW) incineration plant. Results showed that some of the impact categories were underestimated during the design phase stressing the role of the assumptions made during this stage, which might have compromised the reliability of the operational results. Morselli et al. [35] also showed that updated technologies promote lower environmental impacts, matching the needs of modern legislation. Boer et al. [36] developed a decision-support tool for the waste management system assessment. This tool allowed to create and compare planning scenarios for the urban waste management systems, taking into account the design and analysis options.

Herva et al. [37] performed complementary investigations regarding the same IWMS for the data between 2007 and 2011, but using two different methodologies—Energy and Material Flow Analysis (EMFA) and Ecological Footprint (EF). Although this study allowed the delineation of an efficient management strategy, some drawbacks were identified namely the non-assessment of the gaseous emissions from the ERP, which were not included in the chosen indicators. Therefore, concerns such as the yields of dioxins, furans and other toxic substances were not quantified, raising the need for a different approach to be undertaken once they are extremely important due to public health issues, especially for the neighbouring population.

This study assesses the environmental impact of the energy valorisation process of an integrated waste management system during 2015, using a LCA methodology in order to evaluate the performance of each participating facility and their contribution on the total weighted impact. To the best of the author's knowledge, this is the first time this type of study is performed for a waste management institution in this area hence, awareness of the assessed outputs achieved with the actual practices may help to understand results in other business dimensions (like financial, social or technical) and serve as foundations for the development of efforts in finding better management solutions. In regards to the EU legislation, the results from this work were also compared to European average situations in order to understand the trends and evolution of the Portuguese situation in the waste management segment, supporting a follow-up for the EU-proposed frameworks, in order to monitor the progress of this topic within the participating countries.

2. Methods

In this section, the integrated waste management system description and waste characterization will be elucidated. Also, boundaries, functional unit and life cycle inventory for the evaluated system will be defined as well as the LCA scenarios described.

2.1. Integrated Waste Management System for the Portuguese Case Study

Regarding the integrated management system, four main stages are considered from the residues generation until its final valorisation, as can be seen in Figure 1. Stage I refers to the waste creation whether it occurs in homes, small businesses and enterprises or public institutions. Then municipalities have a key role in stage II, providing the necessary containers and differentiated vessels for each kind of waste, as well as taking care of its collection and transport to LIPOR facilities, where stage III takes place in sub-processes like energy, organic and multi-material valorisation. Actually, the three central waste management systems in the IWMS include the sorting plant (SP, where materials from eco-points are received and an additional separation is performed according to their nature—metal, plastic, glass, paper and cardboard), the composting plant (CP, where the organic fraction of the collected waste like tree branches, bushes and grass is composted) and the energy recovery plant (ERP, where the waste incineration occurs). Besides, there is a landfill where by-products from the ERP such as slag (previously separated from the ferrous fraction by magnetic segregation), inert ashes and also raw waste that does not comply with any of the treatment processes offered by this unit are disposed. Finally, stage IV corresponds to the obtainment and commercialisation of the valuable products that result from the previous steps, like electricity, compost and fertilizers [38].

For the reference case (scenario 0), the life cycle assessment of the energy recovery plant (stage III of the Figure 1) was performed, within the conditions defined in Section 2.2. The ERP works in continuous operation treating around 1,100 t of waste per day and producing about 170,000 MWh of electricity per year, from which an average of 90% is supplied to the national electric grid. Waste is discharged in a closed depressurized building, where claws and hoppers move residues to the combustion grids. Here, the waste is decomposed at temperatures between 1000 °C and 1200 °C, generating flue gases which are released at 950 °C and also bottom ash. Before its release into the atmosphere, the gaseous fraction is cleaned passing through scrubbers and filters, hazardous substances being removed and some even converted into marketable products. Bottom ash is collected and landfilled and heat is used in a boiler, where steam is produced and then sent to a turbine to generate electricity [38].

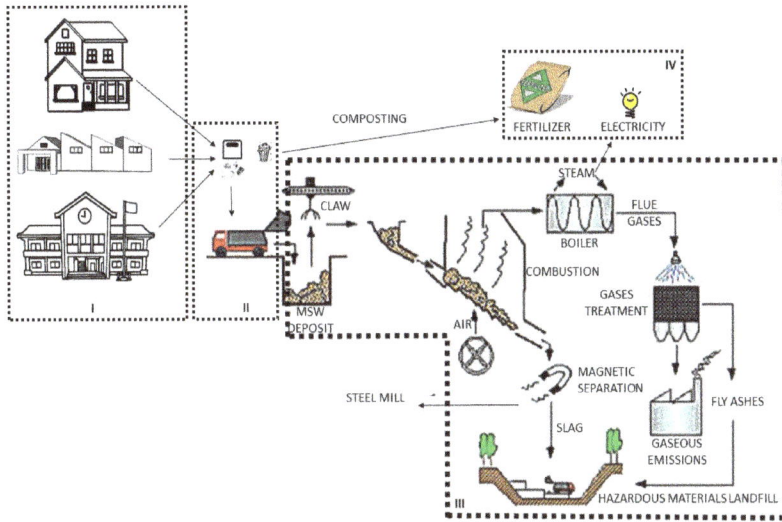

Figure 1. Integrated waste management system operated at the Greater Porto area. ▪▪▪▪▪ (bold dashed lines) representing the chosen boundary for this study.

The waste partition for the collected residues at the referred waste management plant in 2015 is shown in Table 1. As far as the energy recovery plant is concerned, approximately 81% of all the received debris are conveyed towards incineration, which corresponds to roughly 405,000 t/year. From these, the major materials are bio-waste, followed by plastics and health care textiles, as represented in Table 2 [39].

Table 2. Characterization of the material received in the energy recovery plant, in 2015.

Waste Type	Weight %
Bio-waste	37.57
Plastics	12.10
Health care textiles	8.72
Textiles	7.74
Waste < 20 mm	7.59
Composites	6.39
Paper	6.16
Glass	5.53
Cardboard	4.31
Metals	2.45
Others	1.44

2.2. LCA Methodology

2.2.1. Reference Case (Scenario 0)

The scenario depicted in Figure 2 was modelled, taking into account the type of incineration in practice, the landfill usage and also the electricity production. Only major flows are highlighted, the remaining inputs and outputs being reported in a dedicated table (Table 3).

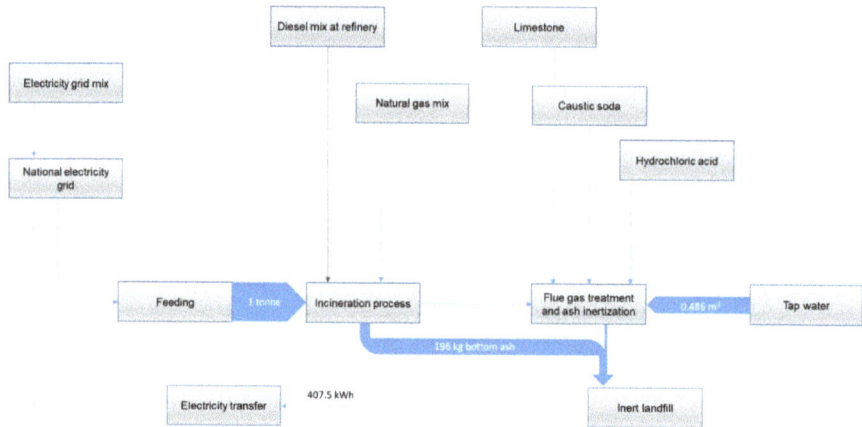

Figure 2. GaBi plan for the reference case (scenario 0)—Portuguese case study.

Table 3. Main lifecycle inventory for the studied ERP (annual data referred for 2015).

Inputs and Outputs	
Waste for incineration, t	407,053
Lower Heating Value, kJ/t	7700
Water, m^3	197,785
Diesel, m^3	1.45
Natural gas, Nm^3	12,556
Auxiliary Materials	
Tripolyphosphate, t	1.06
NaOH, t	16.8
Limestone, t	4703.6
HCl, t	28.8
Activated charcoal, t	202.2
Urea, t	1412.3
Produced steam, t	567
Used in the turbine, t (%)	878,237 (96.9)
Produced electricity, MWh	193,068
Self-consumption, MWh (%)	27,180 (14.1)
Exported to the National grid, MWh (%)	165,888 (85.9)
Exported to the National grid, kWh/t	407.5
Emissions	
HCl, mg/Nm^3	4.8
NOx, mg/Nm^3	164.8
NH_3, mg/Nm^3	7.2
HF, mg/Nm^3	0.1
SO_2, mg/Nm^3	8.3
Total Organic Carbon, mg/Nm^3	0.4
Particulate matter, mg/Nm^3	0.8
CO, mg/Nm^3	4.0
O_2, %	8.8
Hg, $\mu g/Nm^3$	0.3
Neutralized ashes, t	32,427
Removed scraps, t	5646
Removed slags, t	79,627

In this case, the utilised landfill was a process that includes only inert material (neutralized residues as well as other non-hazardous substances), standard end-of-life treatment service for specific waste being considered, landfill gas collection and leachate treatment being excluded (for more details, please see the Supplementary Material section). The electricity production was considered taking into account the self-consumption of the plant and the surplus distribution to the national grid. Therefore, the environmental results of this setup can be taken as the most representative for this case study during the year of 2015, constituting scenario 0. Electricity grid mix refers to the electrical input available at Porto region, mainly composed of renewable energies as described elsewhere [40]. Electricity transfer is a GaBi internal process that grants the outputs related to the production of electricity to be taken into account in the overall sustainability analysis.

2.2.2. Scope, System Limits and Functional Unit

Scenario 0 focused on the waste energy valorisation, other stages and activities (as administrative services or hazardous wastes delivery to other entities, such as ferrous scraps which are sent to a national steel managing entity) being left aside. All the inventory data used and the technical explanations about the operation circuits were provided by the waste management entity through the website [38], oral information during guided tours and internal data described by email and in published reports [39]. System boundaries were established according to the data provided by the company, which led to the omission of some processes, namely regarding waste collection and transportation to the treatment facilities (fulfilled by the associated municipalities), as well as the process of wastewater treatment (currently taken care by the municipal wastewater treatment plant).

Most of the information is given on a yearly basis (in this case referring specifically to the year 2015), all the quantities being then normalized to 1 tonne of residues treated in the ERP, which constitutes the functional unit (fu) employed in the LCA software GaBi, used in this study.

Concerning all the steps involved in MSW incineration, the flowchart depicted in Figure 3 was set according to the stipulated limits for the attributional LCA study performed, where slashed lines indicate the boundaries and boxes represent the processes, which are connected by flows (arrows).

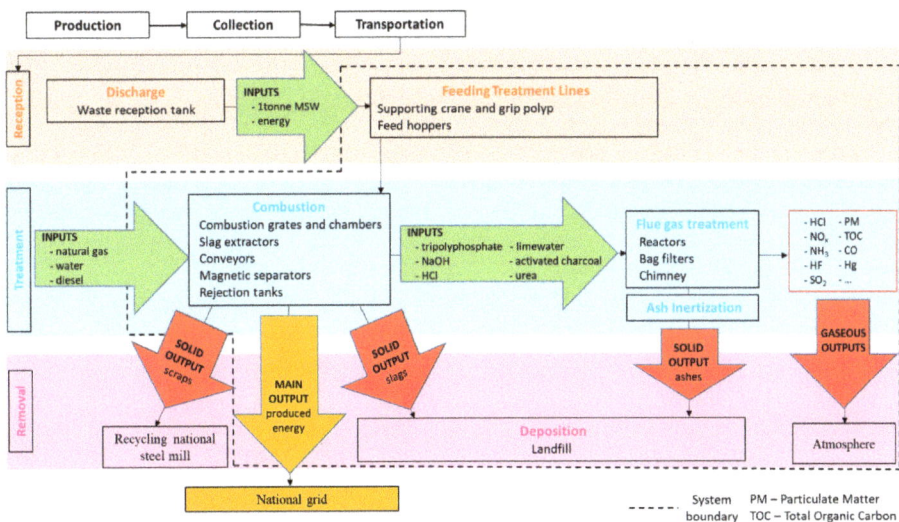

Figure 3. Flowchart of the present LCA study, with defined boundary, inputs and outputs.

2.3. European Average Scenarios

The selected options were a European average incineration (EAI) scenario and a European average landfill (EAL) scenario. The first consists in a plan created on GaBi software (PE International, Stuttgart, Germany) [41] with well-established and documented European processes from the database, simulating the incineration of 1 tonne of residues with average characteristics, inputs and outputs. The former corresponds to the plain situation where waste is sent to landfill, with no previous thermal treatment, once again making use of processes described in the software database [41], adequately standardized for the usual landfill of 1 tonne of municipal solid waste in Europe (for further details please refer to the Supplementary Material section).

2.4. Sensitivity Analysis

Within the energy valorisation of the eligible waste fraction, three main processes occur as can be seen in Figure 2. These are the incineration itself, the flue gas treatment and the neutralization of ashes. Each of these processes has its own inputs and outputs and when differently combined or linked through the available flows, they result in distinct overall impacts. Based on the reference case presented above, four different scenarios were tested to verify the potential effects caused by distinct conditions.

Scenario 1 (Sc1): inclusion of plant construction and waste transportation to the treatment facilities within the system boundaries. This plant was built with the expected duration of 50 years. In this hypothesis, the inputs and outputs related to its construction were taken into account. Regarding waste transportation after collection, the average travel distance to reach the treatment facility is 25 km.

Scenario 2 (Sc2): inclusion of plant construction and wastewater treatment facilities within the system boundaries. Opposite to what happens actually at the evaluated facilities (Sc0), in this scenario the operating process consumes nearly 95% of the produced electricity, only 5% being directed to the national grid.

Scenario 3 (Sc3): using a typical European landfill. The only difference from this scenario to the reference case is the landfill process. In order to better understand the environmental profit of having a restricted type of landfill, a typical landfill for this kind of waste treatment was used in this scenario, representing a solution proposed from the average European situation, several plants being studied to accomplish the net performance of their landfills. This type of disposal includes landfill gas collection, leachate treatment, sludge treatment and deposition (for further details please refer to the Supplementary Material section).

Scenario 4 (Sc4): neglecting electricity production from waste incineration. A virtual plan was created simulating the waste incineration with no electricity production, where electricity production from waste incineration is not accounted neither used to feed the system. For this purpose, electricity production was disregarded, hence self-consumption was admittedly neglected, forcing the system to consider all the energy inputs as if they provided from the national electric grid. This strategy aimed at demonstrating evidence for the chief importance of taking advantage of the waste incineration to produce energy, a highly-demanded asset nowadays.

2.5. Environmental Indicators and Life Cycle Inventory Data

The environmental evaluation methodology chosen within this study was CML 2001, which consists in a database that contains characterisation factors for life cycle impact assessment, such as global warming potential (GWP), acidification potential (AP), human toxicity potential (HTP), abiotic depletion potential (ADP), eutrophication potential (EP) among several others that can be more or less adequate for each assessment, according to the modelled data. These impact categories were evaluated through the use of the product sustainability software GaBi (database version 4.131 distributed by PE International, Stuttgart, Germany), which enabled data modelling and environmental performance estimation rendered by the suitable indicators calculation. These calculations are based on plans, processes, inputs and outputs for the studied system, specific balances being elaborated for each case based on the main operational data, as the life cycle inventory reported in Table 3. For the inventory data of further scenarios, please refer to the Supplementary Material section.

This life cycle inventory enabled the appraisal of the environmental indicators used to assess the energy recovery plant. Table 4 lists the impact categories evaluated, as well as their correspondent units and the applied methodology.

Table 4. Environmental indicators, units and assessment methodology.

Impact Indicator	Units	Assessment Methodology
Abiotic Depletion Potential (ADP)	$kg_{Sb-eq.}$; MJ	CML 2001
Acidification Potential (AP)	$kg_{SO2-eq.}$	CML 2001
Eutrophication Potential (EP)	$kg_{PO4-eq.}$	CML 2001
Global Warming Potential (GWP)	$kg_{CO_2-eq.}$	CML 2001
Freshwater Aquatic Ecotoxicity Potential (FAETP)	$kg_{DCB-eq.}$	CML 2001
Human Toxicity Potential (HTP)	$kg_{DCB-eq.}$	CML 2001
Marine Aquatic Ecotoxicity Potential (MAETP)	$kg_{DCB-eq.}$	CML 2001
Terrestric Ecotoxicity Potential (TETP)	$kg_{DCB-eq.}$	CML 2001

As stated in Section 2.2, the functional unit was chosen on a convenient way for the kind of research this study reports: 1 tonne of wastes treated at the energy recovery plant. Therefore, each contributing step was compared on the same basis, making it easier to take conclusions and understand how each of them affects the global results individually.

3. Results and Discussion

Considering the available data collected, all the quantities and flows were normalized according to the chosen functional unit and LCA plan. Resourcing to the software calculation tools, the LCA study was performed and the waste management system was assessed with the environmental indicators shown in Table 4.

A comparison between the environmental impacts originated by each of the settled situations is herein discussed, as well as the establishment of the individual roles of the main processes is specified for the reference case. A contextualization among reported works was attempted where possible, although it was arduous to accomplish due to the myriad of different boundaries, functional units, utilised software, methodologies and also diverse ways to present the results.

3.1. Portuguese Case Study among the European Scenarios

Each impact category was quantified according to CML 2001 methodology for the reference case (Sc0) as well as for the European average scenarios. This way, results from the reference case may be understood under a range of two extreme possibilities: a typical IWMS and a regular sanitary landfill, both for EU-27 region. Results for environmental impact categories are shown in Table 5. A weak point analysis performed with the software (not shown here) helped to construct a better interpretation of the results.

As can be seen from Table 5, in the European average incineration scenario almost all the impact categories have negative values with higher avoided burdens. In fact, exploring the net LCA results in the software, the most impacting category is GWP (approximately 504 $kg_{CO_2-eq.}$), due to the main output of this scenario: carbon dioxide emissions to air. This has already been observed by other authors for similar studies, values varying between 674 $kg_{CO_2-eq.}$ and 1490 $kg_{CO_2-eq.}$ [34], 637 $kg_{CO_2-eq.}$ and 736 $kg_{CO_2-eq.}$ [42], 58 $kg_{CO_2-eq.}$ and 496 $kg_{CO_2-eq.}$ [43]. In the case of TETP, the positive value achieved is a reflection of the heavy metals released into the atmosphere, namely arsenic (+V), chromium, mercury (+II) and vanadium (+III).

Table 5. Comparison of the environmental impacts for the Portuguese incineration plant and the European standards for incineration and landfill.

Impact Categories	EAI	Sc0	EAL
GWP ($kg_{CO_2\text{-eq.}}$)	503.76	−170.9	655.1
AP ($kg_{SO2\text{-eq.}}$)	−2.426	-242×10^{-3}	167×10^{-3}
EP ($kg_{PO4\text{-eq.}}$)	-38.9×10^{-3}	-38.6×10^{-3}	847×10^{-3}
$ADP_{elem.}$ ($kg_{Sb\text{-eq.}}$)	-8×10^{-3}	-50.1×10^{-6}	9.24×10^{-6}
ADP_{fossil} (MJ)	-4.622×10^3	-2.00×10^3	550.9
FAETP ($kg_{DCB\text{-eq.}}$)	−1.716	-267×10^{-3}	534×10^{-3}
HTP ($kg_{DCB\text{-eq.}}$)	−125.647	−7.45	1.786
MAETP ($kg_{DCB\text{-eq.}}$)	-340×10^3	-26.3×10^3	5.30×10^3
TETP ($kg_{DCB\text{-eq.}}$)	133×10^{-3}	-59×10^{-3}	1.687

When comparing EAI to Sc0 (reference case), albeit EP results are similar in both cases most of the other categories are at least one order of magnitude less impacting in Sc0. The key factors in this plan are the GWP and TETP outputs, due to the reduction of greenhouse gases from the more effective type of landfill. In the reference case, every impact category depicted negative values which means alleviation of the environmental burdens most probably due to the energy recovery held at the IWMS [44], meaning environment is less jeopardised than if no energetic valorisation was achieved. Other studies also confirm that electricity generation in waste treatment facilities is a key advantage in what concerns environmental impacts, once it saves major quantities of natural resources that would be necessary to assure the equivalent amount of energy, as can be seen by the significant value achieved for the abiotic depletion category. Back in 2005, Morselli et al. [45] assessed a municipal solid waste incinerator in Italy with daily waste capacity less than half of the herein reported input. Although several other stages within the applied boundaries were included (such as plant construction/demolition, waste transportation and waste water treatment), the authors concluded that 7 out of 10 impact categories presented avoided impacts, majorly due to the energy recovery step. More recently, Passarini et al. [46] compared the environmental impact of another Italian incineration plant before and after structural upgrades. Enhanced results were achieved after the revamping operations mostly due to the implementation of new procedures related to flue gas treatment, which reduced air emissions. Nevertheless, the authors guarantee that putting the same type of efforts in the energy recovery process would afford even better outcomes, raising plant sustainability. However, due to the national Italian grid mix evolution along the years and to the actual national energy policies, this valuable impact would represent lower net avoided burdens than the new gas treatment techniques. This paradigm highlights the influence of the current transition from fossil to renewable fuels. With these regards, Burnley et al. [47] also studied the factors influencing the burdens associated to energy recovery from municipal wastes namely metal and aggregate recovery, thermal efficiency and the displacement of fossil electricity by the power generated within incineration. Electricity from waste impact has shown to be highly dependent on the type of fuel to be displaced, coal-based electricity affording maximized benefits, and natural gas replacement performing poorly.

In an attempt to briefly compare the achieved results for the reference case (Sc0) with literature, an interesting work by Hong et al. [44] reports a sensitivity analysis using two different calculation methods from which related categories may be compared. In relation to the herein obtained results, improved values for GWP and ADP fossil (non-renewable energy) were presented when IMPACT 2002+ methodology was used, and for GWP, HTP, FAETP when Recipe methodology was used. In this last case, EP depicted similar values to the ones achieved in this work, TETP and MAETP performing poorly than the current study. A possible explanation for the differences encountered may lie in the distinct boundaries applied once Hong et al. included infrastructure construction, leachate treatment and material recovery in their study [44].

In relation to the European average landfill scenario, this undoubtedly constitutes the poorest option, as may be seen by the positive values for all the impact categories, which means grieving of the natural resources. A work conducted by Liamsanguan and Gheewala [42] compares landfilling and incineration in Thailand. The authors found that, for the specific conditions used in the study, landfill only performed better than incineration in regards to the energy recovery impacts achieved in both scenarios, if landfill gas was recovered for electricity production. This is in line with the results of the present work, once the studied ERP does not pursuit electricity produced through biogas exhaled by the landfill. Other authors also stress that improving landfill gas collection would raise the overall incineration efficiency [48]. Likewise, Jeswani and Azapagic [43] compared incineration to biogas recovery from landfill but under two different perspectives: the disposal of 1 tonne of municipal residues and the generation of 1 KWh of electricity. All the environmental categories assessed (except HTP) depicted lower impacts in the case of incineration for both functional units. Improved results were expected if instead of replacing a natural gas-based electricity grid, heavy fuel oil or coal dependent grid was displaced as stated elsewhere [47,49,50]. A more detailed description of what happens in each of the cases in Table 5 will be performed next.

3.2. Environmental Performance of the Reference Case

3.2.1. Global Warming Potential

GWP is assigned to the effect of greenhouse gases (CO_2, CH_4, N_2O, CO, CFCs, HCFC's, HFC's) which are able to absorb heat radiation, increasing atmospheric temperature [51]. All the contributing substances can be modelled and quantified for different time horizons and, in CML 2001, the utilized parameter is GWP for 100 years. Figure 4 shows a detailed balance of the emissions contributing to GWP and, as can be seen, Sc0 constitutes the most sustainable option, preventing more environmental injury.

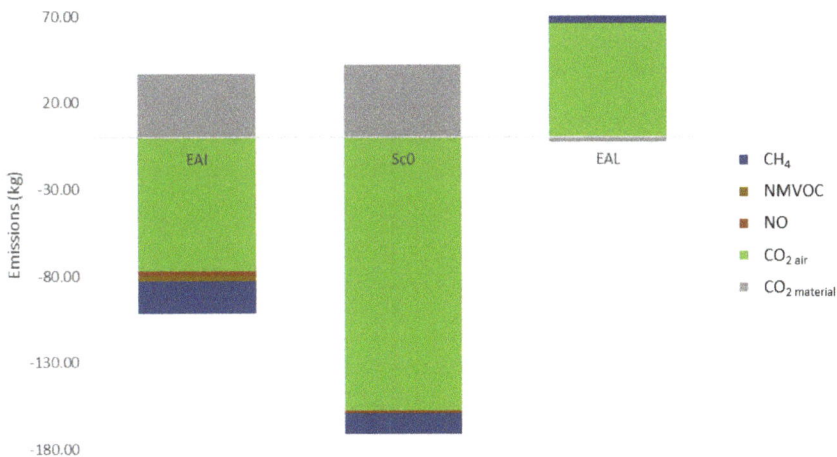

Figure 4. Global warming potential comparison for the reference case (Sc0) among average European scenarios. Results refer to one tonne of treated residues.

This situation is explained through the avoided emissions of CO_2 (-158 $kg_{CO_2\text{-eq.}}$/fu) and CH_4 (-12 $kg_{CO_2\text{-eq.}}$/fu) from the landfill, as reported in another work [52]. Other authors attribute this effect to the CO_2 released from the flue gas to the atmosphere, similar values for the total contribution of Sc0 to GWP being also reported [53]. As shown in Table 2, plastics are the second major type of residues in ERP which constitutes a high calorific value asset contributing to a superior amount of recovered energy from incineration process, explaining the more negative values obtained for Sc0 rather than EAI. In this last scenario, the most problematic contribution to GWP is the CO_2 devastation as material resource (36.7 $kg_{CO_2-eq.}$/fu). GWP result is in the same order of magnitude as reported by other authors [52,54,55], situations where waste collection and transport to the incineration facilities depicted significantly different values [56] although CO_2 is still the major contribution. It must be stressed that the results on biotic CO_2 in all scenarios were not accounted once they are considered part of the carbon cycle, its effect on the GWP being inconsequential [29,57]. This explains the relatively low contribution shown by the incineration process to GWP, Hong et al. [44] obtaining a result similar to Sc0 in this category (-254 $kg_{CO_2-eq.}$/fu). CO_2 emission savings are attained with EAI and Sc0. EAL shows a situation in which there is an overbalance of roughly 70 $kg_{CO_2-eq.}$/fu. This effect was also described in other assessments, Fernández-Nava et al. [52] stating that landfill is the most unfavourable waste management option with even higher CO_2 emissions than this work, and Bezama et al. [58] suggesting upgraded landfill options to reduce this value. Concerning CH_4 emissions, EAL shows that there is only a partial capture, some of it being released to the atmosphere, contrarily to what happens in EAI and Sc0. This was also reported in other cases [50,59].

3.2.2. Acidification Potential

AP is assigned to the release of hydrogen ions into the environment by acidifying substances (SO_2, NO_x, HCl, NH_3), SO_2 being the basis for the determination of this impact. Acid gases that are released into the air or resulting from the reaction of non-acid components of the emissions are taken up by atmospheric precipitations forming "acid rain", which has widespread noxious effects [51]. Figure 5 depicts the contributions of each scenario to this category and it can be seen that the prevailing harmful substances are sulphur dioxide and nitrogen oxides, mainly related to the flue gas treatment process.

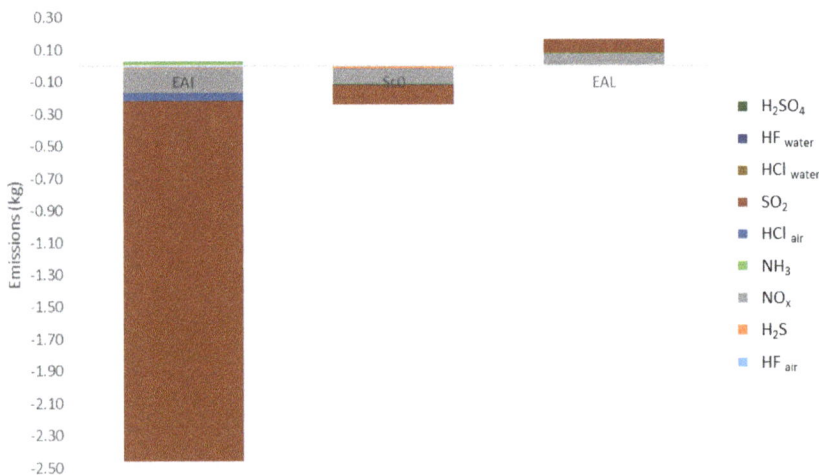

Figure 5. Acidification potential comparison for the reference case (Sc0) among average European scenarios. Results refer to one tonne of treated residues.

In this category EAI portrays the ideal situation, once it keeps back more than 2 $kg_{SO_2-eq.}$/fu, while Sc0 retains only 0.1 $kg_{SO_2-eq.}$/fu, showing also NO_x savings one order of magnitude lower than EAI. Nitrogen oxides are mainly released from the landfill process in the reference case, while sulphur dioxide is mostly emitted during flue gas treatment, although being in compliance to the European guidelines. Regarding EAL, this scenario displays positive values for AP, revealing a heft for the environment, probably due to the absorption of these compounds by plants, soil and surface waters leading to leaf damage and super-acidity of the soil, which in turn affects the solubility and hence the availability of plant nutrients, negatively affecting the ecosystems. This trend is also verified by other authors [46], Mendes et al. [59] confirming NO_2 as the most warning emission from incineration, assuring compliant gas treatment systems that effectively abate other emissions. The production of electric power seems to reduce the positive AP impact as also reported in literature [53], some authors also showing that when the energy recovery doubles, AP will no longer be a hazard, becoming an environmental credit [46]. Meanwhile, Banar et al. [56] conclude that stages such as waste collection and transport afford positive AP values, constituting an environmental burden.

3.2.3. Eutrophication Potential

EP comprises all the potential impacts with high environmental levels of macronutrients (nitrogen and phosphorus) causing increased production of aquatic plants, reducing water quality and oxygen depletion in the bottom layers. In the case of incineration, one of the most worrisome substances is NO_x [51] but there are others as can be seen for Sc0, shown in Figure 6.

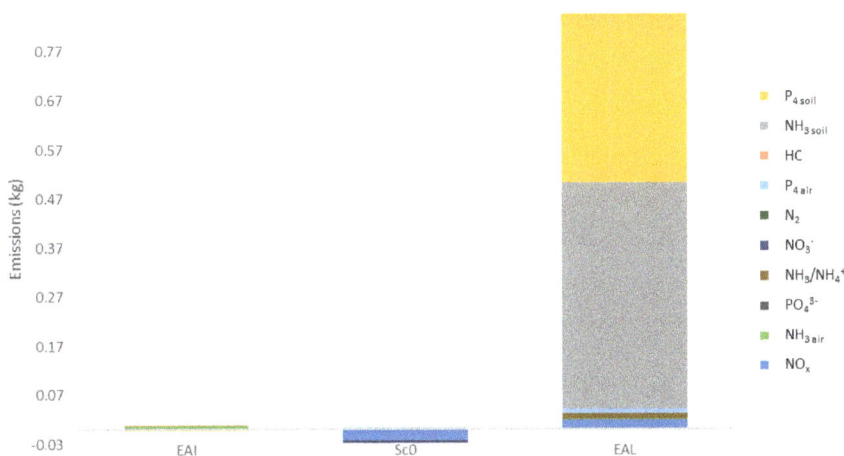

Figure 6. Eutrophication potential comparison for the reference case (Sc0) among average European scenarios. Results refer to one tonne of treated residues.

In the case of EAI the most impacting substance is ammonia (5.98×10^{-3} $kg_{PO_4\text{-eq.}}$/fu), enhanced results being achieved for Sc0 (9.95×10^{-4} $kg_{PO_4\text{-eq.}}$/fu), mostly due to the landfill process. Regarding NO_x, Sc0 depicts negative values caused by the electricity generation. Mendes et al. [59] use "nutrient enrichment potential" to assess what we designate by eutrophication potential, also concluding that nitrogen compounds are the major contributors to this category though the highest level of nitrogen compounds was released to the water. Although the electricity production abates the emission of NO_x, P_4 and NH_3 are the major emissions to the soil, especially in the case of EAL which corresponds to a plan with less restrictive landfilling, permitting the accumulation of these substances also in the soil. The total contribution of this environmental category is somehow comparable to the values found in other reports, Gunamantha and Sarto [60] achieving an EP value double than the Sc0 (-7.87×10^{-2} and -3.86×10^{-2} $kg_{PO_4\text{-eq.}}$/fu, respectively), but far more disruptive in cases where waste collection and transport are considered [56].

3.2.4. Abiotic Depletion Potential

ADP represents the reduction of the total amount of non-renewable natural resources being divided in two sub-categories: elements and fossil resources [28]. This category might be regarded as an indicator of the primary energy usage, to evaluate the efficiency of these resources on the overall system. Figure 7 shows detailed balance for the two ADP categories. In both cases, EAI is the most environmental friendly option, Sc0 showing negative values for all the contributions to ADP while EAL performs poorly.

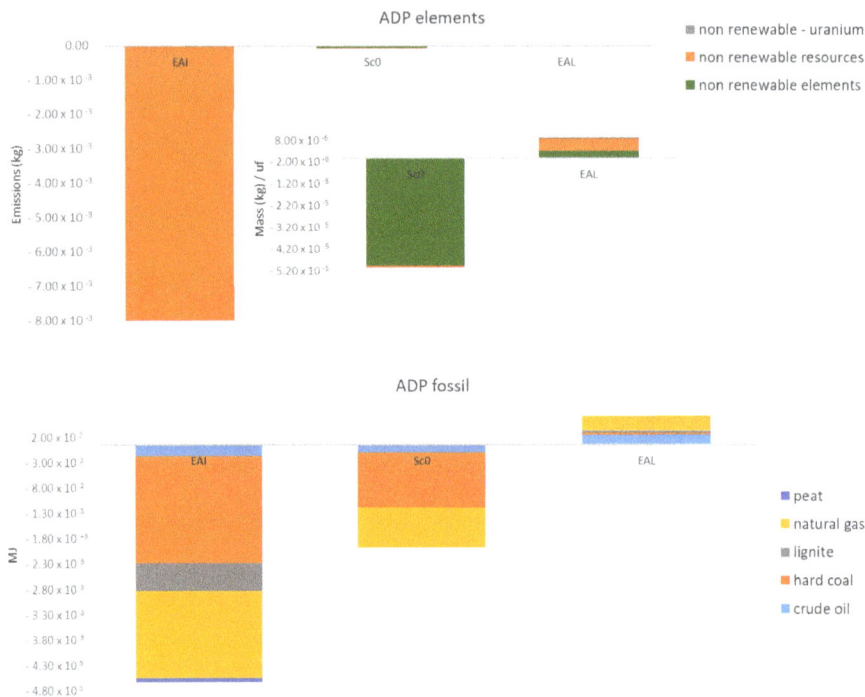

Figure 7. Abiotic depletion potential comparison for the reference case (Sc0) among average European scenarios. **Top**—abiotic depletion of elements, insights for Sc0 and EAL scenarios; **bottom**—abiotic depletion of fossil resources. Results refer to one tonne of treated residues.

Regarding abiotic depletion of elements, EAI contributes to the reconstruction of non-renewable resources like copper, gold, silver, molybdenum, lead and zinc ores (total amount of -7.92×10^{-3} $kg_{Sb\text{-}eq.}/fu$), while Sc0 subscribe mostly to non-renewable elements copper, gold, lead and molibdenum (total amounts of -4.98×10^{-5} $kg_{Sb\text{-}eq.}/fu$), due to the electricity production process. Mineral resources had been reported to be enhanced by the production of electric energy elsewhere [46]. EAL shows positive impacts for ADP elements when compared to scenarios that have a previous waste treatment, as reported in another work [52]. The most contributing process for these results is the landfill itself.

Relative to fossil abiotic depletion, EAI shows a better performance again, sharing negative values for major contributions like hard coal, natural gas and lignite. Hard coal and natural gas lead the avoided burdens as a consequence of the landfill process unit, environmental credits being also achieved by the electricity production. This general avoided consumption of fossil sources was also reported by other authors [42,44], although the present study reveals better efficiency once it utilises only 240 MJ/fu and recovers ca. 2000 MJ/fu (similar to a reported study for a Spanish plant [52]). Chaya and Gheewala [61] performed a similar assessment for an incineration plant treating half the daily waste quantity as compared to the reference case in this study and achieved a total energy resource saving of only (−)563 MJ, most of which comes from the effect of electricity production. Menikpura et al. [54] reported a specific case were incineration promoted a reduction of 190% on net resource consumption. Concerning EAL, positive impacts were achieved as stated in literature [58] especially for crude oil and lignite.

3.2.5. Ecotoxicity

Ecotoxicity may be seen as the effect of the chemical substances in environment. Therefore FAETP, MAETP and TETP define the ecotoxicity in freshwater, marine and terrestrial compartments respectively. The substances contributing to these categories are numerous and difficult to point out even if grouped, their biodegradability, bioaccumulation and distribution being modelled. Figures 8–10 show the results obtained for the three sub-categories.

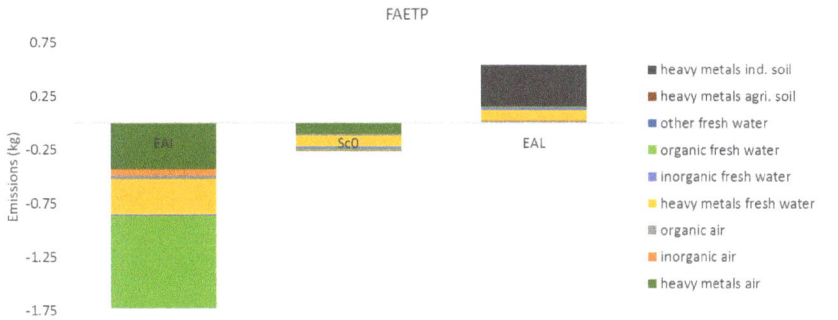

Figure 8. Freshwater ecotoxicity comparison for the reference case (Sc0) among average European scenarios. Results refer to one tonne of treated residues.

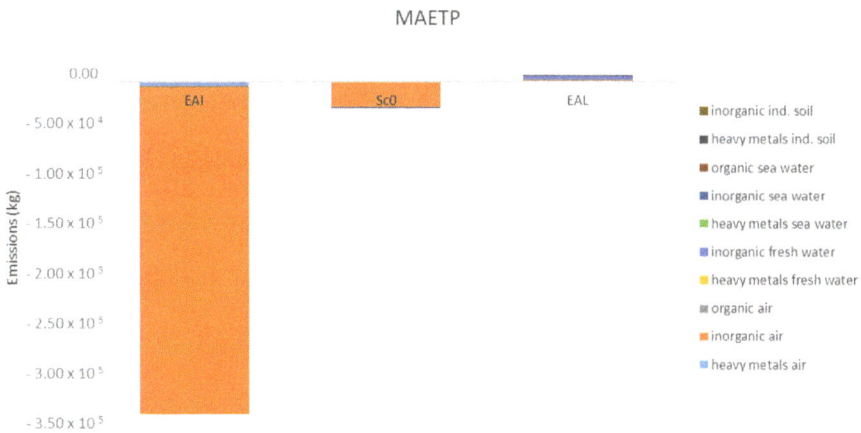

Figure 9. Marine water ecotoxicity comparison for the reference case (Sc0) among average scenarios. Results refer to one tonne of treated residues.

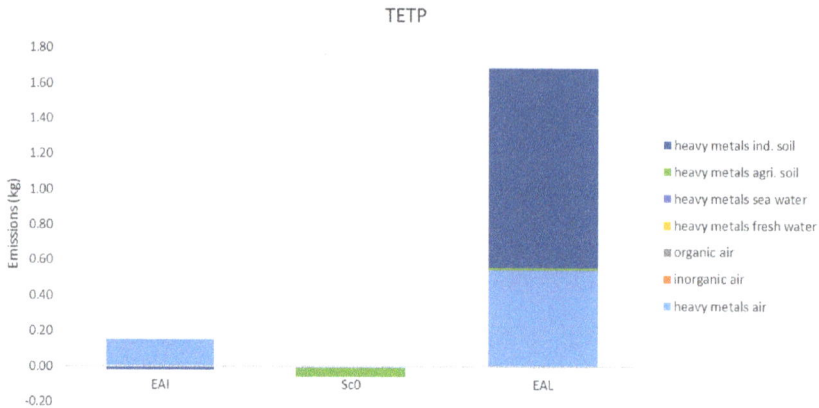

Figure 10. Terrestrial ecotoxicity comparison for the reference case (Sc0) among average European scenarios. Results refer to one tonne of treated residues.

In the case of FAETP, EAI shows the best output alleviating fresh water from $(-)0.862$ kg$_{DCB\text{-}eq.}$/fu due to organic emissions and $(-)0.333$ kg$_{DCB\text{-}eq.}$/fu from heavy metals, and also avoiding the emission of $(-)0.424$ kg$_{DCB\text{-}eq.}$/fu from heavy metals to the atmosphere. Sc0 is the second most compliant scenario, with major contributions of -0.105 kg$_{DCB\text{-}eq.}$/fu and -0.101 kg$_{DCB\text{-}eq.}$/fu from heavy metals to freshwater and air, respectively. These savings can be attributed to the electricity production, and somehow counterbalanced by the landfill process. Both EAI and Sc0 performed better than reported by Toniolo et al. [34], who attained values between 3.09 and 7.27 kg$_{DCB\text{-}eq.}$/fu, thus meaning environmental burdens. For MAETP, EAI has the most environmental friendly results again crediting $(-)3.35 \times 10^5$ kg$_{DCB\text{-}eq.}$/fu of inorganic emissions to the air, Sc0 playing less extensive outputs $(-2.55 \times 10^4$ kg$_{DCB\text{-}eq.}$/fu). The contributing processes to these results are the production of electricity and the flue gas treatment as corroborated by other studies [46], while landfill reduces these achiements showing positive impacts. Once more, when compared to published literature [34] EAI and Sc0 depicted enhanced results (literature values ranging from 0.79 kg$_{DCB\text{-}eq.}$/fu to 4.79 kg$_{DCB\text{-}eq.}$/fu).

As referred earlier, TETP was one of the categories where Sc0 performed better than EAI and this can be proved observing the avoided heavy metals in agricultural soil $(-0.0485$ kg$_{DCB\text{-}eq.}$/fu) for the first and the contribution of heavy metals to air for the second (0.153 kg$_{DCB\text{-}eq.}$/fu). The release of heavy metals in industrial soil is magnified in the case of EAL (1.13 kg$_{DCB\text{-}eq.}$/fu) due to the landfill process. Other authors confirm the release of heavy metals from the landfill [43] (which are reduced when electricity production is considered) or from the incineration process [22] as the major impact for this category. When compared to published works [34], Sc0 assured a good result for this category of impact (literature values ranging from -7.64×10^{-3} kg$_{DCB\text{-}eq.}$/fu to 5.77×10^{-2} kg$_{DCB\text{-}eq.}$/fu).

3.2.6. Human Toxicity Potential

HTP is related to the negative effects of toxic substances (e.g., volatile organic compounds, particulate matter, heavy metals, NO_x, SO_2) on human health. These can be irritative, corrosive, allergenic, irreversible, carcinogenic among others, always excluding indoor exposure [51]. Figure 11 depicts the results of this category in the evaluated scenarios.

In this category, EAI still depicts the best environmental performance when compared to the other scenarios with organic, inorganic and heavy metals emissions to air as well as organic emissions to freshwater being the most pronounced avoided burdens (accounting for a total of nearly -125 kg$_{DCB\text{-}eq.}$/fu). Sc0 follows the trend but with less visible results (total amount of -7.45 kg$_{DCB\text{-}eq.}$/fu), a little under other findings: Toniolo et al. [34] reported values between

-62.9 kg$_{DCB-eq.}$/fu and 156 kg$_{DCB-eq.}$/fu and Banar et al. [56] values ranging from -182 kg$_{DCB-eq.}$/fu to 92 kg$_{DCB-eq.}$/fu. Nevertheless, these are valuable results, consequence of the flue gas treatment and the landfill processes [62]. Comparable conclusions on the scenarios performance were drawn by others [22,52].

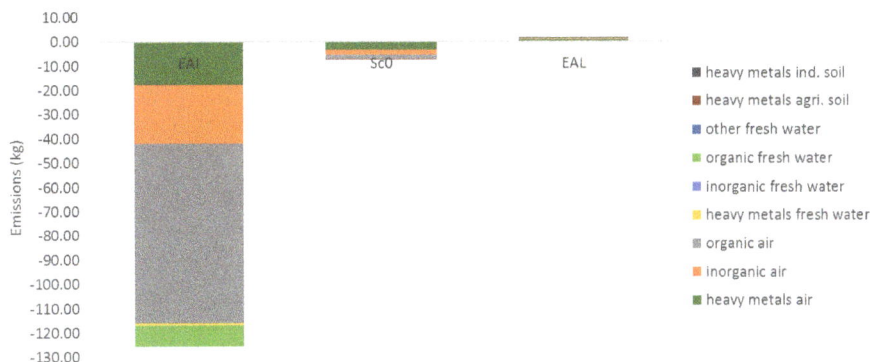

Figure 11. Human toxicity comparison for the reference case (Sc0) among average European scenarios. Results refer to one tonne of treated residues.

3.3. Sensitivity Analysis

After concluding that the actual practices in the ERP give rise to sustainable results, it is interesting to run a sensitivity analysis in order to understand the effect that some methodological changes depict on the environmental impacts. Therefore, the described reference case was compared to four other scenarios and the main results are shown in Table 6. As it is difficult to compare so many different ranges of values, within nine impact categories and five distinct scenarios, a correlation of scenarios 1–4 to the real ERP results is presented in Figure 12. It reports the correlation coefficient between each scenario and the reference case (Sc0), through the comparison of a set of properties. Each scenario is regarded as an environmental matrix composed by the related results for the impact categories and the relationship among the compared scenarios is determined by the differences in their standard deviations.

Table 6. Results for the evaluated scenarios in the sensitivity analysis.

Impact Categories	Sc0	Sc1	Sc2	Sc3	Sc4
GWP (kg$_{CO_2-eq.}$)	-170.9	58	940	49	39.06
AP (kg$_{SO2-eq.}$)	-242×10^{-3}	-7.75×10^{-1}	-2.19	-2.04×10^{-1}	1.50×10^{-2}
EP (kg$_{PO4-eq.}$)	-38.6×10^{-3}	-4.60×10^{-2}	-4.25×10^{-2}	1.84×10^{-1}	1.20×10^{-2}
ADP$_{elem.}$ (kg$_{Sb-eq.}$)	-50.1×10^{-6}	-4.00×10^{-5}	- - -	-4.87×10^{-5}	1.24×10^{-5}
ADP$_{fossil}$ (MJ)	-2.00×10^3	-4.88×10^3	- - -	-1.84×10^3	464.6
FAETP (kg$_{DCB-eq.}$)	-267×10^{-3}	-17	3.09	-1.34×10^{-1}	8.00×10^{-2}
HTP (kg$_{DCB-eq.}$)	-7.45	46	-62.9	-7.161	2.93
MAETP (kg$_{DCB-eq.}$)	-26.3×10^3	-2.16×10^5	7.94×10^{-1}	-2.57×10^4	2.19×10^4
TETP (kg$_{DCB-eq.}$)	-59×10^{-3}	-3.00×10^{-1}	-7.64×10^{-3}	2.12×10^{-1}	1.42×10^{-1}

- - - Values below the methodology error.

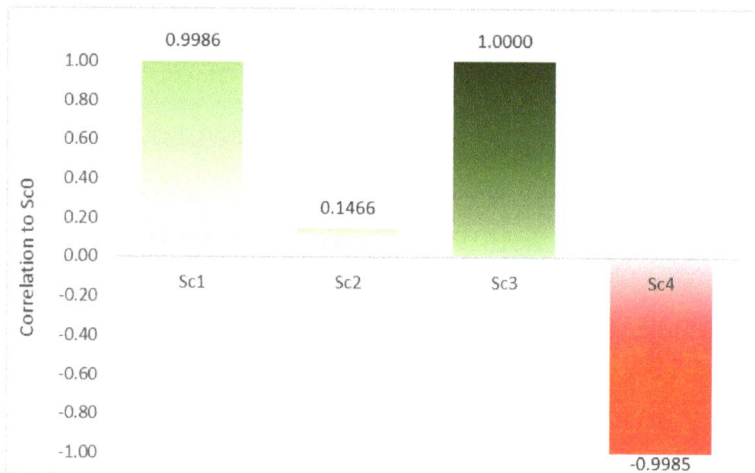

Figure 12. Correlation of different scenarios to the reference case.

Concerning the main differences encountered between Sc0 and Sc1, GWP discrepancy is possibly due to the inclusion of plant construction, waste transport to the incinerator and also bottom ashes processing. Sc1 presented higher CO_2 values (452 kg/t) and higher emissions (600 $g_{DCB-eq.}$/t of NO_x and 75 $g_{DCB-eq.}$/t of SO_2 to air) which contributed to GWP and HTP respectively, contrasting to the values achieved for Sc0 (158 kg/t; -217 $g_{DCB-eq.}$/t of NO_x and -9.92 $g_{DCB-eq.}$/t of SO_2 respectively). An increase in GWP results when plant construction and/or waste collection and transportation are as also shown by other authors. In a thorough LCA work comparing different waste management scenarios in Macau, Song et al. reported waste transportation as one of the most impacting activities, especially when regarding resources depletion [63], contributing to oil consumption as well as NO_x production. This was recently corroborated in studies conducted to assess different possible municipal solid waste management scenarios for the island of Mauritius [64] and also for selected areas in China and Finland [65]. Observing Figure 12 it may be seen that these different boundaries do not interfere significantly in the correlation between both scenarios, meaning that the global performance of the incineration plant is not affected by these modifications.

Sc2 depicts positive impacts for GWP, FAETP and MAETP especially due to stack emissions of CO_2, NO_x and SO_2 whereas Sc0 depicts negative results for these impact categories. This might probably be explained by the study boundaries that include the construction of the plant facilities (accounting for heavier effects on GWP) as well as the wastewater treatment plant [66] (influencing FAETP and MAETP), which are not within the system herein presented for Sc0. Observing Figure 12 these individual differences between impact categories in each case seem to influence the overall performance of the incineration plant, once the correlation coefficient achieved for Sc0 and Sc2 is very low. As both Sc1 and Sc2 comprehend plant construction, comparing their results allows to see that the waste transportation (Sc1) is a more sustainable option than the inclusion of the wastewater treatment facility (Sc2). This may be observed namely for GWP, FAETP, MAETP and TETP which depict lower harmful effects or even avoided burdens for Sc1 rather than Sc2. As a benefit, Sc2 presents a more sustained option in the case of AP and HTP. This is due to the wastewater treatment facility, which reduces the release of acidic and toxic effluents.

Sc3 shows worse results than Sc0 only for GWP, EP and TETP while AP, $ADP_{elements}$ and FAETP are improved and MAETP, HTP and ADP_{fossil} present similar values to the reference case. This seems to create a balance between all the impact categories, and that is why landfill type does not seem to affect the plant performance as may be confirmed by the high correlation coefficient to Sc0 seen in

Figure 12. Liikanen et al. [65] reported the major contribution of landfill to the increase of EP impact due to NO$_x$ emissions, as well as GWP worsening in the absence of landfill gas collection.

Regarding Sc4, all the impact categories show inferior results to the reference case, owing the environmental burdens to the plant energy requirements, once electricity self-consumption will not occur. Other authors had already reported on the importance of recovering energy through incineration concluding that lower carbon footprints and higher electricity savings are attained [55,64,65].

As Figure 12 shows, two of the suggested scenarios are highly correlated to the reference case, Sc1 presenting a correlation coefficient of 0.999 and Sc3 perfectly matching the global environmental results achieved by Sc0. This means that the changes imposed by these scenarios do not change the final environmental performance of the incineration held under the conditions of this study. Therefore, it is possible to say that the life cycle assessment herein conducted for this IWMS is a robust and reliable evaluation as well as the ERP maintains its environmental sustainability even considering the inclusion of the plant construction plus the waste transport to the treatment facilities and the less restricted type of landfill.

Regarding the inclusion of the wastewater treatment facilities, this will have a visible effect on the plant performance, supported by the low correlation coefficient achieved for Sc2. The possible causes for this may rely on the fact that the wastewater treatment affords sludges, which require processes such as water removal, pathogen destruction and digestion before being disposed. This is in accordance to the fact that this scenario consumes much more energy than the reference one, 95% of the generated energy being reused to operate the plant itself. Relative to Sc4, the result was somehow expected not to correlate so well with the reference case once electricity production is one of the major advantages of the incineration plants, as aforementioned [55]. Therefore, when considering only waste treatment, incineration may be viewed as an unsustainable technique, jeopardizing the environment instead of contributing to its maintenance and equilibrium. This is better explained in the Section 3.4, where a hot-spot analysis allows a more detailed discussion.

In a comprehensive evaluation, Figure 13 shows a comparison of the main resource and emissions for scenario 0 and scenarios 3 and 4 (the strongest and the weakest correlations to Sc0), presented in kg of emissions (or materials) per tonne of treated waste. In the case of Sc0 or Sc3, the waste treatments result in severe avoided impacts for the environment, the most spared segments being material resources, fresh water and air.

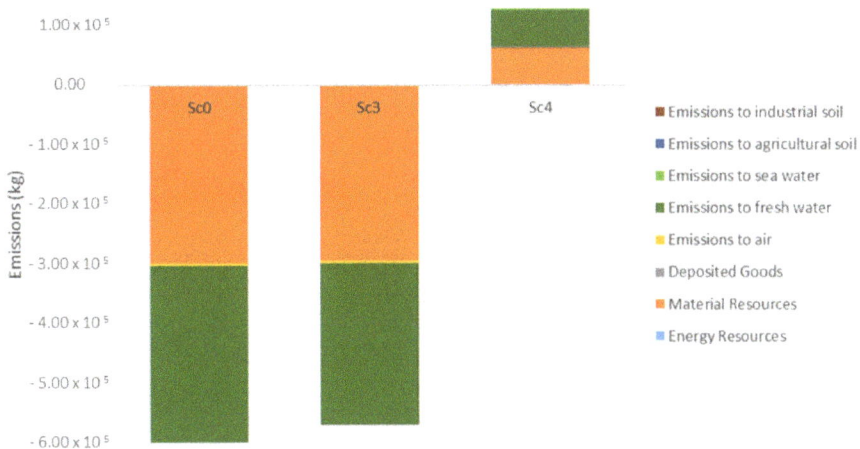

Figure 13. Balance comparison between the reference case (Sc0) and two other alternatives (Sc3 and Sc4). Results refer to one tonne of treated residues.

Figure 13 also shows the added value given by energy production through the incineration process, Sc4 constituting the option with worse results due to this absence: 1 tonne of municipal waste affects more than 120 tonnes of resources available in nature. The production of electrical energy had already been reported as the massive cause for the good performance of incineration plants with this facility, in several impact categories [44,67]. This indicates that using incineration solely as a waste disposal practice would be clearly unsustainable, raising serious environmental issues.

3.4. Hot-Spot Analysis

After a general interpretation of the scenarios with extreme behaviour for the incineration of this waste stream (Sc0 against Sc3 and Sc4), it is interesting to understand what are the main consequences of the type of landfill or the electricity production and reuse within the system for each of the assessed environmental categories, and not only in a global view. For this purpose, a hot-spot analysis was set, Figures 14 and 15 portraying the influence of each of the tested situations.

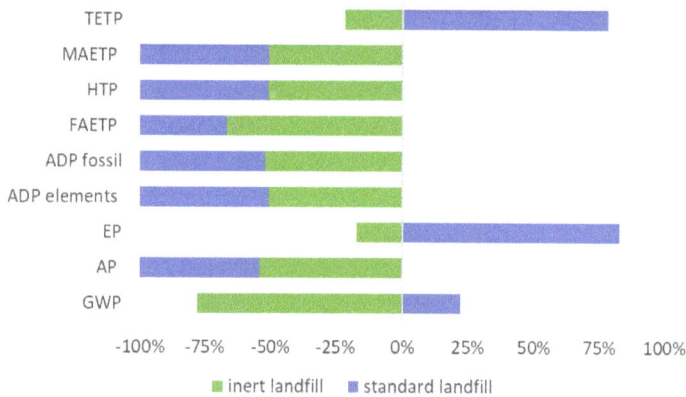

Figure 14. Landfill type effect on the environmental impact categories.

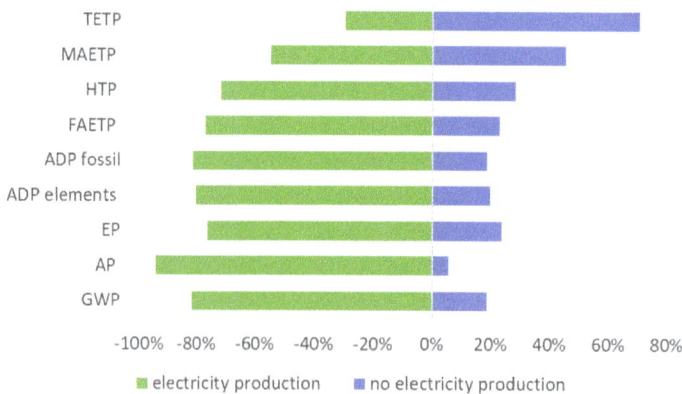

Figure 15. Electricity production and self-consumption effect on the environmental impact categories.

As can be seen from Figure 14, although the correlation coefficient between Sc3 and Sc0 is 1, inert landfill (Sc0) favours a more environment-friendly approach than the typical one (Sc3), all the categories showing negative values for the environmental impacts. This is even more significant in the case of TETP, EP and GWP which, when evaluated under a standard landfill scenario exhibit the opposite behaviour, meaning high environmental damage. MAETP, HTP, ADP (both genres) and AP share relative quotas between the two landfill types, which indicates that they are not influenced by this variable, depicting similar results in both cases, whereas FAETP renders a pronounced effect of approximately 80% towards the inert landfill.

It is relevant to state that in the case of energy recovery held under the conditions presented for Sc0, the inert landfill really is a good asset once it is dedicated to a definite sort of residues, stating a remarkable difference in categories that span from soil, to aquatic and gaseous compartments. Policy makers should be in possession of this kind of information, granting noxious impact remission at the source, instead of having to consider extra means of technical confinement.

In the case of electricity generation within the incineration system (Sc0), Figure 15 shows unquestionable gains for all the assessed categories, TETP and MAETP suffering higher repercussions if energy production is neglected and all the electric inputs have to be supplied by the national grid (Sc4). Regarding the incineration plant herein evaluated, it is not a surprise that it contributes greatly to more sustainable performances, since the plant conversion efficiency is high, with more than 85% of the produced electricity in 2015 being sold to the grid. This benefit has also been reported by Eriksson and Finnveden [49] in recent paper about the key parameters in WtE systems. To enhance this feature, a possible mechanism to be used is the production of electricity from the landfill gas, which can be taken into account in a further stage of development of the facilities, as recommended by other authors [44,48].

3.5. Life Cycle Impact Analysis

After comparing the created scenarios and then their performance in each impact category, the environmental evaluation of the ERP main facilities (Sc0) was done process by process so that weak points were noticed and possibly corrected in the future, if necessary. Figure 16 describes this assessment with an insight for better perception on the flue gas treatment and ash inertization profile, as well as for landfill profile. These results were achieved summing all the contributing inputs and outputs for each of the three main processes presented, as featured by the used software GaBi. For further details, please refer to the Supplementary Material section.

As already commented, the evaluated IWMS is very effective and environmentally sustained, as confirmed here. Furthermore, there are no reports on public health issues related to this facility neither environment emissions above the legal limits. Highlights must be given to the massive avoided burdens in the incineration process unit, mostly due to the electricity production and also to the utilisation of waste as fuel, since this represents a noxious asset for nature and, this way, it is converted into a useful feedstock instead of deposited. In what concerns the electricity production, it must be stressed that this contribution is an approach, once this is not an established process in the plan, rather constituting an output of the incineration process. Hence, the balance was achieved subtracting Sc4 from Sc0, since the electricity production is the only difference between these two scenarios [42], more than 700 $t_{emissions}$/fu being mitigated. The most protected sections are natural resources (saved by incineration in comparison to their grieving if the waste was landfilled), fresh water and minorly air (through the avoided emissions to these environmental compartments when regularly producing electricity from fossil fuels), meaning an overall avoidance of 1300 $t_{emissions}$/fu.

The electric power generation revealed very prone to the success of the incineration facilities (as in general in EU-27 region) and also in other reported studies [45,46,65], but this is not always true. Depending on the type of resources utilized to compose the grid mix electricity available for the consumer, the avoided impacts generated by the electricity provided from incineration will be different and sometimes not so significant [59].

The flue gas treatment and ash inertization process apports resources consumption and emissions to fresh water (summing nearly 1 $t_{emissions}$/fu) and landfill raises this impact to 7.2 $t_{emissions}$/fu, with the contribution of the deposited products and also non-neglegible emissions to the air. These two processes have harmful impacts on environmental compartments, but landfill is the one presenting worse results as reported elsewhere too [59,68].

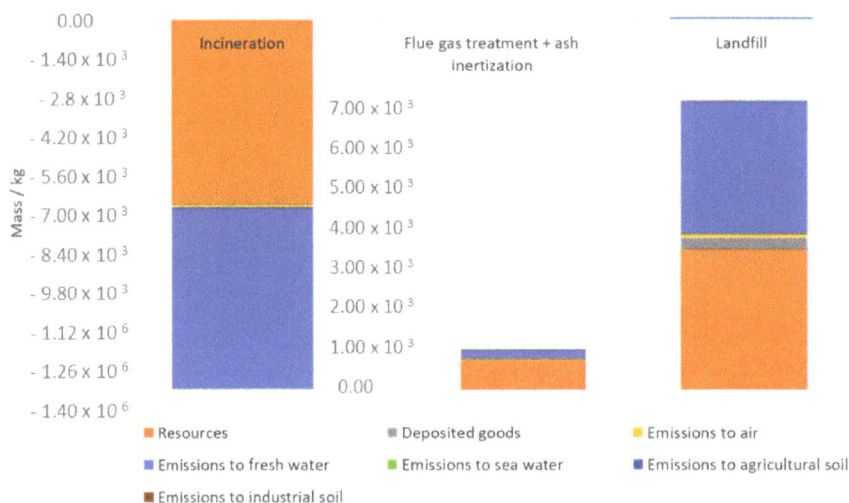

Figure 16. Environmental impact of the incineration processes in the Portuguese case study. Results refer to one tonne of treated residues.

4. Conclusions

A life cycle analysis of an energy recovery plant at the biggest northern city of Portugal was performed for the year 2015. The assessed environmental impacts were compared to European average scenarios in an attempt to position the IWMS amongst a broader panorama, in accordance to all the efforts that are being made worldwide to reduce waste production, to establish new routes for the reuse of everyday products and to embrace several multipurpose opportunities of converting these into more valuable goods. This study aimed at understanding the specific results of this plant, and their effects on the surrounding area and populations, which is a subject that depicts lack of relevant literature. For the majority of the categories, the incineration plant results were within the two European situations ranged, enhanced results being achieved in the case of GWP (-171 kg$_{CO_2\text{-eq.}}$), EP (-39×10^{-3} kg$_{PO_4\text{-eq.}}$) and TETP (-59×10^{-3} kg$_{DCB\text{-eq.}}$), due to the landfill restrictions posed by this facility which reduce noxious emissions to environmental compartments, as confirmed by a hot-spot analysis. This analysis also gave insights that may help policy makers when considering landfill types and, most importantly, energy recovery options for similar WtE facilities.

One of the most resource-demanding process is the requirement for electricity, as it depletes natural reservoirs but, within the scope of the assessed ERP, the incineration plan benefits from energy production, which enables a self-consumption situation, the surplus being sent to the national electrical grid, thus generating revenue and avoiding the grieve of environmental deposits. This is the main factor contributing to the absolute sustainable situation provided by the waste incineration, this waste-to-energy technology saving material resources as well as avoiding emissions to fresh water and air. In fact, 1 tonne of energetically valorised waste saves approximately 1.3 million kg of resources and materials, landfill being established as the weak point of the whole system.

When a comparison of the attained outcomes to recently published literature is made, this plant shows a truly favourable environmental profile, holding a solid position amongst the concurrent results. This validates the LCA approach methodology as a favourable and reproducible procedure to take into account when environmental evaluation of the waste management scenarios is on focus.

Energies **2018**, *11*, 548

Another important conclusion to take from this assessment is that the inclusion of the wastewater treatment facility negatively affects the global incineration plant performance, while including the waste transportation to the incineration facilities as well as using a less restrictive landfill do not influence significantly the outcome.

Supplementary Materials: The following are available online at www.mdpi.com/1996-1073/11/3/548/s1. For further details on the different landfill scenarios, please check the Section Supplementary Information 1. For further details on the European average scenarios, please check the Section Supplementary Information 2. For further details on the life cycle inventory for the sensitivity analysis scenarios, please check Section Supplementary Information 3. For further details on the contributions of each main process to the total mass flows, please check the Section Supplementary Information 4.

Acknowledgments: The authors would like to acknowledge LIPOR for providing the necessary data and information, enabling the LCA study in this work. Funding: This work was supported by the Portuguese Foundation for Science and Technology [grant number SFRH/BD/110787/2015].

Author Contributions: Ana Ramos conceived, designed and performed the experimental work, analysed the data and wrote the paper. Carlos Afonso Teixeira enabled and mentored the use of the LCA software. Abel Rouboa supervised the whole work.

Conflicts of Interest: The authors declare no conflict of interest.

Abbreviations

CP	composting plant
EAI	European average incineration
EAL	European average landfill
EF	Ecological Footprint
EMFA	Energy and Material Flow Analysis
ERP	energy recovery plant
HC	hydrocarbons
IWMS	integrated waste management system
LCA	life cycle assessment
NMVOC	non-methane volatile organic compounds
PERSU	*Plano Estratégico para Resíduos Urbanos* (strategic plan for urban residues, in English)
SP	sorting plant
WtE	Waste-to-Energy

References

1. Chen, H.; Jiang, W.; Yang, Y.; Man, X. Global trends of municipal solid waste research from 1997 to 2014 using bibliometric analysis. *J. Air Waste Manag. Assoc.* **2015**, *65*, 1161–1170. [CrossRef] [PubMed]
2. Eriksson, O. *Energy and Waste Management*; Multidisciplinary Digital Publishing Institute: Basel, Switzerland, 2017.
3. Zhang, D.; Huang, G.; Xu, Y.; Gong, Q. Waste-to-energy in China: Key challenges and opportunities. *Energies* **2015**, *8*, 14182–14196. [CrossRef]
4. Ryu, C.; Shin, D. Combined heat and power from municipal solid waste: Current status and issues in South Korea. *Energies* **2012**, *6*, 45–57. [CrossRef]
5. Wagner, T.; Arnold, P. A new model for solid waste management: An analysis of the Nova Scotia MSW strategy. *J. Clean. Prod.* **2008**, *16*, 410–421. [CrossRef]
6. Pires, A.; Martinho, G.; Chang, N.-B. Solid waste management in European countries: A review of systems analysis techniques. *J. Environ. Manag.* **2011**, *92*, 1033–1050. [CrossRef] [PubMed]
7. Ruj, B.; Ghosh, S. Technological aspects for thermal plasma treatment of municipal solid waste-A review. *Fuel Process. Technol.* **2014**, *126*, 298–308. [CrossRef]
8. Inoue, K.; Yasuda, K.; Kawamoto, K. Report: Atmospheric pollutants discharged from municipal solid waste incineration and gasification-melting facilities in Japan. *Waste Manag. Res.* **2009**, *27*, 617–622. [CrossRef] [PubMed]
9. Brems, A.; Baeyens, J.; Dewil, R. Recycling and recovery of post-consumer plastic solid waste in a European context. *Therm. Sci.* **2012**, *16*, 669–685. [CrossRef]

10. Cucchiella, F.; D'Adamo, I.; Gastaldi, M. Sustainable management of waste-to-energy facilities. *Renew. Sustain. Energy Rev.* **2014**, *33*, 719–728. [CrossRef]

11. Cherubini, F.; Stromman, A.H. Life cycle assessment of bioenergy systems: State of the art and future challenges. *Bioresour. Technol.* **2011**, *102*, 437–451. [CrossRef] [PubMed]

12. Brunner, P.H.; Rechberger, H. Waste to energy-key element for sustainable waste management. *Waste Manag.* **2015**, *37*, 3–12. [CrossRef] [PubMed]

13. Hellweg, S.; Doka, G.; Finnveden, G.; Hungerbühler, K. Assessing the Eco-efficiency of End-of-Pipe Technologies with the Environmental Cost Efficiency Indicator. *J. Ind. Ecol.* **2005**, *9*, 189–203. [CrossRef]

14. Xie, M.; Qiao, Q.; Sun, Q.; Zhang, L. Life cycle assessment of composite packaging waste management—A Chinese case study on aseptic packaging. *Int. J. Life Cycle Assess.* **2013**, *18*, 626–635. [CrossRef]

15. Schwarzböck, T.; Rechberger, H.; Cencic, O.; Fellner, J. Determining national greenhouse gas emissions from waste-to-energy using the Balance Method. *Waste Manag.* **2016**, *49*, 263–271. [CrossRef] [PubMed]

16. Lausselet, C.; Cherubini, F.; Del, A.S.G.; Becidan, M.; Strømman, A.H. Life-cycle assessment of a Waste-to-Energy plant in central Norway: Current situation and effects of changes in waste fraction composition. *Waste Manag.* **2016**, *58*, 191–201. [CrossRef] [PubMed]

17. Magrinho, A.; Didelet, F.; Semiao, V. Municipal solid waste disposal in Portugal. *Waste Manag.* **2006**, *26*, 1477–1489. [CrossRef] [PubMed]

18. Ferreira, S.; Moreira, N.A.; Monteiro, E. Bioenergy overview for Portugal. *Biomass Bioenergy* **2009**, *33*, 1567–1576. [CrossRef]

19. Margallo, M.; Aldaco, R.; Irabien, A.; Carrillo, V.; Fischer, M.; Bala, A.; Fullana, P. Life cycle assessment modelling of waste-to-energy incineration in Spain and Portugal. *Waste Manag. Res.* **2014**, *32*, 492–499. [CrossRef] [PubMed]

20. Agência Portuguesa do Ambiente. *PERSU 2020—Plano Estratégico Para os Resíduos Urbanos*; APA: Lisbon, Portugal, 2014; p. 137.

21. Arena, U.; Mastellone, M.L.; Perugini, F. The environmental performance of alternative solid waste management options: A life cycle assessment study. *Chem. Eng. J.* **2003**, *96*, 207–222. [CrossRef]

22. Tarantini, M.; Loprieno, A.D.; Cucchi, E.; Frenquellucci, F. Life Cycle Assessment of waste management systems in Italian industrial areas: Case study of 1st Macrolotto of Prato. *Energy* **2009**, *34*, 613–622. [CrossRef]

23. Liu, G.; Yang, Z.; Chen, B.; Zhang, Y.; Su, M.; Zhang, L. Emergy evaluation of the urban solid waste handling in Liaoning province, China. *Energies* **2013**, *6*, 5486–5506. [CrossRef]

24. Menikpura, S.N.M.; Gheewala, S.H.; Bonnet, S. Framework for life cycle sustainability assessment of municipal solid waste management systems with an application to a case study in Thailand. *Waste Manag. Res.* **2012**, *30*, 708–719. [CrossRef] [PubMed]

25. Cleary, J. Life cycle assessments of municipal solid waste management systems: A comparative analysis of selected peer-reviewed literature. *Environ. Int.* **2009**, *35*, 1256–1266. [CrossRef] [PubMed]

26. Clift, R.; Doig, A.; Finnveden, G. The application of Life Cycle Assessment to Integrated Solid Waste Management—Part 1—Methodology. *Process Saf. Environ. Prot.* **2000**, *78*, 279–287. [CrossRef]

27. McDougall, F.R.; White, P.R.; Dranke, M.; Hindle, P. *Integrated Solid Waste Management: A Life Cycle Inventory*; John Wiley & Sons: New York, NY, USA, 2008.

28. Ghinea, C.; Petraru, M.; Bressers, H.T.A.; Gavrilescu, M. Environmental evaluation of waste management scenarios—Significance of the boundaries. *J. Environ. Eng. Landsc. Manag.* **2012**, *20*, 76–85. [CrossRef]

29. Thomas, B.; McDougall, F. International expert group on life cycle assessment for integrated waste management. *J. Clean. Prod.* **2005**, *13*, 321–326. [CrossRef]

30. Arushanyan, Y.; Björklund, A.; Eriksson, O.; Finnveden, G.; Söderman, M.L.; Sundqvist, J.-O.; Stenmarck, A. Environmental assessment of possible future waste management scenarios. *Energies* **2017**, *10*, 247. [CrossRef]

31. Arafat, H.A.; Jijakli, K.; Ahsan, A. Environmental performance and energy recovery potential of five processes for municipal solid waste treatment. *J. Clean. Prod.* **2015**, *105*, 233–240. [CrossRef]

32. Astrup, T.F.; Tonini, D.; Turconi, R.; Boldrin, A. Life cycle assessment of thermal Waste-to-Energy technologies: Review and recommendations. *Waste Manag.* **2015**, *37*, 104–115. [CrossRef] [PubMed]

33. Parkes, O.; Lettieri, P.; Bogle, I.D.L. Life cycle assessment of integrated waste management systems for alternative legacy scenarios of the London Olympic Park. *Waste Manag.* **2015**, *40*, 157–166. [CrossRef] [PubMed]

34. Toniolo, S.; Mazzi, A.; Garato, V.G.; Aguiari, F.; Scipioni, A. Assessing the "design paradox" with life cycle assessment: A case study of a municipal solid waste incineration plant. *Resour. Conserv. Recycl.* **2014**, *91*, 109–116. [CrossRef]
35. Morselli, L.; Luzi, J.; De Robertis, C.; Vassura, I.; Carrillo, V.; Passarini, F. Assessment and comparison of the environmental performances of a regional incinerator network. *Waste Manag.* **2007**, *27*, S85–S91. [CrossRef] [PubMed]
36. Den Boer, J.; den Boer, E.; Jager, J. LCA-IWM: A decision support tool for sustainability assessment of waste management systems. *Waste Manag.* **2007**, *27*, 1032–1045. [CrossRef] [PubMed]
37. Herva, M.; Neto, B.; Roca, E. Environmental assessment of the integrated municipal solid waste management system in Porto (Portugal). *J. Clean. Prod.* **2014**, *70*, 183–193. [CrossRef]
38. LIPOR. May 2016. Available online: https://www.lipor.pt/en/municipal-solid-waste/energy-recovery/process-description/ (accessed on 1 May 2016).
39. LIPOR. *Sustainability Report*; LIPOR: Porto, Portugal, 2015.
40. Portugal, E.-E.d. EDP. 2009. Available online: https://www.edp.pt/particulares/apoio-cliente/origem-energia/ (accessed on 14 February 2018).
41. Stuttgart, L.-U.O. *GaBi Software—System and Database for Life Cycle Engineering*; Thinkstep AG: Stuttgart, Germany, 2016; pp. 1992–2016.
42. Liamsanguan, C.; Gheewala, S.H. LCA: A decision support tool for environmental assessment of MSW management systems. *J. Environ. Manag.* **2008**, *87*, 132–138. [CrossRef] [PubMed]
43. Jeswani, H.K.; Azapagic, A. Assessing the environmental sustainability of energy recovery from municipal solid waste in the UK. *Waste Manag.* **2016**, *50*, 346–363. [CrossRef] [PubMed]
44. Hong, J.; Li, X.; Zhaojie, C. Life cycle assessment of four municipal solid waste management scenarios in China. *Waste Manag.* **2010**, *30*, 2362–2369. [CrossRef] [PubMed]
45. Morselli, L.; Bartoli, M.; Bertacchini, M.; Brighetti, A.; Luzi, J.; Passarini, F.; Masoni, P. Tools for evaluation of impact associated with MSW incineration: LCA and integrated environmental monitoring system. *Waste Manag.* **2005**, *25*, 191–196. [CrossRef] [PubMed]
46. Passarini, F.; Nicoletti, M.; Ciacci, L.; Vassura, I.; Morselli, L. Environmental impact assessment of a WtE plant after structural upgrade measures. *Waste Manag.* **2014**, *34*, 753–762. [CrossRef] [PubMed]
47. Burnley, S.; Coleman, T.; Peirce, A. Factors influencing the life cycle burdens of the recovery of energy from residual municipal waste. *Waste Manag.* **2015**, *39*, 295–304. [CrossRef] [PubMed]
48. Chi, Y.; Dong, J.; Tang, T.; Huang, Q.; Ni, M. Life cycle assessment of municipal solid waste source-separated collection and integrated waste management systems in Hangzhou, China. *J. Mater. Cycles Waste Manag.* **2015**, *17*, 695–706. [CrossRef]
49. Eriksson, O.; Finnveden, G. Energy Recovery from Waste Incineration—The Importance of Technology Data and System Boundaries on CO_2 Emissions. *Energies* **2017**, *10*, 539. [CrossRef]
50. Tsiliyannis, C. Report: Comparison of environmental impacts from solid waste treatment and disposal facilities. *Waste Manag. Res.* **1999**, *17*, 231–241. [CrossRef]
51. Stranddorf, H.; Hoffmann, L.; Schmidt, A. *Impact Categories, Normalisation and Weighting in LCA (Påvirkningskategories, Normalisering og Vægtning i LCA–in Danish)*; Environmental News No. 78; The Danish Ministry of the Environment. Environmental Protection Agency: Copenhagen, Denmark, 2005.
52. Fernández-Nava, Y.; del Rio, J.; Rodríguez-Iglesias, J.; Castrillón, L.; Marañón, E. Life cycle assessment of different municipal solid waste management options: A case study of Asturias (Spain). *J. Clean. Prod.* **2014**, *81*, 178–189. [CrossRef]
53. Al-Salem, S.M.; Evangelisti, S.; Lettieri, P. Life cycle assessment of alternative technologies for municipal solid waste and plastic solid waste management in the Greater London area. *Chem. Eng. J.* **2014**, *244*, 391–402. [CrossRef]
54. Menikpura, S.N.M.; Sang-Arun, J.; Bengtsson, M. Assessment of environmental and economic performance of Waste-to-Energy facilities in Thai cities. *Renew. Energy* **2016**, *86*, 576–584. [CrossRef]
55. Tabata, T. Waste-to-energy incineration plants as greenhouse gas reducers: A case study of seven Japanese metropolises. *Waste Manag. Res.* **2013**, *31*, 1110–1117. [CrossRef] [PubMed]
56. Banar, M.; Cokaygil, Z.; Ozkan, A. Life cycle assessment of solid waste management options for Eskisehir, Turkey. *Waste Manag.* **2009**, *29*, 54–62. [CrossRef] [PubMed]

57. US EPA, Office of Atmospheric Programs. *Climate Change Division, Accounting Framework for Biogenic CO_2 Emissions from Stationary Source*; US EPA: Washington, DC, USA, 2011; p. 127.

58. Bezama, A.; Douglas, C.; Méndez, J.; Szarka, N.; Muñoz, E.; Navia, R.; Schock, S.; Konrad, O.; Ulloa, C. Life cycle comparison of waste-to-energy alternatives for municipal waste treatment in the Chilean Patagonia. *Waste Manag. Res.* **2013**, *31*, 67–74. [CrossRef] [PubMed]

59. Mendes, M.R.; Aramaki, T.; Hanaki, K. Comparison of the environmental impact of incineration and landfilling in São Paulo City as determined by LCA. *Resour. Conserv. Recycl.* **2004**, *41*, 47–63. [CrossRef]

60. Gunamantha, M.; Sarto. Life cycle assessment of municipal solid waste treatment to energy options: Case study of KARTAMANTUL region, Yogyakarta. *Renew. Energy* **2012**, *41*, 277–284. [CrossRef]

61. Chaya, W.; Gheewala, S.H. Life cycle assessment of MSW-to-energy schemes in Thailand. *J. Clean. Prod.* **2007**, *15*, 1463–1468. [CrossRef]

62. Chen, D.Z.; Christensen, T.H. Life-cycle assessment (EASEWASTE) of two municipal solid waste incineration technologies in China. *Waste Manag. Res.* **2010**, *28*, 508–519. [CrossRef] [PubMed]

63. Song, Q.; Wang, Z.; Li, J. Environmental performance of municipal solid waste strategies based on LCA method: A case study of Macau. *J. Clean. Prod.* **2013**, *57*, 92–100. [CrossRef]

64. Rajcoomar, A.; Ramjeawon, T. Life cycle assessment of municipal solid waste management scenarios on the small island of Mauritius. *Waste Manag. Res.* **2017**, *35*, 313–324. [CrossRef] [PubMed]

65. Liikanen, M.; Havukainen, J.; Hupponen, M.; Horttanainen, M. Influence of different factors in the life cycle assessment of mixed municipal solid waste management systems—A comparison of case studies in Finland and China. *J. Clean. Prod.* **2017**, *154*, 389–400. [CrossRef]

66. O'Connor, M.; Garnier, G.; Batchelor, W. Life cycle assessment comparison of industrial effluent management strategies. *J. Clean. Prod.* **2014**, *79*, 168–181. [CrossRef]

67. Morselli, L.; Luzi, J.; Bartoli, M.; Vassura, I.; Passarini, F. LCA and an Integrated Environmental Monitoring System as joint tools for incinerator environmental impact assessment. *WIT Trans. Ecol. Environ.* **2004**, *78*, 301–309.

68. Pikon, K.; Gaska, K. Greenhouse Gas Emission Mitigation Relevant to Changes in Municipal Solid Waste Management System. *J. Air Waste Manag. Assoc.* **2010**, *60*, 782–788. [CrossRef] [PubMed]

energies

MDPI

Article

Sucrose Is a Promising Feedstock for the Synthesis of the Platform Chemical Hydroxymethylfurfural

David Steinbach [1,2,*], Andrea Kruse [2], Jörg Sauer [1] and Philipp Vetter [1]

1   Institute for Catalysis Research and Technology, Karlsruhe Institute of Technology (KIT),
    Hermann-von-Helmholtz-Platz 1, 76344 Eggenstein-Leopoldshafen, Germany; j.sauer@kit.edu (J.S.);
    Philipp.Vetter@partner.kit.edu (P.V.)
2   Institute of Agricultural Engineering, University of Hohenheim, Garbenstrasse 9, 70599 Stuttgart, Germany;
    Andrea_Kruse@uni-hohenheim.de
*   Correspondence: David.Steinbach@uni-hohenheim.de; Tel.: +49-7121-989724

Received: 16 February 2018; Accepted: 12 March 2018; Published: 14 March 2018

Abstract: Hydroxymethylfurfural (HMF) has an outstanding position among bio-based platform chemicals, because high-value polymer precursors and fuel additives can be derived from HMF. Unfortunately, the large-scale industrial production of HMF is not yet realized. An open research question is the choice of hexose feedstock material. In this study, we used the highly available disaccharide sucrose for HMF synthesis. The conversion of sucrose was catalyzed by sulfuric acid in water media. Experiments were conducted at temperatures of 180, 200, and 220 °C with reaction times of 2–24 min. A carbon balance showed that the yield of unwanted side products rose strongly with temperature. We also developed a kinetic model for the conversion of sucrose, involving nine first-order reactions, to uncover the kinetics of the main reaction pathways. Within this model, HMF is produced exclusively via the dehydration of fructose. Glucose isomerizes slowly to fructose. Side products arise simultaneously from glucose, fructose, and HMF. A pathway from hexoses to xylose via reverse aldol reaction was also included in the model. We believe that sucrose is the ideal feedstock for large-scale production of HMF because it is more abundant than fructose, and easier to process than sugars obtained from lignocellulosic biomass.

Keywords: kinetics; glucose; fructose; sugar dehydration; HMF; hydrothermal; hydrolysis; acid-catalysed

---

1. Introduction

The production of platform chemicals from renewable biomass is experiencing enormous interest as a result of (1) the latest developments in research, and (2) government funding for sustainable production processes and CO_2 reductions [1,2]. The platform chemical hydroxymethylfurfural (HMF) has an outstanding position among bio-based platform chemicals because it has two different functional groups, and therefore offers a wide range of applications [3]. It is able to be a substitute for fossil-based products in various new materials and products, such as biopolymers, resins, coatings, paints, varnishes, and fuel additives [4]. For example, HMF can be oxidized to furandicarboxylic acid (FDCA), which can be a substitute for fossil-based terephthalic acid in the production of polyesters. HMF has been named as one of the top ten value-added bio-based chemicals by the U.S. Department of Energy [5]. It can be produced exclusively by thermochemical conversion routes, and is highly carbon-efficient during its synthesis [6]. In principle, HMF is available from all hexoses and their polymers.

The dehydration of hexoses to form HMF has been performed in different solvent systems, which can be categorized as aqueous, organic, or biphasic. In addition, a great variety of acid catalysts have been investigated [7,8], which can be distinguished into homogeneous and heterogeneous catalysts. A well-known and established synthetic strategy for HMF is the homogeneous catalyzed dehydration of hexoses in aqueous solutions. Sulfuric acid is a homogeneous catalyst which is comparatively cheap, and thus an expensive acid-recycling step is unnecessary. The capital cost of processes based on diluted sulfuric acid as a catalyst is, therefore, relatively low [9]. The advantages of HMF synthesis in aqueous media are as follows: (1) water is a cheap and environmentally friendly solvent; (2) the process is simple and easy to scale up; (3) water dissolves polar sugars and HMF in high concentrations; and (4) the water elimination of fructose to form HMF is supported by the high level of the ionic product in hot compressed water and the higher thermodynamic stability of double bonds in these conditions [10,11]. Unfortunately, HMF rehydrates in a water-based medium under acidic conditions to form the consecutive products levulinic acid and formic acid. Furthermore, the self-condensation of HMF and condensation with other compounds lead to the formation of high-molecular-weight polymeric substances (humins) [6]. The maximum yields of HMF in water are reported to be around 50–60 mol% when fructose is used as the feedstock [3].

Unfortunately, the large-scale industrial production of HMF has not yet been realized, due to unfavorable reaction kinetics and the lack of progress in process development [6]. Feedstock costs are a key parameter in improving the economics of future processes. Ketoses (e.g., fructose) are much more efficiently dehydrated to form HMF than aldoses (e.g., glucose) [12]. Therefore, fructose is the ideal feedstock from a reaction-engineering point of view. However, fructose feedstock costs are higher as compared with glucose. Generally, when glucose is the feedstock, significantly lower HMF yields have been obtained.

Lignocellulosic biomass is another feedstock containing hexoses. Cellulose consists of glucose building blocks, and hemicelluloses can also contain glucose [13]. A previous isomerization of glucose to fructose is a strategy to overcome low yields from the direct conversion of lignocellulose-derived glucose to HMF [6]. By using isomerase enzymes, the fructose yield can reach the thermodynamic equilibrium (about 50 wt % [14]). However, the yields of fructose are consistently lower when using heterogeneous catalysts [15,16]. Thus, a mixture of glucose and fructose is always obtained after isomerization.

A mixture of fructose and glucose is also obtained when sucrose is subjected to hot water. This is because the glycosidic linkage, which connects a glucose and fructose unit, can be easily hydrolyzed. Sucrose is an available and cheap feedstock; about 171 million tons were produced worldwide in 2016–2017 [17]. However, only a few studies deal with the conversion of sucrose to hexoses and HMF in hot water [18–22]. Khajavi et al. [19] and Gao et al. [21] investigated the decomposition of sucrose in hot water without the addition of a catalyst. They determined the kinetics of the sucrose decomposition, but not the kinetics of HMF formation [19,21]. Bower and coworkers [20] converted sucrose catalyzed by sulfuric acid. They investigated the decomposition kinetics of formed fructose and glucose, but not the HMF formation [20]. A recent work by Tan-Soetedjo and coworkers [22] deals with the acid-catalysed conversion of sucrose to levulinic acid. They developed a detailed kinetic model which includes HMF formation as well as the rehydration of HMF to form levulinic acid [22].

In general, the scope of kinetic studies is to understand the reaction mechanism or to identify the optimal reaction conditions for maximum product yields. Many such studies converting either glucose or fructose to HMF have been performed, and have been reviewed intensively by van Putten et al. [3]. The group led by Heeres, for example, used modified Arrhenius equations with (1) power law dependence of acid catalyst concentration, and (2) power law dependence of reaction order. They investigated the kinetics of the acid-catalyzed conversion of fructose to HMF [23], HMF to levulinic acid [24], and glucose to levulinic acid [25]. However, kinetic parameters which are determined experimentally depend to a great extent on the setup and the procedure of the experiments themselves [26]. Attention must be paid to the following issues for the direct comparison of kinetic data obtained by different studies: (1) which reactor system is applied, (2) modifications to the Arrhenius equation, (3) modifications to reaction order, (4) the parameter range of reaction conditions, (5) anion effects of the acid utilized, and (6) which reaction scheme is applied.

1. The reactor type can influence the determination of kinetic data. Smith et al. [27], for example, investigated glucose decomposition in a batch and tubular reactor using sulfuric acid. Glucose decomposed 4.4 times faster in the continuous system. The activation energy was much lower (88 kJ/mol) compared with that of the continuous reactor (129 kJ/mol) [27].
2. Either a linear or a power law dependence of acid concentration is included in the Arrhenius equation in some models. Depending on the study, the mass percentage of acid [27,28], hydronium concentration at ambient conditions [20], hydronium concentration at reaction conditions [23,25], an activity term [29], or multiple factors [30] may be used in modified Arrhenius equations.
3. The reaction order is often set to one for the conversion of hexoses in aqueous media [3,12,20]. Nevertheless, slightly different reaction orders have been determined by some studies [23,25].
4. Kinetic rate constants are normally determined within small ranges for substrate concentration, catalyst concentration, or temperature [25]. An extrapolation to other reaction conditions is problematic.
5. The type of acid anion influences the kinetics of fructose decomposition, even if the pH is the same [31]. Sulfate ions, for example, have an inhibitor effect on the HMF rehydration reaction [31].
6. Only one reaction is assumed for hexose conversion in some studies. Other studies use networks of more than seven individual reactions [3].

In this work, we converted a 2 wt % sucrose solution in a tubular reactor. Experiments were conducted at temperatures of 180, 200, and 220 °C with reaction times of 2–24 min. Sulfuric acid was used as a catalyst at a concentration of 0.005 mol/L. Furthermore, we compared sucrose, glucose, and fructose concerning their product distribution at 220 °C and 10 min. A detailed kinetic network for the acid-catalyzed conversion of sucrose to HMF was developed. Thereby, the kinetics of main reaction pathways were uncovered.

2. Materials and Methods

2.1. Feedstock

Experiments were performed with a feedstock solution of 2 wt % sucrose in deionized water and 0.005 mol/L sulfuric acid (pH = 2.0). Sucrose, sulfuric acid, and water were mixed together directly before each experiment to minimize degradation reactions of the saccharides. A relatively small concentration of sulfuric acid catalyst was chosen in this study, to minimize the potential demand of a neutralization agent in the wastewater stream. Moreover, under less acidic conditions, unwanted rehydration of HMF to form levulinic acid and formic acid is decreased [31]. For economic reasons, the feedstock concentration should be as high as possible to reduce costs in downstream separation steps. On the other hand, higher hexose feedstock concentrations decrease HMF yield [31], because side reactions forming high-molecular-weight polymeric substances are accelerated at higher concentrations.

Additional experiments were performed with a feedstock containing either glucose or fructose. In this case, a 1.05 wt % glucose or fructose solution was used (corresponding to glucose or fructose in a 2.0 wt % sucrose solution). The sulfuric acid concentration was kept constant at 0.005 mol/L (pH = 2.0).

All feedstock chemicals were purchased (sucrose grade Reag. Ph. Eur. from Merck (Darmstadt, Germany), D(+)-glucose grade normapur from VWR (Radnor, PA, USA), D(−)-fructose ≥ 99% from Sigma Aldrich (St. Louis, MO, USA), 72 wt % sulfuric acid from Alfa Aesar (Haverhill, MA, USA)).

2.2. Acid-Catalyzed Conversion

The conversion of saccharide feedstock was performed in a self-constructed test rig (see Figure 1). This test rig consisted of a 170 mL tubular reactor made of stainless steel 1.4571. The reactor dimensions were 0.30 m in length and 0.0027 m in inner diameter. The reactor was surrounded by a brass block heated by cartridges. The feedstock solution was pumped via a PrepStar SD-1 from Agilent (Santa Clara, CA, USA) into the reactor. The product was cooled to room temperature in a heat exchanger before the system pressure was released with a back pressure valve.

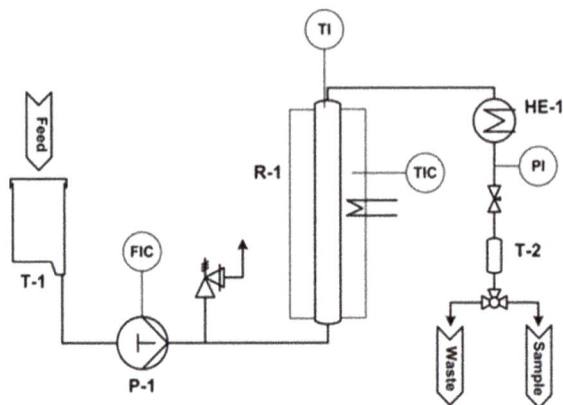

Figure 1. Piping and instrumentationdiagram of test rig. T-1: feedstock tank, P-1: pump, R-1: tubular reactor with surrounding heating block, HE-1: cooler, T-2: small tank.

All experiments were performed at 2.5 MPa reactor pressure to prevent the evaporation of water. The reactor pressure has no effect on the decomposition reactions of hexoses in acidic water media [32]. Reaction temperatures (measured at the reactor exit) of 180, 200, and 220 °C were investigated. The hydrodynamic residence time was varied between 2 and 24 min at 180 °C, and between 2 and 18 min at 200 and 220 °C. Product samples were taken after the twofold residence time was passed to ensure steady state conditions. Samples were stored at 4 °C prior to their analysis. The experiments at 220 °C and 10 min residence time were performed in quadruplicate, and in all other cases as single runs.

2.3. Analytics

Characterization of the liquid samples was performed with several HPLC methods. Preliminary filtration with 0.45 µm GHP syringe filters (Pall, New York, NY, USA) was performed to remove high-molecular-weight products.

The furan compounds HMF, furfural, and methylfurfural were separated at 20 °C in a Lichrospher 100 RP-18 column (Merck, Darmstadt, Germany). Therefore, a water–acetonitrile eluent (9:1 *v/v*) at a flow rate of 1.4 mL/min was used. Furan compounds were quantified by a UV detector at 290 nm. Methylfurfural was only detected in the samples in trace amounts at experimental conditions with higher residence times. Therefore, methylfurfural was not considered further in this article.

The saccharide compounds sucrose, glucose, fructose, and xylose were separated at 35 °C in a Metrosep Carb 2 column (Metrohm, Filderstadt, Germany). An eluent with 0.1 mol/L sodium hydroxide and 0.01 mol/L sodium acetate was used with a flow rate of 0.5 mL/min. Sugars were quantified by an amperometric detector.

Formic acid and levulinic acid were analyzed with an Aminex HPX-87H column (Biorad, Hercules, CA, USA). The column temperature was 25 °C for formic acid separation and 60 °C for levulinic acid. The eluent was 0.004 mol/L sulfuric acid at a flow rate of 0.65 mL/min in both cases. Detection was performed by RI and DAD (210 nm for formic acid, 280 nm for levulinic acid).

2.4. Reaction Model

A chemical reaction model only describes the experimental data correctly if all the relevant reaction pathways are contained in the model [33,34]. However, a formal kinetic reaction model does not show insights into the elemental reactions involved. The formal kinetic reaction model we used in this work is shown in Figure 2. A rate coefficient k_n was defined for each reaction pathway. The reaction order for all reactions was set to one. Different variations of the model shown in Figure 2 were considered, either including additional or omitting reaction pathways. After evaluating the different modeling results, the reaction model presented in Figure 2 described the data most appropriately.

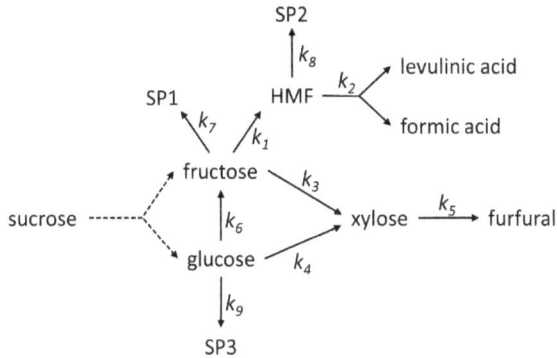

Figure 2. Reaction model to determine the kinetic rate coefficients k_n of the acid-catalyzed conversion of sucrose. HMF = hydroxymethylfurfural; SP = side products.

The hydrolysis reaction of sucrose to equimolar amounts of fructose and glucose was not included in the kinetic model, because sucrose was never detected in the samples in the reaction conditions investigated. Sucrose glycosidic bonds hydrolyze rapidly under acidic conditions at temperatures higher than 160 °C [19,20]. For modelling, it is assumed that sucrose hydrolyzes spontaneously in quantitative yields when entering the reactor.

In our model, glucose is either isomerized to fructose or converted to noncharacterized side products. Even if acids are far less effective catalysts than bases, enolization and isomerization of sugars occurs [30,31,35,36]. We simply implemented the isomerization of glucose to fructose in our model because in other works [37–39], the reverse reaction from fructose to glucose was reported negligible. A pathway from glucose to xylose is also included in the model.

Fructose can undergo dehydration to form HMF or react to form other side products. A pathway from fructose to xylose is also included in the model. Antal et al. [18] assumed that pentoses could be formed together with formaldehyde out of fructose by a reverse aldol reaction, but they did not detect xylose in their studies. Xylose dehydrates to furfural in our model. HMF can react to form other side products, or is rehydrated to form equimolar amounts of levulinic acid and formic acid. Asghari and Yoshida [32] reported in a similar study on fructose decomposition that side products (e.g., soluble polymers) were produced not only from fructose, but also from HMF.

2.5. Kinetic Modelling

A molar ratio $y_{i,meas}$ was calculated from all concentrations c_i experimentally determined (see Equation (1)).

$$y_{i,meas} = \frac{c_i}{c_{sucrose,start}} \tag{1}$$

A set of ordinary differential equations was formulated to describe all rates of change, according to the reaction model (see Equations (2)–(10)).

$$\frac{dy_{fructose}}{dt} = -k_1 \cdot y_{fructose} - k_3 \cdot y_{fructose} - k_7 \cdot y_{fructose} + k_6 \cdot y_{glucose} \tag{2}$$

$$\frac{dy_{glucose}}{dt} = -k_4 \cdot y_{glucose} - k_6 \cdot y_{glucose} - k_9 \cdot y_{glucose} \tag{3}$$

$$\frac{dy_{HMF}}{dt} = +k_1 \cdot y_{fructose} - k_2 \cdot y_{HMF} - k_8 \cdot y_{HMF} \tag{4}$$

$$\frac{dy_{xylose}}{dt} = +k_3 \cdot y_{fructose} + k_4 \cdot y_{gluctose} - k_5 \cdot y_{xylose} \tag{5}$$

$$\frac{dy_{furfural}}{dt} = +k_5 \cdot y_{xylose} \tag{6}$$

$$\frac{dy_{levulinic\ acid}}{dt} = \frac{dy_{formic\ acid}}{dt} = +k_2 \cdot y_{HMF} \tag{7}$$

$$\frac{dy_{SP1}}{dt} = +k_7 \cdot y_{fructose} \tag{8}$$

$$\frac{dy_{SP2}}{dt} = +k_8 \cdot y_{HMF} \tag{9}$$

$$\frac{dy_{SP3}}{dt} = +k_9 \cdot y_{glucose} \tag{10}$$

Experiments were carried out at three different temperatures. The temperature-dependence of the formal kinetic rate coefficients k_n is considered by the Arrhenius equation (see Equation (11)). The activation energy EA_n and the Arrhenius factor A_n are the parameters which were adjusted during optimization.

$$k_n(T) = A_n \cdot e^{\frac{-EA_n}{R \cdot T}} \tag{11}$$

Modelling was conducted in Matlab V 8.4. using the function fminsearch for optimization, which uses the Nelder–Mead simplex search method. The function fminsearch finds a minimum of the scalar function given in Equation (12). The difference between the molar ratio calculated y_i using the kinetic model and the molar ratio measured $y_{i,meas}$ was minimized. The optimization included data of the components l of all experiments z. The maximum molar ratio of each component $y_{l,max}$ was used in Equation (12) for data normalization.

$$f_{min} = \sum_z \sum_l |y_i - y_{i,meas}|^2 \cdot \frac{1}{y_{l,max}} \tag{12}$$

3. Results and Discussion

3.1. Maximum Mass Yield of HMF

The HMF yield at reaction temperatures of 200 and 220 °C shows a maximum within the residence time spectra investigated (see Figure 3). At 180 °C, the lowest temperature considered, however, the HMF yields increased continuously with reaction time. The highest yields of HMF recorded, based on input sucrose mass, were 0.25 g/g (at 180 °C, 22 min), 0.25 g/g (at 200 °C, 14 min), and 0.23 g/g (at 220 °C, 10 min). We expect that temperatures above 220 °C would result in a lower HMF maximum at shorter residence times.

Bowler et al. [20] investigated sucrose hydrolysis at similar reaction conditions (160–200 °C and 3–12 min), but used sulfuric acid in higher concentrations (0.01–0.2 mol/L). They reported a lower maximum HMF yield of 0.18 g/g (25% of the theoretical yield [20]). In addition, Tan-Soetedjo et al. [22] obtained a lower maximum HMF yield of 0.17 g/g at 140 °C and 0.05 mol/L sulfuric acid. Thus, higher sulfuric acid concentrations are not favorable in obtaining high HMF yields.

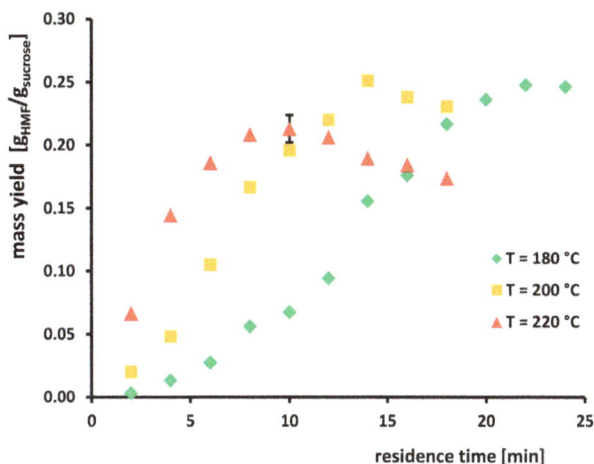

Figure 3. HMF mass yield after acid-catalyzed conversion of sucrose at different residence times and reaction temperatures (T). Feedstock contained 2 wt % sucrose and 0.005 mol/L sulfuric acid.

3.2. Carbon Balance of Sucrose Conversion

A carbon balance is an easy means for evaluating the reactions of organic molecules in water. In that regard, a mass balance often provides not significant results, because most of the mass flow is water. Water is also a reaction partner in the reaction network considered, thus a water-free mass balance is inadvisable.

Figure 4 shows the carbon balance of the sucrose conversion at 180, 200, and 220 °C. A carbon balance of over 100% results from measurement errors of the individual components via HPLC. The carbon from sucrose input was mostly transferred during the reaction to glucose, fructose, or HMF for all residence times considered (see Figure 4). Carbon conversion to xylose and furfural was minor, even at the harshest reaction conditions of 220 °C and 18 min. The hydration products of HMF, levulinic acid and formic acid, become more important at higher reaction temperatures. Sucrose was not detected in the samples, and so was completely hydrolyzed even at the lowest residence times. This is in accordance with Bowler et al. [20], who performed sucrose hydrolysis at similar reaction conditions.

The difference between the sum of all carbon detected and 100% can be assigned to nonquantified side products. This could be either small organic molecules from fragmentation reactions of sugars (such as aldehydes, ketones, and organic acids [31]), sugar reversion products [25], or high-molecular-weight condensation products (humins). The carbon percentage of side products increases strongly with reaction temperature and accounts for 14.6% (180 °C), 22.8% (200 °C), and 38.1% (220 °C) of total carbon at a residence time of 18 min, respectively. We recommend reaction conditions with lower side product formation for a technical process. Firstly, all carbon that is transferred to side products is a waste of feedstock, and must be removed cost-intensively from HMF. Secondly, high-molecular-weight condensation products are poorly soluble in water, and can cause the plugging of pipes and apparatuses [6].

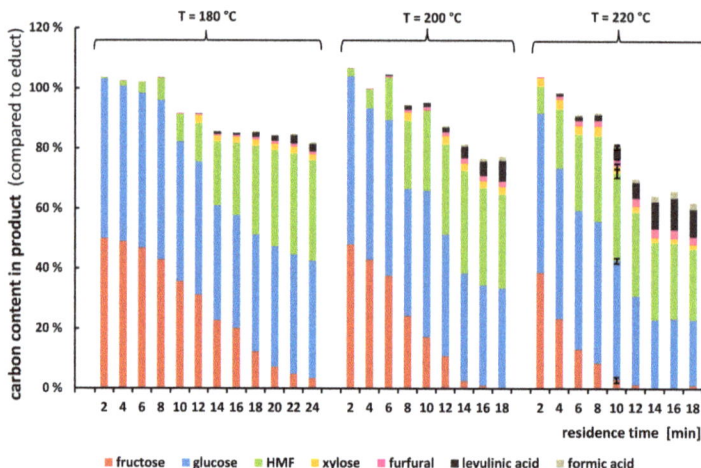

Figure 4. Carbon composition after acid-catalyzed conversion of sucrose at different residence times and reaction temperatures (T). The carbon input of sucrose is represented by 100%. Feedstock contained 2 wt % sucrose and 0.005 mol/L sulfuric acid.

3.3. Carbon Balance of Glucose and Fructose Conversion

The carbon balances of experiments with a feedstock containing either glucose or fructose are shown in Figure 5. Amounts of 1.05 wt % monosaccharide solution were used, corresponding to glucose or fructose in a 2.0 wt % sucrose solution. At 220 °C and 10 min residence time, the conversion of glucose (25%) was much lower than that of fructose (96%). The HMF mass yields accounted for 0.062 $g_{HMF}/g_{glucose}$ and 0.351 $g_{HMF}/g_{fructose}$ at 220 °C and 10 min.

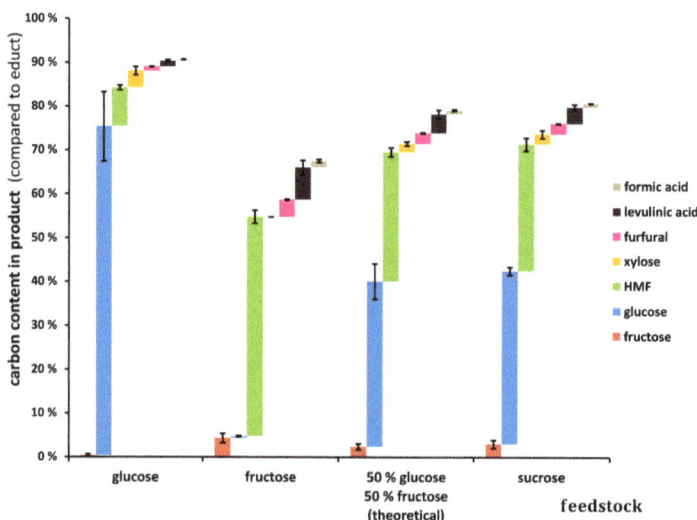

Figure 5. Carbon composition after acid-catalyzed conversion of glucose, fructose, and sucrose at a residence time of 10 min and temperature of 220 °C. The column "50% glucose 50% fructose (theoretical)" shows a calculated 1:1 mixture of glucose and fructose. Feedstock contained 0.005 mol/L sulfuric acid and either 2 wt % sucrose, 1.05 wt % glucose, or 1.05 wt % fructose.

When a glucose feedstock was used, we found traces of the isomerization product fructose in the product samples, and vice versa (see Figure 5). Xylose was only detected after the conversion of glucose, and not in the case of fructose. However, furfural was present in both cases, which is a dehydration product of pentoses. Therefore, we assume that the pathway from fructose to furfural might proceed via another intermediate pentose.

A calculated mixture of glucose and fructose is also depicted in Figure 5, and was compared with experimental data from sucrose feedstock. The conversion of sucrose was well-described by combining the experimental results from fructose and glucose conversion. We used unpaired two-sample Student's *t*-tests to evaluate whether there was a significant difference in the carbon composition. Student's *t*-tests resulted in $p > 0.34$ for all individual components shown in Figure 5, and thus no significant difference could be found in sucrose experiments compared with a calculated mixture of the two hexoses. This implied that the results of glucose and fructose conversion could be used to estimate the yields from sucrose conversion. As a consequence, cross-reactions of glucose products with fructose products were of minor importance in the investigated conditions. This finding is in accordance with Tan-Soetedjo et al. [22], who also observed that individual sugars are not affected in the presence of other sugars.

The side products can be determined by the difference to 100%, as shown in Figure 5. Regarding the literature, the probability of forming high-molecular-weight side products generally increases with the concentrations of HMF, sugars, and intermediates [6], because these condensation reactions are of a higher reaction order [40,41]. The sucrose experiments were conducted with double the carbon input as compared with fructose or glucose experiments (compare Figure 5). However, we do not observe a significant higher side product formation using sucrose as compared with a calculated mixture of glucose and fructose. However, if the sucrose content were to be increased much more, we believe that the formation of high-molecular-weight side products would accelerate. These high-molecular-weight side products can form solid particles at longer reaction times. Velebil et al. [42] reported yields of solid particles increasing with sucrose content.

3.4. Modeling Results

We avoided modeling on a molar yield base, because defining the maximal theoretical yield of each component in the sucrose reaction network is not obvious (compare Figure 2). The modeling is, therefore, performed on a molar ratio base (see Equation (1)). The experimentally determined molar ratios of fructose, glucose, HMF, xylose, furfural, levulinic acid, formic acid, and the sum of the side products can be found in Tables S1 and S2 of the Supplementary Material to this article.

The molar ratio of the noncharacterized side products is not directly measurable. For a rough estimation, we determined the sum of the side products (SP1, SP2, and SP3) via the carbon balance shown in Figure 4. The difference between the sum of all carbon detected and 100% can be assigned to nonquantified side products. An average molar mass of 180.2 g/mol is postulated to calculate the molar ratio of the sum of the side products. However, in reality, the side products can be divided into smaller molecules from defragmentation reactions, such as short-chain aldehydes and organic acids, and high-molecular-weight condensation products.

The experimental results of sucrose conversion are compared with the kinetic model in Figure 6. Since HMF rehydrates to form equimolar amounts of levulinic acid and formic acid, only the data of levulinic acid is displayed in the following diagrams. Within the scope of measuring inaccuracy, the molar ratios of levulinic acid and formic acid are equal within each sample in this study. Glucose molar ratios of greater than one at low residence times in Figure 6 result from measuring inaccuracies; this also leads to negative values for the sum of the side products.

The results of the experiments using glucose or fructose as the feedstock were compared with the results from the kinetic model in Figure 7. These experimental results are also included in the optimization of the kinetic parameters, according to Equation (12).

The model proposed is generally able to describe the experimental data. However, the fructose consumption is overrated at low residence times (2–6 min) by the model, especially at the lowest reaction temperature examined (180 °C). By contrast, the fructose consumption is underrated by the model at higher residence times (14–24 min) (see Figure 6). We are sure that this effect cannot be a result of an insufficient reaction model, but rather arises from experimental conditions in the tubular reactor. Especially at low residence times, which are related to high flow rates, the reactor is not completely isothermal.

The rate coefficients k_n of the kinetic model are summarized in Table 1. The activation energy EA_n and the Arrhenius factor A_n are shown in Table S3 of the Supplementary Material to this article. The rate constant for fructose conversion to HMF (k_1) is roughly an order of magnitude higher than the isomerization of glucose to fructose (k_6). Side-product formation from HMF (k_8) dominates among the reactions that produce noncharacterized side products. The side-product formation from fructose and glucose (k_7, k_9) shows only a small rise with temperature (see Table 1), and the corresponding activation energies EA_7 and EA_9 are the lowest in the reaction network (see Table S3 of the Supplementary Material). We assume this effect occurs because a variety of different elementary reactions are lumped together in k_7 and k_9.

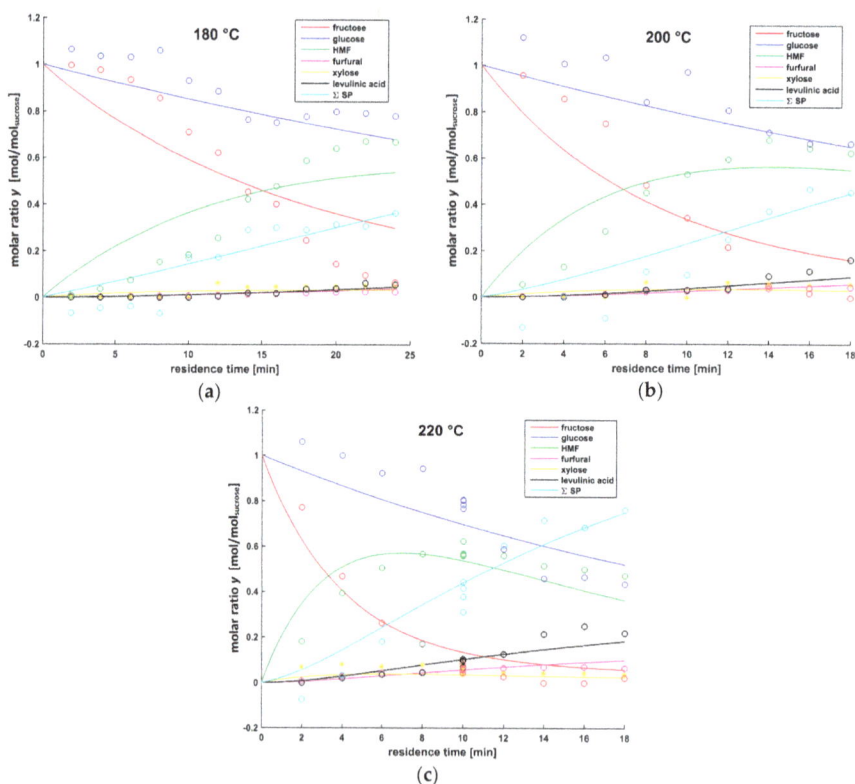

Figure 6. Kinetic plots of the acid-catalyzed conversion of sucrose at different residence times and temperatures of (a) 180 °C, (b) 200 °C, and (c) 220 °C. The molar ratios measured $y_{i,meas}$ are shown as data points and molar ratios calculated y_i as lines. Σ SP accounts for the sum of SP1, SP2, and SP3, as defined in Figure 2. Feedstock contained 2 wt % sucrose and 0.005 mol/L sulfuric acid.

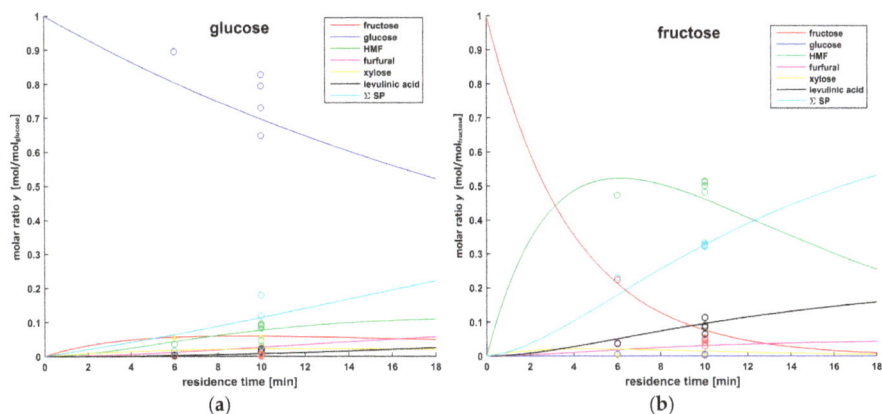

(a) (b)

Figure 7. Kinetic plots of the acid-catalyzed conversion of (**a**) glucose and (**b**) fructose at 10 min residence time at 220 °C. The molar ratios measured $y_{i,meas}$ are shown as data points, and molar ratios calculated y_i as lines. $\sum$ SP accounts for the sum of SP1, SP2, and SP3, as defined in Figure 2. Feedstock contained 0.005 mol/L sulfuric acid and either 1.05 wt % glucose or 1.05 wt % fructose.

Table 1. First-order reactions according to the model (see Figure 2), with their rate coefficients k_n at different reaction temperatures.

Reaction	Rate Coefficient k_n	at 180 °C [min^{-1}]	at 200 °C [min^{-1}]	at 220 °C [min^{-1}]
fructose => HMF	k_1	0.0525	0.116	0.241
HMF => levulinic acid + formic acid	k_2	0.0055	0.0114	0.0224
fructose => xylose	k_3	0.0033	0.0065	0.0123
glucose => xylose	k_4	0.0014	0.0027	0.0050
xylose => furfural	k_5	0.0642	0.111	0.185
glucose => fructose	k_6	0.0061	0.0116	0.0210
fructose => SP1	k_7	0.0042	0.0047	0.0052
HMF => SP2	k_8	0.0156	0.0347	0.0721
glucose => SP3	k_9	0.0086	0.0094	0.0102

According to the proposed model, xylose is produced from fructose as well as from glucose. However, when pure fructose was used as the feedstock, we found no xylose in the product samples (see Figure 5). However, furfural was present, which is a dehydration product of pentoses. Therefore, the pathway from fructose to furfural might proceed via another intermediate pentose, which is not included in the model.

Figure 8 shows the difference between all molar ratios measured and calculated. When the calculation describes the experimental result perfectly, the data point is on the line through the origin with a slope of one. A standard derivation δ of a linear regression function is calculated to evaluate the error (see Equation (13)). Parameters ($a = 1$, $b = 0$) are used to describe the linear equation. The standard deviation δ of the values calculated is about 7.8%. The Pearson correlation coefficient is 0.974.

$$\delta = \sqrt{\frac{1}{N-2} \cdot \sum_{1}^{N}(y_i - (a \cdot y_{i,meas} + b))^2} \tag{13}$$

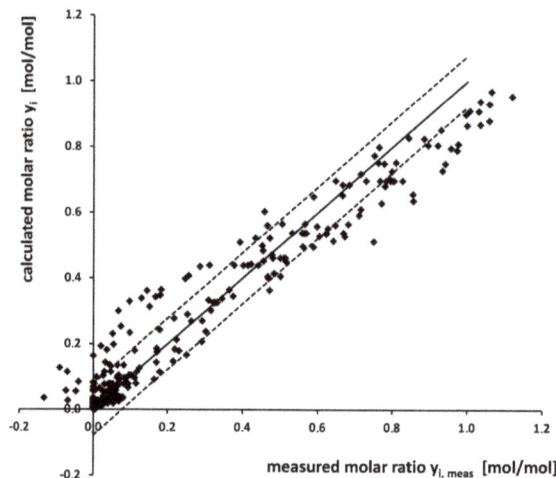

Figure 8. Parity plot of the molar ratio measured $y_{i,meas}$ and the molar ratio calculated y_i. The solid line though the origin has a slope of one. The dashed lines have a slope of one and are shifted by the standard derivation.

Asghari and Yohsida [32] investigated the HCl-catalyzed decomposition of fructose and formulated a kinetic model. They assumed that fructose and HMF could react to form different side products. The formation of side products from fructose was much higher than from HMF [32]. This contrasts to our work, where the side product formation from HMF was dominant ($k_8 > k_7$, see Table 1). The main differences between the study of Asghari and Yohsida [32] and our work are more acidic conditions (pH = 1.8), a smaller feedstock concentration (0.5 wt % fructose), and the use of HCl as the homogeneous catalyst.

Van Putten et al. [12] suggested that the rate-determining step in glucose dehydration to form HMF under acidic conditions is the isomerization to fructose. This is in accordance with our calculations, because the rate constant for the isomerization of glucose to fructose (k_6) is roughly an order of magnitude smaller than the subsequent fructose dehydration to form HMF (k_1) (see Table 1).

4. Conclusions

Sucrose decomposition in water catalyzed by 0.005 mol/L sulfuric acid was studied in a tubular reactor at 180–220 °C. Sucrose hydrolyzed rapidly to fructose and glucose at all the reaction conditions investigated. The formed fructose showed much faster conversion than glucose. The maximum HMF yield was surpassed within the residence time spectra investigated at 200 °C (0.25 g/g based on input sucrose) and at 220 °C (0.23 g/g). A carbon balance showed that the yield of unwanted side products rose strongly with temperature. Therefore, we recommend a mild reaction temperature for industrial HMF production. Nonconverted glucose could be isomerized catalytically to fructose, which is recycled for HMF production. In this regard, efficient purification strategies of HMF from the product mixture are of high interest.

Furthermore, we compared the feedstocks of sucrose, glucose, and fructose concerning their product distribution at 220 °C. The results of glucose and fructose conversion can be used to predict the yields from sucrose conversion.

A reaction network with nine first-order reactions was proposed. Within our network, HMF is only produced via the dehydration of fructose. Glucose isomerizes slowly to fructose. Side products arise simultaneously from glucose, fructose, and HMF, whereby side product formation from HMF is dominant at the conditions investigated. A small part of HMF rehydrates to form levulinic acid and formic acid. A pathway from hexoses to xylose via reverse aldol reaction is also included in the model. However, this postulated pathway needs further experimental investigation.

Supplementary Materials: The following are available online at www.mdpi.com/1996-1073/11/3/645/s1, Table S1: Experimentally determined molar ratios $y_{i,meas}$ of the acid-catalyzed conversion of sucrose. Feedstock contained 2 wt % sucrose and 0.005 mol/L sulfuric acid. Table S2: Experimentally determined molar ratios $y_{i,meas}$ of the acid-catalyzed conversion of glucose or fructose, at 10 min residence times and 220 °C. Feedstock contained 0.005 mol/L sulfuric acid and either 1.05 wt % glucose or 1.05 wt % fructose. Table S3: Activation energy EA_n and Arrhenius factor A_n of the n first-order reactions as defined according to the model.

Acknowledgments: We thank Xavier Sampietro Vallverdú for his experimental work. We acknowledge Nicolaus Dahmen for thesis supervision and Robert Grandl for Matlab support. Matthias Pagel and Thomas Tietz built the reactor. The HPLC measurements were performed by Sonja Habicht and Armin Lautenbach. This work was financially supported by the German Federal Ministry of Food, Agriculture and Consumer Protection (FNR project number 22027811), based on a decision of the German Bundestag. We acknowledge support by Deutsche Forschungsgemeinschaft and Open Access Publishing Fund of Karlsruhe Institute of Technology.

Author Contributions: David Steinbach and Philipp Vetter designed and performed the experiments; David Steinbach, Andrea Kruse and Philipp Vetter analyzed the data; David Steinbach, Andrea Kruse and Jörg Sauer wrote the article.

Conflicts of Interest: The authors declare no conflict of interest. The founding sponsors had no role in the design of the study; in the collection, analyses, or interpretation of data; in the writing of the manuscript, and in the decision to publish the results.

References

1. Harmsen, P.; Hackmann, M.; Bos, H. Green building blocks for bio-based plastics. *Biofuel Bioprod. Biorefin.* **2014**, *8*, 306–324. [CrossRef]
2. Aeschelmann, F.; Carus, M. Biobased building blocks and polymers in the world: Capacities, production, and applications—Status quo and trends towards 2020. *Ind. Biotechnol.* **2015**, *11*, 154–159. [CrossRef]
3. Van Putten, R.J.; van der Waal, J.C.; de Jong, E.; Rasrendra, C.B.; Heeres, H.J.; de Vries, J.G. Hydroxymethylfurfural, a versatile platform chemical made from renewable resources. *Chem. Rev.* **2013**, *113*, 1499–1597. [CrossRef] [PubMed]
4. Kläusli, T. AVA Biochem: Commercialising renewable platform chemical 5-HMF. *Green Process. Synth.* **2014**, *3*. [CrossRef]
5. Bozell, J.J.; Petersen, G.R. Technology development for the production of biobased products from biorefinery carbohydrates—The US Department of Energy's "Top 10" revisited. *Green Chem.* **2010**, *12*, 539–554. [CrossRef]
6. Steinbach, D.; Kruse, A.; Sauer, J. Pretreatment technologies of lignocellulosic biomass in water in view of furfural and 5-hydroxymethylfurfural production—A review. *Biomass Convers. Bior.* **2017**, *7*, 247–274. [CrossRef]
7. Cottier, L.; Descotes, G. 5-Hydroxymethylfurfural syntheses and chemical transformations. *Trends Heterocycl. Chem.* **1991**, *2*, 233–248.
8. Corma, A.; Iborra, S.; Velty, A. Chemical routes for the transformation of biomass into chemicals. *Chem. Rev.* **2007**, *107*, 2411–2502. [CrossRef] [PubMed]
9. Fan, L.; Gharpuray, M.M.; Lee, Y.H. *Cellulose Hydrolysis*; Springer: Berlin, Germany, 1987.
10. Kruse, A.; Dinjus, E. Hot compressed water as reaction medium and reactant—Properties and synthesis reactions. *J. Supercrit. Fluid* **2007**, *39*, 362–380. [CrossRef]
11. Kruse, A.; Dinjus, E. Hot compressed water as reaction medium and reactant—2. Degradation reactions. *J. Supercrit. Fluid* **2007**, *41*, 361–379. [CrossRef]
12. Van Putten, R.J.; Soetedjo, J.N.M.; Pidko, E.A.; van der Waal, J.C.; Hensen, E.J.M.; de Jong, E.; Heeres, H.J. Dehydration of different ketoses and aldoses to 5-hydroxymethylfurfural. *Chemsuschem* **2013**, *6*, 1681–1687. [CrossRef] [PubMed]
13. Ramos, L.P. The chemistry involved in the steam treatment of lignocellulosic materials. *Quim. Nova* **2003**, *26*, 863–871. [CrossRef]
14. Scallet, B.L.; Shieh, K.; Ehrenthal, I.; Slapshak, L. Studies in the isomerization of D-glucose. *Starch Stärke* **1974**, *26*, 405–408. [CrossRef]
15. Yan, K.; Liu, Y.; Lu, Y.; Chai, J.; Sun, L. Catalytic application of layered double hydroxide-derived catalysts for the conversion of biomass-derived molecules. *Catal. Sci. Technol.* **2017**, *7*, 1622–1645. [CrossRef]

16. Delidovich, I.; Palkovits, R. Catalytic Isomerization of Biomass-Derived Aldoses: A Review. *Chemsuschem* **2016**, *9*, 547–561. [CrossRef] [PubMed]

17. Sugar: World Markets and Trade. (2016) United States Department of Agriculture. Available online: https://apps.fas.usda.gov/psdonline/circulars/sugar.pdf (accessed on 13 March 2018).

18. Antal, M.J.; Mok, W.S.L.; Richards, G.N. Mechanism of formation of 5-(hydroxymethyl)-2-furaldehyde from d-fructose and sucrose. *Carbohyd. Res.* **1990**, *199*, 91–109. [CrossRef]

19. Khajavi, S.H.; Kimura, Y.; Oomori, T.; Matsuno, R.; Adachi, S. Kinetics on sucrose decomposition in subcritical water. *Lwt-Food Sci. Technol.* **2005**, *38*, 297–302. [CrossRef]

20. Bower, S.; Wickramasinghe, R.; Nagle, N.J.; Schell, D.J. Modeling sucrose hydrolysis in dilute sulfuric acid solutions at pretreatment conditions for lignocellulosic biomass. *Bioresour. Technol.* **2008**, *99*, 7354–7362. [CrossRef] [PubMed]

21. Gao, D.; Kobayashi, T.; Adachi, S. Kinetics of Sucrose Hydrolysis in a Subcritical Water-ethanol Mixture. *J. Appl. Glycosci.* **2014**, *61*, 9–13. [CrossRef]

22. Tan-Soetedjo, J.N.M.; van de Bovenkamp, H.H.; Abdilla, R.M.; Rasrendra, C.B.; van Ginkel, J.; Heeres, H.J. Experimental and Kinetic Modeling Studies on the Conversion of Sucrose to Levulinic Acid and 5-Hydroxymethylfurfural Using Sulfuric Acid in Water. *Ind. Eng. Chem. Res.* **2017**, *56*, 13229–13240. [CrossRef] [PubMed]

23. Fachri, B.A.; Abdilla, R.M.; van de Bovenkamp, H.H.; Rasrendra, C.B.; Heeres, H.J. Experimental and kinetic modeling studies on the sulfuric acid catalyzed conversion of D-fructose to 5-hydroxymethylfurfural and levulinic acid in water. *ACS Sustain. Chem. Eng.* **2015**, *3*, 3024–3034. [CrossRef]

24. Girisuta, B.; Janssen, L.P.B.M.; Heeres, H.J. A kinetic study on the decomposition of 5-hydroxymethylfurfural into levulinic acid. *Green Chem.* **2006**, *8*, 701–709. [CrossRef]

25. Girisuta, B.; Janssen, L.P.B.M.; Heeres, H.J. A kinetic study on the conversion of glucose to levulinic acid. *Chem. Eng. Res. Des.* **2006**, *84*, 339–349. [CrossRef]

26. Torres, A.P.; Oliveira, F.A.R. Application of the acid hydrolysis of sucrose as a temperature indicator in continuous thermal processes. *J. Food Eng.* **1999**, *40*, 181–188. [CrossRef]

27. Smith, P.C.; Grethlein, H.E.; Converse, A.O. Glucose decomposition at high-temperature, mild acid, and short residence times. *Sol. Energy* **1982**, *28*, 41–48. [CrossRef]

28. Saeman, J.F. Kinetics of wood saccharification—Hydrolysis of cellulose and decomposition of sugars in dilute acid at high temperature. *Ind. Eng. Chem.* **1945**, *37*, 43–52. [CrossRef]

29. McKibbins, S.W.; Harris, J.F.; Saeman, J.F.; Neill, W.K. Kinetics of the acid catalyzed conversion of glucose to 5-hydroxymethyl-2-furaldehyde and levulinic acid. *For. Prod. J.* **1962**, *12*, 17–23.

30. Xiang, Q.; Lee, Y.Y.; Torget, R.W. Kinetics of glucose decomposition during dilute-acid hydrolysis of lignocellulosic biomass. *Appl. Biochem. Biotechnol.* **2004**, *113*, 1127–1138. [CrossRef]

31. Asghari, F.S.; Yoshida, H. Acid-catalyzed production of 5-hydroxymethyl furfural from D-fructose in subcritical water. *Ind. Eng. Chem. Res.* **2006**, *45*, 2163–2173. [CrossRef]

32. Asghari, F.S.; Yoshida, H. Kinetics of the decomposition of fructose catalyzed by hydrochloric acid in subcritical water: Formation of 5-hydroxymethylfurfural, levulinic, and formic acids. *Ind. Chem. Res.* **2007**, *46*, 7703–7710. [CrossRef]

33. Forchheim, D.; Hornung, U.; Kruse, A.; Sutter, T. Kinetic modelling of hydrothermal lignin depolymerisation. *Waste Biomass Valori.* **2014**, *5*, 985–994. [CrossRef]

34. Gasson, J.R.; Forchheim, D.; Sutter, T.; Hornung, U.; Kruse, A.; Barth, T. Modeling the lignin degradation kinetics in an ethanol/formic acid solvolysis approach. Part 1. Kinetic model development. *Ind. Eng. Chem. Res.* **2012**, *51*, 10595–10606. [CrossRef]

35. Clarke, M.A.; Edye, L.A.; Eggleston, G. Sucrose decomposition in aqueous solution, and losses in sugar manufacture and refining. *Adv. Carbohyd. Chem. Biochem.* **1997**, *52*, 441–470. [CrossRef]

36. Usuki, C.; Kimura, Y.; Adachi, S. Isomerization of Hexoses in Subcritical Water. *Food Sci. Technol. Res.* **2007**, *13*, 205–209. [CrossRef]

37. Kabyemela, B.M.; Adschiri, T.; Malaluan, R.M.; Arai, K. Kinetics of glucose epimerization and decomposition in subcritical and supercritical water. *Ind. Eng. Chem. Res.* **1997**, *36*, 1552–1558. [CrossRef]

38. Bonn, G.; Bobleter, O. Determination of the hydrothermal degradation products of D-(U-C14) glucose and D-(U-C-14) fructose by tlc. *J. Radioanal. Chem.* **1983**, *79*, 171–177. [CrossRef]

Energies **2018**, *11*, 645

39. Kabyemela, B.M.; Adschiri, T.; Malaluan, R.M.; Arai, K. Glucose and fructose decomposition in subcritical and supercritical water: Detailed reaction pathway, mechanisms, and kinetics. *Ind. Eng. Chem. Res.* **1999**, *38*, 2888–2895. [CrossRef]
40. Chuntanapum, A.; Matsumura, Y. Formation of tarry material from 5-HMF in subcritical and supercritical water. *Ind. Eng. Chem. Res.* **2009**, *48*, 9837–9846. [CrossRef]
41. Kruse, A.; Grandl, R. Hydrothermal carbonization: 3. Kinetic model. *Chem. Ing. Tech.* **2015**, *87*, 449–456. [CrossRef]
42. Velebil, J.; Malaťák, J.; Bradna, J. Mass yield of biochar from hydrothermal carbonization of sucrose. *Res. Agric. Eng.* **2016**, *62*, 179–184. [CrossRef]

![energies](energies logo)

MDPI

Article

Life Cycle Performance of Hydrogen Production via Agro-Industrial Residue Gasification—A Small Scale Power Plant Study

Sara Rajabi Hamedani [1], Mauro Villarini [2,*], Andrea Colantoni [2], Michele Moretti [3] and Enrico Bocci [4]

[1] Department of Astronautics, Electrical and Energy Engineering, Sapienza University, 00184 Rome, Italy; sara.rajabi@uniroma1.it

[2] Department of Agricultural and Forestry Sciences, Tuscia University of Viterbo—Via San Camillo de Lellis, snc, 01100 Viterbo, Italy; colantoni@unitus.it

[3] Research Group of Environmental Economics, Center for Environmental Sciences, Hasselt University, Agoralaan—Building D, 3590 Diepenbeek, Belgium; michele.moretti@uhasselt.be

[4] Department of Innovation and Information Engineering, Marconi University, 00193 Rome, Italy; e.bocci@unimarconi.it

* Correspondence: mauro.villarini@unitus.it; Tel.: +39-340-226-6196

Received: 28 January 2018; Accepted: 13 March 2018; Published: 16 March 2018

Abstract: This study evaluates the environmental profile of a real biomass-based hydrogen production small-scale (1 MW$_{th}$) system composed of catalytic candle indirectly heated steam gasifier coupled with zinc oxide (ZnO) guard bed, water gas shift (WGS) and pressure swing absorber (PSA) reactors. Environmental performance from cradle-to-gate was investigated by life cycle assessment (LCA) methodology. Biomass production shows high influence over all impact categories. In the syngas production process, the main impacts observed are global warming potential (GWP) and acidification potential (AP). Flue gas emission from gasifier burner has the largest proportion of total GWP. The residual off gas use in internal combustion engine (ICE) leads to important environmental savings for all categories. Hydrogen renewability score is computed as 90% due to over 100% decline in non-renewable energy demand. Sensitivity analysis shows that increase in hydrogen production efficiency does not necessarily result in decrease in environmental impacts. In addition, economic allocation of environmental charges increases all impact categories, especially AP and photochemical oxidation (POFP).

Keywords: hydrogen production; biomass gasification; life cycle assessment; environmental impact

---

1. Introduction

Growing concerns about climate change, rising costs of fossil fuel, and the geopolitical uncertainty associated with an uninterrupted energy supply have motivated individuals, organizations and nations to look for substitutes that are clean and renewable [1,2]. Hydrogen (H$_2$) is, currently, considered as one of the leading candidates in the search alternatives to fossil fuels (FF) [3]. Nevertheless, H$_2$ is only an energy carrier like electricity and not a primary energy source. It can be produced from a wide variety of energy sources, such as natural gas, coal, biomass, solar (thermal and photovoltaic), etc. [4] and act as a unique energy hub providing low or zero emission in energy use to all energy consuming sectors [5]. However, the environmental performance of hydrogen-production systems highly depends on the type of primary energy and conversion technology used [6]. Among several renewable energy sources, bio-hydrogen is gaining a lot of attraction because of its conversion efficiency with less pollutant generation [7]. In particular, hydrogen production through lignocellulosic biomass residues gasification is a promising pathway in terms of global warming impacts and energy security [8–10]. However, although lignocellulosic biomass is a renewable resource, its use does not guarantee an appropriate environmental performance [11] and, even under strategies of using agriculture and forestry residues to avoid emissions from direct and indirect land use change, the feedstock use can potentially cause negative environmental impacts [12]. Therefore, it is requisite to analyze entire biomass to hydrogen production process considering the impacts from the cultivation to the conversion plant.

LCA is a well-established methodology to evaluate the environmental impacts of a product over its life cycle. It is often used to determine the greenhouse gas footprint and minimum selling price in several kinds of processes and in particular regarding biomasses [13]. This is achieved by: (i) gathering all relevant inputs and outputs of the considered system; (ii) evaluating the potential environmental impacts related to these inputs and outputs; and (iii) interpreting the results obtained in the impact assessment phase [14]. Hence, LCA is a viable tool to identify environmental hotspots and find approaches that enhance environmental results related to the products [15,16]. The LCA application to energy is becoming increasingly important for evaluating which technology is more affordable and more sustainable [17]. LCA often demonstrated the benefits of replacement of oil with bioenergy systems [18] not only for energy but also for refinery scopes [19]. Despite the potential role of LCA in sustainable development of energy systems [20], few LCA studies have been carried out on H_2 production from biomass [8,9,21,22]. In these papers, agricultural and forestry waste that feed oxygen/steam fixed or circulating fluidized bed coupled with tar cracker/reformer/scrubber, sulfur removal systems, steam reformer and water gas shift reactors have been assessed. Kalinci et al. [9] evaluated two different large scale gasification systems, circulating fluidized bed and fixed bed gasifiers coupled with tar reformer, water gas shift (HTS (High Temperature Shift), MTS (Medium Temperature Shift), and LTS (Low Temperature Shift)) and PSA reactors. In this study, impacts were just evaluated in term of kg CO_2. Moreno and Dufour [21] investigated environmental efficiency of a 131 MW_{th} power plant that applied a fixed bed gasifier with steam reformer, WGS and PSA. Susmozas et al. [8] assessed biomass gasification via circulating fluidized bed gasifier and SMR technology for hydrogen production in a medium size power plant. ZnO reactor and steam reformer were not included. Finally, Suwatthikul et al. [22], after the model implementation, by means of ASPEN Plus, a simulation software for chemical process, of a steam gasification system and its validation, developed a LCA comparing conventional gasification and energy self-sufficient gasification in terms of GWP and marine aquatic ecotoxicity.

Since biomass resources are typical of a distributed nature and are limited by low energy density, perishability and complexity of the supply chain, distributed power plants offer significantly better energy security and possibility to exploit the full potential than the equivalent centralized plant [23,24]. To sustain development of these distributed units, reliable, high efficient and low environmental impact small scale power plants have to be developed [24]. Thus, it is important to conduct a LCA of hydrogen production at small scale systems. Previous literature regarding this specific topic is missing and, therefore, the contribution of the present study is relevant because it deals with LCA of hydrogen production via small scale biomass plant realized during four-year project of UNIfHY (UNIque gasifier for HYdrogen production) [25]. In this case study, almond shell, which represents agro-industrial residue, is used as feedstock, and a catalytic candle indirectly heated steam gasifier coupled with zinc oxide (ZnO) guard bed, water gas shift (WGS) and pressure swing absorber (PSA) reactors is considered as conversion technology. Almond shell has been chosen because it has the most suitable characteristics to this thermochemical process, as explained by Bocci et al. [24]. The chemical characterization of any feedstock to be used in a gasification process is an important step for qualifying the process itself in terms of conversion efficiency (mass and energy) and presence of potential poisoning elements. In this paper, almond shells have been selected as representative of agro industrial residue because, as a food industry derived biomass source, they are largely available, have a limited cost of supply, do not have potential poisoning elements and have already been tested in the UNIfHY project [25].

The article is organized as follows: after the Introduction, Section 2 describes the methodology used; Section 3 reports a description of the conceptual organization of the gasification system within the model; Sections 4 and 5, respectively, present and discuss the main results of the work; and Section 6 reports the conclusions.

2. Methodology

2.1. The Life Cycle Assessment

LCA is a methodology for the comprehensive evaluation of the impact that a product (good or service) has on the environment throughout its life cycle [14]. This method presents a holistic approach for a comprehensive environmental assessment, following a standardized method which guarantees reproducibility of results [14,26].

2.1.1. Goal and Scope Definition

The environmental impacts and energy requirements of the whole process aimed at hydrogen production from the almond shell by means of gasification process and the subsequent use for electricity have been determined. In addition, environmental hotspots were identified to reduce the impact and improve the environmental and energy profiles. StimaPro 8.1 software, developed by PRé Consultants, was used for the environmental evaluation of the process.

2.1.2. Functional Unit and System Boundaries

The functional unit (FU) expresses the function of the system in quantitative terms and provides the reference to which all the inputs and outputs of the product system are calculated [14]. The functional unit selected for this assessment was the production of 1 MJ of hydrogen (purity: 99.99%).

A cradle-to-gate LCA has been performed including feedstock production, gasification, H_2 purification and electricity generation from off-gas.

2.1.3. Inventory Data Acquisition

The most effort-consuming step in the execution of LCA studies is the collection of inventory data to build the Life Cycle Inventory (LCI). High quality data (input and output) are essential for a reliable evaluation [27].

Inventory data for the bioenergy production plant are collected from foreground data of a pilot plant and form questionnaire filled out by designer, producer and operators of the equipment and gasifier plant. In feedstock production phase, on farm emissions derived from fertilizer application have also been modeled following the IPCC (Intergovernmental Panel on Climate Change) guidelines [28]. Temporary storage of carbon in tree biomass, and carbon storage in orchard soils, is accounted for based on net annual changes in carbon stored or emitted. The average soil carbon accumulation in almond orchards was estimated as 624 kg of carbon ha^{-1} year^{-1}. More detailed information concerning these calculations can be found in [29]. Emissions from fuel combustion in both bioenergy plant and agricultural machinery have been calculated according to Tier 1 method described in IPCC guidelines. The fuel consumption does not include the extraction stage. The inventory involves the distribution of fuel to the final consumer including all necessary transports. In the off-gas management subsystem presented in the following paragraph, the derived emissions from CHP were calculated with the emission limits reported by [30].

2.1.4. Impact Assessment

According to some criteria defined for selecting the impact assessment methods, such as the scientific robustness, which considers the level of uncertainty, development that occurs over time, their application in LCA practice and the European environmental policy goals, the midpoint CML method is chosen to assess hydrogen production system (Fuel cell and Hydrogen-Joint Undertaking 2011). In line with [31–33], four midpoint categories were taking into account: acidification potential (AP), eutrophication potential (EP), global warming potential (GWP) and photochemical oxidation potential

(POFP). The cumulative (non-renewable) fossil and nuclear energy demand (CED) of the whole life cycle was also quantified [34].

2.2. Hydrogen Renewability

The concept of renewability has firstly been introduced by Neelis et al. [35] and, as an indicator, it can assist decision makers to choose fuel with superior renewability character. According to cumulative energy demand (CED) method, renewable and non-renewable energy involved in whole life cycle can be counted. The index represented for this cycle is shown in Equation (1), [35]:

$$Hydrogen\ renewability[\%] = \frac{renewable\ energy\ input}{(renewable\ energy + nonrenewable\ energy\ inputs)} \quad (1)$$

2.3. Sensitivity Analysis

In addition, sensitivity analysis has been performed to assess the influence of change in hydrogen production efficiency and environmental charge allocation on the model results. Among the factors that affect hydrogen production, the steam to biomass ratio (S/B) has been chosen [36,37]. Therefore, study results have been compared with those considering S/B equal to 1 and 1.5, respectively. Experimental conditions for these different ratios are found in [38]. Alternatively, an economic allocation between hydrogen and electricity produced in the bio-energy plant was assumed. According to [39,40], allocation percentages for hydrogen and electricity in economic basis were 96% and 4%, respectively.

3. Description and Model of the Gasification System

The present system is based on a 1 MW$_{th}$ innovative gasification system [41–43]. Figure 1 shows the CHEMCAD, a simulation software for chemical process, flowsheet of the system with the indirectly heated gasifier here considered. To realize the 1 MW$_{th}$ system with Steam–Oxygen gasifier, the depicted plant in Figure 2 is used.

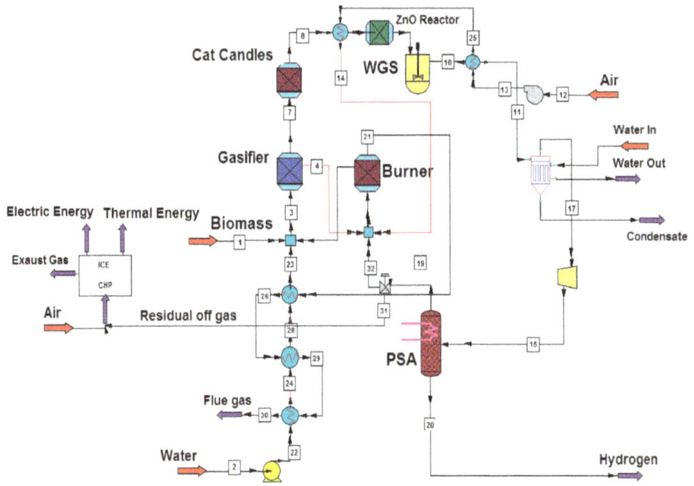

Figure 1. Flow sheet of the plant evaluated in this study.

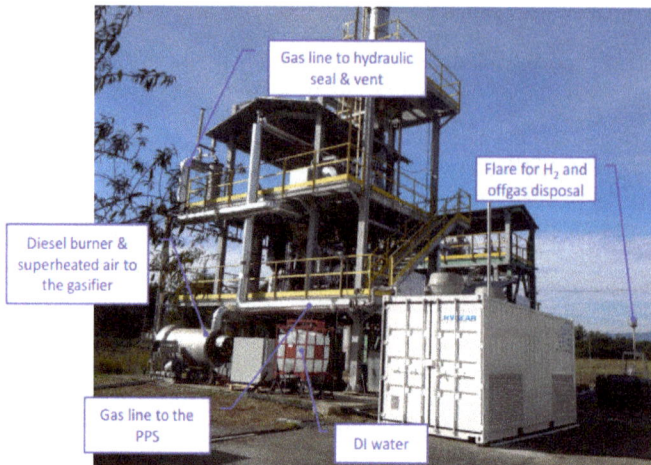

Figure 2. A picture of the plant.

The system is divided into four subsystems. First is the Biomass production (SS1). The bioenergy plant operation has been divided into two main stages: from biomass to syngas representing the syngas production stage (SS2) and from syngas to hydrogen representing the syngas conversion stage (SS3). All inputs and outputs demanded for the plant operations are included to consider the related environmental burdens. The last subsystem is composed of the off-gas management (SS4). Figure 3 illustrates the system boundaries and the considered processes.

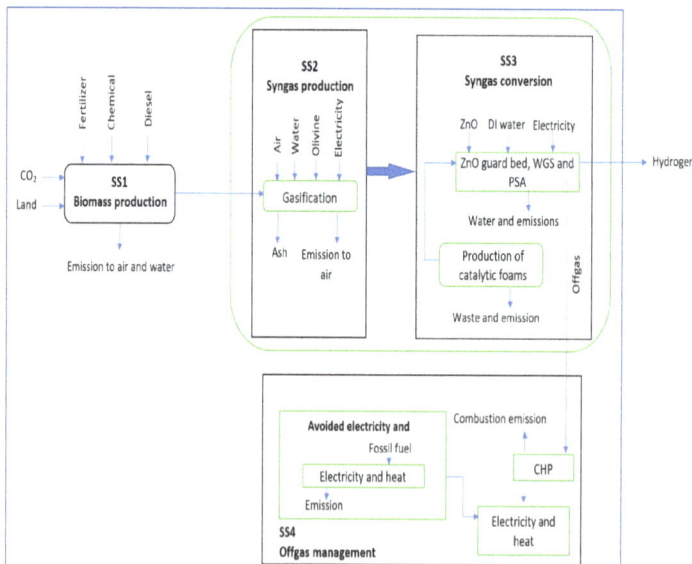

Figure 3. Life cycle boundaries for hydrogen production system.

3.1. Subsystem 1: Biomass Production

In Subsystem 1, all the biomass production processes were considered, comprising main crop planting, fuels consumed by agricultural machinery, requirements of fertilizer, pesticides and water for irrigation, transportation, fuel consumed in factory and biomass collection [29,44,45]. The global inventory data for this subsystem are reported in Table 1. Since waste production is not the goal of farming, it is required to designate all outputs (main products and waste) of cultivation and adapt allocation method to account only for the environmental burdens allocated to waste [21,27]. In this study, environmental charges of cultivation stage are allocated using market prices (1% of charge associated to waste).

Table 1. Global inventory data (per 1 MJ$_{H2}$) for Subsystem 1.

Input from Technosphere		Output to Technosphere	
Materials and Fuels		**Products and Coproducts**	
Diesel	0.017 kg	Kernel (main product)	0.257 kg
Nitrogen fertilizer	0.025 kg	Almond shell	0.144 kg
Energy		Other Co-products	0.74 kg
Electricity	0.037 kWh		
Input from nature			
Water	800 kg	Output to environment	
		Emissions to air	
		Nitrogen oxides	0.45×10^{-3} kg
		Methane	0.35×10^{-6} kg
		Carbon dioxide	0.0087 kg
		Emissions to water	
		Nitrogen oxides	2×10^{-4} kg

3.2. Subsystem 2: Syngas Production

The main operating conditions of the entire power plant are listed in Table 2 [38]. In addition, almond shell characterization and syngas composition are shown in Tables 3 and 4, respectively.

Table 2. Operating conditions of the gasifier.

Parameters (Unit)	Value
Power (kW$_{th}$)	1000
Biomass feeding rate (kg/h)	200
Steam feeding rate (kg/h)	100
Electricity consumption (kW)	
Start up	4
water pump	0.61
Air blower	13
Syngas blower	3
Syngas Compressor	36
Gasification T (°C)	850
S/B	0.5
Combustor T (°C)	950
Olivine between combustor/gasifier (kg/h)	1000

Table 3. Almond shell characterization [46].

Bulk Density (kg/m^3)	Humidity (%wt)	Ash (%wt)	Volatile Matter (%wt)	Fixed Carbon (%wt)	C (%)	H (%)	N (%)	O (%)	Cl (%)	S (%)
450	12	1.2	80.6	18.2	47.9	6.3	0.32	44.27	0.012	0.015

Table 4. Syngas composition.

Compositions (Unit)	Value
Gas yield ($Nm^3_{dry}/kg_{bio,daf}$)	1.45
H_2 ($\%_{dry}$)	44
CH_4 ($\%_{dry}$)	10
CO ($\%_{dry}$)	28
CO_2 ($\%_{dry}$)	18
H_2O ($\%_{wet}$)	16
C_6H_6 (g/Nm^3_{dry})	4.11
C_7H_8 (g/Nm^3_{dry})	0.68
$C_{10}H_8$ (g/Nm^3_{dry})	0.68
MJ/Nm^3_{dry}	11.4

Syngas production stage involves the gasification process. Gasification is performed in a 1 MW_{th} Indirectly heated Bubbling Fluidized Bed gasifier (IBFB), working under mentioned conditions in Table 2 for 7000 operating hours (even if during the project only 120 h of pure hydrogen production has been achieved). Indirectly heated bubbling fluidized bed with tar reforming inside the gasifier was chosen in this study since this configuration provides a synthesis gas with higher hydrogen content than fluidized and fixed bed gasifiers [6,36,41,47–49]. The 1 MW_{th} gasifier built and experimented in the research project is an oxygen bubbling fluidized bed gasifier. Thus, we considered the same technology running in indirectly heated configuration.

First, tar and methane are treated and the cleaning process of synthesis gas from particles is carried out by 60 Catalytic Filter Candles (CFC) allocated in the gasifier freeboard. This is an innovative method adapted and tested during the two European research projects called UNIQUE [50] and UNIfHY [25]. In the gasification process, both electricity and heat are requisite. The electricity consumed in the syngas compressor and in the air blower is taken directly from the Italian national grid. Thermal energy is required to preheat the air, heat the water, and generate and overheat the steam, and is provided by syngas and flue gas cooling. Owing to the indirectly heated configuration, hot bed material (olivine) from the combustor is circulated back to the gasifier supplying the thermal power needed for the gasification reactions. Carbon dioxide emissions from flue gas have been considered.

3.3. Subsystem 3: Syngas Conversion

In the syngas conversion stage, the produced syngas has been used in a portable purification station (PPS), which generates the H_2. This unit combines a ZnO reactor (to remove the sulfur compounds within syngas), a WGS reactor (to convert CO into further H_2) and a PSA membrane reactor (to separate H_2 from the other syngas components). Since atmospheric pressure gasification is more suitable for small-scale applications, the conventional WGS that operates at high pressure has been substitute with ceramic foams catalytic WGS operating at atmospheric pressure. The use of a ceramic foam with a wide catalytic surface area impregnated with catalyst eliminates the need to operate at high pressure for high conversion efficiency [36].

The ZnO data have been considered but this unit is not present in the flowsheet because the trace elements have not been considered in the simulation. The heat from the heat exchangers in PPS is recovered to meet low temperature thermal need.

All inputs and outputs needed for the hydrogen production unit (PPS), such as electricity, ZnO, de-ionized water and WGS Cu foams, are included. Additionally, the derived emissions from burning off-gas (such as nitrogen oxides, carbon dioxide, and sulfur dioxide) were considered. Construction of gasifier plant with 20-year life time has also been considered within the subsystem boundaries. The details of catalysts production are extracted from a questionnaire filled out by the University of Strasburg, a partner of UNIfHY project. The materials and energy inputs required for the catalysts production to be used in a 1 MW_{th} power plant are presented in Table 5.

Table 5. Material and energy employed for catalysts production.

Parameters (Unit)	Value
Number	20
Size (cm)	D: 70 H: 10
Material and Energy	
Water (kg)	800
Cerium nitrate (kg)	367
Heat (MJ)	12,000
Lifetime (year)	3
Emission to water	
Cerium nitrate solution (L)	500
Copper nitrate solution (L)	500
Emission to air (kg)	
NO_x (kg)	6

During the 20-year lifetime of power plant, the catalysts with three-year lifetime need to be replaced seven times. To integrate this production phase into the model, the energy and materials required for these catalysts (Table 5) have been multiplied by seven and divided by the total energy output produced during 20 years.

The main inventory data for hydrogen production via almond shell gasification are shown in Table 6.

Table 6. Main inventory data for hydrogen production via almond shell gasification per FU.

Input from Technosphere		Output to Technosphere	
Materials and fuels		Products and coproducts	
Almond shell (SS1)	0.144 kg	Hydrogen	1 MJ
Diesel (SS2)	3.55×10^{-5} kg		
Water (SS2)	0.144 kg		
De-ionized water (SS3)	0.128 kg	Avoided product	
Zinc oxide (SS3)	1.3×10^{-4} kg	Electricity	0.26 kWh
Cu foam (SS3)	3×10^{-7} P		
Energy			
Electricity (SS2)	0.014 kWh		
Electricity (SS3)	0.027 kWh	Emissions to air	
Input from nature		Nitrogen oxide (SS2)	1×10^{-9} kg
		Methane (SS2)	1.7×10^{-8} kg
Olivine (SS2)	0.00027 kg	Carbon dioxide (SS2)	0.16 kg
		Water (SS2)	0.15 kg
		Sulfur dioxide (SS4)	2×10^{-6} kg
		Nitrogen oxide (SS4)	2.5×10^{-5} kg
		Emission to water	
		Wastewater (SS3)	0.05 kg
		Solid waste flows	
		Ash (SS2)	1.6×10^{-3} kg
		Olivine (SS2)	2.7×10^{-4} kg
		Insulation (SS2)	5.5×10^{-7} kg
		Hydrogen Sulfide (SS3)	5×10^{-5} kg

3.4. Subsystem 4: Off-Gas Management

In the syngas conversion process, off-gas is co-produced from PSA and is mainly used in the burner (syngas production process). This subsystem involves the use of the residual off gas into a conventional Combined Heat and Power system (CHP, via an Internal Combustion Engine, ICE). Under the conditions mentioned in Table 2, with respect to the total energy output: the energy content of residual off-gas represents 12%; the H_2 represents 43%; and the thermal energy losses are 45% (8% in flue gas and another 37% mainly due to the heat of the cooling water and condensate) (Figure 1).

According to previous energy study of the authors [38] and Fremaux et al. [37], it is inferred that the residual off-gas decreases with higher S/B ratio (e.g., with S/B greater than 1.5–2, depending on the PSA efficiency, the off-gas is completely used in the burner). In particular, the global efficiency is limited by the thermal losses and by the PSA compressor consumption. The hydrogen efficiency is limited by the PSA efficiency and by the absence of a methane reformer. Thus, to increase the hydrogen efficiency, high temperature hydrogen membrane separation that works at low pressure (as Palladium membranes) including methane reforming are considered but, up to now, are still in a development phase. Indeed, the hydrogen efficiency is about 38%, while the global efficiency (considering hydrogen, residual off gas and the useful heat) is about 70%.

The production of off-gas, under the fixed gasification and conditioning parameters (residence time, temperature, catalysts, etc.) mainly depends on S/B ratio and PSA efficiency. Being the PSA efficiency a technological parameter rather than an operational parameter, only the variation of S/B has been considered in the following sensitivity analysis.

The off-gas use for CHP was considered within the system boundaries together with the consequently avoided conventional electricity production, while heat generation has been disregarded. Indeed, electricity can be used for own demand and injected to the grid, meanwhile the heat produced by CHP can only be used if there is a near low temperature heat demand. In accordance to ISO standards, allocation procedure is avoided by system expansion in this LCA study.

During biomass-based hydrogen production, hydrogen and off-gas are coproduced. The hydrogen was considered as the main product, although the produced electricity from off-gas can be sold to the national grid. In fact, it is considered as an avoided input in the base case of this study. In other words, the system is expanded rather than allocated. Therefore, hydrogen production was only addressed as the base case. In addition, impacts derived from the production and transmission of the avoided electricity were also included within the subsystem boundaries in this case.

4. Results

Table 7 summarizes the LCA characterization results for the different subsystems under study per functional unit. Positive values are environmental burdens while negative values signify environmental credits or benefits accrued from carbon dioxide uptake and the substitution of electricity from national grid.

Table 7. Characterization results corresponding to the production of 1 MJ_{H2}.

Category	Unit	Total	SS1	SS2	SS3	SS4
GWP	Kg CO_2 eq	0.042	−0.12	0.19	0.11	−0.14
EP	Kg PO_4 eq	2.6×10^{-4}	5×10^{-4}	3.6×10^{-5}	3.3×10^{-6}	-2.8×10^{-4}
AP	Kg SO_2 eq	-4.5×10^{-4}	2×10^{-4}	1.3×10^{-4}	1.5×10^{-5}	-8×10^{-4}
POFP	Kg C_2H_4 eq	-1.8×10^{-5}	6×10^{-6}	2.7×10^{-6}	3.3×10^{-6}	-3×10^{-5}
CED	MJ eq	−0.46	0.39	0.14	0.3	−1.3

Figure 4 displays the relative contributions of each subsystem to the environmental results for the studied system. According to these results, the biomass production (SS1) shows high influence over the environmental profile, especially in term of EP with a contribution of 60%. This result is mainly related to the emissions derived from fertilizer application. In addition, the large amount of energy required to produce fertilizer causes intense fossil energy consumption in feedstock production phase. Contrarily, this subsystem achieves a positive effect on GWP owing to the CO_2 uptake by photosynthesis. Therefore, the emission of 120 g CO_2 eq per functional unit is avoided. The main environmental impacts of the syngas production (SS2) are observed on GWP (34%) and AP (15%), mostly related with flue gas emitted from gasifier combustor and feedstock supply. The syngas conversion (SS3) has relatively lower impact on GWP (19%) than the other categories.

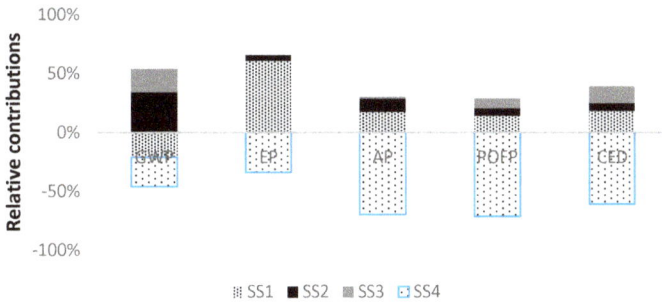

Figure 4. Relative contributions from subsystems involved to each impact category.

SS4 covers the energy generation from off-gas and its utilization as electricity. Thus, the production of electricity from the grid is avoided. As shown in Table 7, it leads to environmental benefits in terms of all the analyzed categories. In the following, a detailed assessment per subsystem has been carried out to recognize in detail the main contribution to these environmental outcomes.

As displayed in Figure 5, impact of fertilizers in SS1 is notable in EP, AP and POCP categories with a contribution greater than 84% because of drastic energy consumption in production of nitrogen fertilizer required. Irrigating orchard, owing to diesel and electricity requirements in irrigation, has also impact on AP and POFP with 5% and 15%, respectively. Regarding GWP, the fixation CO_2 by photosynthesis can offset the GHG emitted throughout the cultivation system which are chiefly derived from fertilizer production.

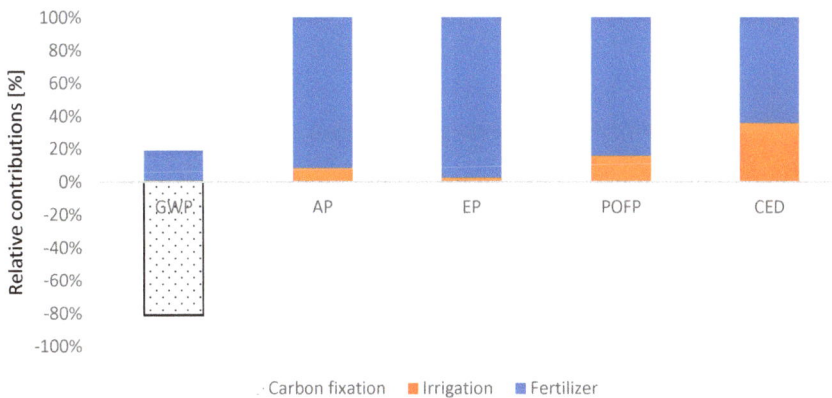

Figure 5. Breakdown of contributions from process involved in Subsystem 1.

With regard to the syngas production process (SS2), flue gas emission from gasifier burner has the largest proportion of total GWP (55%). By virtue of carbon fixation of feedstock, this impact eventually experiences a massive decrease.

The other categories (AP, POFP and EP) are significantly influenced by feedstock supply process. Moreover, the electricity required to run the gasification has been taken from the Italian grid; its production turns out as second large contributor for AP and POFP categories with 16% and 23%, respectively. Eighty percent of the electricity consumption in SS2 is due to the air blower to provide air needed for gasification process.

In SS3, the syngas production phase impacts on all the environmental categories. In particular, for EP, it impacts with 97% related to feedstock and fertilizer consumption. The materials used for syngas conversion to hydrogen have minor impact on this subsystem. Among them, the Cu foam production has higher environmental impact than the other materials accounting for almost the 3.6% of the impact on AP and POFP.

In SS4, the emissions from the CHP (SO$_2$ and NO$_x$) have been estimated according to emission limits legislated by European Commission. Carbon dioxide released by off-gas burning is biogenic and it is not treated as a net source of carbon dioxide by IPCC. The detailed information is indicated in Table 8. The avoided emissions from electricity production and transmission, as shown in Figure 6, have the greatest negative impact on all categories.

Table 8. Global inventory data (per 1 MJ$_{H2}$) for Subsystem 4.

Input from Technosphere		Output to Technosphere	
Materials and fuels		Avoided products	
Off-gas (from SS3)	1.33 MJ	Electricity	0.26 kWh
		Output to environment	
		Emissions to air	
		Sulfur dioxide	2×10^{-6} kg
		Nitrogen oxide	2.5×10^{-5} kg

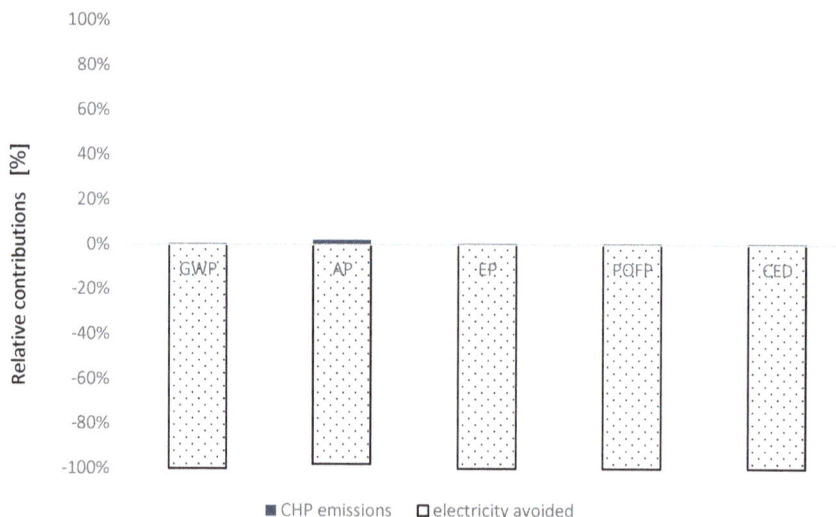

Figure 6. Breakdown of contributions from process involved in Subsystem 4.

The renewability score of hydrogen under the conditions considered in our study has been fixed at 90%, which is close to the values defined for wind and solar based hydrogen production, 96% and 93%, respectively [51]. In fact, high energy saving in SS4 leads to achieving this appreciative value.

To calculate the renewability score in the current study, the energy content of biomass as the renewable energy was considered 18 MJ/kg $_{biomass}$ [46]. Therefore, the consumption of 0.144 kg biomass per 1 MJ$_{H2}$ equals to 2.6 MJ biomass-renewable energy consumption. According to the cumulative energy demand indicator (Table 7), 0.46 MJ energy per 1 MJ$_{H2}$ would be saved in the whole life cycle. In fact, the electricity produced from off-gas leads to 1.3 MJ energy saving per 1 MJ_{H2}. The non-renewable energy of whole life cycle has been calculated as 0.22 MJ/MJ_{H2} considering 43% contribution of off-gas treatment subsystem (SS4) to save energy. On the other hand, if SS4 were excluded from the model, the H_2 renewability could be expressed as 75%. In other words, off gas treatment can improve hydrogen renewability by 15%. According to [6] and Table A1 in Appendix A, other approaches for hydrogen production are compared in Figure 7.

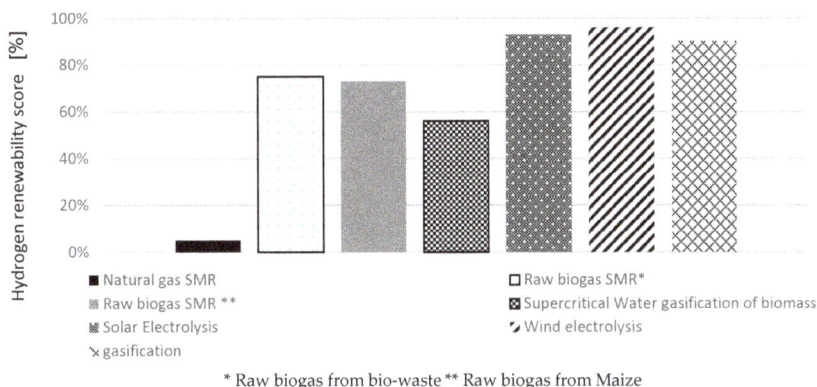

* Raw biogas from bio-waste ** Raw biogas from Maize

Figure 7. Hydrogen renewability for each technology [6].

5. Discussion

In this section, the results obtained in our study for the GWP, AP and EP impact categories are compared with other LCA studies. To simplify these comparisons, our results are presented using the same functional units as other studies (0.042 kg CO_2 eq per MJ_{H2}, 0.45 kg CO_2 eq per Nm^3_{H2} or 15.96 CO_2 eq g/s).

Kalinci et al. [9] reported values of GWP as 17.13 CO_2 eq g/s in the CFBG system and 0.175 CO_2 eq g/s in the DG system. In this study, the production of pine wood as feedstock for H_2 production in Proton Exchange Membrane (PEM) fuel cell vehicle were defined into system boundary. H_2 compression and transportation have been found as the main sources of environmental impact. Although the authors consider more subsystems and a wider system boundary, their result for the CFBG (comparable with our system) is just marginally higher than the one of our system. This is justified by the kind of biomass used, which does not need cultivation process and has a greater carbon fixation.

Moreno and Dufour [21] reported values for the GWP concerning different biomasses. Economic allocation has been considered to designate relevant emissions to waste biomass in orchard products. They found that use of allocation approach leads to decrease in CO_2 fixation and carbon credits of waste, since it is distributed between fruit with 90–99% of total price and waste. Therefore, the GWP, for almond pruning, by considering or not the allocation has been determined as 1.5 and 0.18 kg CO_2 eq per Nm^3 of hydrogen, respectively and GWP for vine pruning by considering or not the allocation 1.1 and 0.2 kg CO_2 eq per Nm^3 of hydrogen, respectively. In this estimation, non-converted CH_4 has been recovered into system to provide energy needs. The emissions estimated in this study are slightly lower than the ones in our study due to the use different type of gasifier. In fact, in line with results obtained by [9], fixed bed gasifier released lower emission than fluidized bed gasifier.

Thus, differences on the feedstock and the system boundaries have influence on the results and explain the variations in relation with other studies.

In other research [52], the CO_2 emitted from raw biogas reforming under two different feedstocks (maize and bio waste) was assessed. The GWP value is estimated to be 0.046 kg CO_2 eq. per 1 MJ_{h2}

for maize-based biogas and 0.037 kg CO_2 eq. if bio waste is used as feedstock. The emissions from fertilizer supply and use of diesel for farming machines in maize cultivation are mainly responsible for these values. These results are quite similar to the system investigated in this study. However, the value reported to generate H_2 by natural gas SMR was 0.1 kg CO_2 per 1 MJ_{H2} [6,53,54]. Therefore, 0.058 kg CO_2 per 1 MJ_{H2} can be reduced if waste biomass gasification is utilized to produce hydrogen.

Regarding the other impact categories, remarkable differences have been identified. In terms of AP and EP, Moreno and Dufour [21] reported 0.006 kg SO_2 eq and 0.008 kg SO_2 eq and 0.03 kg PO_4 and 0.045 kg PO_4 eq per Nm^3 of hydrogen for vine and almond pruning, respectively. According to these authors, these categories were significantly affected by the emissions of nitrate and ammonia used as fertilizer.

Thus, Table 9 summarizes the differences with the reviewed studies.

Table 9. Comparative synoptic of this and previous researches.

GWP	AP	EP	Feedstock	Handling Multifunctional Process	Reference
kg CO_2/Nm^3_{H2}	kg SO_2/Nm^3_{H2}	kg PO_4/Nm^3_{H2}			
0.45	−0.0048	0.0028	Almond shell	System expansion	Our result
1.5	0.0009	0.0054	Almond pruning	Economic allocation	[21]
0.18	0.008	0.045	Almond pruning	-	[21]
1.1	0.001	0.005	Vine pruning	Economic allocation	[21]
0.2	0.006	0.03	Vine pruning	-	[21]
kg CO_2/MJ_{H2}	kg SO_2/MJ_{H2}	kg PO_4/MJ_{H2}			
0.042	−4.5 × 10^{-4}	2.6 × 10^{-4}	Almond shell	System expansion	Our result
0.046	-	-	Maize-biogas	-	[52]
0.037	-	-	waste-biogas	-	[52]
0.1	-	-	Methane	-	[6,53,54]
g CO_2/s	-	-			
15.96	-	-	Almond shell	System expansion	Our result
17.13	-	-	Pine wood	-	[9]

Sensitivity Analysis

Sensitivity analysis has been performed to assess influence of H_2 productivity and the allocation method used in the model. As mentioned in Section 2, among several factors affecting H_2 productivity, the steam to biomass ratio (S/B), (see Section 3.4) has been chosen to assess hydrogen production. In Figure 8, the current status characterizes the gasification system presented in Section 3 (S/B = 0.5). This status has been compared with the LCA output considering the S/B ratio equal to 1 and 1.5 (Table 10). Therefore, the results of this study have been compared with results related to S/B 1 and 1.5. The experimental conditions for these different ratios are found in [38].

Table 10. Syngas composition in outlet of gasifier and inlet of WGS.

Composition (Unit)	Outlet Gasifier		Outlet WGS	
	S/B 1	S/B 1.5	S/B 1	S/B 1.5
Gas yield (Nm^3_{dry}/$kg_{bio,daf}$)	1.63	1.72	1.9	2
H_2 (%dry)	49	51	58	60
CH_4 (%dry)	7	6	7	5
CO (%dry)	24	23	3	2
CO_2 (%dry)	20	20	32	33
H_2O (%wet)	34	46	40	50
C_6H_6 (g/Nm^3_{dry})	5.51	6.97	4.7	6
C_7H_8 (g/Nm^3_{dry})	1.23	1.53	1.05	1.3
$C_{10}H_8$ (g/Nm^3_{dry})	0.33	0.77	0.29	0.66

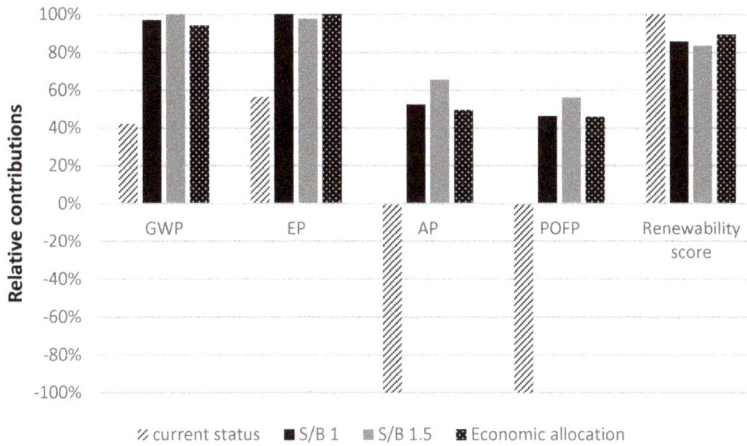

Figure 8. Sensitive analysis results for increase in hydrogen production.

According to Table 10, increasing the steam to biomass ration improves gas yield and H_2 production but also the CO_2 and H_2O content. The gas yield and H_2 production increase is more relevant than the increase of inert gas, because the LHV of the syngas increases [36].

Figure 8 shows that, in cases of S/B 1 and 1.5, impacts are higher (from 60% of GWP to around 100% for AP and POFP). Numerical values related to Figure 8 are presented in Table 11. This is due to the decrease in electricity production and subsequently decrease in environmental benefits (see Section 4). In addition, regarding renewability score, the case of S/B 0.5 shows the best performance. The economic allocation case, where environmental impacts are allocated to H_2 and electricity, shows environmental improvements compared with the case of S/B 1 and 1.5 but always higher than the case of S/B 0.5. This means that increase in hydrogen production not necessarily results in diminution of environmental impacts on the contrary these impacts can grow. Rise in hydrogen produced leads to fall in off-gas volume, electricity obtained and its avoided impacts. These results clarify role importance of by-products in environmental efficiency of hydrogen production. In addition, as the Italian electric profile includes high fraction of non-renewable sources in [27], which results that more than 65% is generated from non-renewable sources, mainly natural gas, oil and hard coal, renewability of hydrogen can be affected by decline in avoided grid derived electricity demand, which entails consumption of fossil-based electricity due to drop in off-gas volume.

Table 11. Comparison of different scenarios.

Impact Category	Unit	Scenarios			
		Current Status	S/B 1	S/B 1.5	Economic Allocation
GWP	Kg CO_2 eq	0.042	0.097	0.1	0.094
EP	Kg PO_4 eq	2.6×10^{-4}	4.6×10^{-4}	4.4×10^{-4}	4.5×10^{-4}
AP	Kg SO_2 eq	-4.5×10^{-4}	2.3×10^{-4}	2.9×10^{-4}	2.2×10^{-4}
POFP	Kg C_2H_4 eq	-1.8×10^{-5}	8.3×10^{-6}	1×10^{-5}	8.2×10^{-6}
Renewability score	%	90	77	75	80

Decentralized power plants entail environmental benefits despite the technical and economic challenges for this scale of power plant. Indeed, investment cost per power unit of small scale power plants is higher than large scale plants and there are higher requirements for the technological development of these small-scale power plants (thermal balance, etc.). Moreover, hydrogen production cost can be higher (10 euro/kg) compared with large scale production cost (1–2 euro/kg).

On the other hand, the energy, economic and environmental impacts of the biomass transportation are relatively lower due to avoiding distribution and because CHP and CCHP systems can also be more applicable for small scale power plants as heating and cooling production is more effortlessly distributed and usable. In fact, currently, hydrogen production systems at large scale can be more technically and economically applicable, but they impose higher impacts.

6. Conclusions

This study evaluates the environmental characteristics of hydrogen and electricity production in an innovative small-scale biomass gasification plant. The LCA methodology has been applied to assess the environmental performance of this production scenario.

Real input and output flows for the whole system have been identified and managed in detail from a cradle-to-gate perspective. This study showed that the biomass production phase influences all impact categories, which is a unanimous result of previous research. Furthermore, due to multifunctional nature of this process and considering byproduct converted into electricity as well as no auxiliary energy consumption, these environmental impacts massively decrease. Negative values in AP, POFP and CED illustrate this improvement. Following this result, a sensitivity analysis was conducted to assess the influence of the variations in hydrogen and byproduct volume on environmental impacts. The S/B ratio has been chosen as parameter affecting both off-gas and hydrogen production because it is commonly recognized as the main factor in the analysis of performance of gasification process. Results indicate that, although increase in hydrogen production can directly reduce all environmental impacts, this implies a fall in off-gas volume, which indirectly leads to increased impacts for all categories. As a result, in the considered production system, the environmental effects cannot be univocally represented by hydrogen production rate because relevance of byproducts produced plays an even more important role. In this line of research, development of system boundary into bio-refinery products can also lead to higher reduction in environmental footprints of hydrogen production system. In addition, the sensitivity analysis shows that allocation method also highly affects the system profile. Allocation of environmental charges can considerably overestimate results. Results of this work have a relevant impact with respect to previous literature works regarding hydrogen production by means of biomass gasification because it has analyzed the effect of some design parameters on the environmental impact of the overall system, quantifying the contribution, for each category, of the different configurations.

Finally, despite the limits of small scale hydrogen production system, many potential advantages of this category can be enumerated. Therefore, a highly efficient small-scale power plant with low environmental burden needs to be developed to tackle the low energy density and perishability of biomass.

Acknowledgments: This research has been supported by the following projects: the Italian national *Ricerca di Sistema* research project HBF2.0 funded by *Italian Ministry for Economic Developments* and European project UNIfHY (UNIque gasifier for HYdrogen production) funded by *FP7 European Commission Programme*. This research has been supported by the following projects: *HBF2.0* funded by *Italian Ministry of Economic Development* within the *Italian National System Research*, in the Programme called *Ricerca di Sistema*, and the European project UNIfHY (*UNIque gasifier for HYdrogen production*) funded by FP7 *EuropeanCommissionProgramme*. Within these projects we thank especially Andrea Di Carlo of University of L'Aquila, for the plant flowsheet and part of the plant data and Donatella Barisano of ENEA for the data and the picture of the UNIFHY 1000 kWth plant.

Author Contributions: Enrico Bocci and Mauro Villarini conceived and designed the model of the plant, took care of the system energy balance and, with Andrea Colantoni, revised the functional unit data and the paper; Sara Rajabi Hamedani and Michele Moretti developed the LCA analysis; Sara Rajabi Hamedani and Andrea Colantoni analyzed the data; contributed reagents/materials/analysis tools; Sara Rajabi Hamedani wrote the paper with Mauro Villarini.

Conflicts of Interest: The authors declare no conflicts of interest.

Nomenclature

AP	acidification potential
CED	Cumulative Energy Demand
CFC	Catalytic Filter Candles
CHP	Combined Heat and Power
CCHP	Combined Cooling Heating and Power
EP	eutrophication potential
FU	Functional Unit
GWP	global warming potential
IBFB	Bubbling Fluidized Bed gasifier
ICE	Internal Combustion Engine
LCA	Life Cycle Assessment
LCI	Life Cycle Inventory
LHV	Low Heating Value
POFP	photochemical oxidation potential
PSA	Pressure Swing Adsorber
SMR	Steam Methane Reforming
SS1	Subsystem1
SS2	Subsystem2
SS3	Subsystem3
SS4	Subsystem4
WGS	Water Gas Shift

Appendix

Table A1. Energy demand per kWh_{H2} and H_2 renewability score [6].

Scenario	Fossil Energy (kWh Primary Energy)	Nuclear Energy (kWh Primary Energy)	Renewable Energy (kWh Primary Energy)	H_2 Renewability Score (%)
Natural gas	2.3	0.5	0.14	5
Wind electrolysis	0.06	0.03	2	96
SCWG	1.2	0.4	2.1	56
Raw biogas SMR	0.6	0.4	3	75
Raw biogas SMR-maize	0.6	0.5	3	73

References

1. O'Keefe, P.; Geoff, O.; Nicola, P. *The Future of Energy Use*, 2nd ed.; Routledge: Abingdon, UK, 2010.
2. IIASA. I institude for applied systems analysis. In *Energy Perspectives for Eurasia and the Kyoto Protocol*; IIASA: Laxenburg, Austria, 1998.
3. Khila, Z.; Baccar, I.; Jemel, I.; Hajjaji, N. Thermo-environmental life cycle assessment of hydrogen production by autothermal reforming of bioethanol. *Energy Sustain. Dev.* **2017**, *37*, 66–78. [CrossRef]
4. Ball, M.; Wietschel, M. The future of hydrogen-opportunities and challenges. *Int. J. Hydrog. Energy* **2009**, *34*, 615–627. [CrossRef]
5. Dunn, S. Hydrogen futures: Toward a sustainable energy system. *Int. J. Hydrog. Energy* **2002**, *27*, 235–264. [CrossRef]
6. Fuell cells and hydrogen joint undertaking. In *Study on Hydrogen from Renewable Resources in the EU, Final Report*; FCH: Brussels, Belgium, 2015. [CrossRef]
7. Singh, A.; Sevda, S.; Abu Reesh, I.; Vanbroekhoven, K.; Rathore, D.; Pant, D. Biohydrogen Production from Lignocellulosic Biomass: Technology and Sustainability. *Energies* **2015**, *8*, 13062–13080. [CrossRef]
8. Susmozas, A.; Iribarren, D.; Dufour, J. Life-cycle performance of indirect biomass gasification as a green alternative to steam methane reforming for hydrogen production. *Int. J. Hydrog. Energy* **2013**, *38*, 9961–9972. [CrossRef]

9.   Kalinci, Y.; Hepbasli, A.; Dincer, I. Life cycle assessment of hydrogen production from biomass gasification systems. *Int. J. Hydrog. Energy* **2012**, *37*, 14026–14039. [CrossRef]

10. Bocci, E.; Di Carlo, A.; McPhail, S.J.; Gallucci, K.; Foscolo, P.U.; Moneti, M.; Villarini, M.; Carlini, M. Biomass to fuel cells state of the art: A review of the most innovative technology solutions. *Int. J. Hydrog. Energy* **2014**, *39*, 21876–21895. [CrossRef]

11. Koroneos, C.; Dompros, A.; Roumbas, G. Hydrogen production via biomass gasification-A life cycle assessment approach. *Chem. Eng. Process. Process Intensif.* **2008**, *47*, 1267–1274. [CrossRef]

12. IEA Bioenergy. *Bioenergy—A Sustainable and Reliable Energy Source a Review of Status and Prospects*; IEA Bioenergy: Wageningen, The Netherlands, 2009; p. 108.

13. Staples, M.D.; Malina, R.; Olcay, H.; Pearlson, M.N.; Hileman, J.I.; Boies, A.; Barrett, S.R.H. Lifecycle greenhouse gas footprint and minimum selling price of renewable diesel and jet fuel from fermentation and advanced fermentation production technologies. *Energy Environ. Sci.* **2014**, *7*, 1545–1554. [CrossRef]

14. ISO 14040. *Environmental Management-Life Cycle Assessment-Principles and Framework 2006*; ISO: Geneva, Switzerland, 2006.

15. Meul, M.; Van Middelaar, C.E.; de Boer, I.J.M.; Van Passel, S.; Fremaut, D.; Haesaert, G. Potential of life cycle assessment to support environmental decision making at commercial dairy farms. *Agric. Syst.* **2014**, *131*, 105–115. [CrossRef]

16. Moretti, M.; De Boni, A.; Roma, R. Economic and environmental sustainability of forestry measures in Apulia Region Rural Development Plan: An application of life cycle approach. *Land Use Policy* **2014**, *41*, 284–289. [CrossRef]

17. Murphy, D.; Carbajales-Dale, M.; Moeller, D. Comparing Apples to Apples: Why the Net Energy Analysis Community Needs to Adopt the Life-Cycle Analysis Framework. *Energies* **2016**, *9*, 917. [CrossRef]

18. Cherubini, F.; Strømman, A.H. Life cycle assessment of bioenergy systems: State of the art and future challenges. *Bioresour. Technol.* **2011**, *102*, 437–451. [CrossRef] [PubMed]

19. Cherubini, F. The biorefinery concept: Using biomass instead of oil for producing energy and chemicals. *Energy Convers. Manag.* **2010**, *51*, 1412–1421. [CrossRef]

20. Lelek, L.; Kulczycka, J.; Lewandowska, A.; Zarebska, J. Life cycle assessment of energy generation in Poland. *Int. J. Life Cycle Assess.* **2015**, *21*, 1–14. [CrossRef]

21. Moreno, J.; Dufour, J. Life cycle assessment of hydrogen production from biomass gasification. Evaluation of different Spanish feedstocks. *Int. J. Hydrog. Energy* **2013**, *38*, 7616–7622. [CrossRef]

22. Suwatthikul, A.; Limprachaya, S.; Kittisupakorn, P.; Mujtaba, I. Simulation of Steam Gasification in a Fluidized Bed Reactor with Energy Self-Sufficient Condition. *Energies* **2017**, *10*, 314. [CrossRef]

23. Buchholz, T.; Volk, T.A. Considerations of Project Scale and Sustainability of Modern Bioenergy Systems in Uganda. *J. Sustain. For.* **2012**, *31*, 154–173. [CrossRef]

24. Bocci, E.; Sisinni, M.; Moneti, M.; Vecchione, L.; Di Carlo, A.; Villarini, M. State of art of small scale biomass gasification power systems: A review of the different typologies. *Energy Procedia* **2014**, *45*, 247–256. [CrossRef]

25. EU. UNIfHY Collaborative Project, Project ID 299732 7FP n.d. Available online: http://www.fch.europa.eu/project/unique-gasifier-hydrogen-production (accessed on 14 January 2018).

26. ISO 14044. *Environmental Management-Life Cycle Assessment-Requirements and Guidelines 2006*; ISO: Geneva, Switzerland, 2006.

27. Lijó, L.; González-garcía, S.; Bacenetti, J.; Fiala, M.; Feijoo, G.; Lema, J.M.; Moreira, M.T. Life Cycle Assessment of electricity production in Italy from anaerobic co-digestion of pig slurry and energy crops. *Renew. Energy* **2014**, *68*, 625–635. [CrossRef]

28. Intergovernmental Panel on Climate Change (IPCC). *Guidelines for National Greenhouse Gas Inventories*; IPCC: Geneva, Switzerland, 2006.

29. Marvinney, E.; Kendall, A.; Brodt, S. Life Cycle—Based Assessment of Energy Use and Greenhouse Gas Emissions in Almond Production, Part II Uncertainty Analysis through Sensitivity Analysis and Scenario Testing. *Res. Anal.* **2015**, *19*, 1019–1029. [CrossRef]

30. European Union. *PE-CONS 42/15 DGE 1 EB/vm*; European Union: Brussels, Belgium, 2015; Volume 2015.

31. Fuell Cell and Hydrogen-Joint Undertaking. *Guidance Document for Performing LCAs on Fuel Cells and H2 Technologies*; Fuell Cell and Hydrogen-Joint Undertaking: Brussels, Belgium, 2011.

32. Guinée, J.B.; Gorrée, M.; Heijungs, R.; Huppes, G.; Kleijn, R.; de Koning, A.; van Oers, L.; Wegener Sleeswijk, A.; Suh, S.; Udo de Haes, H.A.; et al. *Handbook on Life Cycle Assessment. Operational Guide to the ISO Standards. Part III: Scientific Background*; Kluwer Academic Publishers: Dordrecht, The Netherlands, 2002.

33. PRé. SimaPro Database Manual Methods 2015. Available online: https://www.pre-sustainability.com/download/DatabaseManualMethods.pdf (accessed on 14 January 2018).

34. Hischier, R.; Weidema, B.; Althaus, H.J.; Bauer, C.; Doka, G.D.R. *Implementation of Life Cycle Impact Assessment Methods, Ecoinvent Report*; Swiss: Dubendorf, Switzerland, 2010.

35. Neelis, M.-L.; van der Kooi, H.J.; Geerlings, J.J.C. Exergetic life cycle analysis of hydrogen production and storage systems for automotive applications. *Int. J. Hydrogen Energy* **2004**, *29*, 537–545. [CrossRef]

36. Pallozzi, V.; Di Carlo, A.; Bocci, E.; Villarini, M.; Foscolo, P.U.; Carlini, M. Performance evaluation at different process parameters of an innovative prototype of biomass gasification system aimed to hydrogen production. *Energy Convers. Manag.* **2016**, *130*, 34–43. [CrossRef]

37. Fremaux, S.; Beheshti, S.-M.; Ghassemi, H.; Shahsavan-Markadeh, R. An experimental study on hydrogen-rich gas production via steam gasification of biomass in a research-scale fluidized bed. *Energy Convers. Manag.* **2015**, *91*, 427–432. [CrossRef]

38. Hamedani Rajabi, S.; Bocci, E.; Villarini, M.; Di Carlo, A.; Naso, V. Techno-economic Analysis of Hydrogen Production Using Biomass Gasification-A Small Scale Power Plant Study. *Energy Procedia* **2016**, *101*, 806–813.

39. GSE. *Prezzi Minimi Grantiti Per l'anno 2015*; GSE: Chicago, IL, USA, 2015.

40. Alto Garda Sevizi (AGS). *Tariffe Di Riferimento IV Trimestre 2015 Ottobre/Dicembre*; AGS: Riva del Garda, Italy, 2015.

41. Moneti, M.; Di Carlo, A.; Enrico, B.; Foscolo, P.U.; Villarini, M.; Carlini, M. Influence of the main gasifier parameters on a real system for hydrogen production from biomass. *Int. J. Hydrog. Energy* **2016**, *41*, 11965–11973. [CrossRef]

42. Villarini, M.; Bocci, E.; Di Carlo, A.; Savuto, E.; Pallozzi, V. The case study of an innovative small scale biomass waste gasification heat and power plant contextualized in a farm. *Energy Procedia* **2015**, *82*, 335–342. [CrossRef]

43. Bocci, E.; Di Carlo, A.; Vecchione, L.; Villarini, M.; De Falco, M.; Dell'Era, A. Technical-economic analysis of an innovative cogenerative small scale biomass gasification power plant. *Lect. Notes Comput. Sci.* **2013**, *7972*, 256–270.

44. Kendall, A.; Marvinney, E.; Brodt, S.; Zhu, W. Life Cycle—Based Assessment of Energy Use and Greenhouse Gas Emissions in Almond Production, Part I Analytical Framework and Baseline Results. *J. Ind. Ecol.* **2015**, *19*, 1008–1018. [CrossRef]

45. Kendall, A.; Marvinney, E.; Brodt, W.; Zhu, S.; Yuan, J. *Greenhouse Gas and Energy Footprint (Life Cycle Assessment) of California Almond Production*; Almond board of California: Modesto, CA, USA, 2012.

46. Nanna, F.; Barisano, D.; Villone, A. *Deliverable 2.1 Chemical Characterisation of Feedstock. UNIfHY UNIQUE Gasifier for Hydrogen Production*; FCH: Brussels, Belgium, 2013.

47. Savuto, E.; Di Carlo, A.; Bocci, E.; D'Orazio, A.; Villarini, M.; Carlini, M.; Foscolo, P.U. Development of a CFD model for the simulation of tar and methane steam reforming through a ceramic catalytic filter. *Int. J. Hydrogen Energy* **2015**, *40*, 7991–8004. [CrossRef]

48. Hofbauer, H.; Rauch, R.; Loeffler, G.; Kaiser, S.; Fercher, E.; Tremmel, H. Six years experience with the FICFB-gasification process. In Proceedings of the 12th European Conference and Technology Exhibition on. Biomass for Energy, Industry and Climate Protection, Amsterdam, The Netherlands, 17–21 June 2002; pp. 982–985.

49. Hofbauer, H.; Knoef, H. Success stories on biomass gasification. In *Handbook Biomass Gasification*; BTG Biomass Technology: Enschede, The Netherlands, 2005.

50. EU. UNIQUE Cooperative Research Project, Contract N.211517 7FP. 2013-01-18 n.d. Available online: http://cordis.europa.eu/result/rcn/55532_en.html (accessed on 14 January 2018).

51. Dufour, J.; Serrano, D.P.; Gálvez, J.L.; González, A.; Soria, E.; Fierro, J.L.G. Life cycle assessment of alternatives for hydrogen production from renewable and fossil sources. *Int. J. Hydrogen Energy* **2012**, *37*, 1173–1183. [CrossRef]

52. Deutsches Biomasseforschungszentrum (DBFZ). *Evaluierung der Verfahren und Technologien für die Bereitstellung von Wasserstoff auf Basis von Biomasse*; DBFZ: Leipzig, Germany, 2014.

53. Edwards, R.; Hass, H.; Larivé, J.-F.; Lonza, L.; Mass, H.; Rickeard, D. Well-to-Wheel Analysis of Future Automotive fuels and Powertrains in the European Context, Version 4.a. 2014. Available online: http://iet.jrc.ec.europa.eu (accessed on 1 March 2018).

54. FCH JU. *Fuel Cells and Hydrogen Joint Undertaking (FCH JU), Annual Implamantaion Plant*; FCH JU: Bruxelles, Belgium, 2011.

energies

Article

Thermogravimetric, Devolatilization Rate, and Differential Scanning Calorimetry Analyses of Biomass of Tropical Plantation Species of Costa Rica Torrefied at Different Temperatures and Times

Johanna Gaitán-Álvarez [1], Róger Moya [1,*], Allen Puente-Urbina [2] and Ana Rodriguez-Zúñiga [1]

1 Escuela de Ingeniería Forestal, Instituto Tecnológico de Costa Rica, Apartado 159-7050, Cartago, Costa Rica; jgaitan@itcr.ac.cr (J.G.-Á.); ana.rodriguez@itcr.ac.cr (A.R-Z.)
2 Instituto Tecnológico de Costa Rica, Centro de Investigación y de Servicios Químicos y Microbiológicos (CEQUIATEC), Escuela de Quimica, Apartado, 159-7050 Cartago, Costa Rica; apuente@itcr.ac.cr
* Correspondence: rmoya@itcr.ac.cr; Tel.: +506-2550-9092

Received: 18 February 2018; Accepted: 19 March 2018; Published: 21 March 2018

Abstract: We evaluated the thermogravimetric and devolatilization rates of hemicellulose and cellulose, and the calorimetric behavior of the torrefied biomass, of five tropical woody species (*Cupressus lusitanica*, *Dipteryx panamensis*, *Gmelina arborea*, *Tectona grandis* and *Vochysia ferruginea*), at three temperatures (T_T) and three torrefaction times (t_T) using a thermogravimetric analyzer. Through a multivariate analysis of principal components (MAPC), the most appropriate torrefaction conditions for the different types of woody biomass were identified. The thermogravimetric analysis-derivative thermogravimetry (TGA-DTG) analysis showed that a higher percentage of the hemicellulose component of the biomass degrades, followed by cellulose, so that the hemicellulose energy of activation (Ea) was less than that of cellulose. With an increase in T_T and t_T, the Ea for hemicellulose decreased but increased for cellulose. The calorimetric analyses showed that hemicellulose is the least stable component in the torrefied biomass under severe torrefaction conditions, and cellulose is more thermally stable in torrefied biomass. From the MAPC results, the best torrefaction conditions for calorimetric analyses were at 200 and 225 °C after 8, 10, and 12 min, for light and middle torrefaction, respectively, for the five woody species.

Keywords: thermogravimetric analysis; differential scanning calorimetry; hemicellulose; cellulose; torrefaction; thermostability

---

1. Introduction

Biomass is widely available worldwide, and often used in biofuel production to help reduce the use of fossil energy reserves and mitigate problems the environmental problems caused by petroleum derived fuels [1]. In addition, biomass produces lower carbon dioxide (CO_2) emissions, as it maintains the carbon cycle by freeing the carbon that was previously fixed during photosynthesis [2]. However, despite the importance of biomass, it has not been developed into other types of energy, such as hydroelectricity, eolic, or solar energies, which are being highly exploited [3].

Despite the increase in the use of biomass as an energy source, some disadvantages still limit its optimum performance, such as difficulties in collection due to disperse distribution, irregular shape, high volume, low energy density, and storage and transportation problems [2]. These problems affect the variability of the biomass's physical properties [4]. Other challenges posed by biomass include its low calorific power, high moisture content, and hygroscopic nature that cause economic problems and deficiencies during transportation, handling, storage, and conversion of the material [5].

Technologies that use thermal treatment of biomass could be applied to solve the mentioned difficulties and convert biomass into energy via combustion [6]. Of the thermal treatments, torrefaction appears to be an effective solution [5]. Studies have shown that torrefaction increases the energy density of the biomass and reduces its hygroscopicity [7,8]. Biomass torrefaction occurs at temperatures from 200 and 300 °C, at atmospheric pressure, and in the presence of inert atmosphere, meaning a limited oxygen presence during the process. The advantages of torrefaction include calorific power increase, reduction of O-H and H-C, lower, moisture content, higher hydrophobicity, and better grinding capacity [9–12].

The success and quality of the torrefied biomass depend on the temperature and time of torrefaction, mainly due to the heterogeneity of the chemical composition of the biomass [13]. Among the chemical components of biomass, lignin is the most thermally stable, followed by cellulose and then by [5,12,14]. For average ranges of torrefaction, the component that degrades the most is hemicellulose as well as some non-structural components, such as extractives [9,12,15].

Important changes in biomass composition were observed using thermogravimetric analysis (TGA) after torrefaction [16]. The curves in this analysis demonstrate the thermal stability of the components of the biomass, including the mass loss and residual mass [14,17]. Previous studies confirmed that hemicellulose decreased and, consequently, the proportion of cellulose and lignin increased in the species after torrefaction [12]. However, the characteristics of the torrefied biomass of tropical species have rarely been studied [12,18]. Large volumes of biomass residues of tropical species in Costa Rica are constantly produced in the timber industry, so torrefaction is an option to process this raw material [12,19]. The biomass of tropical timber species and the different torrefaction processes have been characterized [18–21]. This study continued these studies on the biomass torrefaction of the most used species with energy potential in Costa Rica [22–24].

The present study aimed to evaluate the thermogravimetric behaviour, devolatilization of hemicellulose and cellulose, and the calorimetric behaviour of the torrefied biomass of five tropical woody species (*Cupressus lusitanica, Dipteryx panamensis, Gmelina arborea, Tectona grandis* and *Vochysia ferruginea*), at three temperature conditions (light, middle and severe) and three torrefaction times using simultaneous thermogravimetric and differential scanning calorimetry analyses. Then, we aimed to find the most appropriate torrefaction conditions for the different types of woody biomass using multivariate analysis of principal components (MAPC) in relation to the thermo-chemical degradation without significantly affecting the chemical composition of the material. This study will enhance the treatment of biomass to obtain renewable and viable raw material for the generation of clean energy from a lignocellulosic material [25].

2. Material and Methods

2.1. Material Characteristics

The woody waste biomass of *C. lusitanica, D. panamensis, G. arborea, T. grandis* and *V. ferruginea* from fast growth plantations at different sites in Costa Rica was used. The age of the plantations ranged between 8 and 14 years. The details of the materials are available in Moya et al. [18] and Gaitán-Álvarez et al. [12,21]. Sawdust from all the species was directly collected from the sawing process, conditioned to a 7% moisture content and then sieved. After sieving, the sawdust particles used were 70% of 2.00–4.00 mm and 30% 0.42–2.00 mm. The chemical compositions of the five species are shown in Table 1.

Table 1. Chemical composition of five fast-growth plantation species in Costa Rica.

Properties	*Cupressus lusitanica*	*Dipterix panamensis*	*Gmelina arborea*	*Tectona grandis*	*Vochysia ferruginea*
Cellulose (%)	64.7	49.9	55.6	54.4	50.9
Lignin (%)	31.4	20.3	24.2	21.90	11.2
Ash (%)	0.18	3.04	0.96	2.81	0.99
Carbon (%)	50.18	48.64	48.39	49.77	49.32
Nitrogen (%)	0.27	0.24	0.20	0.20	0.27

2.2. Torrefaction Process

Three 500 g samples of sawdust were obtained from each species. The material was then divided to apply three different torrefaction durations (8, 10 and 12 min), and the three different torrefaction temperatures (200, 225, and 250 °C), resulting in nine treatments per species. Figure 1 shows the nine treatment and their abbreviations. These durations and temperatures were selected according to a previous study [5]. A modified Thermolyne Furnace 48,000 (Thermolyne, Waltham, MA, USA) was used for the torrefaction process. The furnace was sealed to prevent airflow from the manual system to maintain pressure. Every 4–5 min, the air was freed, allowing the development of the torrefaction process within an environment with limited oxygen content, adding N_2 to the furnace [19].

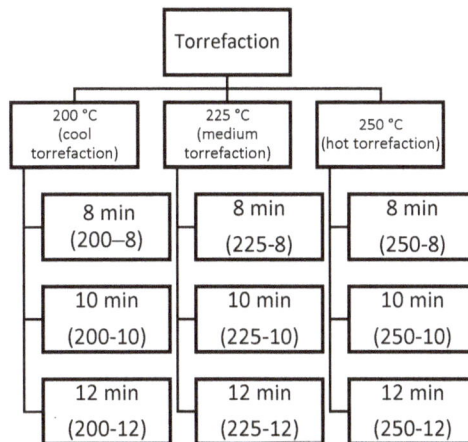

Figure 1. Temperature and time for the torrefaction of the biomass of five fast-growth plantation species of Costa Rica. Note: the numbers in parentheses indicate the abbreviation of this torrefaction condition.

2.3. Thermogravimetric Analysis (TGA)

To obtain the degradation curves, TGA was performed at atmospheric pressure under inert ambient nitrogen, using about 5 mg of sawdust of each species. The heating rate was 20 °C/min in a nitrogen atmosphere of ultra-high purity N_2 at 100 mL/min, reaching a temperature of 800 °C. A TA Instruments (New Castle, DE, USA) thermogravimetric analyzer, model SDT Q600, was used. The TGA provided values for mass loss in relation to temperature, from which the derivative thermogravimetry (DTG) was obtained, allowing us to determine the position and temperature at which sample degradation occurred. The TGA data and their first and second derivatives (DTG and D^2TG) were analyzed using TA Instruments Universal Analysis 2000 software. The parameters are presented in Figure 2a,b: (i) the temperature at the beginning of degradation (T_i) and the percentage of residual mass at Ti (W_{Ti}); (ii) the temperature corresponding to the maximum degradation of hemicellulose (T_{sh}) and the percentage of residual mass at T_{sh} (WT_{sh}); (iii) the temperature corresponding to the maximum cellulose mass loss rate (T_m) and the percentage of the residual mass at Tm (W_{Tm}) and (iv) the temperature corresponding to the end of degradation (T_f) and the percentage of residual mass at T_f (W_{Tf}), when mass loss began to stabilize as the temperature increased. Additional parameters were obtained from the derivative thermogravimetry (DTG): (v) the temperature of hemicellulose degradation onset ($T_{onset(hc)}$) and residual mass at $T_{onset(hc)}$ ($W_{Tonset\ (hc)}$); (vi) the end temperature of the hemicellulose degradation ($T_{offset(hc)}$) and the residual mass at $T_{offset(hc)}$ ($W_{Toffset\ (hc)}$); (vii): the temperature of cellulose degradation onset ($T_{onset(c)}$) and the residual mass at T onset(c) ($W_{Tonset(c)}$); (viii): the end temperature of cellulose degradation ($T_{offset(c)}$)and the residual mass at $T_{offset(c)}$ ($W_{Toffset(c)}$). Figure 2c shows the DTG curve representing the different temperature. MAgicPlot 2.5.1 software was used to obtain these values.

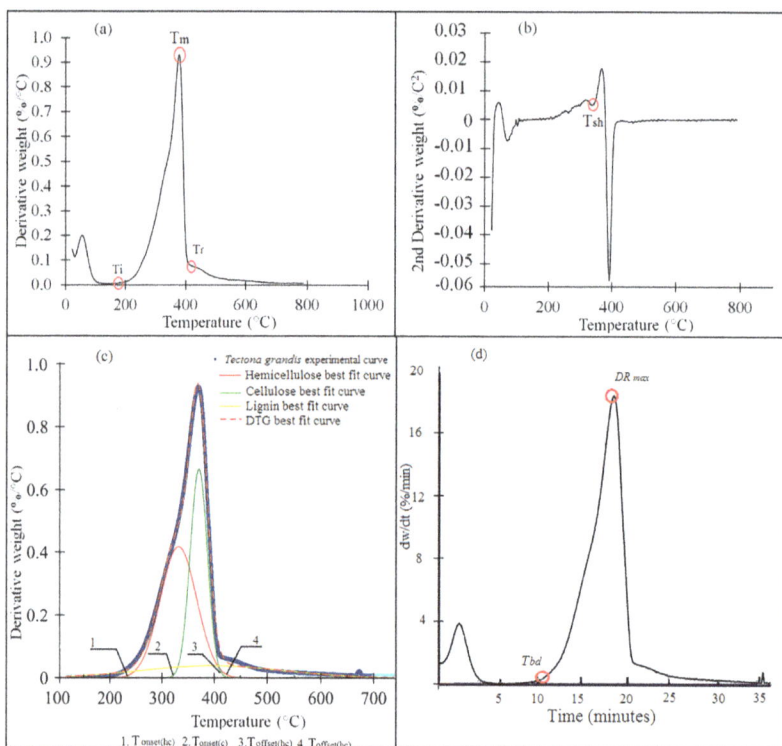

Figure 2. (**a**,**c**) Derivative thermogravimetry (DTG) and (**b**) second derivative (D²TG) parameters for the different woody biomasses analyzed; (**d**) Devolatilization rate measured by first time derivates of the mass frcationas a function of time. Note: t_{bd} is the start time of the maximum devolatilization rate and D_{max} is the maximum devolatilization rate [18].

Once the decomposition start points for hemicellulose and cellulose were obtained, the thermostability of these components was evaluated using the model described in Equation (2), which was obtained from the linearized model in Equation (1) according to Sbirrazzuoli et al. [26]. The differential was the conversional method used by Friedman. The objective was to calculate the activation energy of the decomposition for each component of the materials being studied (hemicellulose and cellulose):

$$K = A * e^{\left(\frac{-Ea}{RT}\right)} \tag{1}$$

$$\ln\left(\frac{d\alpha}{dt}\right) = \ln K_0 + \left(\frac{-Ea}{RT}\right) + n * \ln(1 - \alpha) \tag{2}$$

where α is the degraded mass, $\frac{d\alpha}{dt}$ is the percentage of the degraded sample per unit time, A is the pre-exponential factor, Ea is the energy of activation, and T is temperature.

2.4. Devolatilization Variation

Several methods can be used to measure the degree of biomass devolatilization [18,27]. According to Grønli et al. [27], the total volatiles released during devolatilizatiion include mass fractions, whose dynamics are described by first-order kinetics. In this research, the devolatilization behavior during the thermal degradation of the biomass components in different samples was evidence by the percentage of devolatilized mass relative to time, and a subsequent analysis of the devolatilization rate (D_{rate}). The D_{rate} behavior with different T_T and t_T was first described. Next, we determined the maximum devolatilization rate (D_{max}) and the time at which D_{rate} was obtained. Figure 2d shows where these parameters were determined in the first time derivatives of the mass fraction with respect to time.

2.5. Statistical Analysis

The experiment had a two-level factorial design. Level one corresponded to the t_T of the biomass at three different times: 8, 10, and 12 min. The second factorial level was T_T at three temperatures: 200, 225, and 250 °C. This design was applied to each species studied (*C. lusitanica, D. panamenisis, G. arborea, T. grandis* and *V. ferruginea*). We worked with three samples for each treatment per species. Secondly, a multivariate analysis of the principal components (MAPC) was performed, including all the variables of the TGA and the determined devolatilization parameters. Two main components were selected. This analysis was performed with SAS software (SAS Inc., Cary, NC, USA). Significance level was established at 5%.

3. Results

3.1. TGA-DTG Analysis

The thermogravimetric decomposition behavior of the torrefied biomass for the five species showed the same pattern with different T_T and t_T (Figure 3a–h). However, the DTG curve showed some differences in biomass decomposition (Figure 3a–h). For the TGA curve, five important stages were observed. A predominant signal appeared in the first stage prior to 100 °C. The second stage showed a pronounced peak between 290 °C and 330 °C, the third stage occurred between 340–380 °C. And the fourth stage appeared between 400–500 °C, where the speed of mass loss mas lower compared with the two previous decomposition stages. Finally, few changes in the sample occurred as temperature continued to increase.

Overall, the TGA and DTG curves (Figure 3a–h) showed small visually noticeable differences in the thermal behaviour of the torrefied biomass at various T_T. For all species studied, the biomass torrefied at 250 °C at the three t_T were thermally different. First, the TGA curves show that the biomass torrefied at 250 °C, or severe torrefaction, behaved differently compared to the rest of the T_T. After 340–380 °C, mass loss was less than for biomass torrefied at 250 °C (Figure 3a–h). Second, the DTG curves showed a strong signal at 290 °C, but it appeared as a small shoulder along that at 250 °C (Figure 3a–h). This signal was more visible in *D. panamensis* (Figure 3c) and *V. ferruginea* (Figure 3i), whereas this shoulder was not present in *C. lusitanica* (Figure 3a), *G. arborea* (Figure 3e), and *T. grandis* (Figure 3g) in biomass torrefied at 250-12.

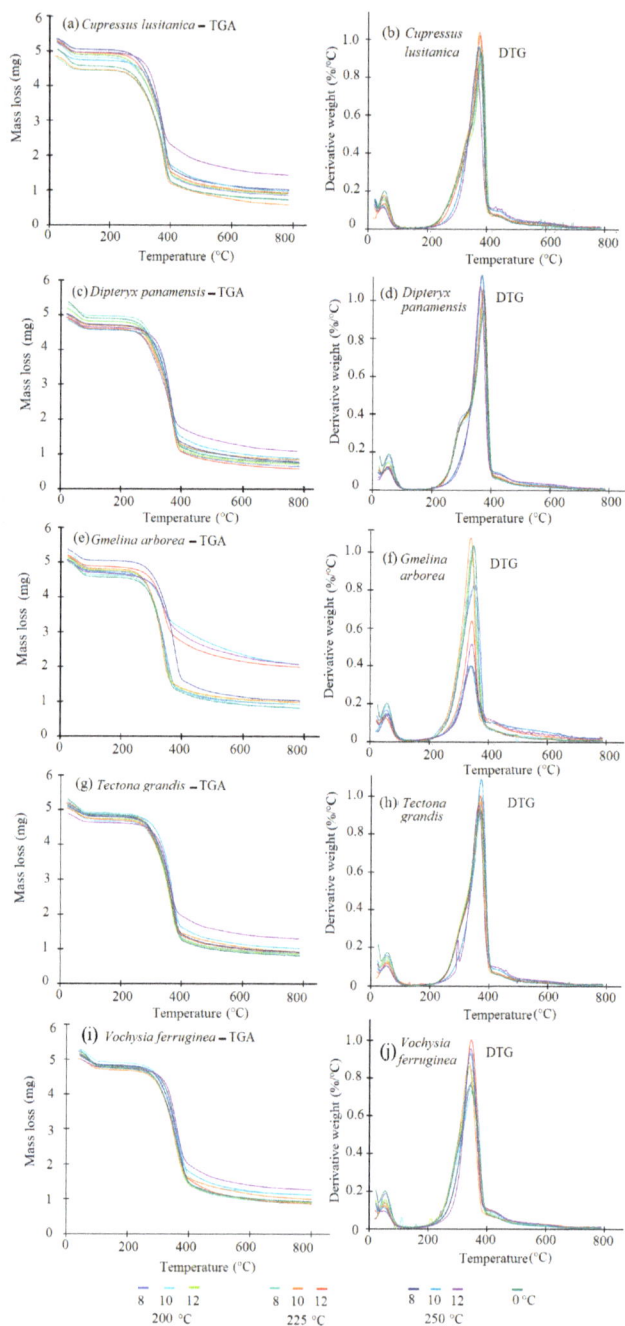

Figure 3. Thermogravimetric analysis (TGA) and DTG of biomasses for *Cupressus lusitanica* (**a–b**), *Dipteryx panamensis* (**c–d**), *Gmelina arborea* (**e–f**) and *Tectona grandis* (**g–h**) and *Vochysia guatemalensis* (**i–j**), torrefied at different times and temperatures.

Tables 2 and 3 show the detailed analyses of the temperatures and mass loss for the various species, where the main changes in the degradation of the different chemical components of the torrefied biomass during the TGA occurred. In the evaluation of the decomposition T_i in *C. lusitanica*, the T_i of biomass torrefaction increased with respect to untorrefied biomass, except for the 250-8 condition. For the remaining four species under all torrefaction conditions, decomposition T_i increased (Table 2). As for W_{Ti} at 250 °C, torrefaction was greater in the torrefied biomass compared with untorrefied biomass for all species (Table 3). Conversely, T_i tended to increase in torrefied biomass under the light torrefaction condition (200-8) to the middle torrefaction condition between 225-10 and 225-12, depending on the species, and decreased under 250-10 or 250-12 conditions. For *D. panamensis*, *G. arborea*, *T. grandis*, and *V. ferruginea*, the decomposition T_f was lower for the torrefied biomass than the untorrefied biomass under any condition of T_T and t_T. T_f increased at 200 °C in the torrefied biomass of *C. lusitanica* compared to untorrefied biomass. The remaining conditions (225 and 250 °C) displayed lower T_f compared to the untorrefied biomass (Table 2). The behavior of W_{Ti}, W_{Tf}, and the residual mass differed among T_i and T_f conditions for all species, as W_{Ti} and W_{Tf} increased with increasing T_T and t_T (Tables 2 and 3).

C. lusitanica behaved differently with respect to hemicellulose parameters compared with the other species. $T_{onset(hc)}$ and T_{sh} were higher in the torrefied biomass compared to the untorrefied biomass (Table 2), whereas the $T_{offset(hc)}$ was lower in all torrefied biomasses compared to the untorrefied biomass (Table 2). $T_{onset(hc)}$ was lower in the torrefied biomass of *D. panamensis*, *G. arborea*, *T. grandis*, and *V. ferruginea* compared to untorrefied biomass. The T_{sh} and $T_{offset(hc)}$ of the torrefied biomass of these species were higher than the untorrefied biomass (Table 2).

The different torrefaction conditions had varying effects on the hemicellulose of *C. lusitanica* compared to the other four species. The $T_{onset(hc)}$ of the torrefied biomass of *C. lusitanica* increased as T_T and t_T increased, whereas T_{sh} and $T_{offset(hc)}$ decreased with increasing T_T and t_T. The $T_{onset(hc)}$ also increased in the torrefied biomass of the remaining species (*D. panemensis*, *G. arborea*, *T. grandis* and *V. feruginea*) under light 200-8) to medium (between 225-10 or 225-12 depending on the species), torrefaction conditions, then decreasing under the 250-10 or 250-12 conditions. For the T_{sh} and $T_{offset(hc)}$ parameters, their values decreased with increasing T_T and t_T, whereas some irregularities were observed in this behaviour in *C. lusitanica* and *T. grandis*.

In biomass torrefied at any T_T or t_T, $W_{Tonset(hc)}$, W_{Tsh}, and $W_{Toffset(hc)}$ had higher values than in untorrefied biomass in all species (Table 3). The values of T_T and t_T varied under different torrefaction conditions. In general, the values of $W_{Tonset(hc)}$, W_{Tsh} and $W_{Toffset(hc)}$ for all species increased with increasing T_T, particularly in biomass torrefied at 250 °C. Few changes were observed in t_T at the same T_T in the $W_{Tonset(hc)}$ and W_{Tsh} values for all t_T of the different species. However, for $W_{Toffset(hc)}$, for 8 and 10 min, the parameter values were similar, whereas under condition 250-12, a significant increase in $W_{Toffset(hc)}$ was observed in all species (Table 3).

For the cellulose decomposition parameters, biomass torrefaction increased $T_{onset(c)}$ compared to the untorrefied biomass in *C. lusitanica*, *G. arborea*, *T. grandis*, and *V. ferruginea*, whereas in *D. panamensis*, $T_{onset(c)}$ increased from the 200-8 to the 250-10 condition, and then decreased under the most severe condition (250-12) (Table 3). Conversely, T_m increased in the torrefaction of the biomass of *C. lusitanica* and *D. panamensis* from the least severe condition (200-8) to condition 225-10. Also, under conditions 225-12 and T_T at 250 °C, the torrefied biomass had a lower T_m than the untorrefied biomass. The torrefied biomass of *G. arborea* had a higher T_m value compared to the untorrefied biomass, except under condition 250-8. In the biomass of *T. grandis*, torrefaction increased T_m compared to untorrefied biomass, except for condition 250-12. In the biomass of *V. ferruginea*, torrefaction reduced T_m under conditions 200-12, 225-8, and 225-10, whereas T_m was higher under the rest of the torrefaction conditions (Table 3) compared with untorrefied biomass. Lastly, torrefaction of the biomass of the five species decreased $T_{offset(c)}$ compared with untorrefied biomass (Table 3).

Table 2. Thermogravimetric analysis (TGA) temperatures of biomasses of five fast-growth plantation species in Costa Rica torrefied at different times and temperatures.

Species	Temperature (°C)	Time (min)	T_i (°C)	T_f (°C)	Residual Mass (%)	$T_{onset(hc)}$ (°C)	$T_{offset(hc)}$ (°C)	T_{sh} (°C)	$T_{onset(c)}$ (°C)	$T_{offset(c)}$ (°C)	T_m (°C)
Cupressus lusitanica	0	0	172.1	448.7	21.0	221.2	455.4	339.3	253.8	465.8	378.8
	200	8	177.2	449.8	22.6	231.2	454.4	345.3	345.5	418.7	383.5
		10	181.7	438.0	24.1	234.0	452.3	340.1	346.4	418.1	383.5
		12	191.7	453.5	23.3	231.4	454.5	345.3	346.1	418.4	382.6
	225	8	215.4	436.2	22.8	230.9	452.4	346.5	344.6	416.4	380.8
		10	183.5	412.6	23.6	237.1	438.2	329.7	340.1	409.7	375.3
		12	173.5	445.3	25.0	245.0	439.1	337.5	341.2	413.6	378.0
	250	8	161.7	437.1	28.3	247.9	440.9	342.7	343.4	415.7	379.8
		10	172.6	425.3	31.3	266.5	425.7	263.5	333.5	413.8	371.7
		12	213.8	476.6	36.1	273.5	415.8	275.2	323.3	410.7	364.4
Dipteryx panamensis	0	0	146.3	460.9	18.7	239.5	375.4	315.0	238.2	468.2	372.1
	200	8	194.7	431.1	19.6	206.9	438.6	310.2	330.3	413.0	373.8
		10	188.2	441.3	21.5	206.4	443.0	307.6	330.4	416.7	375.1
		12	201.2	425.7	20.2	205.2	443.1	306.3	332.4	415.6	375.1
	225	8	212.8	430.9	20.8	211.6	440.9	316.7	333.0	411.3	373.8
		10	205.1	433.5	23.4	211.0	437.9	314.1	331.2	409.2	372.5
		12	206.4	446.5	18.5	212.4	441.3	311.5	331.2	411.9	372.5
	250	8	212.8	433.5	24.1	237.6	428.4	308.9	325.7	409.3	367.3
		10	206.4	438.7	27.2	263.6	420.2	272.6	329.8	407.9	368.6
		12	196.0	424.4	33.5	266.1	413.8	272.6	232.8	404.0	362.1
Gmelina arborea	0	0	172.1	471.5	20.8	249.1	385.0	305.8	258.7	417.9	349.9
	200	8	194.7	410.2	24.4	196.7	392.7	247.9	305.6	395.6	351.7
		10	197.3	401.1	24.4	225.1	410.4	247.9	301.7	401.6	346.5
		12	221.9	419.2	25.5	217.9	417.2	302.4	295.9	391.7	341.4
	225	8	201.2	399.8	26.6	227.0	411.4	303.7	305.3	381.4	341.4
		10	199.9	406.3	26.4	236.3	391.9	240.1	303.9	376.0	338.8
		12	207.7	424.4	50.0	238.5	418.1	262.2	305.2	380.7	344.0
	250	8	173.9	454.3	25.8	247.9	440.9	267.4	343.4	415.7	341.4
		10	198.6	442.6	56.9	236.6	416.5	242.7	295.0	384.6	337.5
		12	179.1	412.8	56.9	188.0	495.9	255.7	294.7	386.6	344.0

Table 2. *Cont.*

Species	Temperature (°C)	Time (min)	T_i (°C)	T_f (°C)	Residual Mass (%)	$T_{onset(hc)}$ (°C)	$T_{offset(hc)}$ (°C)	T_{sh} (°C)	$T_{onset(c)}$ (°C)	$T_{offset(c)}$ (°C)	T_m (°C)
Tectona grandis	0	0	164.5	473.0	19.2	262.6	398.5	321.0	257.8	454.6	368.2
	200	8	233.6	425.7	22.1	226.6	435.2	312.8	322.0	418.8	371.2
		10	219.3	430.9	23.2	233.0	433.8	320.6	330.2	410.6	369.9
		12	227.1	425.7	23.1	225.7	434.4	316.7	320.4	413.4	366.0
	225	8	198.6	438.7	22.1	227.2	441.1	312.8	320.4	413.4	373.8
		10	220.6	423.1	26.5	233.0	430.8	312.8	329.1	409.6	369.9
		12	233.6	443.9	24.2	232.0	434.8	316.7	328.6	411.9	369.9
	250	8	212.8	432.2	25.1	240.2	434.9	308.9	330.8	414.9	373.8
		10	241.4	427.0	28.8	266.0	422.4	286.8	331.9	412.0	369.9
		12	272.6	430.9	36.9	256.6	416.1	298.5	317.6	408.6	363.4
Vochysia ferruginea	0	0	161.5	439.6	22.3	233.7	369.6	301.3	227.2	442.9	339.3
	200	8	219.3	427.0	22.5	220.1	414.8	301.1	301.9	407.7	350.4
		10	233.6	430.9	26.2	222.0	419.1	306.3	306.3	408.1	355.6
		12	223.2	424.4	22.9	217.1	384.1	299.8	282.2	405.5	338.8
	225	8	224.5	420.5	23.7	222.3	415.6	255.7	297.9	408.4	337.5
		10	224.5	401.1	29.6	214.5	407.1	298.5	286.1	394.1	336.2
		12	251.8	432.2	24.8	237.9	401.2	245.3	294.2	400.5	346.5
	250	8	245.3	415.4	24.7	224.4	414.8	297.2	303.4	406.7	350.4
		10	245.1	420.5	30.4	228.4	401.8	253.1	288.4	398.4	340.1
		12	227.1	414.1	35.3	251.7	395.4	268.7	294.8	395.7	342.7

Table 3. TGA residual masses of the biomasses of five fast-growth plantation species in Costa Rica torrefied at different times and temperatures.

Specie	Temperature (°C)	Time (min)	W_{Ti} (%)	W_{Tf} (%)	$W_{Tonset(hc)}$ (%)	$W_{Toffset\,(hc)}$ (%)	W_{Tsh} (%)	$W_{Tonset(c)}$ (%)	$W_{Toffset(c)}$ (%)	W_{Tm} (%)
Cupressus lusitanica	0	0	90.3	21.0	89.7	20.6	65.4	88.1	20.0	38.5
	200	8	91.2	22.6	90.4	22.3	65.3	65.3	24.7	39.8
		10	91.8	24.1	91.0	23.3	68.5	65.7	25.6	40.5
		12	92.0	23.3	91.3	23.2	66.2	65.7	25.7	41.7
	225	88	92.0	22.8	91.7	21.7	65.1	66.1	24.2	40.9
		10	92.0	23.6	91.1	21.8	71.9	66.5	23.9	40.4
		12	93.3	25.0	92.3	25.5	72.9	71.0	27.5	43.8
	250	8	94.0	28.3	92.8	26.8	72.1	71.6	28.8	42.5
		10	93.7	31.3	92.1	31.3	92.3	79.4	32.5	52.7
		12	94.1	36.1	92.7	42.1	92.6	86.1	42.7	62.2
Dipteryx panamensis	0	0	90.9	18.7	89.6	35.5	72.6	89.7	18.4	38.3
	200	8	91.4	19.6	91.3	19.3	74.0	65.6	20.6	36.5
		10	92.4	21.5	92.3	21.4	78.1	69.4	22.8	39.6
		12	92.6	20.2	92.6	19.2	77.8	67.6	20.8	37.7
	225	8	93.1	20.8	93.1	20.2	21.7	68.4	22.1	39.6
		10	93.8	23.4	93.8	23.2	77.0	70.0	25.0	41.1
		12	93.4	18.5	93.3	18.8	77.4	69.1	20.8	39.5
	250	8	93.4	24.1	92.8	24.4	82.9	76.8	25.9	46.4
		10	93.8	27.2	92.3	28.8	91.7	83.8	30.1	50.9
		12	93.9	33.5	92.3	34.5	92.0	93.4	35.3	56.4
Gmelina arborea	0	0	89.8	20.8	88.0	26.0	76.1	87.2	23.5	44.1
	200	8	91.4	24.4	91.4	25.6	89.2	76.4	25.4	45.9
		10	90.8	24.4	90.0	23.8	88.4	75.7	24.4	46.1
		12	92.3	25.5	92.4	25.6	78.4	81.3	27.5	51.0
	225	8	92.2	26.6	91.7	25.7	77.2	76.3	28.2	49.5
		10	92.5	26.4	91.5	27.5	91.3	76.9	29.1	49.0
		12	93.0	50.0	92.3	50.6	91.1	85.1	54.8	67.5
	250	8	93.9	25.8	92.8	26.8	91.6	71.6	28.8	72.5
		10	92.1	56.9	91.1	59.4	90.9	86.5	62.7	74.4
		12	92.4	56.9	92.3	50.0	90.5	87.5	60.0	72.2

Table 3. *Cont.*

Specie	Temperature (°C)	Time (min)	W_Ti (%)	W_Tf (%)	W_Tonset(hc) (%)	W_Toffset (hc) (%)	W_Tsh (%)	W_Tonset(c) (%)	W_Toffset(c) (%)	W_Tm (%)
Tectona grandis	0	0	91.0	19.2	88.8	23.9	74.4	89.2	19.9	43.2
	200	8	91.1	22.1	91.3	21.5	78.5	74.5	22.5	41.5
		10	91.5	23.2	91.2	23.0	76.0	71.7	24.5	44.1
		12	91.9	23.1	91.9	22.5	77.1	75.4	23.9	44.2
	225	8	92.7	22.1	92.4	22.0	79.7	76.7	23.7	43.1
		10	93.1	26.5	92.9	26.1	80.4	73.3	27.4	45.2
		12	92.7	24.2	92.7	24.7	79.5	74.7	26.3	46.2
	250	8	93.7	25.1	93.2	25.0	84.3	76.3	26.5	45.5
		10	93.6	28.8	92.6	29.3	91.0	80.6	30.2	52.9
		12	92.6	36.9	93.2	38.3	89.6	86.5	39.1	58.8
Vochysia ferruginea	0	0	89.6	22.3	88.1	32.3	74.9	88.3	22.2	52.1
	200	8	90.2	22.5	90.2	23.4	76.0	76.0	23.9	45.7
		10	90.8	26.2	91.2	26.9	76.1	76.1	27.6	46.0
		12	91.0	22.9	91.2	26.4	76.9	83.2	24.3	52.0
	225	8	91.4	23.7	91.4	24.1	89.3	78.2	24.7	53.6
		10	92.3	29.6	92.6	29.2	79.6	84.0	30.2	54.6
		12	91.7	24.8	92.4	27.7	92.0	85.9	27.9	52.8
	250	8	91.5	24.7	92.5	24.8	81.6	79.3	25.4	48.9
		10	92.2	30.4	92.9	32.1	91.7	87.2	32.4	58.7
		12	93.5	35.3	92.7	37.4	91.8	88.7	37.4	62.4

With respect to the different torrefaction conditions, the increase in T_T and t_T decreased $T_{onset(c)}$ in *C. lusitanica* and *D. panamensis*. Conversely, in *G. arborea*, the increase in T_T and t_T decreased $T_{onset(c)}$, except under condition 225-8. As for *T. grandis* and *V. ferruginea*, no trend was found in $T_{onset(c)}$ with either an increase or decrease of T_T or t_T (Table 3). T_m and $T_{offset(c)}$ decreased in all species as T_T or t_T increased (Table 3).

The evaluation of the residual mass of the different biomasses showed that torrefaction decreased the $W_{Tonset(c)}$ value in *C. lusitanica* and *T. grandis*, whereas in *D. panamensis*, *G. arborea* and *V. ferruginea*, torrefaction decreased the $W_{Tonset(c)}$ value, except under the most severe condition (250-12) (Table 3). Torrefaction increased T_m in all species, except under condition 200-8 for *D. panamensis* and *T. grandis* and conditions 200-8 and 200-10 for *V. ferruginea*, where T_m decreased. Lastly, $W_{Toffset(c)}$ was higher in torrefied biomass than in untorrefied biomass for all species (Table 3). The evaluation of the residual mass under the different torrefaction conditions showed that all parameters related to residual mass ($W_{Tonset(c)}$, W_{Tm}, and $W_{Toffset(c)}$) increased their values with increasing T_T and t_T of torrefaction in all species (Table 3).

Table 4 shows the kinetic parameters of hemicellulose and cellulose decomposition in torrefied biomass observed with TGA. In the torrefied biomass of *C. lusitanica*, the activation energy (Ea) value of hemicellulose increased with the increase in T_T and t_T up to 225 °C; however, in torrefaction at 250 °C for 10 and 12 min, the Ea values were lower. The Ea cellulose values for the biomass of *C. lusitanica* increased with the increase in T_T and t_T, requiring more energy to degrade the cellulose in the biomass. For Ea, the torrefied biomasses at 200 °C and 225-8 had lower values than the untorrefied biomass; then Ea increased with T_T and t_T, decreasing again under the 250-12 condition.

The torrefaction increased the pre-exponential factor (A) and Ea of the hemicellulose in the *D. panamensis* biomass compared to the untorrefied biomass. The A and Ea increased from 200-8 to 225-12, and decreased at 250 °C. For cellulose, the A and Ea in the torrefied biomass decreased with the increase in T_T and t_T, but any torrefaction produced lower values of A and Ea compared to the untorrefied biomass.

The *G. arborea* torrefied biomass had lower A and Ea hemicellulose values than the untorrefied biomass. Torrefaction at 200 °C had high Ea hemicellulose values, whereas the Ea decreased significantly with T_T and t_T above 225 °C. For cellulose, Ea increased as T_T and t_T increased, especially under condition 225-10 (Table 4). The torrefied biomass had higher Ea values compared to untorrefied biomass, except for conditions 200-8 and 200-10. In general, torrefaction Ea increased with increasing T_T and t_T.

With *T. grandis*, the hemicellulose A and Ea increased up to torrefaction condition 225-12, with values greater than those for untorrefied biomass. Beyond these torrefaction conditions, A and Ea decreased in the torrefied biomass at 250 °C, with values lower than found for the untorrefied biomass. For cellulose, Ea was lower for the different types of torrefied biomass, except under condition 250-12. Ea was also lower in torrefied biomass under 225-8, 225-10, and 250-8 conditions. Ea increased with T_T for the other temperatures. However, Ea was greater for all torrefied biomasses, increasing as T_T and t_T increased (Table 4).

The torrefied biomass of *V. ferruginea* had lower Ea values for hemicellulose than the untorrefied biomass. Ea in hemicellulose decreased with decreasing T_T for the different torrefaction conditions, whereas at the same temperature, Ea decreased at 10 and 12 min. For cellulose, the Ea value was lower as T_T decreased except for conditions 225-10 and 250-12, in which Ea was higher in the torrefied biomass. For the EA value for cellulose for different T_T, Ea increased with T_T, excluding conditions 200-10 and 225-8, which had a low Ea value.

Notably, the correlation coefficients (R^2) for all torrefaction conditions remained close to 0.99, with the exception of the cellulose models for *G. arborea*, which were low.

Table 4. Activation energies and pre-exponential factors for the thermal decomposition of hemicellulose and cellulose observed in biomasses of five fast-growth plantation species in Costa Rica torrefied at different times and temperatures.

Species	Temperature (°C)	Time (min)	Hemicellulose			Cellulose		
			A	Ea	R^2	A	Ea	R^2
Cupressus lusitanica	0	0	2×10^9	77.9	0.999	4×10^{19}	158.3	0.955
	200	8	3×10^9	78.5	0.995	8×10^7	68.2	0.999
		10	4×10^9	79.8	0.993	9×10^7	68.5	0.999
		12	5×10^9	80.8	0.994	5×10^7	65.7	0.999
	225	8	2×10^{10}	87.0	0.996	1×10^9	81.4	0.999
		10	1×10^{10}	85.1	0.996	3×10^{17}	177.7	0.997
		12	6×10^9	84.3	0.998	9×10^{15}	160.3	0.999
	250	8	3×10^9	82.1	0.999	1×10^{16}	161.1	0.998
		10	5×10^8	76.0	0.998	2×10^{25}	267.6	0.973
		12	6×10^7	68.2	0.996	6×10^{19}	201.1	0.999
Dipteryx panamensis	0	0	2×10^8	66.3	0.979	2×10^8	324.7	0.977
	200	8	2×10^{12}	105.1	0.997	1×10^{12}	113.4	0.997
		10	4×10^{13}	118.2	0.997	3×10^{12}	119.1	0.997
		12	3×10^{13}	117.6	0.997	6×10^{13}	133.8	0.997
	225	8	1×10^{14}	124.5	0.998	7×10^{15}	157.1	0.994
		10	1×10^{14}	123.1	0.998	1×10^{19}	193.4	0.985
		12	1×10^{13}	113.5	0.998	4×10^{16}	164.9	0.991
	250	8	3×10^{10}	89.7	0.999	6×10^{18}	189.2	0.989
		10	4×10^8	75.0	0.999	9×10^{21}	227.4	0.982
		12	2×10^7	63.5	0.998	2×10^{28}	299.9	0.946
Gmelina arborea	0	0	3×10^{12}	109.6	0.914	9×10^{15}	146.8	0.777
	200	8	7×10^8	72.3	0.989	2×10^9	77.6	0.993
		10	8×10^9	81.8	0.994	2×10^{11}	96.8	0.986
		12	1×10^{11}	94.1	0.999	4×10^{22}	223.1	0.899
	225	8	3×10^{10}	89.1	1.000	4×10^{24}	244.7	0.810
		10	4×10^{10}	90.2	0.999	2×10^{32}	327.9	0.729
		12	6×10^6	56.7	0.997	2×10^{29}	298.7	0.760
	250	8	3×10^8	73.1	0.997	1×10^{16}	161.1	0.998
		10	1×10^6	51.5	0.999	2×10^{12}	112.7.7	0.999
		12	9×10^2	24.5	0.998	5×10^{27}	280.2	0.856
Tectona grandis	0	0	2×10^9	79.3	0.997	4×10^{22}	143.6	0.991
	200	8	4×10^{11}	100.7	1.000	9×10^{14}	144.9	0.994
		10	3×10^{11}	100.1	1.000	6×10^{16}	167.4	0.990
		12	1×10^{12}	105.5	1.000	5×10^{19}	198.6	0.971
	225	8	3×10^{12}	109.4	0.999	4×10^{12}	118.9	0.996
		10	2×10^{12}	108.0	0.999	7×10^{12}	121.4	0.994
		12	2×10^{12}	108.5	0.999	4×10^{15}	153.0	0.993
	250	8	4×10^{11}	101.83	0.999	4×10^{13}	130.32	0.998
		10	4×10^8	75.58	0.999	7×10^{16}	168.65	0.997
		12	9×10^8	79.99	0.949	9×10^{28}	306.93	0.935
Vochysia ferruginea	0	0	9×10^{10}	93.16	0.998	3×10^{26}	225.20	0.901
	200	8	6×10^8	71.85	0.996	7×10^{11}	104.78	1.000
		10	6×10^9	81.67	0.999	2×10^9	78.04	0.999
		12	4×10^8	69.73	0.971	4×10^{24}	244.37	0.996
	225	8	4×10^{10}	88.96	0.999	1×10^9	71.99	0.995
		10	5×10^9	80.92	0.999	1×10^{30}	303.55	0.985
		12	3×10^7	63.22	0.998	1×10^{20}	198.03	0.993
	250	8	2×10^{11}	83.23	0.997	1×10^{13}	105.55	0.990
		10	1×10^8	68.60	1.000	1×10^{27}	274.73	0.981
		12	2×10^6	53.10	0.999	1×10^{36}	376.89	0.958

3.2. Devolatilization

Figure 4 displays the devolatilization rate of the torrefied and untorrefied biomasses. Table 5 shows when D_{max} was reached and the D_{max} values. For the *C. lusitanica* biomass, torrefaction at 250-10 and 250-12 had lower D_{rate} values (Figure 5a) and reached D_{max} more quickly (Table 5), whereas D_{max} increased between 200-8 and 225-12, and then decreased in biomass torrefied at 250 °C.

The *D. panamensis* biomass torrefied at 250 °C at the three temperatures had the lowest devolatilization rate. In addition, the shoulder in the devolatilization curve at 13 min disappeared in the biomass torrefied at 250 °C (Figure 4b). The time to reach D_{max} showed no significant variation, except for condition 250-12 where the time required was shorter and D_{max} increased with T_T and t_T, except for condition 250-12, where again the value was low (Table 5). *G. arborea* had the lowest devolatilization rate for all torrefactions, and especially for 225-10, 225-12, and 250 °C in the three t_T (Figure 4c). The time to reach D_{max} was approximately 16 min in the different types of biomass. However, the shortest time was obtained with 250-10 (Table 5). The D_{max} value increased at 225 °C, but decreased at 250 °C.

Table 5. Time to reach the maximum devolatilization rate determined by thermogravimetric analysis (TGA) experiments of the biomasses of five fast-growth plantation species in Costa Rica torrefied at different times and temperatures.

Species	Temperature (°C)	Time (min)	Time Max. (min)	D_{max} (% wt/min)
	0	0	18.33	17.3
	200	8	18.7	14.7
		10	18.8	14.1
		12	18.7	14.8
Cupressus lusitanica	225	8	18.6	16.1
		10	18.3	19.2
		12	18.3	18.8
	250	8	18.3	18.4
		10	17.0	8.2
		12	17.6	15.1
	0	0	18.0	18.6
	200	8	18.3	18.0
		10	18.3	18.2
		12	18.3	19.5
Dipteryx panamensis	225	8	18.3	20.7
		10	18.3	20.7
		12	18.2	21.1
	250	8	18.0	22.7
		10	18.0	20.5
		12	17.8	16.1
	0	0	16.7	16.9
	200	8	17.0	16.6
		10	16.8	14.5
		12	16.5	17.3
Gmelina arborea	225	8	16.6	17.9
		10	16.4	20.2
		12	16.6	14.0
	250	8	16.5	18.1
		10	14.4	13.7
		12	16.6	8.4

Table 5. *Cont.*

Species	Temperature (°C)	Time (min)	Time Max. (min)	D_{max} (% wt/min)
Tectona grandis	0	0	17.7	15.8
	200	8	17.9	14.6
		10	17.9	15.0
		12	17.7	15.8
	225	8	18.2	19.6
		10	18.1	19.3
		12	18.0	20.1
	250	8	18.1	20.2
		10	18.0	21.8
		12	17.6	16.5
Vochysia ferruginea	0	0	16.5	14.5
	200	8	17.0	15.7
		10	17.1	14.8
		12	16.4	16.2
	225	8	16.4	14.5
		10	16.4	16.6
		12	16.8	16.4
	250	8	17.0	17.3
		10	16.6	16.9
		12	16.4	17.9

The *T. grandis* torrefied biomass had a D_{rate} above 20 dw/dt, whereas torrefactions at 250-10 and 250-12 displayed no inflexion at 13 min (Figure 3d). The maximum devolatilization was reached at 17 min for the untorrefied biomass and 200 °C and 250-12 for torrefied biomass. Under other torrefaction conditions, the time exceeded 18 min. D_{max} increased with increasing T_T and t_T of torrefaction, with the exception of 250-12, which had a low D_{max} value (Table 5).

Under conditions 200-8, 200-10, and 225-8, the torrefied *V. ferruginea* had a lower D_{rate} relative to the torrefied and untorrefied biomass under the other conditions (Figure 4e). The time to reach D_{max} was close to 17 min, with a slight increase in the value of D_{max} with increasing T_T and t_T of torrefaction (Table 5).

Notably, in all species, once D_{max} was reached, the slope of the curve became more severe, with a steeper slope (Figure 4a–e).

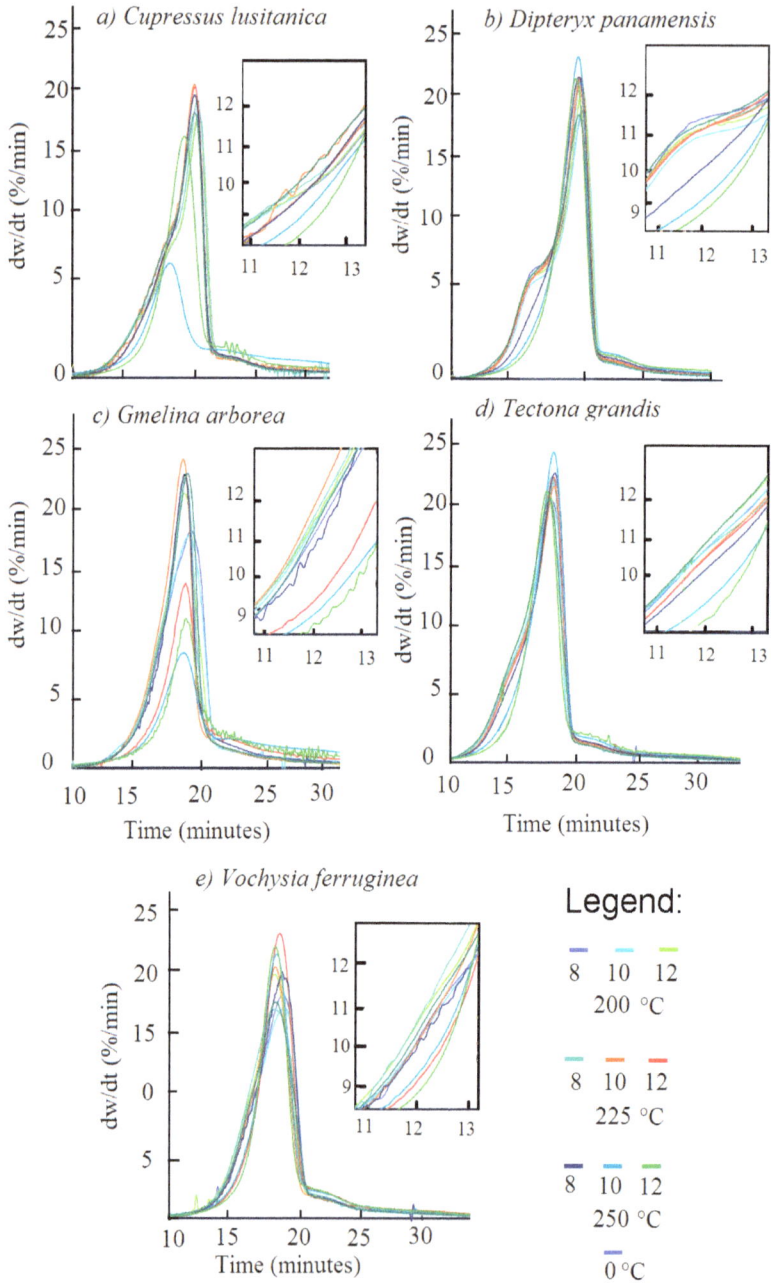

Figure 4. Devolatilization rate measured by the first derivative of the mass fraction with respect to time for the biomasses of for *Cupressus lusitanica* (**a**), *Dipteryx panamensis* (**b**), *Gmelina arborea* (**c**) and *Tectona grandis* (**d**) and *Vochysia guatemalensis* (**e**), torrefied at different times and temperatures.

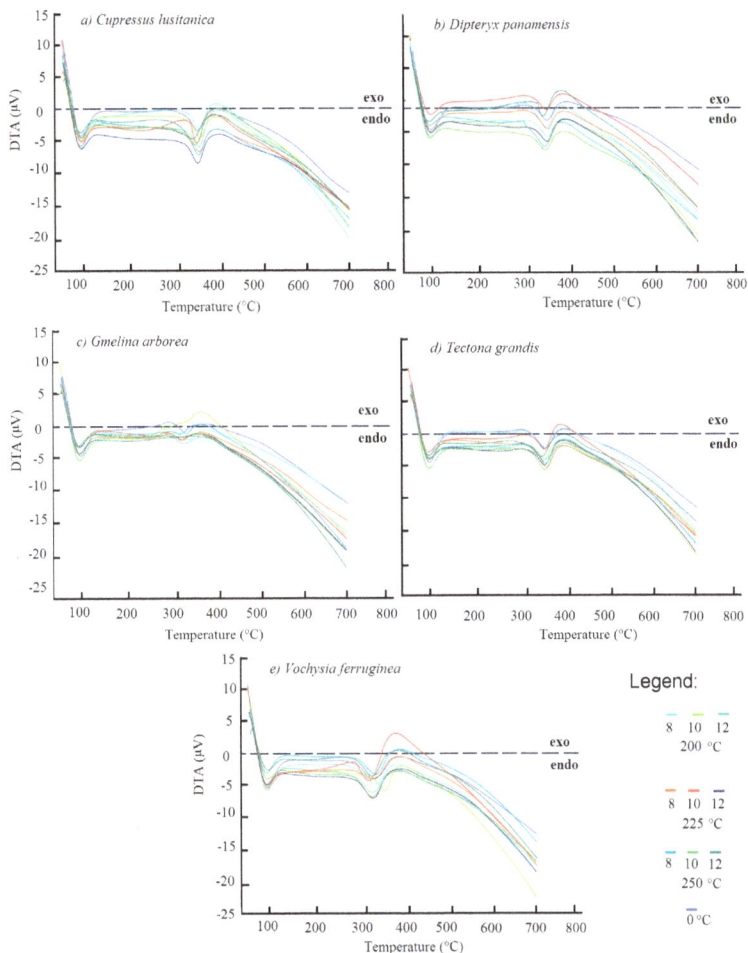

Figure 5. Differential Scanning Calorimetry (DSC) analysis of the torrefaction at different times and temperatures for for *Cupressus lusitanica* (**a**), *Dipteryx panamensis* (**b**), *Gmelina arborea* (**c**), *Tectona grandis* (**d**) and *Vochysia guatemalensis* (**e**).

3.3. Differential Scanning Calorimetry Analyses

Figure 5 displays the DTG curves of the calorimetric analysis of the reactions that occurred during the TGA. In all torrefaction conditions and for all woody species, the first endothermic peaks occurred at 100 and 300 °C, whereas exothermic peaks were observed between 350 and 400 °C, with some variations among the species and torrefaction conditions. All torrefaction conditions demonstrated endothermic processes in the *C. lusitanica* biomass. However, untorrefied biomass for condition 200-8 had a more pronounced exothermic peak between 350 and 450 °C (Figure 5a). In the *D. panamensis* biomass, torrefaction at 200-8, 200-12, 225-8, and 250-12 demonstrated exothermic processes between 350 and 450 °C, whereas endothermic peaks of greater magnitude were observed in biomass torrefied at 225-12, 250-10, and 250-12 (Figure 5c). Exothermic reactions occurred between 300 and 400 °C in *G. arborea* untorrefied biomass and with torrefaction at 200-8 and 225-10, whereas the biomass under the other torrefaction conditions only showed endothermic reactions (Figure 4c). For the *T. grandis* biomass, the exothermic reactions between 350 and 450 °C appeared in untorrefied biomass and under conditions 200-8 and 225-10 (Figure 5d). The opposite occurred in the biomass of *V. ferruginea*, as torrefied biomass presented exothermic peaks under severe torrefaction conditions (225-10, 250-10, and 250-12) (Figure 6e).

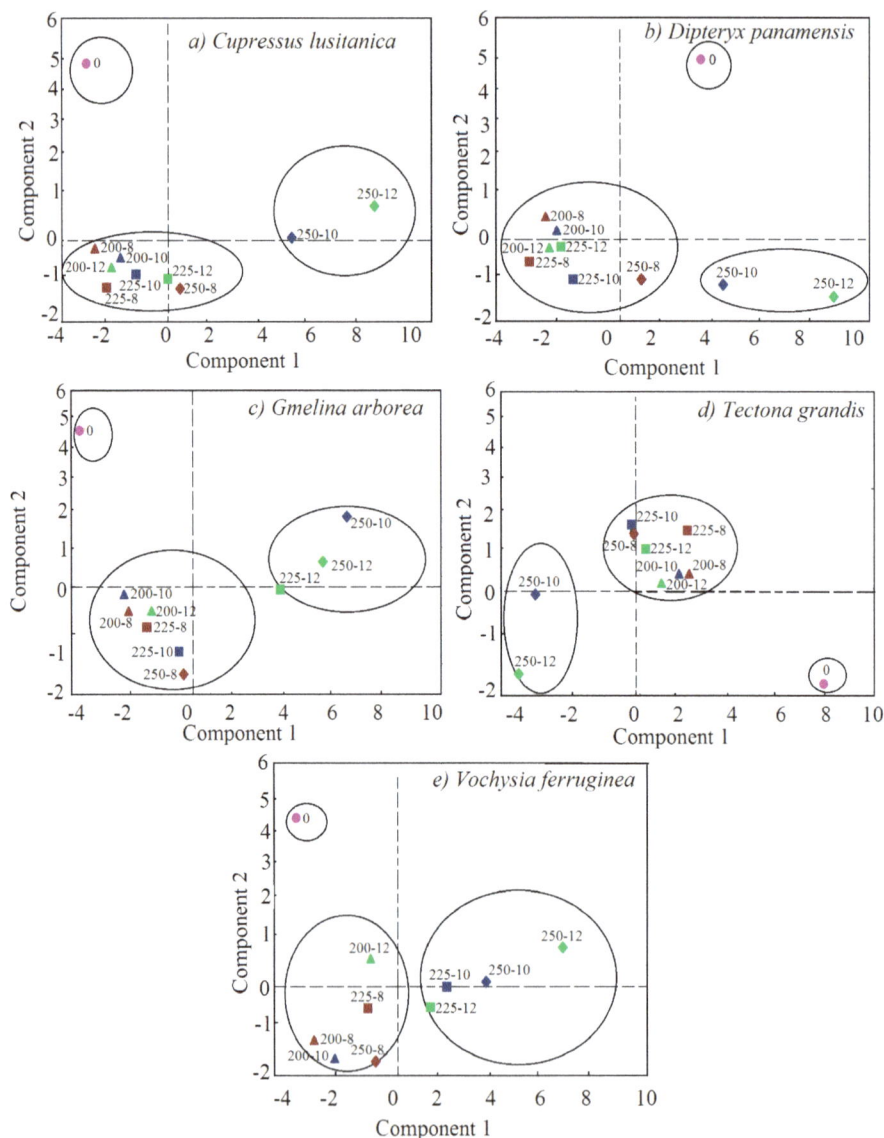

Figure 6. Relationship between the auto-vector of components 1 and 2 of the multivariate analysis by means of principal components of for *Cupressus lusitanica* (**a**), *Dipteryx panamensis* (**b**), *Gmelina arborea* (**c**) and *Tectona grandis* (**d**) and *Vochysia guatemalensis* (**e**), torrefied at different times and temperatures of five fast-growth plantations in Costa Rica.

3.4. Multivariate Analysis

Table 6 shows the MAPC that determined that the first two components represented approximately 60% of the total variation in the evaluated variables, of which 45% was explained by principle component one (PC1). In general, the variables influencing these components are related to hemicellulose for PC1, and cellulose for PC2. However, a slight variation was found between the different species (Table 6). For *C. lusitanica*, PC1 mainly included the hemicellulose-related variables, such as $T_{onset(hc)}$, $T_{offset(hc)}$, $W_{Tonset(hc)}$, $W_{Toffset(hc)}$, T_{sh}, and W_{Tsh}, and PC2 included cellulose variables such as $T_{onset(c)}$, $T_{offset(c)}$, and $W_{Tonset(c)}$. For *D. panamensis*, the variables that most influenced PC1 were the same as for *C. lusitanica* long with the Ea of hemicellulose and cellulose. In PC2, $T_{offset(c)}$, T_i, T_f, $W_{Toffset(c)}$, and W_{Ti} were the most influential. For *G. arborea*, the variables representing PC1 were percentage of residual mass, W_{Ti}, W_{Tm}, and W_{Tf}, whereas PC2 included $W_{Tonset(hc)}$, $W_{Tonset(c)}$, and W_{Ti}. The components of *T. grandis* included T_{sh}, T_i, mass residual, W_{Tsh}, W_{Tm}, and W_{Tf} for PC1 and by $T_{onset(hc)}$, $T_{offset(hc)}$, T_{sh}, and $W_{Tonset(c)}$ for PC2. In *V. ferruginea*, PC1 included the percentage of residual mass, $W_{Tonset(hc)}$, $W_{Toffset(c)}$, W_{Ti}, W_{Tm}, and W_{Tf}, and PC2 included $T_{offset(hc)}$, $T_{onset(c)}$, T_i, and $W_{Tonset(c)}$.

By plotting the auto-vector for PC1 and PC2 for each species (Figure 6), we identified three different groups. In *C. lusitanica*, *D. panamensis* and *T. grandis*, the first group included torrefactions under 200 °C, 225 °C, and 250-8 °C; the second group included conditions 250-10 and 250-12; whereas the untorrefied biomass behaved differently compared with the other groups (Figure 5a–d). In *G. arborea*, the first group included torrefactions under 200 °C, 225-8, 225-10, and 250-8, whereas the second group included 225-12, 250-10, and 250-12. Similarly, the untorrefied biomass behaved differently compared with the other torrefactions (Figure 5c). The first *V. ferruginea* group was formed by torrefactions under 200 °C, 225-8, and 250-8; whereas the second group included 225-10, 225-50, 250-10, and 250-12. Untorrefied biomass had no similarities to any of the torrefactions (Figure 6e).

Table 6. Matrix of the multivariate analysis correlations for all variables evaluated of biomass torrefied at different times and temperatures of five fast-growth plantations species in Costa Rica.

Variable	Cupressus Lusitanica		Dipteryx panamensis		Gmelina arborea		Tectona grandis		Vochysia ferruginea	
	C1	C2	C1	C2	C1	C2	C1	C2	C1	C2
T_i (°C)	-	-	-	-0.92 **	-	-	-0.83 **	-	-	-0.80 **
T_m (°C)	-0.90 **	-	-0.89 *	-	-	-	-	-	-	-
T_f (°C)	-	-	-	0.79 **	-	-	-	-0.71 *	-	-
T_{sh} (°C)	-0.91 **	-	-0.78 **	-	-	-	0.83 **	-	-	-
$T_{offset(hc)}$ (°C)	-0.96 **	-	-0.69 **	-0.69 *	-	-	-	0.94 **	-	-0.93 **
$T_{onset(c)}$ (°C)	-	-0.97 **	-0.78 **	-	-	-	-	0.88 **	-	-0.94 **
$T_{offset(c)}$ (°C)	-	0.89 **	-	0.97 **	-	-0.87 **	-	-0.78 **	-0.69 *	-
$T_{onset(hc)}$ (°C)	0.97 **	-	0.95 **	-	-	-	-0.91 **	-0.69 *	-	-
WT_{sh} (%)	0.97 **	-	-	-0.87 **	-	-	-	-	0.76 *	-
WT_i (%)	0.81 *	-	-	-	-	-0.70 *	-	-	0.96 **	-
WT_m (%)	0.98 **	-	0.85 **	-	0.87 **	-	-0.96 **	-	0.87 **	-
WT_f (%)	0.98 **	-	0.80 **	-	0.96 **	-	-0.98 **	0.71 *	0.91 **	-
$WT_{onset(hc)}$ (%)	0.69 *	-	-	-0.85 **	-	-0.76 *	-0.66 *	-	0.78 **	-
$WT_{offset(hc)}$ (%)	0.97 **	-	0.91 **	-	0.93 **	-	-0.93 **	-	0.68 *	-
$WT_{onset(c)}$ (%)	-	0.79 *	0.96 **	-	-	0.79 *	-	-0.88 **	-	0.74 *
$WT_{offset(c)}$ (%)	0.96 **	-	0.75 *	-0.64 *	0.96 **	-	-0.98 **	-	0.94 **	-
Ea Hemicellulose	-0.72 *	-	-0.92 **	-	-0.93 **	-	-	0.75 *	-0.73 *	-
Ea Cellulose	0.73 *	-	0.86 **	-	0.69 *	-	-0.74 *	-	0.75 *	-
Residual mass (%)	0.98 **	-	0.80 *	-	0.96 **	-	-0.98 **	-	0.91 **	-
Time max (min)	-0.77 *	-	-0.95 **	-	-0.64 *	-	-	0.74 *	-	-
Rate max (wt/%)	-	-	-	-	-0.83 **	-	-	-	0.79 *	-
Percentage of variance	60.88	16.46	52.18	31.49	44.52	18.28	46.36	32.63	47.45	26.66
Cumulative variance	60.88	77.35	52.18	83.67	44.52	62.80	46.36	78.99	47.45	74.11

Note: C1: correlations of component 1; C2: correlations of component 2. * Significance at 95%, ** significance at 99%, - not present significance.

4. Discussion

4.1. TGA-DTG Analysis

In general, TGA trends for torrefied and untorrefied biomass of the different woody species were similar, which is consistent with previous reports [18,20,21]. The DTG curves support this finding, where important stages were defined (Figure 3a–e) and differences were clarified.

During thermogravimetric analyses, the initial decrease in mass is attributed to the release of the moisture in the samples [28]. This water release is lower for biomasses torrefied under more severe conditions, consistent with a higher drying temperature and an increase in the hydrophobicity related to such conditions [12]. Higher temperatures enable the decomposition of the polymers present in the samples. Hemicellulose degradation occurs between 230 and 330 °C [29]. This degradation mainly tends to disappear under severe torrefaction conditions for all five species because a higher percentage of hemicellulose has already been eliminated during torrefaction [4,28,30,31]. Then, cellulose decomposition occurs at temperatures between 305 and 380 °C [28,32], which appears in all the biomasses analyzed considering that torrefaction processes affect this biopolymer less than hemicellulose. Temperatures between 400 and 500 °C cause the final decomposition of cellulose and most of the lignin [33]. During this stage, the decomposition rate slows and then continues to a period of limited change as temperature increases.

The torrefied biomass displays the four decomposition stages of the well-defined components (Figure 3a–e). The first signal in the DTG curves before 150 °C is attributable to the removal of moisture in the samples, since moisture decreases with T_T [12]. The next signal or decomposition stage between 230 and 330 °C is due to hemicellulose degradation [29]; however, contrary to the untorrefied biomass, this signal tends to disappear under severe torrefaction in all five species (Figure 3). This is because high percentages of hemicelluloses have already been eliminated in the process prior to torrefaction [28,30]. This result agrees with the work reported by Bach et al. [31] and Ren et al. [4], who torrefied the biomass of conifers under temperatures above 250 °C and found that the signal decreased in the TGA curve. The next peak in the curve is related to cellulose decomposition, which occurs in the range of 305 to 380 °C [28,32]. This curve occurs in all torrefaction conditions and in untorrefied biomass, with differences in the magnitude of the peak, evidenced by weight loss. Lastly, in the final stage between 400 and 500 °C, the rate of decomposition slows, which is attributable to the final decomposition of cellulose and most of the lignin [33].

Using the parameters for material degradation (T_i, W_i) and hemicellulose degration ($T_{onset(hc)}$, $W_{Tonset\ (hc)}$, T_{sh}, $T_{offset(hc)}$, and $W_{Toffset(hc)}$)), the evaluation of stages two and three of the TGA curve shows that an increase of T_T and t_T increase T_i (Tables 2 and 3), indicating that torrefied biomass is more thermally stable than untorrefied biomass, which agrees with results found by Lee et al. [34] and Islam et al. [35] when evaluation some tropical species (*Dyera costulata*, *Esdospermun diadenum*, *Paraserianthes moluccana*, *Hevea brasiliensis*, and *Alstonia pneumatophora*). This result also indicates that an increase in T_T and t_T stabilizes the biomass, leading to a reduction in the mass loss values ($W_{Tonset(hc)}$ and $W_{Toffset(hc)}$) of this component (Table 4). Nevertheless, this behaviour should be viewed cautiously, as some authors indicated that this relationship is the result of the content of extractives in the wood, and the volatile material [27], affecting the combustion process [36]. In fact, Gaitán-Alvarez et al. [12] showed that, with these same woody species, weight loss during torrefaction is correlated with the type and content of extractives.

As for the cellulose degradation parameters, the evaluation showed that the most important differences were in temperature and residual mass at different T_T and t_T (Tables 2 and 3). The temperature parameters ($T_{onset(c)}$, T_m, and $T_{offset(c)}$) increased as T_T and t_T increased, except in *G. arborea* (Table 2), again indicating that this biomass component has higher thermal stability than in torrefied biomass, causing a reduction in weight loss values in the different stages of the evaluated cellulose decomposition evaluated ($W_{Tonset(c)}$, W_{Tm}, and $W_{Toffset(c)}$).

Some of the differences found in the species, or in the behaviour of the parameters evaluated in the decomposition of cellulose in torrefied biomass between species (Tables 2 and 3), show that the decomposition process of cellulose is complex in both torrefied and untorrefied biomass. The thermal stability of cellulose is related to the natural variation in the material, having a more chemically complex structure than hemicellulose [32,37]. Therefore, the parameters evaluated in the five species differ among them, even under the torrefaction conditions (Tables 3 and 4).

The activation energy was expected to increase with increasing T_T [38]. However, each structural component of the biomass has its own behaviour due to its chemical nature [39]. In the first stage, early degradation of hemicellulose appears during the torrefaction process [39]. Here, low activation energy is required to initiate hemicellulose degradation compared to activation energy of cellulose [38]. Hemicellulose starts to decompose at low temperatures, between 180 and 350°C [40]. Ramos [41] indicated that xylan, a type of hemicellulose, depolymerizes and reduces hemicelluloses into smaller molecules with lower molecular weight, which are more sensitive to pyrolysis [42]. Thus, with an increase in the torrefaction conditions in T_T and t_T, the hemicellulose decomposes in monosaccharides and volatilizes more rapidly [42]. For this reason, torrefied biomass has a low percentage of hemicellulose. Within the same woody species, Gaitán-Alvarez et al. [12] found that the weight loss of hemicellulose increases with T_T and t_T. Then, given the low hemicellulose content in the biomass torrefied under severe conditions, the activation energy is lower as T_T and t_T increase (Table 3). This result coincides with the studies of Bach et al. [43] on Norway spruce, Bobleter [44] on plants, and Wyman et al. [45] on biomass. These authors found that an increase in T_T significantly decreases the Ea in hemicellulose.

Cellulose degradation requires higher energy [38]. Biomass torrefaction increases the Ea value for cellulose (Table 4), since the thermal process increases the order of the regions of cellulose [40]. This means that heat transportation is more difficult [46], so the thermal stability of the biomass is greater [38]. This behaviour was observed in the biomasses studied, where the Ea for cellulose increased with increasing T_T and t_T, particularly under severe torrefaction conditions (Table 4).

4.2. Devolatilization

D_{max} is associated with the activation energy of cellulose decomposition. Higher Ea makes the polymer decomposition process more difficult, which is reflected in the lower D_{max} values and vice versa (Tables 4 and 5). Likewise, the decrease in devolatilization rates at higher temperatures at 250-10 and 250-12 (Figure 4a–e, Table 5) is attributed to the fact that at these T_T, a high proportion of hemicellulose has been degraded [27,46], leaving a low percentage of hemicellulose and less cellulose to devolatilize when the biomass is used for energy production. Likewise, a reduction in the devolatilization rate at the higher temperatures of 250-10 and 250-12 (Figure 4a–e, Table 5) is attributed to the degradation of a high proportion of hemicellulose at these T_T [27,46], leaving a low percentage of hemicellulose and less cellulose for devolatilization when the biomass is used for energy production.

The differences found in the devolatilization and D_{max} values among the various species (Figure 4a–e, Table 5) are associated with the proportion and nature of the hemicellulose and cellulose contained in the biomass, since each species has its unique behaviour and chemical structure, and therefore its own pyrolysis characteristics [18].

Chen et al. [5] showed that an increase in T_T and t_T affects D_{max}, without affecting the time to reach maximum devolatilization, with differences of approximately 2 min (Table 5). This behaviour indicates that, in torrefied biomass, the decomposition of cellulose (the component associated with maximum devolatilization) and the time to reach maximum devolatilization are maintained, whereas thermal stability of the torrefied biomass causes values of D_{max} to vary.

4.3. Differential Scanning Calorimetry Analyses

At temperatures below 200 °C, all DSC curves of the five species show endothermic values, which is linked to the energy biomass needs to absorb to remove moisture [32]. Later, the exothermic peaks at 275 °C correspond to degradation of hemicellulose, while yhe peak at 365 °C corresponds to ligninn [32,47]. The endothermic peak close to 355 °C corresponds to degradation of cellulose [32]. Figure 4a–e clearly shows the processes previously described.

Although all torrefied biomasses of the different species display the exothermic processes of hemicellulose and lignin and the endothermic process of cellulose [48,49], the different behavior of each species with respect to torrefaction conditions are evident. For the torrefied biomass of *C. lusitanica* (Figure 4a), the endothermic peaks at 375 °C are more pronounced than for the other species, indicating the greater stability of the cellulose in this species [38]. Conversely, in the torrefied biomass of *D. panamensis* and *T. grandis* (Figure 5b,d), the endothermic peaks at 375 °C are small or less pronounced, occurring mainly in the torrefaction conditions above 225-12, meaning that under these torrefaction conditions, cellulose is less stable [38].

The exothermic peaks at 275 °C corresponding to hemicellulose [48,49], are less pronounced or absent in some torrefied biomasses (Figure 5a–e), and especially in the biomass of *G. arborea* under all torrefaction conditions (Figure 5c). In the remainder of the species, this behaviour mainly appears under torrefaction conditions above 225-10 (Figure 5a,b,d,e). This is because at those T_T, part of the hemicellulose was removed during the torrefaction process [5]. Therefore, the exothermic peak with severe torrefaction is unclear [35].

4.4. Multivariate Analysis

The variables related to hemicellulose ($T_{onset(hc)}$, $T_{offset(hc)}$, $W_{Tonset(hc)}$, $W_{Toffset(hc)}$, T_{sh}, and W_{Tsh}) form PC1, whereas PC2 is related to the cellulose parameters (Table 6). This demonstrates that the behaviour of torrefied biomass at different T_T and t_T can be classified relative to the content of these components. In addition, these two components reflect the thermal stability of the torrefied biomass, as these were statistically and significantly reflected in the principal components (Table 3). However, the relationships between the principal components and the hemicellulose or cellulose parameters may not always be significant under some torrefied biomass conditions, likely due to the nature and quantity of these components in the biomass [18,21].

The scores of the components of the different types of biomass under torrefaction conditions display the different T_T and t_T conditions of hemicellulose and cellulose (Figure 6a,e, respectively). *C. lusitanica*, *D. panamensis*, and *T. grandis*, likely due to greater thermal stability under severe torrefaction conditions (250-10 and 250-12), form a unique group, different from the group formed with biomasses torrefied under light and middle torrefaction, which have similar conditions amongst the two groups. The torrefied biomass of *G. arborea* and *V. ferruginea* of form a group with biomasses torrefied under these conditions, different from biomass torrefied at 200 and 225 °C probably due to lesser thermal stability under condition 225-10 [5,18,21]. Then, the group formed by the different types of biomass torrefied under light and middle conditions, at 200 and 225 °C, respectively, indicate that these are the appropriate torrefaction conditions for those species, since they have the most appropriate parameters for combustion, such as positive correlation with D_{max} thermal stability (Table 6).

5. Conclusions

Based on our results, we conclude that the best torrefaction temperatures and times for the tested species are 200 °C for 8, 10, and 12 min and 225 °C for 8, 10, and 12 min, classified as light and medium torrefaction. Under these conditions, optimum thermo-chemical degradation is achieved for using biomass as an energy source, without significantly affecting the chemical composition of the material. In all species, severe torrefaction at 250 °C produced important degradation of the material, especially hemicellulose and part of the cellulose. As such, we do not recommend the use of this temperature in the biomass torrefaction of tropical species. However, behaviour among species presents some differences *C. lusitanica*, *D. panamensis* and *T. grandis* showed higher thermal stability that *G. arborea* and *V. ferruginea*.

Acknowledgments: The author thank the Vicerrectoría de Investigacion y Extension of the Instituto Tecnologico de Costa Rica and the Comsejo Nacional de Rectores (CONARE) for their financial support.

Author Contributions: Ana Rodriguez-Zúñiga and Róger Moya conceived and designed the experiments; Ana Rodriguez-Zúñiga and Róger Moya performed the experiments; Johanna Gaitán-Álvarez and Allen Puente-Urbina analyzed the data; Ana Rodriguez-Zúñiga and Róger Moya contributed reagents/materials/analysis tools; Johanna Gaitán-Álvarez and Róger Moya wrote the paper.

Conflicts of Interest: The authors declare no conflict of interest.

References

1. Yue, Y.; Singh, H.; Singh, B.; Mani, S. Torrefaction of sorghum biomass to improve fuel properties. *Bioresour. Technol.* **2017**, *232*, 372–379. [CrossRef] [PubMed]
2. Puig-Arnavat, M.; Shang, L.; Sárossy, Z.; Ahrenfeldt, J.; Henriksen, U.B. From a single pellet press to a bench scale pellet mill—Pelletizing six different biomass feedstocks. *Fuel Process. Technol.* **2016**, *142*, 27–33. [CrossRef]
3. Eseltine, D.; Thanapal, S.S.; Annamalai, K.; Ranjan, D. Torrefaction of woody biomass (Juniper and Mesquite) using inert and non-inert gases. *Fuel* **2013**, *113*, 379–388. [CrossRef]
4. Ren, S.; Lei, H.; Wang, L.; Bu, Q.; Chen, S.; Wu, J. Thermal behaviour and kinetic study for woody biomass torrefaction and torrefied biomass pyrolysis by TGA. *Biosyst. Eng.* **2013**, *116*, 420–426. [CrossRef]
5. Chen, W.H.; Peng, J.; Bi, X.T. A state-of-the-art review of biomass torrefaction, densification and applications. *Renew. Sustain. Energy Rev.* **2015**, *44*, 847–866. [CrossRef]
6. Medic, D.; Darr, M.; Shah, A.; Potter, B.; Zimmerman, J. Effects of torrefaction process parameters on biomass feedstock upgrading. *Fuel* **2012**, *91*, 47–154. [CrossRef]
7. Wang, G.; Luo, Y.; Deng, J.; Kuang, J.; Zhang, Y. Pretreatment of biomass by torrefaction. *Chin. Sci. Bull.* **2011**, *56*, 1442–1448. [CrossRef]
8. Peng, J.H.; Bi, X.T.; Sokhansanj, S.; Lim, C.J. Torrefaction and densification of different species of softwood residues. *Fuel* **2013**, *111*, 411–421. [CrossRef]
9. Van der Stelt, M.J.C.; Gerhauser, H.; Kiel, J.H.A.; Ptasinski, K.J. Biomass upgrading by torrefaction for the production of biofuels: A review. *Biomass Bioenergy* **2011**, *35*, 3748–3762. [CrossRef]
10. Chew, J.J.; Doshi, V. Recent advances in biomass pretreatment–Torrefaction fundamentals and technology. *Renew. Sustain. Energy Rev.* **2011**, *15*, 4212–4222. [CrossRef]
11. Ciolkosz, D.; Wallace, R. A review of torrefaction for bioenergy feedstock production. *Biofuel Bioprod. Biorefin.* **2011**, *5*, 317–329. [CrossRef]
12. Gaitán-Álvarez, J.; Moya, R.; Rodríguez-Zúñiga, A.; Puente-Urbina, A. Characterization of torrefied biomass of five reforestation species (*Cupressus lusitanica*, *Dipteryx panamensis*, *Gmelina arborea*, *Tectona grandis* and *Vochysia ferruginea*) in Costa Rica. *Bioresources* **2017**, *12*, 7566–7589. [CrossRef]
13. Da Silva, C.M.S.; Carneiro, A.D.; Pereira, B.L.C.; Vital, B.R.; Alves, I.C.N.; de Magalhaes, M.A. Stability to thermal degradation and chemical composition of woody biomass subjected to the torrefaction process. *Eur. J. Wood Prod.* **2016**, *74*, 845–850. [CrossRef]
14. Korošec, R.C.; Lavrič, B.; Rep, G.; Pohleven, F.; Bukovec, P. Thermogravimetry as a possible tool for determining modification degree of thermally treated Norway spruce wood. *J. Therm. Anal. Calorim.* **2009**, *98*, 189. [CrossRef]
15. Aydemir, D.; Gunduz, G.; Altuntas, E.; Ertas, M.; Sahin, H.T.; Alma, M.H. Investigating changes in the chemical constituents and dimensional stability of heat-treated hornbeam and uludag fir wood. *BioResources* **2011**, *6*, 1308–1321.
16. Vassilev, S.V.; Baxter, D.; Andersen, L.K.; Vassileva, C.G. An overview of the chemical composition of biomass. *Fuel* **2010**, *89*, 913–933. [CrossRef]
17. Poudel, J.; Ohm, T.I.; Oh, S.C. A study on torrefaction of food waste. *Fuel* **2015**, *140*, 275–281. [CrossRef]
18. Moya, R.; Rodríguez-Zúñiga, A.; Puente-Urbina, A. Thermogravimetric and devolatilisation analysis for five plantation species: Effect of extractives, ash compositions, chemical compositions and energy parameters. *Thermochim. Acta* **2017**, *647*, 36–46. [CrossRef]
19. Gaitán-Álvarez, J.; Moya, R. Characteristics and properties of pellet fabricated with torrefaccioned biomass of *Gmelina arborea* and *Dipterix panamensis* at different time. *Revista Chapingo* **2016**, *23*, 325–337.

20. Tenorio, C.; Moya, R. Thermogravimetric characteristics, its relation with extractives and chemical properties and combustion characteristics of ten fast-growth species in Costa Rica. *Thermochim. Acta* **2013**, *563*, 12–21. [CrossRef]

21. Gaitán-Álvarez, J.; Moya, R.; Puente-Urbina, A.; Rodríguez-Zúñiga, A. Physical and compression properties of pellets manufactured with the biomass of five woody tropical species of Costa Rica torrefied at different temperatures and times. *Energies* **2017**, *10*, 1205. [CrossRef]

22. Aragón-Garita, S.; Moya, R.; Bond, B.; Valaert, J.; Tomazello Filho, M. Production and quality analysis of pellets manufactured from five potential energy crops in the Northern Region of Costa Rica. *Biomass Bioenergy* **2016**, *87*, 84–95. [CrossRef]

23. Tenorio, C.; Moya, R.; Tomazello Filho, M.; Valaert, J. Application of the X-ray densitometry in the evaluation of the quality and mechanical properties of biomass pellets. *Fuel Process. Technol.* **2015**, *132*, 62–73. [CrossRef]

24. Tenorio, C.; Moya, R.; Tomazello-Filho, M.; Valaert, J. Quality of pellets made from agricultural and forestry crops in Costa Rican tropical climates. *BioResources* **2014**, *9*, 482–498. [CrossRef]

25. Moya, R.; Rodriguez-Zuñiga, A.; Puente-Urbina, A.; Gaitan-Alvarez, J. Study of light, middle and severe torrefaction and effects of extractives and chemical compositions on torrefaction process by thermogravimetric analyses in five fast-growing plantation of Costa Rica. *Energy* **2018**, *149*, 152–160. [CrossRef]

26. Sbirrazzuoli, N.; Vyazovkin, S.; Mititelu, A.; Sladic, C.; Vincent, L. A Study of Epoxy-Amine Cure Kinetics by Combining Isoconversional Analysis with Temperature Modulated DSC and Dynamic Rheometry. *Macromol. Chem. Phys.* **2003**, *204*, 1815–1821. [CrossRef]

27. Grønli, M.G.; Várhegyi, G.; Di Blasi, C. Thermogravimetric analysis and devolatilization kinetics of wood. *Ind. Eng. Chem. Res.* **2002**, *41*, 4201–4208. [CrossRef]

28. Prins, M.J.; Ptasinski, K.J.; Janssen, F.J. Torrefaction of wood: Part 2. Analysis of products. *J. Anal. Appl. Pyrolysis* **2006**, *77*, 35–40. [CrossRef]

29. Arias, B.; Pevida, C.; Fermoso, J.; Plaza, M.G.; Rubiera, F.; Pis, J.J. Influence of torrefaction on the grindability and reactivity of woody biomass. *Fuel Process. Technol.* **2008**, *89*, 169–175. [CrossRef]

30. Ramiah, M.V. Thermogravimetric and differential thermal analysis of cellulose, hemicellulose, and lignin. *J. Appl. Polym. Sci.* **1970**, *14*, 1323–1337. [CrossRef]

31. Bach, Q.V.; Skreiberg, Ø. Upgrading biomass fuels via wet torrefaction: A review and comparison with dry torrefaction. *Renew. Sustain. Energy Rev.* **2016**, *54*, 665–677. [CrossRef]

32. Yang, H.; Yan, R.; Chen, H.; Lee, D.H.; Zheng, C. Characteristics of hemicellulose, cellulose and lignin pyrolysis. *Fuel* **2007**, *86*, 1781–1788. [CrossRef]

33. Avni, E.; Coughlin, R.W. Kinetic analysis of lignin pyrolysis using non-isothermal TGA data. *Thermochim. Acta* **1985**, *90*, 157–167. [CrossRef]

34. Lee, S.Y.; Kang, I.A.; Doh, G.H.; Kim, W.J.; Kim, J.S.; Yoon, H.G.; Wu, Q. Thermal, mechanical and morphological properties of polypropylene/clay/wood flour nanocomposites. *Express Polym. Lett.* **2008**, *2*, 78–87. [CrossRef]

35. Islam, M.S.; Hamdan, S.; Rahman, M.R.; Jusoh, I.; Ahmen, A.S. Dynamic Young's modulus, morphological, and thermal stability of 5 tropical light hardwoods modified by benzene diazonium salt treatment. *BioResources* **2011**, *6*, 737–750.

36. Arteaga-Pérez, L.E.; Grandón, H.; Flores, M.; Segura, C.; Kelley, S.S. Steam torrefaction of *Eucalyptus globulus* for producing black pellets: A pilot-scale experience. *Bioresour. Technol.* **2017**, *238*, 194–204. [CrossRef] [PubMed]

37. McKendry, P. Energy production from biomass (part 2): Conversion technologies. *Bioresour. Technol.* **2002**, *83*, 47–54. [CrossRef]

38. Doddapaneni, T.R.; Konttinen, J.; Hukka, T.I.; Moilanen, A. Influence of torrefaction pretreatment on the pyrolysis of Eucalyptus clone: A study on kinetics, reaction mechanism and heat flow. *Ind. Crops Prod.* **2016**, *92*, 244–254. [CrossRef]

39. Chen, W.H.; Kuo, P.C. Torrefaction and co-torrefaction characterization of hemicellulose, cellulose and lignin as well as torrefaction of some basic constituents in biomass. *Energy* **2011**, *36*, 803–811. [CrossRef]

40. Park, J.; Meng, J.; Lim, K.H.; Rojas, O.J.; Park, S. Transformation of lignocellulosic biomass during torrefaction. *J. Anal. Appl. Pyrolysis* **2013**, *100*, 199–206. [CrossRef]

41. Ramos, L.P. The chemistry involved in the steam treatment of lignocellulosic materials. *Quim. Nova* **2003**, *26*, 863–871. [CrossRef]

42. Biswas, A.K.; Umeki, K.; Yang, W.; Blasiak, W. Change of pyrolysis characteristics and structure of woody biomass due to steam explosion pretreatment. *Fuel Process. Technol.* **2011**, *92*, 1849–1854. [CrossRef]

43. Bach, Q.V.; Tran, K.Q.; Skreiberg, Ø.; Khalil, R.A.; Phan, A.N. Effects of wet torrefaction on reactivity and kinetics of wood under air combustion conditions. *Fuel* **2014**, *137*, 375–383. [CrossRef]

44. Bobleter, O. Hydrothermal degradation of polymers derived from plants. *Prog. Polym. Sci.* **1994**, *19*, 797–841. [CrossRef]

45. Wyman, C.E.; Decker, S.R.; Himmel, M.E.; Brady, J.W.; Skopec, C.E.; Viikari, L. Hydrolysis of cellulose and hemicellulose. *Polysaccharides* **2005**, *1*, 1023–1062.

46. Skreiberg, A.; Skreiberg, Ø.; Sandquist, J.; Sørum, L. TGA and macro-TGA characterisation of biomass fuels and fuel mixtures. *Fuel* **2011**, *90*, 2182–2197. [CrossRef]

47. Ball, R.; McIntosh, A.C.; Brindley, J. Feedback processes in cellulose thermal decomposition: Implications for fire-retarding strategies and treatments. *Combust. Theory Model.* **2004**, *8*, 281–291. [CrossRef]

48. Shen, J.; Igathinathane, C.; Yu, M.; Pothula, A.K. Biomass pyrolysis and combustion integral and differential reaction heats with temperatures using thermogravimetric analysis/differential scanning calorimetry. *Bioresour. Technol.* **2015**, *185*, 89–98. [CrossRef] [PubMed]

49. Stenseng, M.; Jensen, A.; Dam-Johansen, K. Investigation of biomass pyrolysis by thermogravimetric analysis and differential scanning calorimetry. *J. Anal. Appl. Pyrolysis* **2001**, *58*, 765–780. [CrossRef]

![energies logo]

MDPI

Article

Optimization and Scale-Up of Coffee Mucilage Fermentation for Ethanol Production

David Orrego [1,2,†], **Arley David Zapata-Zapata** [3,†] and **Daehwan Kim** [1,4,*]

1   Laboratory of Renewable Resources Engineering, Purdue University, West Lafayette, IN 47907, USA; dorregol@purdue.edu
2   Department of Agricultural and Biological Engineering, Purdue University, West Lafayette, IN 47907, USA
3   School of Chemistry, Faculty of Science, National University of Colombia, Calle 59A N, Medellin 63-20, Colombia; adzapata@unal.edu.co
4   Department of Biology, Hood College, 401 Rosemont Avenue, Frederick, MD 21701, USA
*   Correspondence: kim1535@purdue.edu; Tel.: +1-765-637-8603
†   Authors contributed equally to the study.

Received: 31 January 2018; Accepted: 5 March 2018; Published: 29 March 2018

Abstract: Coffee, one of the most popular food commodities and beverage ingredients worldwide, is considered as a potential source for food industry and second-generation biofuel due to its various by-products, including mucilage, husk, skin (pericarp), parchment, silver-skin, and pulp, which can be produced during the manufacturing process. A number of research studies have mainly investigated the valuable properties of brewed coffee (namely, beverage), functionalities, and its beneficial effects on cognitive and physical performances; however, other residual by-products of coffee, such as its mucilage, have rarely been studied. In this manuscript, the production of bioethanol from mucilage was performed both in shake flasks and 5 L bio-reactors. The use of coffee mucilage provided adequate fermentable sugars, primarily glucose with additional nutrient components, and it was directly fermented into ethanol using a *Saccharomyces cerevisiae* strain. The initial tests at the lab scale were evaluated using a two-level factorial experimental design, and the resulting optimal conditions were applied to further tests at the 5 L bio-reactor for scale up. The highest yields of flasks and 5 L bio-reactors were 0.46 g ethanol/g sugars, and 0.47 g ethanol/g sugars after 12 h, respectively, which were equal to 90% and 94% of the theoretically achievable conversion yield of ethanol.

Keywords: bioethanol; coffee mucilage; fermentation; *Saccharomyces cerevisiae*; second generation biofuels

1. Introduction

According to previous research studies, coffee is one of the most traded commodities, along with crude oils, and is consumed by millions of people on a daily basis. It is known that more than 70 countries in South American and tropical nations, including Brazil, Vietnam, Colombia, Indonesia, and Ethiopia extensively produce coffee, and they play major roles in the global coffee trades, developments, and the country's economy [1,2]. Furthermore, the current trend of coffee consumption and its relative global market has rapidly grown over recent decades due to an intriguing ingredient—caffeine as a stimulant. Because of the increase of consumption of coffee with consumers' interest in health and functional foods, a number of works have investigated the cognitive, physical, and functional attributes of coffee (mainly caffeine) [3,4].

The coffee fruit is mainly composed of bean, silver-skin, parchment, mucilage, pulp, and pericarp [5,6]. The endosperm (coffee bean) is attached and covered by silver-skin, which is enclosed in parchment (yellowish thin endocarp). The thin, colorless, and hydrated pectin layer, which is called the mucilage, covers the parchment, and finally a soft and stiff outer pericarp protects the fibrous pulp and parchment. The brief structure of the coffee bean is shown in Figure 1.

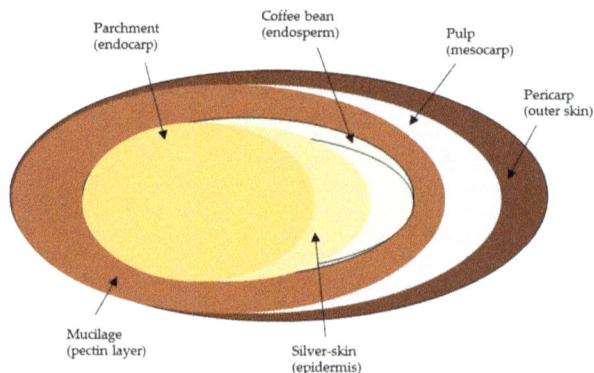

Figure 1. The structure of the coffee fruit.

As the production of coffee has increased, the resulting by-products of coffee generated during wet or dry processing also increased [7]. In the common coffee production process, the green coffee bean is mainly used while other components of silver-skin, parchment, mucilage, pulp, and pericarp, which are more than 50% of the coffee fruit, are removed and discarded during processing [8,9]. Bressani [10] reported that the recovery of the coffee residuals after wet process contained 43.2% (w/w) skin and pulp, 11.8% (w/w) mucilage and soluble sugars, and 6.1% (w/w) parchment. Recently, it has been found that the coffee residual by-products made up approximately 15 million tons per year [11], and they generated a large amount of agricultural waste. This includes the coffee mucilage, which represents approximately 22% of the grain wet weight [12]. Some coffee by-products are currently used in manure [12], adsorption molecules [13–15], ethanol [16–18], gibberellic acid [19], and α-amylase [20]. Further studies of the coffee silver-skin showed that it could be utilized as a functional material due to its high dietary fiber component (60%), low contents of fat and carbohydrate, and associated antioxidant phenolic compounds [21,22]. Although coffee by-product wastes could be utilized in agricultural fertilizer and animal feedstock, the high composition of lignin (25% w/w in coffee ground) and anti-nutritional components (phenolics and caffeine) in the coffee wastes are restrictive factors of direct use for animal feedstock [23–28]. Un-utilized coffee industry residuals are discarded into land or water resources, which could generate environmental issues with their high biochemical oxygen demand and toxic compounds [29,30]; therefore, further utilization and practical application of these residues are required.

Coffee mucilage, rich in carbohydrates and nitrogen, is considered to be a potential substrate in bio-based industry to produce value-added molecules and commodities, such as ethanol and lactic acid because it consists of 85–91% (w/w) water, 6.2–7.4% (w/w) sugars and small amounts of protein, non-reducing sugar, and pectin [12,30]. Moreover, the sugars present in the mucilage are mainly composed of reducing sugars (63%, w/w) that can be biochemically transformed to other molecules [23,31]. Although several research studies have been published regarding the chemical/structural composition of coffee mucilage [32], functional properties [33] and production of value-added products such as hydrogen [34], ethanol [17,18], and lactic acid [35], to the best of our knowledge there is no research study for the scale-up of ethanol production using coffee mucilage under optimal conditions. This study reports the use of coffee mucilage waste as a substrate for bioethanol production without any pretreatment or addition of nutrition sources prior to microbial fermentation. Initially, the evaluation of microbial fermentations using a *Saccharomyces cerevisiae* NRRL (Northern Regional Research Laboratory) Y-1546 strain was carried out at the lab scale with flasks to find an optimal condition, and the designed experimental condition was applied to the 5 L bio-reactors, which resulted in 0.47 g ethanol/g fermentable sugars, showing almost theoretically achievable yield (94%). Different variable parameters such as temperature, pH, and inoculum concentrations were tested and compared. Our work presents the potential to directly utilize reducing sugars (mainly glucose) in the coffee mucilage for ethanol production through microbial fermentation.

2. Results and Discussion

2.1. Compositional Analysis of Coffee Mucilage

Chemical composition of the initial coffee mucilage was determined by International Organization for Standardization (ISO) standard analytical methods as described in the methods section. The coffee mucilage was provided from a production farm in Antioquia, Colombia and its initial chemical composition is summarized in Table 1. Previous studies on the chemical composition of the coffee mucilage reported that it had a composition of 84.2% water, 8.9% protein, 4.1% sugar, 0.91% pectin, and 0.7% ash [5]. To identify the cell growth and fermentability of reducing sugars in the coffee mucilage, initial runs with nutrient supplements of yeast extract (1 g L^{-1}) or diammonium hydrogen phosphate ($(NH_4)_2HPO_4$, 0.5 g L^{-1}) in the mucilage were examined and compared to the tests without additional supplements. The presence of nitrogen source in the mucilage had little or no difference compared to the results from control, and all tests resulted in similar final ethanol yields (unpublished observations). Other previous studies support this phenomenon, since the mucilage contains various nutrients, protein, and valuable sources that could meet the requirements of cell growth and conversion of fermentable sugars to ethanol [17,35]. In addition, there was no substrate inhibition from the coffee mucilage that was consistent with another work done by Margaritis and Bajpai [36] for the ethanol production from artichoke extract, where substrate inhibition was observed at 250 g L^{-1} of reducing sugar concentration. Considering previous references and our preliminary data, we used the coffee mucilage without additional supplements in further microbial fermentation tests.

Table 1. Chemical composition of coffee mucilage.

(A) Component	Concentration (g L^{-1})
Glucose	37.1
Galactose	14.7
Lactose	0.8
Acetic acid	1.2
Protein	0.3
(B) Mineral	**Parts-Per-Million (ppm)**
Calcium	337
Iron	73
Magnesium	81
Potassium	116 (mg L^{-1})
Phosphorus	115
Sodium	29

2.2. Optimization of Coffee Mucilage Fermentation

In order to determine the optimal experimental condition for ethanol fermentation from the coffee mucilage, a full factorial design method was considered and designed with different levels of temperature, pH, and initial inoculum concentration. Generally, the *Saccharomyces cerevisiae* strain has its optimal growth condition at a temperature range of 25–30 °C [37] or 30–33 °C [38] with acid pH ranging from 3.0 to 7.0, and the optimal pH condition being at 5.0 [38]. Another study found that *S. cerevisiae* was able to grow at 4 °C with the minimum growth while the maximum cell growth was achieved at 38–39 °C [39]. Recent research of the utilization of coffee mucilage reported that the reducing sugars in the coffee mucilage could effectively be fermented to ethanol at 32 °C, pH 5.1, and initial reducing sugar concentration of 61.8 g L^{-1} with a cell density of 1.0×10^6 cfu/mL [17]. In the light of previous publications, different levels of temperature (28–35 °C), pH (4.0–7.0), and initial cell concentration (3.0–6.0 g L^{-1}) were selected to determine optimal experimental conditions. Because substrate inhibition (reducing sugar) would occur at >250 g L^{-1} [36], we have not considered the reducing sugar concentration and directly used the initial mucilage as a substrate.

To fulfill tests, a two-level full factorial experimental fermentation was designed by Minitab 17 software; the total regular experiments were 9, including + 1 central point experiment, which was for the estimation of the experimental variance. The total experiments were also carried out in random order to avoid lurking variables, and ethanol yield was calculated each run. The designed experiments, results of ethanol production and percent ethanol yields are summarized in Table 2. The theoretical achievable ethanol yield was estimated with the calculation of 0.511 times the total initial fermentable sugars (glucose and galactose) in the coffee mucilage medium [40,41]. All glucose and galactose were metabolized and converted to ethanol within 24 h, however, the final yields were different, depending on experimental factors of temperature, pH, and initial cell density.

Table 2. Two-level full factorial experimental design and brief summary of the results. Ethanol fermentation tests were carried out in 250 mL Erlenmeyer flasks at the given experimental conditions for 24 h with agitation of 150 rpm in a shaking incubator. All tests were performed in triplicate.

Experiment	Temperature (°C)	pH	Inoculum (g L^{-1})	Ethanol Production (g L^{-1})	Standard Deviation (g L^{-1})	Percent Ethanol Yield (%)
1	28.0	4.0	3.0	22.2	0.8	83.9
2	35.0	4.0	3.0	20.3	1.2	76.8
3	28.0	7.0	3.0	21.6	1.3	81.8
4	35.0	7.0	3.0	20.0	0.8	75.5
5	28.0	4.0	6.0	21.6	0.7	81.5
6	35.0	4.0	6.0	18.4	1.2	69.3
7	28.0	7.0	6.0	19.2	1.1	72.5
8	35.0	7.0	6.0	17.4	0.9	65.8
9	31.5	5.5	4.5	19.6	1.2	73.9

At a given experimental condition, the highest ethanol production of 22.2 g L^{-1} was obtained at 28°C (pH 4.0) with an initial cell density of 3.0 g L^{-1}, which was equivalent to 83.9% of the theoretically achievable yield. On the other hand, the lowest yield (17.4 g L^{-1}) was achieved at 35 °C (pH 7.0) with an initial inoculum of 6 g L^{-1}. To better understand the main factors for the microbial fermentations, the analysis of each variance was carried out by ANOVA tests. As shown in Figure 2A, the incubation temperature and initial cell concentration significantly contributed toward the ethanol production (sharp slope). The coefficient correlation (R^2) of the best model was 0.9449, which indicated the variability of response in this experimental design was 94.49%. In order to verify the superior model with a less than 0.05 probability value, the minor reciprocal interaction factors were excluded, and the model was adjusted with the Pareto chart for individual and two-factor interaction effects. Similar to previous analytic results, the resulting data presented that the levels of temperature and initial cell density had remarkable effects on the ethanol fermentation, while the variable of pH had less effect (Figure 2B). Our previous studies highlighted that the acetic acid concentration and pH on microbial fermentation were thought to be the major contributors for cell growth and metabolic ethanol yield [40,42]. For example, a 67% reduction in ethanol productivity was observed in the presence of 15 g L^{-1} acetic acid at pH 5.0 compared to the control run without acetic acid or under the lower acetic acid concentration of 3 g L^{-1}. The un-dissociated and/or un-charged form of weak acids (mainly acetic acid) could disrupt the proton influx system in the cell membrane [40,43,44]; however, with a small amount of acetic acid (1.2 g L^{-1}), no dramatic response was observed in the pH range of 4.0 and 7.0.

So far, as temperature and initial cell inoculum were considered as dominant parameters, we were interested in exploring their interaction and optimal conditions for the scale-up at the 5 L bio-reactors. Response surface (optimal condition) was performed for the maximum ethanol yield using a contour plot experimental design (Figure 3A–C). The critical point was determined regarding the modification of the obtained experimental results, which resulted in the following equation with >95% correlation coefficients (R^2). The data was fitted with the key factors of temperature and initial cell inoculum; the resulting equation represented the dependent response with regard to the relation to the variable factors under the experimental conditions: ethanol yield (g L^{-1}) = 20.68 − 0.017 temperature + 0.81 pH + 2.07 inoculum − 0.011 temperature * pH − 0.0586 temperature * inoculum − 0.1422 pH * inoculum − 0.66 center point. The optimal ethanol estimation is shown in the surface plot model in Figure 3D.

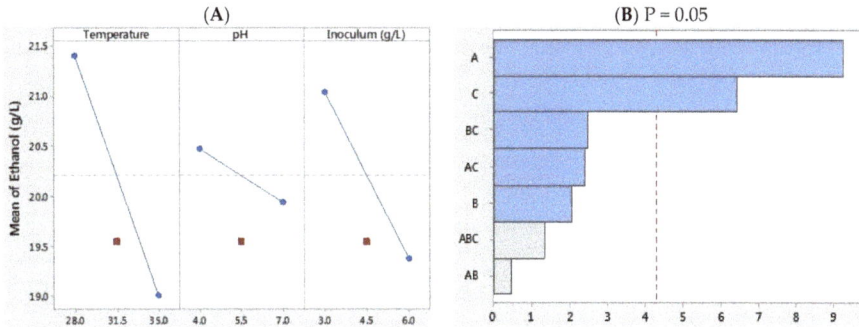

Figure 2. (**A**) The main effect of temperature, pH, and inoculum concentration on the ethanol production from coffee mucilage; (**B**) Pareto chart of individual and reciprocal interaction effects of temperature, pH, and inoculum concentration. A: temperature, B: pH, C: inoculum concentration.

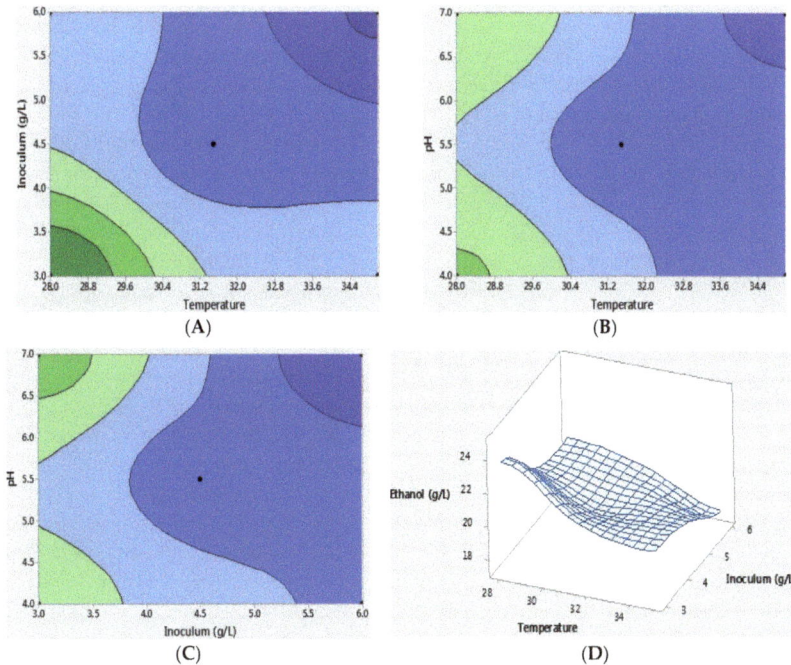

Figure 3. Contour plots present effect of varying (**A**) temperature and inoculum; (**B**) temperature and pH; and (**C**) inoculum and pH. The dark green and dark blue colors denote a high ethanol concentration and a low ethanol concentration, respectively. The fitted optimal conditions with two dominant factors of temperature and initial cell concentration are presented by the surface plot (**D**).

The maximum yield of ethanol was obtained at a low temperature of 28 °C, pH 4.0, with an initial cell density of 3 g L^{-1}, which reached around 23 g L^{-1} of ethanol production, corresponding to 86.9% theoretically achievable yield. The statistical analysis of the resulting equation was employed by a *T*-test analysis, which showed more than 95% significant difference for ethanol production from the mucilage at a given experimental condition. Interestingly, the effect of initial pH on ethanol production was less than the other two factors, possibly due to the low acid concentration in the substrate medium and the acceptable pH changes during the fermentation performances [40,42]. For the next step, the application of the designed optimal condition to the lab scale (flasks) and the 5 L bio-reactor fermentation were studied, respectively.

2.3. Mucilage Fermentation Both in the Flasks and 5 L Bio-Reactors

In order to access the optimal condition of temperature, pH, and initial cell density, further tests were carried out both in the lab scale level and scale-up at the 5 L bio-reactors, respectively. In initial runs with the flasks, most of the glucose and galactose were consumed at 9 h and the final ethanol concentration (23.8 g L^{-1}) was achieved at 12 h, which was equal to approximately 89.9% theoretically achievable yield (Figure 4A). In the scale-up test at the 5 L bio-reactor at similar experimental conditions, a slight increase of the final ethanol concentration (24.8 g L^{-1}) was observed with a 93.7% theoretically achievable value (Figure 4B). It might have happened that the remaining sugars in the small coffee mucilage slurries provided more fermentable sugars and/or better mixing with the dual impeller that resulted in the higher ethanol production compared to the results from the flasks. Both runs in the flasks and bio-reactors had a similar fermentation pattern, where glucose was primarily utilized and converted into ethanol within 9 h, while the galactose was slowly metabolized and totally consumed in 12 h (Figure 4A,B). *S. cerevisiae* has the ability to utilize and metabolize different C6 sugars, including glucose, fructose, mannose, and galactose; however, the yeast prefers to use simple structure sugars such as glucose and fructose rather than galactose because the metabolic pathway of galactose requires a combined step of the Leloir pathway and glycolysis for alcohol fermentation [45].

Our fermentation results are somewhat different from previous observations for ethanol production from coffee mucilage. Pérez-Sariñana et al. [17] have reported that the best ethanol concentration (16.29 g L^{-1}) was obtained at 32 °C, pH 5.1, cell inoculum of 1.0×10^6 cfu/mL, and initial sugar concentration of 61.8 g L^{-1}. Distinctions in the coffee mucilage material, sugar concentration, and yeast strain may help to explain this. Some inhibitory compounds such as acetic acid and phenolic compounds were detected in the previous study [17] while fewer or no potential inhibitors were identified in our work. Phenolic molecules are one of the dominant contributors to inhibit enzymes [46–49] and microbial fermentation [50–52]; thus, remarkable inhibition could have occurred even at a low concentration, which results in the reduction of sugar utilization and diminished final product yield. Also, the use of different yeast strain and initial cell concentration may reflect the different ethanol yields under optimal conditions. Further studies are required in order to gain a better knowledge and understanding of the coffee mucilage and its practical utilization.

Further comparison between our data and earlier study [23] with the fermentative ethanol production from the sulfuric acid treated coffee silver-skin and spent coffee grounds hydrolysates is summarized in Table 3. When the reducing sugars (mainly galactose and mannose) in the coffee silver-skin and spent coffee grounds hydrolysates were fermented into ethanol by different yeast strains, a maximum ethanol yield of 0.13 g ethanol/g substrate and 0.26 g ethanol/g substrate were obtained, respectively (Table 3). The ethanol production of coffee mucilage was evaluated at around 0.46 g ethanol/g substrate (lab scale), which was 1.77 times higher compared to the previous research data from the coffee silver-skin and spent coffee grounds. Furthermore, the ethanol yield slightly increased when the mucilage fermentation was performed at the 5 L bio-reactors, which gave 1.84 times higher yield than the results from silver-skin and coffee grounds (Table 3). It can be explained that the different chemical composition of the coffee by-products (mucilage vs. silver-skin vs. grounds) and the acid pretreatment before the fermentation step generated toxic compounds (mainly phenolic molecules), and inhibited the fermentation performance [41,48,49,53].

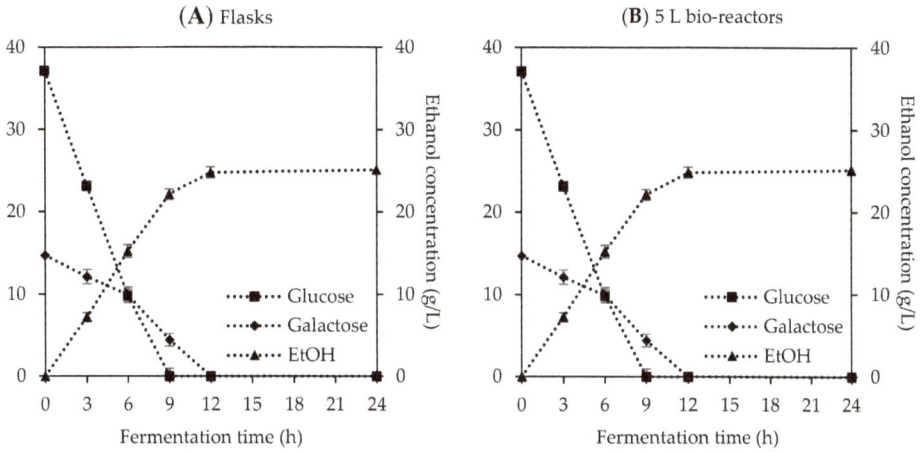

(A) Flasks   **(B)** 5 L bio-reactors

Figure 4. Consumption of glucose and galactose and production of ethanol during microbial fermentation in the (**A**) flasks level and (**B**) scale-up at 5 L bio-reactors. All experimental tests were performed at 28 °C, pH 4.0, and initial cell concentration of 3 g L^{-1}. Orbital shake agitation of 150 rpm was utilized for the flasks fermentation while a double Rushton impeller at 350 rpm was used for the 5 L bio-reactors fermentation.

Table 3. Comparison of ethanol productions from coffee by-products of mucilage, silver-skin, and grounds using different yeast strains.

Yeast Strain	Ethanol Yield (g ethanol/g Substrate)			Reference
	Coffee Mucilage	Coffee Silver-Skin	Spent Coffee Grounds	
S. cerevisiae (NRRL Y-1546)	0.46 ± 0.02	-	-	Current Study
P. stiptis (NRRL-Y-7124)	-	0.11 ± 0.02	0.26 ± 0.01	[23]
S. cerevisiae (RL-11)	-	0.13 ± 0.03	0.26 ± 0.02	[23]
K. fragilis (Kf1)	-	0.01 ± 0.00	0.13 ± 0.02	[23]

It is worth stating that the sugar metabolism and ethanol production in this study did not require any pretreatment step. Additional supplement of nutrition sources, and further processes could avoid the higher costs of other methods and can be suggested as a potential application for industrial fields. In addition, this research on fermentative utilization of the coffee mucilage is a first attempt to optimize and scale-up ethanol production. The resulting design model and ethanol yield here would be enhanced if other valuable factors were considered and performed at the best performing conditions (oxygen availability, genetically modified yeast strain, impeller type, mixing condition, and others). Therefore, the present study with the coffee mucilage is thought to be a promising alternative agricultural waste feedstock for further re-use and applications with a high ethanol yield.

3. Materials and Methods

3.1. Coffee Mucilage

Coffee mucilage was collected from the San Rafael farm located in the municipality of Envigado (Antioquia, Colombia) at 1575 MASL (meters above sea level) and an average temperature throughout the year of 21 °C. As soon as the coffee mucilage was provided by the San Rafael farm, the materials were sterilized at 121 °C for 15 min and stored at 10 °C. In order to separate the biggest solids in the coffee mucilage, a sample of 500 mL mucilage was sieved over a stainless-steel screen with a 20-mesh opening size (Tyler, USA standard testing sieve, ASTM E11 specification, VWR, Philadelphia, PA, USA). The resulting slurry was centrifuged at 8000 rpm for 5 min to eliminate the remaining solids and identified in regard to the concentration of fermentable sugars, including glucose and galactose by HPLC. The prepared mucilage medium was adjusted to the pH range 4.0 to 7.0 by adding NaOH (pellets) and stored at 4°C until used in fermentation tests.

3.2. Microorganism and Fermentation Experiments

A yeast strain of *Saccharomyces cerevisiae* NRRL Y-1546 was utilized for all fermentations of the coffee mucilage. This *S. cerevisiae* is able to metabolize C6 sugars (glucose, fructose, mannose, and galactose) for ethanol fermentation [41,54,55]. Our previous research studies highlighted that *S. cerevisiae* NRRL Y-1546 was suitable to ferment the sugars in the pretreated-lignocellulosic material hydrolysates and other agricultural residues, which had some inhibitory compounds [41,50,53]; thus, we selected this strain for current work and evaluated the fermentation tests. 2 mL of the stock solution was used for the inoculum preparation in 250 mL Erlenmeyer flasks containing 100 mL YEPD medium (1% yeast extract, 2% peptone, and 2% glucose) [41]. The prepared flasks were cultured in an orbital shaking incubator (Centricol, Colombia) at 30°C, overnight with agitation of 150 rpm. When the cell density was reached to a desired concentration, cells were centrifuged (at 5000 rpm for 5 min), recovered, and re-suspended with YEP medium (no glucose).

For the lab scale tests, batch fermentation tests were performed in 250 mL Erlenmeyer flasks containing 100 mL of the mucilage samples at desired experimental conditions: temperature range of 28–35 °C, pH range of 4.0–7.0, and initial cell concentration range of 3.0–9.0 g L^{-1} (Table 2). For further tests of the scale-up, 5 L bio-reactors (Bioengineering AFL, Wald, Switzerland) were used in the presence of a 3 L mucilage medium. Scale-up fermentations were carried out at optimal experimental conditions (28 °C, pH 4.0, and an initial cell density of 3 g L^{-1}) obtained from the lab scale tests. Instead of orbital agitation, mixing was performed at 350 rpm using two Rushton impellers. During all fermentation experiments in flasks and bio-reactors, samples were taken regularly for the consumption of fermentable sugars and ethanol production. All the fermentation experiments were carried out in triplicate.

3.3. Analytic Assays

A two-level factorial experimental design and statistical analysis was conducted by Minitab 17 software program. The *T*-test was performed with >95% significant difference for ethanol production from the coffee mucilage. The fermentable sugars and ethanol were analyzed by high performance liquid chromatography (HPLC) before and after fermentation performances followed by previous works [41,48,49] using a Waters alliance 2695 system coupled to a FID detector with an Aminex HPX-87H ion-exchange column (300 nm × 7.8 mm, Bio-Rad Laboratories, Hercules, CA, USA). The HPLC system was maintained at 60 °C with a 5 mM diluted sulfuric acid in distilled water as eluent mobile phase at 0.6 mL/min flow rate. Other mineral components, including calcium, iron, magnesium, potassium, sodium, and zinc were determined by atomic absorption spectroscopy (ISO 6869:2012) [56]. The analysis of phosphorous and crude protein content were identified by UV-VIS spectrophotometry (ISO method 6498:1998) [57] and Kjeldahl method (ISO method 5983:2005) [58], respectively. The theoretical maximum ethanol yield (26.47 g L^{-1}) was calculated based on the equation; Y_{max} (g L^{-1}) = G × 0.511, where the value of G represents the initial total fermentable sugars of glucose (37.1 g L^{-1}) and galactose (14.7 g L^{-1}) in the coffee mucilage. A percent of ethanol yield was estimated as per equation; Y_{EtOH} (%) = E/Y_{max} × 100, where E denotes ethanol production during microbial fermentation [59].

4. Conclusions

Coffee mucilage, un-utilized agro-industrial residue from coffee processing, was successfully re-used as a potential substrate for ethanol production. Through the two-level factorial experimental design, the optimal condition for microbial fermentation was obtained at 28 °C, pH 4.0, and initial cell concentration of 3 g L^{-1}, which resulted in 0.46 (g ethanol/g sugars) ethanol concentration with a 90% theoretically achievable value. Such designed conditions applied to 5 L scale-up tests, which improved the final ethanol yield ($Y_{p/s}$ = 0.47 g/g) with a volumetric productivity (Q_p) of 2.26 g/Lh. The incubation temperature and initial cell density were considered to be key factors for the conversion of fermentable sugars in the coffee mucilage to ethanol, while the pH value between 4.0 and 7.0 had less effect on the fermentations. The ethanol yield (Yp/s, g ethanol/g substrate) of coffee mucilage is higher than the results from other agro-industrial residues such as barley hay (0.3 g/g), barley straw (0.31 g/g), hay millet (0.27 g/g), sweet sorghum hay (0.31 g/g), triticale hay (0.34 g/g), and wheat straw (0.31 g/g) [59]. Furthermore, the raw coffee mucilage material does not require any pretreatment processes and supplement sources prior to fermentation, which could also be attractive to other applications and utilization in fermentation processes.

Acknowledgments: The authors thank Deisy Sanchez Cuasapud and Doriett Ceilia Prada at the National University of Colombia for their technical assistance; Juan Fernando Botero and Diana Ramirez Lopez of the National University of Colombia for their internal review of this manuscript; Rachel Atherton of Purdue University for her professional assistance in preparing this draft. The authors also thank Michael R. Ladisch and Eduardo A. Ximenes for their assistance and support of David Orrego and Daehwan Kim.

Author Contributions: David Orrego and Arley David Zapata-Zapata started this research at the National University of Colombia and completed with Daehwan Kim at Purdue University. As a principal author of this research, David Orrego performed the designed microbial fermentation experiments and literature research. Arley Zapata-Zapata designed and supported the experiments. Daehwan Kim contributed to summarize, revise, and analyze data for this manuscript.

Conflicts of Interest: The authors declare no conflict of interest.

References

1. Moore, P.H.; Ming, R. (Eds.) *Genomics of Tropical Crop Plants*; Springer: New York, NY, USA, 2008; ISBN 9780387712185.
2. Souza, R.M. (Ed.) *Plant-Parasitic Nematodes of Coffee*; Springer: Dordrecht, The Netherlands, 2008; ISBN 9781402087196.
3. McLellan, T.M.; Caldwell, J.A.; Lieberman, H.R. A review of caffeine's effects on cognitive, physical and occupational performance. *Neurosci. Biobehav. Rev.* **2016**, *71*, 294–312. [CrossRef] [PubMed]
4. Cappelletti, S.; Daria, P.; Sani, G.; Aromatario, M. Caffeine: Cognitive and Physical Performance Enhancer or Psychoactive Drug? *Curr. Neuropharmacol.* **2015**, *13*, 71–88. [CrossRef] [PubMed]
5. Belitz, H.D.; Grosch, W.; Schieberle, P. *Food Chemistry*; Springer: Berlin, Germany, 2009; ISBN 9783540699330.
6. Berbert, P.A.; Queiroz, D.M.; Sousa, E.F.; Molina, M.B.; Melo, E.C.; Faroni, L.R.D. Dielectric properties of parchment coffee. *J. Agric. Eng. Res.* **2001**, *80*, 65–80. [CrossRef]
7. Prata, E.R.B.A.; Oliveira, L.S. Fresh coffee husks as potential sources of anthocyanins. *LWT Food Sci. Technol.* **2007**, *40*, 1555–1560. [CrossRef]
8. Kondamudi, N.; Mohapatra, S.K.; Misra, M. Spent Coffee Grounds as a Versatile Source of Green Energy. *J. Agric. Food Chem.* **2008**, *56*, 11757–11760. [CrossRef] [PubMed]
9. Saenger, M.; Hartge, E.U.; Werther, J.; Ogada, T.; Siagi, Z. Combustion of coffee husks. *Renew. Energy* **2001**, *23*, 103–121. [CrossRef]
10. Bressani, R.; Braham, J.E. (Eds.) *Coffee Pulp: Composition, Technology, and Utilization*; International Development Research Centre: Ottawa, ON, Canada, 1978.
11. Bakker, R.R.C. *Availability of Lignocellulosic Feedstocks for Lactic Acid Production. Feedstock Availability, Lactic Acid Production Potential and Selection Criteria*; Wageningen UR Food & Biobased Research: Wageningen, The Netherlands, 2013; pp. 1–62.

12. Murthy, P.S.; Madhava Naidu, M. Sustainable management of coffee industry by-products and value addition—A review. *Resour. Conserv. Recycl.* **2012**, *66*, 45–58. [CrossRef]

13. Franca, A.S.; Oliveira, L.S.; Nunes, A.A.; Alves, C.C.O. Microwave assisted thermal treatment of defective coffee beans press cake for the production of adsorbents. *Bioresour. Technol.* **2010**, *101*, 1068–1074. [CrossRef] [PubMed]

14. Oliveira, W.E.; Franca, A.S.; Oliveira, L.S.; Rocha, S.D. Untreated coffee husks as biosorbents for the removal of heavy metals from aqueous solutions. *J. Hazard. Mater.* **2008**, *152*, 1073–1081. [CrossRef] [PubMed]

15. Oliveira, L.S.; Franca, A.S.; Alves, T.M.; Rocha, S.D.F. Evaluation of untreated coffee husks as potential biosorbents for treatment of dye contaminated waters. *J. Hazard. Mater.* **2008**, *155*, 507–512. [CrossRef] [PubMed]

16. Gouvea, B.M.; Torres, C.; Franca, A.S.; Oliveira, L.S.; Oliveira, E.S. Feasibility of ethanol production from coffee husks. *Biotechnol. Lett.* **2009**, *31*, 1315–1319. [CrossRef] [PubMed]

17. Pérez-Sariñana, B.Y.; De León-Rodriguez, A.; Saldaña-Trinidad, S.; Joseph, S.P. Optimization of bioethanol production from coffee mucilage. *BioResources* **2015**, *10*, 4326–4338. [CrossRef]

18. Pérez-Sariñana, B.Y.; Saldaña-Trinidad, S.; Guerrero-Fajardo, C.A.; Santis-Espinosa, L.F.; Sebastian, P.J. A simple method to determine bioethanol production from coffee mucilage, verified by HPLC. *BioResources* **2015**, *10*, 2691–2698. [CrossRef]

19. Machado, C.M.M.; Soccol, C.R.; De Oliveira, B.H.; Pandey, A. Gibberellic acid production by solid-state fermentation in coffee husk. *Appl. Biochem. Biotechnol. Part A Enzym. Eng. Biotechnol.* **2002**, *102–103*, 179–191. [CrossRef]

20. Murthy, P.S.; Naidu, M.M.; Srinivas, P. Production of α-amylase under solid-state fermentation utilizing coffee waste. *J. Chem. Technol. Biotechnol.* **2009**, *84*, 1246–1249. [CrossRef]

21. Borrelli, R.C.; Esposito, F.; Napolitano, A.; Ritieni, A.; Fogliano, V. Characterization of a New Potential Functional Ingredient: Coffee Silverskin. *J. Agric. Food Chem.* **2004**, *52*, 1338–1343. [CrossRef] [PubMed]

22. Napolitano, A.; Fogliano, V.; Tafuri, A.; Ritieni, A. Natural occurrence of ochratoxin A and antioxidant activities of green and roasted coffees and corresponding byproducts. *J. Agric. Food Chem.* **2007**, *55*, 10499–10504. [CrossRef] [PubMed]

23. Mussatto, S.I.; Machado, E.M.S.; Carneiro, L.M.; Teixeira, J.A. Sugars metabolism and ethanol production by different yeast strains from coffee industry wastes hydrolysates. *Appl. Energy* **2012**, *92*, 763–768. [CrossRef]

24. Brand, D.; Pandey, A.; Rodriguez-Leon, J.A.; Roussos, S.; Brand, I.; Soccol, C.R. Packed bed column fermenter and kinetic modeling for upgrading the nutritional quality of coffee husk in solid-state fermentation. *Biotechnol. Prog.* **2001**, *17*, 1065–1070. [CrossRef] [PubMed]

25. Brand, D.; Pandey, A.; Roussos, S.; Soccol, C.R. Biological detoxification of coffee husk by filamentous fungi using a solid state fermentation system. *Enzyme Microb. Technol.* **2000**, *27*, 127–133. [CrossRef]

26. Orozco, A.L.; Pérez, M.I.; Guevara, O.; Rodríguez, J.; Hernández, M.; González-Vila, F.J.; Polvillo, O.; Arias, M.E. Biotechnological enhancement of coffee pulp residues by solid-state fermentation with Streptomyces. Py-GC/MS analysis. *J. Anal. Appl. Pyrolysis* **2008**, *81*, 247–252. [CrossRef]

27. Pandey, A.; Soccol, C.R.; Nigam, P.; Brand, D.; Mohan, R.; Roussos, S. Biotechnological potential of coffee pulp and coffee husk for bioprocesses. *Biochem. Eng. J.* **2000**, *6*, 153–162. [CrossRef]

28. Ulloa Rojas, J.B.; Verreth, J.A.J.; Van Weerd, J.H.; Huisman, E.A. Effect of different chemical treatments on nutritional and antinutritional properties of coffee pulp. *Anim. Feed Sci. Technol.* **2002**, *99*, 195–204. [CrossRef]

29. Heimbach, J.T.; Marone, P.A.; Hunter, J.M.; Nemzer, B.V.; Stanley, S.M.; Kennepohl, E. Safety studies on products from whole coffee fruit. *Food Chem. Toxicol.* **2010**, *48*, 2517–2525. [CrossRef] [PubMed]

30. Murthy, P.S.; Naidu, M.M. Recovery of Phenolic Antioxidants and Functional Compounds from Coffee Industry By-Products. *Food Bioprocess Technol.* **2012**, *5*, 897–903. [CrossRef]

31. Clifford, M.N.; Willson, K.C. (Eds.) *Coffee: Botany, Biochemistry and Production of Beans and Beverage*; AVI Publishing Company, Inc.: Westport, CT, USA, 1985; ISBN 9781461566595.

32. Avallone, S.; Guyot, B.; Brillouet, J.M.; Olguin, E.; Guiraud, J.P. Microbiological and biochemical study of coffee fermentation. *Curr. Microbiol.* **2001**, *42*, 252–256. [CrossRef] [PubMed]

33. Esquivel, P.; Jiménez, V.M. Functional properties of coffee and coffee by-products. *Food Res. Int.* **2012**, *46*, 488–495. [CrossRef]

34. Hernández, M.A.; Rodríguez Susa, M.; Andres, Y. Use of coffee mucilage as a new substrate for hydrogen production in anaerobic co-digestion with swine manure. *Bioresour. Technol.* **2014**, *168*, 112–118. [CrossRef] [PubMed]

35. Neu, A.K.; Pleissner, D.; Mehlmann, K.; Schneider, R.; Puerta-Quintero, G.I.; Venus, J. Fermentative utilization of coffee mucilage using Bacillus coagulans and investigation of down-stream processing of fermentation broth for optically pure l(+)-lactic acid production. *Bioresour. Technol.* **2016**, *211*, 398–405. [CrossRef] [PubMed]

36. Margaritis, A.; Bajpai, P. Effect of sugar concentration in Jerusalem artichoke extract on Kluyveromyces marxianus growth and ethanol production. *Appl. Environ. Microbiol.* **1983**, *45*, 723–725. [PubMed]

37. Spencer, J.F.T.; Spencer, D.M. (Eds.) *Yeasts in Natural and Artificial Haritats*; Springer: Berlin/Heidelberg, Germany, 1997; ISBN 9783642081606.

38. Campbell, I. *Yeast and Fermentation*; Elsevier Ltd.: Amsterdam, The Netherlands, 2003; ISBN 9780126692020.

39. John, I.P.; Ailsa, D.H. *Fungi and Food Spoilage*; Springer: Berlin/Heidelberg, Germany, 2009; ISBN 9780387922065.

40. Casey, E.; Sedlak, M.; Ho, N.W.Y.; Mosier, N.S. Effect of acetic acid and pH on the cofermentation of glucose and xylose to ethanol by a genetically engineered strain of Saccharomyces cerevisiae. *FEMS Yeast Res.* **2010**, *10*, 385–393. [CrossRef] [PubMed]

41. Kim, D.; Ximenes, E.A.; Nichols, N.N.; Cao, G.; Frazer, S.E.; Ladisch, M.R. Maleic acid treatment of biologically detoxified corn stover liquor. *Bioresour. Technol.* **2016**, *216*, 437–445. [CrossRef] [PubMed]

42. Casey, E.; Mosier, N.S.; Adamec, J.; Stockdale, Z.; Ho, N.; Sedlak, M. Effect of salts on the Co-fermentation of glucose and xylose by a genetically engineered strain of Saccharomyces cerevisiae. *Biotechnol. Biofuels* **2013**, *6*, 1–10. [CrossRef] [PubMed]

43. Narendranath, N.V.; Thomas, K.C.; Ingledew, W.M. Effects of acetic acid and lactic acid on the growth of Saccharomyces cerevisiae in a minimal medium. *J. Ind. Microbiol. Biotechnol.* **2001**, *26*, 171–177. [CrossRef] [PubMed]

44. Pampulha, M.E.; Loureiro-Dias, M.C. Combined effect of acetic acid, pH and ethanol on intracellular pH of fermenting yeast. *Appl. Microbiol. Biotechnol.* **1989**, *31*, 547–550. [CrossRef]

45. Van Maris, A.J.A.; Abbott, D.A.; Bellissimi, E.; van den Brink, J.; Kuyper, M.; Luttik, M.A.H.; Wisselink, H.W.; Scheffers, W.A.; van Dijken, J.P.; Pronk, J.T. Alcoholic fermentation of carbon sources in biomass hydrolysates by Saccharomyces cerevisiae: Current status. *Antonie van Leeuwenhoek* **2006**, *90*, 391–418. [CrossRef] [PubMed]

46. Ximenes, E.; Kim, Y.; Mosier, N.; Dien, B.; Ladisch, M. Deactivation of cellulases by phenols. *Enzyme Microb. Technol.* **2011**, *48*, 54–60. [CrossRef] [PubMed]

47. Ximenes, E.; Kim, Y.; Mosier, N.; Dien, B.; Ladisch, M. Inhibition of cellulases by phenols. *Enzyme Microb. Technol.* **2010**, *46*, 170–176. [CrossRef]

48. Cao, G.; Ximenes, E.; Nichols, N.N.; Frazer, S.E.; Kim, D.; Cotta, M.A.; Ladisch, M. Bioabatement with hemicellulase supplementation to reduce enzymatic hydrolysis inhibitors. *Bioresour. Technol.* **2015**, *190*, 412–415. [CrossRef] [PubMed]

49. Kim, D.; Orrego, D.; Ximenes, E.A.; Ladisch, M.R. Cellulose conversion of corn pericarp without pretreatment. *Bioresour. Technol.* **2017**, *245*, 511–517. [CrossRef] [PubMed]

50. Kim, D. Physico-Chemical Conversion of Lignocellulose: Inhibitor Effects and Detoxification Strategies: A Mini Review. *Molecules* **2018**, *23*, 309. [CrossRef] [PubMed]

51. Adeboye, P.T.; Bettiga, M.; Aldaeus, F.; Larsson, P.T.; Olsson, L. Catabolism of coniferyl aldehyde, ferulic acid and p-coumaric acid by Saccharomyces cerevisiae yields less toxic products. *Microb. Cell Fact.* **2015**, *14*, 149. [CrossRef] [PubMed]

52. Klinke, H.B.; Thomsen, A.B.; Ahring, B.K. Inhibition of ethanol-producing yeast and bacteria by degradation products produced during pre-treatment of biomass. *Appl. Microbiol. Biotechnol.* **2004**, *66*, 10–26. [CrossRef] [PubMed]

53. Kim, Y.; Kreke, T.; Hendrickson, R.; Parenti, J.; Ladisch, M.R. Fractionation of cellulase and fermentation inhibitors from steam pretreated mixed hardwood. *Bioresour. Technol.* **2013**, *135*, 30–38. [CrossRef] [PubMed]

54. Kim, D.; Ku, S. Beneficial Effects of Monascus sp. KCCM 10093 Pigments and Derivatives: A Mini Review. *Molecules* **2018**, *23*, 98. [CrossRef] [PubMed]

55. Kim, D.; Ku, S. Bacillus Cellulase Molecular Cloning, Expression, and Surface Display on the Outer Membrane of *Escherichia coli*. *Molecules* **2018**, *23*, 503. [CrossRef] [PubMed]

56. International Oranization for Standardization. *ISO 6869 Animal Feeding Stuffs—Method Using Atomic Absorption Spectrometry*; ISO: Geneva, Switzerland, 2000.

57. International Oranization for Standardization. *Validating UV/Visible Spectrophotometers*; ISO: Geneva, Switzerland, 1998; pp. 1–10.

58. International Oranization for Standardization. *International Standard*; ISO: Geneva, Switzerland, 2005.

59. Chen, Y.; Sharma-Shivappa, R.R.; Keshwani, D.; Chen, C. Potential of agricultural residues and hay for bioethanol production. *Appl. Biochem. Biotechnol.* **2007**, *142*, 276–290. [CrossRef] [PubMed]

energies

MDPI

Article

Carbon Debt Payback Time for a Biomass Fired CHP Plant—A Case Study from Northern Europe

Kristian Madsen [†] and **Niclas Scott Bentsen** *

Department of Geosciences and Natural Resource Management, University of Copenhagen, Roligheds vej 23, 1958 Frederiksberg C, Denmark; kma@viegandmaagoe.dk
* Correspondence: nb@ign.ku.dk
† Current Affiliation: Viegand Maagøe A/S, Nørre Farimagsgade 37, 1364 Copenhagen K, Denmark. Research was conducted, while affiliated with the Department of Geosciences and Natural Resource Management, University of Copenhagen.

Received: 22 January 2018; Accepted: 23 March 2018; Published: 31 March 2018

Abstract: The European Union (EU) has experienced a large increase in the use of biomass for energy in the last decades. In 2015, biomass used to generate electricity, heat, and to a limited extent, liquid fuels accounted for 51% of the EU's renewable energy production. Bioenergy use is expected to grow substantially to meet energy and climate targets for 2020 and beyond. This development has resulted in analyses suggesting the increased use of biomass for energy might initially lead to increased greenhouse gas (GHG) emissions to the atmosphere, a so-called carbon debt. Here, we analyze carbon debt and payback time of substituting coal with forest residues for combined heat and power generation (CHP). The analysis is, in contrast to most other studies, based on empirical data from a retrofit of a CHP plant in northern Europe. The results corroborate findings of a carbon debt, here 4.4 kg CO_2eq GJ^{-1}. The carbon debt has a payback time of one year after conversion, and furthermore, the results show that GHG emissions are reduced to 50% relative to continued coal combustion after about 12 years. The findings support the use of residue biomass for energy as an effective means for climate change mitigation.

Keywords: carbon debt; bioenergy; forest residues; payback time; combined heat and power (CHP)

---

1. Introduction

The recent decades, have seen a massive increase in the use of biomass for energy [1]. The majority of this increase has occurred in the European Union (EU), where bioenergy currently (2015) accounts for 51% of renewable energy production [2] and it is expected to increase further to meet the targets on renewable energy by 2020 [3,4]. A number of studies find that increased use of bioenergy creates a so-called carbon debt [5–8], which implies a period with larger greenhouse gas (GHG) emissions as compared to a continued fossil fuel use scenario. Assuming regrowth of harvested biomass, estimates of the payback time of this carbon debt have been as high as 400 years [9]. The implication of a carbon debt is that the contribution of bioenergy to climate change mitigation is delayed [10,11]. Buchholz et al. [12] conducted a meta-analysis of 59 carbon debt studies, and showed that the majority (47 studies) was based on hypothetical data and only a dozen were based on field data. Studies reporting long carbon debt payback times in general assume that the biomass is utilized for electricity production with low conversion efficiencies and that the woody biomass originates from the dedicated harvest of trees for energy from long rotation forestry [13]. Looking at the current use of bioenergy in the EU, there is little evidence that such supply chains dominate [14].

Data from EUROSTAT show that less than 1% of electricity production in the EU comes from solid biomass fired power plants [15]. Solid biomass is more prevalent in combined heat and power (CHP) and heat production that are plants feeding into district heating systems. 16.3% of heat production to district heating in the EU comes from solid biomass, while the majority comes from natural gas and coal [15]. Regarding the origin of the woody biomass used for bioenergy data are very sparse. Regarding wood chips, a survey by [14] found that the main source of wood chips for power and heat production in the EU was logging residues (36%) and whole trees from thinning operations (19%). Information from Drax power plant in the UK suggests similar sources [16]. These conclusions are in line with [3,17,18], who also estimate the current source of woody bioenergy as mainly wood waste, harvest residues, and thinnings. The objective of this study is to analyze the carbon dynamics involved in retrofitting a CHP plant from primarily coal firing to primarily wood firing, and to estimate the carbon debt payback time. Contrary to most other studies, which are based on hypothetical scenarios, this analysis benefits from the use of data from an existing power plant retrofit in northern Europe, which is considered to be representative for the use of biomass for CHP in the EU. This analysis is based on the carbon debt concept. Dehue [19] points out that there is no universally applied definition of 'carbon debt' and 'carbon debt payback time', leading authors to apply different definitions in an inconsistent manner. A definition often referred to is by Mitchell et al. [6], where the terms 'carbon debt', 'carbon debt repayment', and 'carbon offset parity point' are introduced. However, this definition only applies to bioenergy scenarios where the source of woody biomass comes from dedicated harvest and forest regrowth is included in the modelling. In contrast, bioenergy sources from wood waste and forest residues are resources that are generated independently of a bioenergy demand. The method that is used here is in line with the typical approach to carbon debt and payback time analyses [20], allowing for a comparison with other studies. In the following Section 2, we present the applied methodology, Section 3 presents the results of the analysis and their sensitivity to key assumptions. Section 4 discusses the findings relative to earlier research, leading to conclusions in Section 5.

2. Material and Methods

The analysis follows a life cycle inventory (LCI) approach quantifying the GHG emissions from both scenarios and ensuring comparability by using produced heat and electricity (1 GJ) as the common functional unit. The analysis estimates the global warming potential, and do not assess other environmental impacts, such as human toxicity, acidification, or resource depletion. The product system analyzed is illustrated in Figure 1.

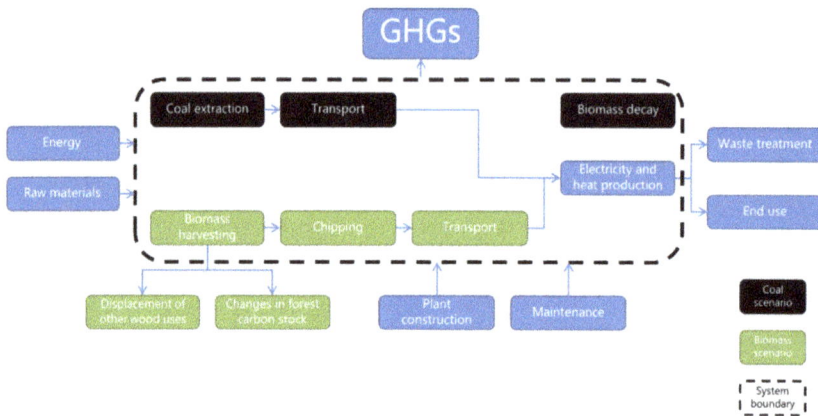

Figure 1. Overview of system boundaries and system components included in this analysis.

The study covers the direct emissions from the extraction and processing, transportation and the combustion of the fuels. It therefore excludes the embodied emissions in the used fuels or materials, e.g., the emissions related to produce diesel used for transportation of biomass or coal.

The system boundary also exclude emissions related to distribution and use of the produced heat and electricity together with emissions that are related to the end of life of the CHP plant. Furthermore, GHG emission related to indirect effects, e.g., indirect land use change or indirect wood use change, of biomass consumption are not considered.

The carbon debt concept is adopted from Mitchell et al. [6], but applied to waste and residue resources, as e.g., demonstrated by Sathre et al. [20] (Equation (1)).

$$NE = E_{bio} - \left(E_{fossil} + E_{decay} \right) \qquad (1)$$

where *NE* is the annual net GHG emission to the atmosphere, E_{bio} is the direct GHG emissions from the bioenergy supply chain including emissions from biomass combustion, E_{fossil} the direct GHG emissions from the counterfactual fossil supply chain, including emissions from fossil fuel combustion, and E_{decay} the GHG emissions from the counterfactual decay of forest residues. The payback time is determined as the time, where the time integrated *NE* = 0. (Figure 2).

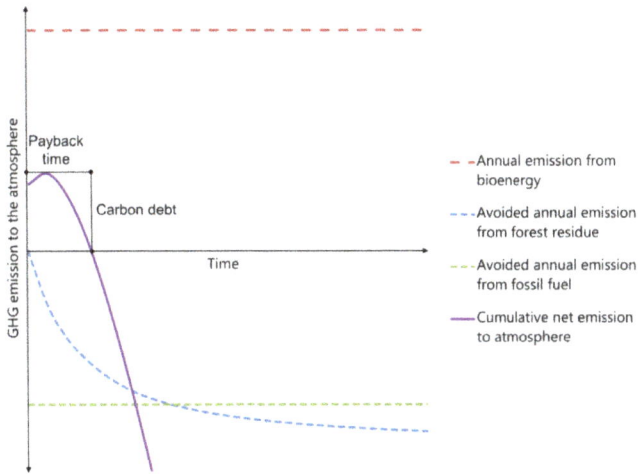

Figure 2. Conceptualized representation of carbon debt and carbon debt payback time with reference to the use of forest residues for energy. Induced greenhouse gas (GHG) emissions from bioenergy are positive, while avoided GHG emissions from the counterfactual fossil supply chain together with avoided GHG emissions from decaying forest residues are illustrated with negative values. Cumulative net emissions correspond to the time integrated difference between induced and avoided GHG emissions.

The conceptual carbon emission profile corresponds to modelled profiles for the use of stumps or branches for energy [20,21]. The payback time is understood as the point in time, where the bioenergy scenario starts to reduce the atmospheric GHG emissions relative to the counterfactual reference scenario. The combined heat and power (CHP) plant that is considered here is a medium sized Northern European utility feeding into a district heating network and the electricity grid. The plant has been converted from primarily coal firing to biomass firing, mainly wood chips. In retrofitting, the plant the boiler was replaced, while the rest of the plant needed little refurbishment. Detailed information about the plant has been anonymized on request. In contrast to assumptions applied by a number of other studies [6,7,22], the biomass scenario shows slightly higher fuel efficiency than the reference scenario, 85.9% as compared to 84.2%. Fuel efficiency is normalized with the power/heat ratio, which is influenced by different heat and electricity demands in 2005 and 2015. According to the head of production at the plant, the alternative to replacing the coal boiler with a biomass boiler would have been to keep the old coal boiler in operation and refurbish it.

Data regarding electricity and heat production has been acquired for the year 2015, where the plant was running exclusively on biomass along with data on the amount, type, and origin of the biomass fuel input. As reference scenario, the year 2005, where the plant was primarily firing coal, has been chosen. Had the plant not been converted into firing biomass, the existing coal boiler would likely have remained in place, with unchanged efficiencies and fuel inputs. It should be noted that during the conversion a flue gas condenser was installed to increase the overall efficiency of the plant. Heat production from this installation has been excluded from the 2015 data to ensure the comparability of general operation conditions between the two scenarios. As for 2015, the same data has been acquired for 2005.

The fuel inputs for 2005 (reference scenario) and for 2015 (biomass scenario) can be seen in Table 1. Regarding the reference scenario, the specifics concerning 'Other biomass', such as origin and type are not known. It is assumed that it shares the same characteristics as the biomass input in 2015. The other fuel inputs are all residue products, and therefore supply chain emissions that are related to forest harvest operations are omitted.

Table 1. Fuel inputs for the biomass year (2005) and the coal year (2015) based on energy content.

	Fuel Input 2005	Fuel Input 2015
	% (Based on Energy Content)	
Coal	74.47	0.41
Wood chips		54.98
Wood logs		24.61
Wood pellets		10.88
Olive seeds	7.60	
Other biomass	17.15	8.56
Meat and bone meal	0.17	
Carbon residues	0.06	
Land fill gas	0.56	0.57

Unlike coal, biomass is delivered in many smaller batches with varying characteristics and are piled up in the storage area, from where they are fired into the boiler. Because of the varying quality of every batch (difference in water content) biomass from several batches are mixed and fired into the boiler to achieve the greatest reliability and consistency. The average storage time at the plant is about 1–2 months, which gives some overlap between years. Data on biomass delivery from 2014 has been acquired to normalize the 2015 data against the total biomass input. The 'Other biomass' fraction is made up of various waste products from different industries consuming woody biomass that is not suitable for other purposes. 'Logs' are delivered at the plant and then chipped and fired into the boiler.

Data available to estimate GHG emissions are the amounts delivered, energy content (carbon content for coal), and the origin of the fuel. The transport distance has been calculated with Google Maps for road transport and with online shipping distance tools for sea transport. Specific data on fuel use for transport, chipping, and harvesting were not available, and standard values from the Biograce II [23] GHG accounting tool were used to estimate GHG emissions from biomass processing and transportation. For emissions related to coal, data from the Ecoinvent LCI database [24] was used, which include all of the emissions related to the extraction of coal from different locations around the world. The GHGs included in the analysis are CO_2, CH_4 and N_2O with the GWP of 1, 28, and 265, respectively. Data are presented in Table 2.

Table 2. GHG emission factors used in the study.

	Process		Unit	GHG Emission	Reference
Transport	Bulk carrier		MJ (tonnes km)$^{-1}$	0.24	Biograce II [23]
	Truck		g CO$_2$eq MJ^{-1}	93.3	Ibid
			g CO$_2$eq (tonnes km)$^{-1}$	22.39	Ibid
			MJ (tonnes km)$^{-1}$	0.84	Ibid
			g CO$_2$eq MJ^{-1}	93.95	Ibid
			g CO$_2$eq (tonnes km)$^{-1}$	79.48	Ibid
Processing	Wood chips	Residue harvest	g CO$_2$eq MJ^{-1} chips	1.48	Ibid
		Road transport (100 km)	g CO$_2$eq MJ^{-1} chips	0.61	Ibid
		Chipping	g CO$_2$eq MJ^{-1} chips	0.38	Ibid
		Total	g CO$_2$eq MJ^{-1} chips	2.47	Ibid
	Logs	Harvest	g CO$_2$eq MJ^{-1} chips	1.09	Ibid
		Road transport (100 km)	g CO$_2$eq MJ^{-1} chips	0.61	Ibid
		Chipping	g CO$_2$eq MJ^{-1} chips	0.38	Ibid
		Total	g CO$_2$eq MJ^{-1} chips	2.08	Ibid
	Wood pellets	Total	g CO$_2$eq MJ^{-1} pellets	17.99	Ibid
	Coal (origin)	Russia	kg CO$_2$eq tonnes^{-1} coal	280	Ecoinvent [24]
		Canada	kg CO$_2$eq tonnes^{-1} coal	865	Ibid
		Australia	kg CO$_2$eq tonnes^{-1} coal	91	Ibid
		Central and eastern Europe	kg CO$_2$eq tonnes^{-1} coal	233	Ibid
		Latin America	kg CO$_2$eq tonnes^{-1} coal	30	Ibid
		North America	kg CO$_2$eq tonnes^{-1} coal	99	Ibid
		Western Europe	kg CO$_2$eq tonnes^{-1} coal	352	Ibid
		South Africa	kg CO$_2$eq tonnes^{-1} coal	108	Ibid
		Mix	kg CO$_2$eq tonnes^{-1} coal	207	Ibid
Combustion	Wood		kg CO$_2$eq GJ^{-1} wood	113.9	IPCC [25]
	Coal		kg CO$_2$eq GJ^{-1} coal	95	Ibid
	Land fill gas		kg CO$_2$eq GJ^{-1} gas	54.7	Ibid

In order to set up a reference scenario, the realistic alternative use of the biomass must be determined. Based on a literature review and interview with some of the biomass suppliers to the plant, the most likely alternative is decomposition on the forest floor, either as logs or as branches. We use an exponential decay function, as described by Gustavsson et al. [26], to model the natural decay of biomass in the reference scenario:

$$M_t = M_o \cdot e^{-kt} \tag{2}$$

M_o is the initial dry mass of organic matter, M_t is the remaining mass at time t and k is the decay constant. The decay constant can be described as a function of the half-life period of a mass:

$$k = \frac{\ln(2)}{l} \tag{3}$$

l is the half-life period.

Figure 3 shows the decay rates for the different types of woody biomass used as fuel input, based on Gustavsson et al. [26]. They analyze a boreal forest system that is representative for most of the biomass supplied to the plant in the biomass scenario. Biomass is, however, also supplied from temperate forest ecosystems, which are assumed to have a higher decay rate. Using decay functions that were validated for boreal forest ecosystems on all biomass supplied to the plant represents a worst case scenario and avoids underestimation of carbon debt payback times. Slower natural decay rates of the biomass increases the payback time [13].

Other fuel inputs in the reference scenario, such as carbon residue and sawdust, are assumed to be used in various products. Carbon residue in the fly ash is a common additive in cement production, which is assumed to have a relatively long lifetime and in this analysis assigned a half-life period of 50 years. Other fuels, such as sawdust, which can be used as animal bedding and subsequently burned are assumed to have very short life times, and therefore assigned a half-life period of one year. These two types of fuels make up a marginal share of the total fuel consumption and do not have a significant impact on the overall GHG balance.

In both scenarios there is a relatively small input of landfill gas (<1%). Landfill gas is extracted from an old landfill and consists mainly of 55% methane (CH_4) and 35% carbon dioxide (CO_2). There are no alternative uses to landfill gas than burning it for energy production. If not used it would be emitted to the atmosphere as methane and CO_2. Since methane is a much more potent GHG than CO_2 (GWP of 28), there is a climate benefit from burning it. The avoided methane emissions have been converted into CO_2 equivalents and are subtracted from the final results.

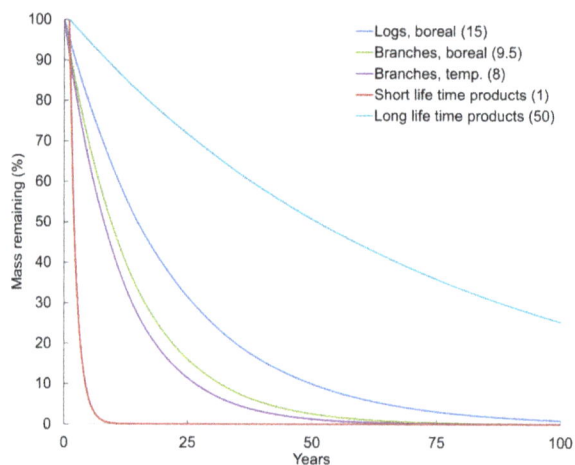

Figure 3. Decay rate of different biomass fractions and products for a 100-year period. Half-life periods in years are given in brackets.

Sensitivity Analysis

To test the robustness of the result, model parameters with the highest uncertainties are adjusted. Included in the sensitivity analysis is: (a) the decay rates (half-life periods) of biomass resources, (b) emissions factor for wood, and (c) the fuel efficiency of the plant (Table 3). The carbon content of the coal used in 2005 was known, allowing for accurate estimates of the CO_2 emissions from combustion. The carbon content of wood was not known, and therefore an average value of 113.9 kg CO_2eq GJ^{-1} has been used as recommended in the IPCC Guidelines for National GHG Inventories [25]. The IPCC Guidelines furthermore emphasize that the emission factor of wood can vary considerably, and therefore contribute to inherent uncertainty. It was not considered meaningful to analyze changing emission factors for transportation and processing of the fuels, since their share of the overall emissions was relatively small.

Table 3. Parameters and parameter values included in the sensitivity analysis.

Parameters	Changes Relative to Base Case
Half-life of decaying biomass	50%, 200%, 400%
Emission factor of wood	−10%, +10%
Biomass fuel conversion efficiency	−10%
Coal fuel conversion efficiency	+10%

3. Results

The total emissions per produced GJ are almost identical for both scenarios. Emissions from the biomass scenario are slightly higher by 4.4 kg CO_2eq GJ^{-1}, which represents the carbon debt, equaling 3.2% of the total emissions. Emissions from fuel combustion constitute the largest share with 92% in the reference scenario and 96% in the biomass scenario. Emissions from other sources than combustion, also referred to as supply chain emissions, constitute a minor share of the total emissions. The carbon debt more or less equals the emissions related to transportation in the biomass scenario. Figure 4 shows total CO_2eq emission per produced GJ allocated to combustion, processing and transportation. Avoided methane emissions from landfill gas have been included in both scenarios.

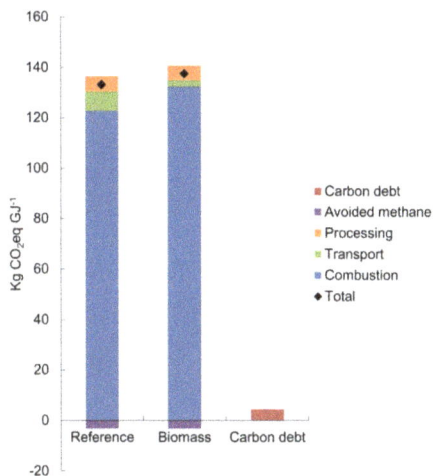

Figure 4. Direct GHG emissions in CO_2 equivalents per GJ produced for the biomass scenario and reference scenario respectively. The difference between the two scenarios is termed the carbon debt, which is paid back through avoided GHG emissions from decaying forest residues.

Emissions from processing are roughly identical for both scenarios; however, transport emissions are approximately three times higher in the reference scenario than the biomass scenario. This is in line with earlier research and is mainly attributable to longer transport distances for coal [11,27]. The carbon debt incurred in the transition from coal to biomass is primarily related to the higher carbon intensity of biomass when compared to coal due to a lower carbon to oxygen ratio in biomass. Lower supply chain emissions in the biomass scenario, on the other hand, reduces the carbon debt.

In this analysis, the carbon debt payback time is dependent on the decay rates of the different biomass resources supplied to the plant. The results are shown in Figure 5, which illustrates the difference in emissions between the two scenarios. It covers a 100-year period following the direct emissions in year 0.

We find that the payback of the carbon debt occurs almost immediately—within the first year. This means that after less than one year, the biomass-fired plant has a lower impact on global warming than the continued operation on coal would have had. The contribution to global warming mitigation increases further in the following years and throughout the modelling period. After 12 years, the cumulative GHG emissions in the biomass scenario will be half that of the reference scenario (Table 4). In other words, the global warming impact of the biomass scenario is reduced to half that of the reference scenario after 12 years.

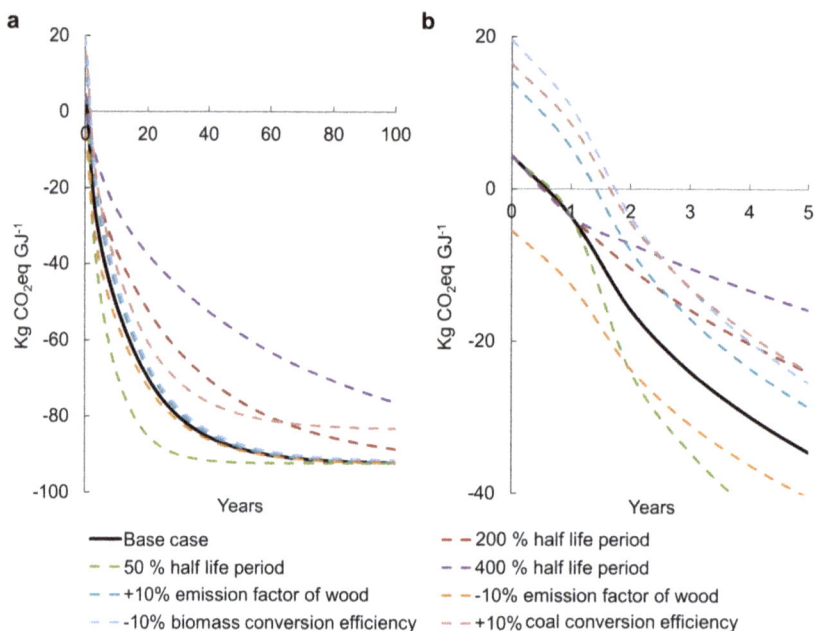

Figure 5. (a) Cumulative GHG emissions in a 100-year perspective from biomass use at the plant relative to continued use of coal (*x*-axis). The influence of varying biomass half-life periods, emission factors and plant efficiencies on the 100-year cumulative GHG emission profile of shifting from coal to biomass are illustrated with dotted lines. (b) Highlights the GHG emission profiles the first five years.

Table 4. Influence of the sensitivity analysis on estimated carbon debt and the period it takes to reach 50% reduction of GHG emissions in the biomass scenario compared to the reference scenario.

Sensitivity Parameter	Carbon Debt	50% Reduction Relative to Reference Scenario
	kg CO_2eq GJ^{-1}	Years
Base case	4.4	12
200% half-life period	4.4	22
50% half-life period	4.4	7
400% half-life period	4.4	43
10% increased emission factor of wood	14.2	14
10% decreased emission factor of wood	−5.4	10
10% decrease in biomass plant fuel efficiency	19.6	14
10% increase in coal plant fuel efficiency	16.5	14

Sensitivity

The test the sensitivity of the carbon debt payback time reported above key parameter values are changed returning alternative carbon debts (Table 4) and payback times (Figure 5). A slower decay rate of biomass left in the forest limits the benefits of the bioenergy scenario when compared to the reference scenario, at least in the short to medium term. Nevertheless, the biomass scenario will still achieve a rapid climate benefit compared to the reference scenario. The GHG emission factor for wood that is used in this analysis is generic. However, different tree species have different composition and ratios of carbon (C), hydrogen (H), and oxygen (O), and consequently different GHG emissions per unit of energy. If the emission factor of wood is increased by 10%, the carbon debt increases to 14.2 kg

CO_2eq GJ^{-1}, however, the payback time is only increased by one year. By decreasing the emission factor of wood by 10%, the carbon debt becomes negative to -5.4 kg CO_2eq GJ^{-1}. This implies that the direct emissions from the biomass supply chain are lower than those from the coal supply chain, and that the biomass scenario will be immediately beneficial to climate change mitigation when compared to continued use of coal. Adjusting the power plant efficiency influences the carbon debt and payback time to a similar degree. Changing biomass decay rates (by changing the half-life period) does not affect the carbon debt, but only the payback time. Doubling the half-life period slows down the decay rate and increase the carbon debt payback time. The time it takes to achieve 50% reduction of GHG emissions from heat and electricity production is increased from 12 to 22 years (Table 4). With a 50% reduction in half-life periods a 50% reduction in GHG emissions is achieved after seven years. In the extreme case, with a four times increase of biomass half-life periods, the 50% reduction is achieved after 43 years.

4. Discussion

This study contributes to the body of literature on the climate mitigation potential of biomass as a replacement for coal in CHP production. Zetterberg & Chen [28] and Cintas et al. [29] analyzed scenarios for the use of forest residues to displace coal in CHP production in Scandinavian countries. Both of the studies reported carbon payback times of 0 years, corroborating the findings reported here. Colnes et al. [30] looked at the use of mixed resources (residues, thinning wood, round wood) from south-eastern USA to displace coal in CHP production and found a payback time of 35 years. The longer payback time reported for SE USA may be attributable to the fact that not only residue resources was included in the supply chain, but also dedicated harvests, which would not have been left to decay, but left to continued growth and carbon sequestration. Repo et al. [31] reported on electricity production in Finland using forest residues to displace coal with a carbon payback time of 0 years, while Cherubini et al. [32] for heat production in Norway using mixed wood resources to displace coal found a carbon payback time of 20 years. In correspondence with Colnes et al. [30], Cherubini et al. [32] reports longer payback times for a supply chain, including mixed wood resources. Also, Gustavsson et al. [26], Repo et al. [33], Sathre & Gustavsson [34], and Pingoud et al. [35] report carbon payback times close to 0 years in corroboration with this analysis suggesting almost immediate climate benefits from using thinning and residue wood instead of coal for heat and electricity production. A large number of studies have been carried out in other parts of the world on scenarios with comparable fuel source transformation reported here. Zanchi et al. [8], on forest residues to displace coal for electricity production in Austria, reported a payback time of 0 years. Lamers et al. [36], McKechnie et al. [37], and Ter-Mikaeliean et al. [38], studied forest residues to displace coal for electricity production in North America and reported carbon payback times between 0 and 30 years. Walker et al. [7], Daigneault et al. [39] also reported for North America carbon payback times between 0 and 75 years using mixed wood resources to displace coal in electricity production. The use of mixed wood resources seems to extend the payback period and also the production of electricity only significantly reduces the overall conversion efficiency. Gustavsson et al. [26], furthermore, coupled the emissions data with a Cumulative Radiative Forcing (CRF) model, which suggest a slower increase in climate benefits than a simple GHG emission balance portrayed in this study does. This is due to the immediate emission pulse and the long atmospheric residence time of a CO_2 molecule.

Studies comparing a biomass scenario to other fossil fuel scenarios, such as natural gas and oil [26,33–35], find considerable 'carbon debts' with longer payback times; however in the long term they all end up being more beneficial for the climate than the fossil alternative. In this case study, it is not relevant to look at alternative fuels or technologies, since its purpose is to evaluate the climatic benefits of a decision already made. Other alternatives may have been ruled out on the basis of economical, technological, political, or geographical limitations. These can however change in the future, which can alter the results. The reference scenario is key to these types of analyses and the reference scenario for the used biomass can also very quickly change due to market dynamics. If, as an example, the biomass used in the plant, instead of decaying on the forest floor, would have been used for other wood products, the resultant carbon payback time would change considerably. Changes in fuel conversion efficiency have proved to have a large influence on carbon debt and payback times; see e.g., Mitchell et al. [6]. Here, the biomass scenario exhibited slightly higher fuel efficiency than the reference scenario, 85.9% when compared to 84.2%. If the coal boiler had been replaced with a more modern coal boiler instead of a biomass boiler, the fuel efficiency of the reference scenario would likely have been higher than the biomass scenario. The sensitivity analysis demonstrates that fuel conversion efficiency has an influence on the carbon debt, but the payback time is not influenced much.

The major uncertainty in this study lies in the assumed biomass decay rates, since the wood is sourced from a great number of different locations in different climate regions. The local conditions, especially temperature and humidity, have a strong influence on decay rates. Over the long term (>100 years), slower decay rates are less influential. More complex modelling of decay rates, such as the Yasso07 model [40], suggest a quicker decline in decay rates than the exponential functions, which could influence climate benefits in the medium term (50 to 100 years).

5. Conclusions

In conclusion, this study shows that the use of biomass for energy is not carbon neutral in absolute terms. In the transition from fossil to biomass resources increased GHG emissions to the atmosphere can be expected. These are, however, quickly countered by fossil fuel savings and ecosystem responses. For the transition from coal to forest residues that would otherwise decay, for combined heat and power production in northern Europe, we find that a carbon debt of 4.4 kg CO_2eq GJ^{-1} was paid back within one year. Furthermore, the results show that GHG emissions are reduced by 50% relative to continued coal combustion after about 12 years. While the results are sensitive to underlying assumptions, the sensitivity analysis showed little effect on carbon debts and payback times. For all scenario variations analyzed, the carbon debt is paid back within two years. The findings support the use of residue biomass for energy as an effective means for climate change mitigation, and for a transition away from fossil energy resources.

Author Contributions: K.M. and N.S.B. conceived and designed the study; K.M. performed the analysis; K.M. and N.S.B. wrote the paper.

Conflicts of Interest: The authors declare no conflict of interest.

References

1. Lamers, P.; Junginger, M.; Hamelinck, C.; Faaij, A. Developments in international solid biofuel trade—An analysis of volumes, policies, and market factors. *Renew. Sustain. Energy Rev.* **2012**, *16*, 3176–3199. [CrossRef]
2. European Commission. *Renewable Energy Progress Report*; European Commission: Brussels, Belgium, 2015; p. 4.
3. European Commission. *State of Play on the Sustainability of Solid and Gaseous Biomass Used for Electricity, Heating and Cooling in the EU*; European Commission: Brussels, Belgium, 2014; p. 34.
4. Bentsen, N.; Felby, C. Biomass for energy in the European Union—A review of bioenergy resource assessments. *Biotechnol. Biofuels* **2012**, *5*, 25. [CrossRef] [PubMed]
5. Holtsmark, B. Harvesting in boreal forests and the biofuel carbon debt. *Clim. Chang.* **2012**, *112*, 415–428. [CrossRef]
6. Mitchell, S.R.; Harmon, M.E.; O'Connell, K.E.B. Carbon debt and carbon sequestration parity in forest bioenergy production. *GCB Bioenergy* **2012**, *4*, 818–827. [CrossRef]
7. Walker, T.; Cardellichio, P.; Gunn, J.S.; Saah, D.S.; Hagan, J.M. Carbon Accounting for Woody Biomass from Massachusetts (USA) Managed Forests: A Framework for Determining the Temporal Impacts of Wood Biomass Energy on Atmospheric Greenhouse Gas Levels. *J. Sustain. For.* **2012**, *32*, 130–158. [CrossRef]
8. Zanchi, G.; Pena, N.; Bird, N. Is woody bioenergy carbon neutral? A comparative assessment of emissions from consumption of woody bioenergy and fossil fuel. *GCB Bioenergy* **2012**, *4*, 761–772. [CrossRef]
9. Lamers, P.; Junginger, M. The 'debt' is in the detail: A synthesis of recent temporal forest carbon analyses on woody biomass for energy. *Biofuels Bioprod. Biorefin.* **2013**, *7*, 373–385. [CrossRef]

10. Cowie, A.; Berndes, G.; Smith, T. *On the Timing of Greenhouse Gas Mitigation Benefits of Forest-Based Bioenergy*; IEA Bioenergy Executive Committee Statement: Dublin, Ireland, 2013; Volume 4.

11. Laganière, J.; Paré, D.; Thiffault, E.; Bernier, P. Range and uncertainties in estimating delays in greenhouse gas mitigation potential of forest bioenergy sourced from Canadian forests. *GCB Bioenergy* **2015**, *9*, 358–369. [CrossRef]

12. Buchholz, T.; Hurteau, M.D.; Gunn, J.; Saah, D. A global meta-analysis of forest bioenergy greenhouse gas emission accounting studies. *GCB Bioenergy* **2016**, *8*, 281–289. [CrossRef]

13. Bentsen, N.S. Carbon debt and payback time—Lost in the forest? *Renew. Sustain. Energy Rev.* **2017**, *73*, 1211–1217. [CrossRef]

14. Díaz-Yáñez, O.; Mola-Yudego, B.; Anttila, P.; Röser, D.; Asikainen, A. Forest chips for energy in Europe: Current procurement methods and potentials. *Renew. Sustain. Energy Rev.* **2013**, *21*, 562–571. [CrossRef]

15. Eurostat. *Electricity and Heat Statistics—Statistics Explained*; Euroatat: Amsterdam, The Netherlands, 2015.

16. DRAX. *Biomass Supply*; DRAX: Selby, UK, 2016.

17. Agostini, A.; Giuntoli, J.; Boulamanti, A. *Carbon Accounting of Forest Bioenergy: Conclusions and Recommendations from a Critical Literature Review*; Publications Office of the European Union: Ispra (Va), Italy, 2013.

18. Matthews, R.; Sokka, L.; Soimakallio, S.; Mortimer, N.; Rix, J.; Schelhaas, M.; Jenkins, T.; Hogan, G.; Mackie, E.; Morris, A.; et al. *Review of Literature on Biogenic Carbon and Life Cycle Assessment of Forest Bioenergy*; Final Task 1 Report; Forest Research: Farnham, UK, 2014.

19. Dehue, B. Implications of a 'carbon debt' on bioenergy's potential to mitigate climate change. *Biofuels Bioprod. Biorefin.* **2013**, *7*, 228–234. [CrossRef]

20. Sathre, R.; Gustavsson, L.; Truong, N.L. Climate effects of electricity production fuelled by coal, forest slash and municipal solid waste with and without carbon capture. *Energy* **2017**, *122*, 711–723. [CrossRef]

21. Stendahl, J.; Repo, A.; Hammar, T.; Liski, J. *Climate Impact Assessments of Forest Bioenergy Affected by Decomposition Modelling—Comparison of the Q and Yasso Models*; IEA Bioenergy: Dublin, Ireland, 2017.

22. Holtsmark, B. Quantifying the global warming potential of CO_2 emissions from wood fuels. *GCB Bioenergy* **2013**, *7*, 195–206. [CrossRef]

23. RVO. *Biograce II—Harmonised Greenhouse Gas Calculations for Electricity, Heating and Cooling from Biomass*; RVO: Utrecht, The Netherlands, 2016.

24. Ecoinvent. *Ecoinvent 3.2*; Ecoinvent: Zurich, Switzerland, 2015.

25. IPCC. *2006 IPCC Guidelines for National Greenhouse Gas Inventories, Prepared by the National Greenhouse Gas Inventories Programme*; Eggleston, H.S., Miwa, K., Srivastava, N., Tanabe, K., Eds.; IGES: Hayama, Japan, 2006.

26. Gustavsson, L.; Haus, S.; Ortiz, C.A.; Sathre, R.; le Truong, N. Climate effects of bioenergy from forest residues in comparison to fossil energy. *Appl. Energy* **2015**, *138*, 36–50. [CrossRef]

27. Weisser, D. A guide to life-cycle greenhouse gas (GHG) emissions from electric supply technologies. *Energy* **2007**, *32*, 1543–1559. [CrossRef]

28. Zetterberg, L.; Chen, D. The time aspect of bioenergy—Climate impacts of solid biofuels due to carbon dynamics. *GCB Bioenergy* **2015**, *7*, 785–796. [CrossRef]

29. Cintas, O.; Berndes, G.; Cowie, A.L.; Egnell, G.; Holmström, H.; Ågren, G.I. The climate effect of increased forest bioenergy use in Sweden: Evaluation at different spatial and temporal scales. *Wiley Interdiscip. Rev. Energy Environ.* **2016**, *5*, 351–369. [CrossRef]

30. Colnes, A.; Doshi, K.; Emick, H.; Evans, A.; Perschel, R.; Robards, T.; Saah, D.; Sherman, A. *Biomass Supply and Carbon Accounting for Southeastern Forests*; Biomass Energy Resource Center: Montpelier, VT, USA, 2012; p. 123.

31. Repo, A.; Tuomi, M.; Liski, J. Indirect carbon dioxide emissions from producing bioenergy from forest harvest residues. *GCB Bioenergy* **2011**, *3*, 107–115. [CrossRef]

32. Cherubini, F.; Bright, R.M.; Strømman, A.H. Global climate impacts of forest bioenergy: What, when and how to measure? *Environ. Res. Lett.* **2013**, *8*, 014049. [CrossRef]

33. Repo, A.; Känkänen, R.; Tuovinen, J.; Antikainen, R.; Tuomi, M.; Vanhala, P.; Liski, J. Forest bioenergy climate impact can be improved by allocating forest residue removal. *GCB Bioenergy* **2012**, *4*, 202–212. [CrossRef]

34. Sathre, R.; Gustavsson, L. Time-dependent climate benefits of using forest residues to substitute fossil fuels. *Biomass Bioenergy* **2011**, *35*, 2506–2516. [CrossRef]

35. Pingoud, K.; Ekholm, T.; Soimakallio, S.; Helin, T. Carbon balance indicator for forest bioenergy scenarios. *GCB Bioenergy* **2016**, *8*, 171–182. [CrossRef]

36. Lamers, P.; Junginger, M.; Dymond, C.C.; Faaij, A. Damaged forests provide an opportunity to mitigate climate change. *GCB Bioenergy* **2014**, *6*, 44–60. [CrossRef]
37. McKechnie, J.; Colombo, S.; Chen, J.; Mabee, W.; MacLean, H.L. Forest Bioenergy or Forest Carbon? Assessing Trade-Offs in Greenhouse Gas Mitigation with Wood-Based Fuels. *Environ. Sci. Technol.* **2011**, *45*, 789–795. [CrossRef] [PubMed]
38. Ter-Mikaelian, M.T.; McKechnie, J.; Colombo, S.; Chen, J.; MacLean, H. The carbon neutrality assumption for forest bioenergy: A case study for northwestern Ontario. *For. Chron.* **2011**, *87*, 644–652. [CrossRef]
39. Daigneault, A.; Sohngen, B.; Sedjo, R. Economic Approach to Assess the Forest Carbon Implications of Biomass Energy. *Environ. Sci. Technol.* **2012**, *46*, 5664–5671. [CrossRef] [PubMed]
40. Liski, J.; Palosuo, T.; Peltoniemi, M.; Sievänen, R. Carbon and decomposition model Yasso for forest soils. *Ecol. Model.* **2005**, *189*, 168–182. [CrossRef]

![energies logo] *energies*

MDPI

Article

Utilization of Energy Crops and Sewage Sludge in the Process of Co-Gasification for Sustainable Hydrogen Production

Adam Smoliński [1,*], Natalia Howaniec [2] and Andrzej Bąk [3]

1   Central Mining Institute, 40-166 Katowice, Poland
2   Department of Energy Saving and Air Protection, Central Mining Institute, 40-166 Katowice, Poland;
    n.howaniec@gig.eu
3   Department of Synthesis Chemistry, University of Silesia, 40-007 Katowice, Poland; andrzej.bak@us.edu.pl
*   Correspondence: smolin@gig.katowice.pl; Tel.: +48-32-259-2252

Received: 28 February 2018; Accepted: 29 March 2018; Published: 31 March 2018

Abstract: The increasing world energy demand driven by economic growth and technical development contributes to the severe depletion of conventional energy resources and various environmental issues. The need for the employment of low-emission, highly efficient technologies of thermochemical conversion, flexible in terms of both raw resources and product applications is declared, when the utilization of solid, alternative fuels is considered. Gasification is the proven technology of lower unit emission of contaminants and higher efficiency than combustion systems, as well as versatile applicability of the synthesis gas, as its main product. While the conversion of fossil fuels in gasification systems is technically mature, the co-utilization of biomass and waste still requires research and optimization in various technical and economic aspects. In this paper, the results of experimental work on co-gasification of energy crops biomass and sewage sludge with steam to produce hydrogen-rich gas are presented. The process is performed at 700, 800 and 900 °C under atmospheric pressure. The experimental results are analyzed with the application of the Hierarchical Clustering Analysis. The optimal results in terms of hydrogen production in co-gasification of selected biomass and sewage sludge are observed for *Helianthus tuberosus L.* blends of 10% w/w of sewage sludge content at 900 °C.

Keywords: biomass; sewage sludge; gasification; hydrogen; hierarchical clustering analysis (HCA)

---

1. Introduction

The main path of economic utilization of excess sewage sludge has been by now the land application as a fertilizer. Phosphorus and nitrogen content in sewage sludge is important for cultivation of crops of a long growing season, and the organic matter improves the soil structure. The environmental concerns, e.g., content of toxic metals, however give some limitations to its use as a source of organic and inorganic components valuable in terms of plant cultivation [1,2]. The shortage of conventional energy resources as well as costs and environmental issues related to the landfilling of the considerable amounts of excess sewage sludge raised the interest in the development of methods of its utilization for energy purposes [3–5]. Among these methods, special attention is given to thermochemical co-processing of sewage sludge in combustion, pyrolysis and gasification systems. Previous works reported results on co-combustion of sewage sludge with fossil fuels and biomass [6–10] and pyrolysis of sewage sludge and its blends [11–13]. Gasification is considered to be the technology giving the product, synthesis gas, of the widest application range among the thermochemical conversion methods. The recent works concern the production of syngas [14–17] or hydrogen-rich gas [18–20] in gasification and co-gasification of sewage sludge, as well as hybrid systems combining thermochemical and biotechnological processes [21,22]. The process of steam co-gasification of sewage sludge with coal is of particular interest for several reasons [23–25]. First, it gives the economic and operating benefits of stable supplies of high-energy density fuel, coal. Second, it makes use of waste material hardly applicable at large scale in any other economic process. Third, it enables the production of the prospective, clean energy carrier. However, to make the process less carbon-intensive, biomass should be considered as a component of a fuel blend instead of fossil fuels. Previous studies on thermal utilization of energy crops or biowaste in the gasification process using various gasification agents proved their applicability in syngas or hydrogen-rich-gas production [26,27]. In this paper, the experimental study on thermochemical processing of energy crops biomass *Helianthus tuberosus L.* and *Miscanthus x giganteus* and dried sewage sludge in steam co-gasification is presented. The main objective of the study is the assessment of the process performed in terms of the hydrogen-rich gas production efficiency. The experimental dataset is analyzed with the application of Hierarchical Clustering Analysis (HCA), which is one of the classical methods of data visualization and interpretation. The exploratory analysis of a studied dataset often starts with hierarchical clustering, which reveals the internal structure, i.e., its clustering tendency. HCA usually leads to sub-optimal clustering of objects (due to its hierarchical nature) and largely depends on the method used for clusters' linkage. Often, different linkage methods are applied to the same dataset and their performance is determined mainly by interpretability of the results. However, interpretability of clustering is not an easy task, especially when clustering is performed in high-dimensional space of parameters. HCA, employed in the study presented, proves to be an efficient and useful tool for the extraction of the essential information from the experimental dataset.

2. Materials and Methods

The samples of energy crops biomass: *Helianthus tuberosus L.* (HTL) and *Miscanthus x giganteus* (MXG) were provided by the Department of Agriculture, Faculty of Agricultural Sciences in Zamość (University of Life Sciences in Lublin, Poland) and by the experimental crops plantation in Föhren (Germany), respectively. The sewage sludge (SS) samples were collected in the municipal waste water treatment plant located in the Silesia region, Poland. The ultimate and proximate analyses of samples were performed in accredited laboratories of the Central Mining Institute according to the relevant standards and testing procedures (Table 1). The content of moisture, ash and volatiles were determined with the application of automatic thermogravimetric analyzers LECO: TGA 701 or MAC 500 according to PN-G-04560:1998 and PN-G-04516:1998 standards. The heat of combustion and calorific value were analyzed with the use of calorimeters LECO: AC-600 and AC-350 according to PN-G-04513:1981. The content of carbon, hydrogen and nitrogen were determined employing TruSpecCHN analyzer according to PN-G-04571:1998. The content of sulfur was analyzed on TruSpecS analyzer according to PN-G-04584:2001. The contents of oxygen and fixed carbon were calculated as differential factors according to formulas 100%-moisture-ash-carbon-hydrogen-sulfur-nitrogen (PN-G-04510:1991), and 100%-moisture-ash-volatiles (PN-G-04516:1998), respectively.

Table 1. Physical and chemical properties of samples tested.

Parameter	Helianthus tuberosus L.	Miscanthus x giganteus	Sewage sludge
Total moisture, %	8.81	6.78	2.24
Ash, %	3.18	1.60	35.39
Volatiles, %	69.24	76.00	54.96
Heat of combustion, kJ/kg	15,989	16,546	14,230
Calorific value, kJ/kg	14,543	14,942	13,410
Sulfur, %	0.04	0.05	1.2
Carbon, %	46.62	53.71	34.52
Hydrogen, %	5.64	6.59	4.98
Nitrogen, %	<0.01	<0.01	8.80
Oxygen, %	35.71	31.27	12.87
Fixed carbon, %	18.77	15.62	7.41

The experiments of steam co-gasification of energy crops biomass with sewage sludge were performed in a fixed bed reactor, under atmospheric pressure and at temperatures of 700, 800 and 900 °C. The fixed bed reactor is of 0.8 L working volume and is heated with a resistance furnace. Further details on the experimental stand may be found in [28]. The fuel blends tested were composed of *Helianthus tuberosus L.* or *Miscanthus x giganteus* biomass with 10 or 20% *w/w* of sewage sludge. The samples were placed at the bottom of the reactor and heated to the process temperature (700, 800 or 900 °C) in the inert gas atmosphere. Next, steam was injected to the gasifier with a flow rate 5×10^{-2} mL/s. The amount of the product gas and its composition were measured with a mass flow meter and gas chromatograph Agilent 3000A, respectively.

3. Results and Discussion

The process of sewage sludge utilization in steam co-gasification with *Helianthus tuberosus L.* or *Miscanthus x giganteus* biomass was investigated with a focus on hydrogen production. The total volumes of product gas and hydrogen were quantified (see Figure 1).

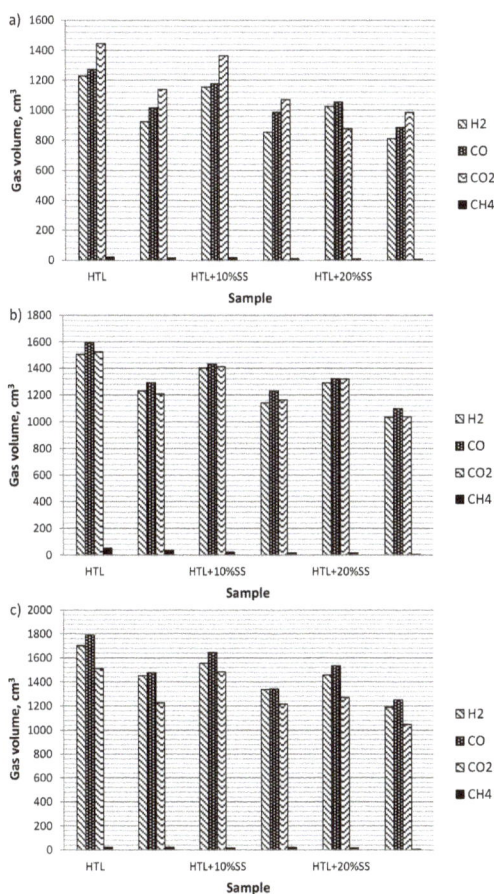

Figure 1. Volumes of the main gaseous components produced in steam co-gasification of *Helianthus tuberosus L.* or *Miscanthus x giganteus* biomass with sewage sludge at: (**a**) 700 °C; (**b**) 800 °C; and (**c**) 900 °C.

The amounts of gaseous product increased with temperature. Higher total volumes of produced gas and hydrogen were observed in gasification of biomass samples than in the co-gasification of the respective fuel blends irrespective of process temperature. The highest volume of hydrogen was reported for steam gasification of *Helianthus tuberosus L.* biomass at 900 °C. Addition of sewage sludge as a fuel blend component resulted in a deterioration of the gasification effects in terms of the total gas yield; the relevant decrease was from 5.6% in co-gasification of *Miscanthus x giganteus* biomass with 10% w/w of sewage sludge at 700 °C to 8.7% for co-gasification of *Helianthus tuberosus L.* biomass with 10% w/w of sewage sludge at 800 °C. Higher share of sewage sludge in fuel blends (20% w/w) caused more pronounced reduction in the total gas volume when compared to gasification of energy crops biomass: from 13.2% in co-gasification of *Miscanthus x giganteus* to 16.9% in co-gasification of *Helianthus tuberosus L.* blends at 700 °C. The focus of the experimental study performed was to determine the parameters for the most effective utilization of sewage sludge in hydrogen-rich gas production in the zero-emission co-gasification approach adopted. For this purpose, the Hierarchal Clustering Analysis (HCA) was applied [29–32]. HCA is one of the most effective methods of data visualization and interpretation. It is characterized by a similarity measure used and the way the similar objects are linked (single linkage, complete linkage, average linkage, centroid linkage or Ward linkage). The results of HCA are presented in the form of dendrograms. The experimental dataset explored within the study presented was organized into matrix **X** (18 × 5), in which rows correspond to studied fuel samples (listed in Table 2) processed at various temperatures, while columns represent the volumes of hydrogen, carbon monoxide, carbon dioxide, methane and the total product gas (Parameter Nos. 1–5).

Table 2. Objects of the Hierarchical Clustering Analysis—fuel samples gasified at various temperatures.

No	Sample	Temperature, °C
1	HTL	700
2	MXG	700
3	HTL	800
4	MXG	800
5	HTL	900
6	MXG	900
7	HTL + 10%SS	700
8	MXG + 10%SS	700
9	HTL + 10%SS	800
10	MXG + 10%SS	800
11	HTL + 10%SS	900
12	MXG + 10%SS	900
13	HTL + 20%SS	700
14	MXG + 20%SS	700
15	HTL + 20%SS	800
16	MXG + 20%SS	800
17	HTL + 20%SS	900
18	MXG + 20%SS	900

The HCA allowed investigating the similarities and dissimilarities between steam gasification and co-gasification experiments performed at various process temperatures. The resultant HCA dendrograms revealed the internal data structure in terms of objects in the parameter space and parameters in the object space, respectively, but did not allow interpreting these relationships simultaneously (Figure 2). That is why a color map of the experimental standardized data sorted according to the order of objects and parameters observed on the dendrograms was employed (see Figure 2c). The dendrogram presented in Figure 2a groups studied fuel samples (energy crops biomass and their blends with sewage sludge) gasified at 700, 800 and 900 °C into two main clusters. Cluster A collects *Helianthus tuberosus L.* biomass samples gasified at all studied temperatures, *Miscanthus x giganteus* samples processed at 800 and 900 °C (Object Nos. 1 and 3–6, respectively),

Helianthus tuberosus L. biomass blends with 10% w/w of sewage sludge processed at 700, 800 and 900 °C (Object Nos. 7, 9 and 11, respectively), blends of *Helianthus tuberosus L.* and *Miscanthus x giganteus* with 10% w/w of sewage sludge processed at 900 °C (Object Nos. 11 and 12, respectively) and *Helianthus tuberosus L.* biomass samples of 20% w/w content of sewage sludge gasified at 800 and 900 °C (Object Nos. 15 and 17, respectively). Cluster B groups *Miscanthus x giganteus* biomass sample processed at 700 °C (Object No. 2), *Miscanthus x giganteus* blends of 10% w/w of sewage sludge content processed at 700 and 800 °C (Object Nos. 8 and 10, respectively), *Helianthus tuberosus L.* blends of 20% w/w of sewage sludge processed at 700 °C (Object No. 13) and *Miscanthus x giganteus* blends containing 20% w/w of sewage sludge processed at all studied temperatures (Object Nos. 14, 16 and 18, respectively). Furthermore, within the main clusters, sub-clusters may further be distinguished. Namely, within Cluster A, there is Sub-Cluster A1 composed of *Helianthus tuberosus L.* biomass gasified at 700 °C, *Miscanthus x giganteus* biomass processed at 800 and 900 °C (Object Nos. 1, 4 and 6, respectively), *Helianthus tuberosus L.* blends of 10% w/w of sewage sludge content processed at 700 and 800 °C (Object Nos. 7 and 9, respectively), *Miscanthus x giganteus* blends of 10% w/w of sewage sludge content processed at 900 °C (Object No. 12) and *Helianthus tuberosus L.* blends containing 20% w/w of sewage sludge gasified at 800 and 900 °C (Object Nos. 15 and 17, respectively). The Sub-Cluster A2 contains *Helianthus tuberosus L.* biomass gasified at 800 and 900 °C (Object Nos. 3 and 5, respectively) and *Helianthus tuberosus L.* blends of 10% w/w of sewage sludge content processed at 900 °C (Object No. 11). Within Cluster B, Sub-Cluster B1 collecting *Miscanthus x giganteus* biomass and blends of 10% w/w of sewage sludge content processed at 700 °C (Object Nos. 2 and 8, respectively), *Helianthus tuberosus L.* blends containing 20% w/w of sewage sludge processed at 700 °C (Object No. 13) and *Miscanthus x giganteus* blends of 20% w/w of sewage sludge content gasified at 700 and 800 °C (Object Nos. 14 and 16, respectively) may be distinguished. Sub-Cluster B2 groups *Miscanthus x giganteus* blend of 10% w/w of sewage sludge content gasified at 800 °C (Object No. 10) and *Miscanthus x giganteus* blend of 20% w/w of sewage sludge content processed at 900 °C (Object No. 18).

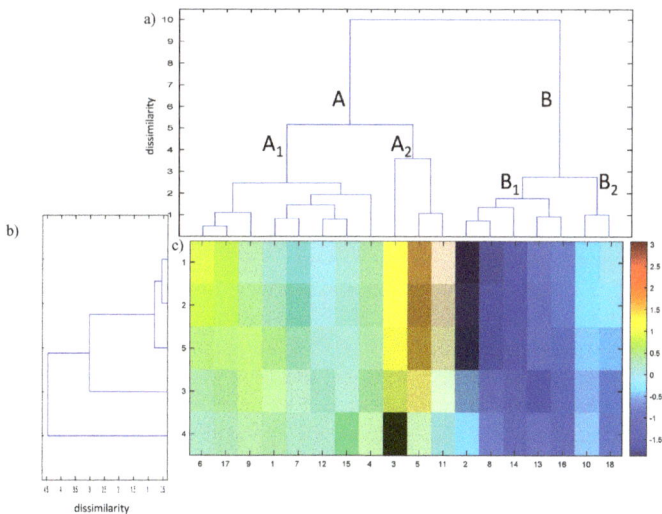

Figure 2. Dendrograms of: (**a**) studied biomass samples (objects listed in Table 2) in the space of measured parameters; and (**b**) parameters in the objects space; and (**c**) the color map of the studied data sorted according to the Ward linkage method.

The color map of studied data was employed in a more in-depth analysis of the clustering tendency between studied fuel samples gasified at various temperatures, sorted according to the Ward linkage method (see Figure 2c). The objects collected in Cluster A were characterized by relatively higher amounts of hydrogen and carbon monoxide produced in gasification and co-gasification tests than fuel samples collected in Cluster B. *Helianthus tuberosus L.* samples processed at 800 and 900 °C (Object Nos. 3 and 5) and *Helianthus tuberosus L.* blends of 10% *w/w* of sewage sludge content gasified at 900 °C (Object No. 11) (see Sub-Cluster A2) were unique because of the highest hydrogen, carbon monoxide, carbon dioxide and total gas volumes (Parameter Nos. 1, 2, 3 and 5, respectively) among all studied samples. This corresponds to the highest fixed carbon content in these fuels and high process temperature providing the energy for endothermic gasification reactions. Furthermore, the uniqueness of *Helianthus tuberosus L.* sample gasified at 800 °C (Object No.3) was observed due to the highest amount of methane produced among all studied samples. The biomass samples collected in Sub-Cluster B1 were characterized by relatively low amounts of gaseous products. Moreover, *Miscanthus x giganteus* blend of 20% *w/w* of sewage sludge content processed at 700 °C gave the lowest amounts of hydrogen, carbon monoxide and total amount of product gases among all studied samples. *Helianthus tuberosus L.* blend containing 20% *w/w* of sewage sludge and processed at 700 °C (Object No. 13) was unique due to the lowest amount of carbon dioxide generated. These results may be attributed to both the relatively lower fixed carbon and higher volatiles content than in the case of elements of Cluster A, and lower process temperature applied which was less favorable in terms of steam gasification efficiency. Based on the simultaneous interpretation of the dendrograms of studied fuels gasified at various temperatures in the space of studied parameters with the color map of studied data the optimal conditions for production of hydrogen may be determined. The highest amount of hydrogen was reported in *Helianthus tuberosus L.* gasification at the highest tested temperature. This observation suggests that the effective utilization of sewage sludge for hydrogen-rich gas production in co-gasification with selected biomass should be focused on *Helianthus tuberosus L.* Indeed, based on the results obtained, it was confirmed that the highest amount of hydrogen in co-gasification was reported for *Helianthus tuberosus L.* blends of 10% *w/w* of sewage sludge at 900 °C.

4. Conclusions

Application of sewage sludge as a fuel blend component results in a deterioration of the gasification process effects in terms of the total gas yield and hydrogen yield when compared to biomass gasification, but gives the benefits of zero-emission utilization of waste material for energy purposes.

The Hierarchical Clustering Analysis employed in the exploration of experimental dataset shows that, for the objects of Cluster A, higher amounts of hydrogen and carbon monoxide produced in gasification and co-gasification are reported than for objects of Cluster B.

Helianthus tuberosus L. samples processed at 800 and 900 °C and *Helianthus tuberosus L.* blends of 10% *w/w* of sewage sludge content processed at 900 °C give the highest hydrogen, carbon monoxide, carbon dioxide and total gas volumes.

Miscanthus x giganteus blend of 20% *w/w* of sewage sludge content processed at 700 °C gives the lowest amounts of hydrogen, carbon monoxide and total gas volume.

The optimal results in terms of hydrogen production in co-gasification of selected biomass and sewage sludge, determined based on HCA analysis of experimental dataset, are reported for *Helianthus tuberosus L.* blends of 10% *w/w* of sewage sludge content processed at 900 °C.

Acknowledgments: This work was supported by the Ministry of Science and Higher Education, Poland (Grant No 11157018).

Author Contributions: Adam Smoliński and Natalia Howaniec conceived, designed and performed the experiments; Adam Smoliński and Andrzej Bąk analyzed the data; Adam Smoliński, Natalia Howaniec and Andrzej Bąk wrote the paper.

Energies **2018**, *11*, 809

Conflicts of Interest: The authors declare no conflict of interest. The founding sponsors had no role in the design of the study; in the collection, analyses, or interpretation of data; in the writing of the manuscript, and in the decision to publish the results.

References

1. Singh, R.P.; Agrawal, M. Potential benefits and risks of land application of sewage sludge. *Waste Manag.* **2008**, *28*, 347–358. [CrossRef] [PubMed]
2. Camargo, F.P.; Tonello, P.S.; dos Santos, A.C.A.; Silveira Duarte, I.C. Removal of Toxic Metals from Sewage Sludge Through Chemical, Physical, and Biological Treatments—A Review. *Water Air Soil Pollut.* **2016**, *227*, 433. [CrossRef]
3. Garrido-Baserba, M.; Molinos-Senante, M.; Abelleira-Pereira, J.M.; Fdez-Güelfo, L.A.; Poch, M.; Hernandez-Sancho, F. Selecting sewage sludge treatment alternatives in modern wastewater treatment plants using environmental decision support systems. *J. Clean. Prod.* **2015**, *107*, 410–419. [CrossRef]
4. Kokalj, F.; Arbiter, B.; Samec, N. Sewage sludge gasification as an alternative energy storage model. *Energy Convers. Manag.* **2017**, *149*, 738–747. [CrossRef]
5. Ramos, A.; Monteiro, E.; Silva, V.; Rouboa, A. Co-gasification and recent developments on waste-to-energy conversion: A Review. *Renew. Sustain. Energy Rev.* **2018**, *81*, 380–398. [CrossRef]
6. Skoglund, N.; Bäfver, L.; Fahlströmd, J.; Holmén, E.; Renström, C. Fuel design in co-combustion of demolition wood chips and municipal sewage sludge. *Fuel Process. Technol.* **2016**, *141*, 196–201. [CrossRef]
7. Liu, J.; Huang, L.; Buyukada, M.; Evrendilek, F. Response surface optimization, modeling and uncertainty analysis of mass loss response of co-combustion of sewage sludge and water hyacinth. *Appl. Therm. Eng.* **2017**, *125*, 328–335. [CrossRef]
8. Xiao, Z.; Yuan, X.; Jiang, L.; Chen, X.; Li, H.; Zeng, G.; Leng, L.; Wang, H.; Huang, H. Energy recovery and secondary pollutant emission from the combustion of co-pelletized fuel from municipal sewage sludge and wood sawdust. *Energy* **2015**, *91*, 441–450. [CrossRef]
9. Lin, Y.; Liao, Y.; Yu, Z.; Fang, S.; Ma, X. The investigation of co-combustion of sewage sludge and oil shale using thermogravimetric analysis. *Thermochim. Acta* **2017**, *653*, 71–78. [CrossRef]
10. Niu, S.; Chen, M.; Li, Y.; Song, J. Co-combustion characteristics of municipal sewage sludge and bituminous coal. *J. Therm. Anal. Calorim.* **2018**, *131*, 1821–1834. [CrossRef]
11. Deng, S.; Tan, H.; Wang, X.; Yang, F.; Cao, R.; Wang, Z.; Ruan, R. Investigation on the fast co-pyrolysis of sewage sludge with biomass and the combustion reactivity of residual char. *Bioresour. Technol.* **2017**, *239*, 302–310. [CrossRef] [PubMed]
12. Ma, W.; Du, G.; Li, J.; Fang, Y.; Hou, L.; Chen, G.; Ma, D. Supercritical water pyrolysis of sewage sludge. *Waste Manag.* **2017**, *59*, 371–378. [CrossRef] [PubMed]
13. Zhang, W.; Yuan, C.; Xu, J.; Yang, X. Beneficial synergetic effect on gas production during co-pyrolysis of sewage sludge and biomass in a vacuum reactor. *Bioresour. Technol.* **2015**, *183*, 255–258. [CrossRef] [PubMed]
14. Hu, M.; Gao, L.; Chen, Z.; Ma, C.; Zhou, Y.; Chen, J.; Ma, S.; Laghari, M.; Xiao, B.; Zhang, B.; et al. Syngas production by catalytic in-situ steam co-gasification of wet sewage sludge and pine sawdust. *Energy Convers. Manag.* **2016**, *111*, 409–416. [CrossRef]
15. Gai, C.; Chen, M.; Liu, T.; Peng, N.; Liu, Z. Gasification characteristics of hydrochar and pyrochar derived from sewage sludge. *Energy* **2016**, *113*, 957–965. [CrossRef]
16. Zhu, J.G.; Yao, Y.; Lu, Q.G.; Gao, M.; Ouyang, Z.Q. Experimental investigation of gasification and incineration characteristics of dried sewage sludge in a circulating fluidized bed. *Fuel* **2015**, *150*, 441–447. [CrossRef]
17. Pinto, F.; Andre, R.N.; Lopes, H.; Dias, M.; Gulyurtlu, I.; Cabrita, I. Effect of experimental conditions on gas quality and solids produced by sewage sludge cogasification. Sewage sludge mixed with biomass. *Energy Fuel* **2008**, *22*, 2314–2325. [CrossRef]
18. Li, H.; Chen, Z.; Huo, C.; Hu, M.; Guo, D.; Xiao, B. Effect of bioleaching on hydrogen-rich gas production by steam gasification of sewage sludge. *Energy Convers. Manag.* **2015**, *106*, 1212–1218. [CrossRef]
19. Chao, G.; Guo, Y.; Liu, T.; Peng, N.; Liu, Z. Hydrogen-rich gas production by steam gasification of hydrochar derived from sewage sludge. *Int. J. Hydrogen Energy* **2016**, *41*, 3363–3372.

20. Chiang, K.Y.; Lu, C.H.; Liao, C.K.; Hsien-Ruen, R. Characteristics of hydrogen energy yield by co-gasified of sewage sludge and paper-mill sludge in a commercial scale plant. *Int. J. Hydrog. Energy* **2016**, *41*, 21641–21648. [CrossRef]
21. Methling, T.; Armbrust, N.; Haitz, T.; Speidel, M.; Poboss, N.; Braun-Unkhoff, M.; Dieter, H.; Kempter-Regel, B.; Kraaij, G.; Schliessmann, U.; et al. Power generation based on biomass by combined fermentation and gasification—A new concept derived from experiments and modelling. *Bioresour. Technol.* **2014**, *169*, 510–517. [CrossRef] [PubMed]
22. Speidel, M.; Kraaij, G.; Wörner, A. A new process concept for highly efficient conversion of sewage sludge by combined fermentation and gasification and power generation in a hybrid system consisting of a SOFC and a gas turbine. *Energy Convers. Manag.* **2015**, *98*, 259–267. [CrossRef]
23. Smoliński, A.; Howaniec, A. Co-gasification of coal/sewage sludge blends to hydrogen-rich gas with the application of simulated high temperature reactor excess heat. *Energy* **2016**, *41*, 8154–8158. [CrossRef]
24. Cormos, C.C. Hydrogen and power co-generation based on coal and biomass/solid wastes co-gasification with carbon capture and storage. *Int. J. Hydrogen Energy* **2012**, *37*, 5637–5648. [CrossRef]
25. Smoliński, A.; Howaniec, N. Thermal Utilization of Sewage Sludge in the Process of Steam Co-Gasification with Coal to Hydrogen-Rich Gas. In Proceedings of the 17th International Multidisciplinary Scientific GeoConference SGEM 2017, SGEM2017 Vienna GREEN Conference proceedings, Vienna, Austria, 27–29 November 2017; Volume 17, pp. 829–836.
26. Howaniec, N.; Smoliński, A. Effect of fuel blend composition on the efficiency of hydrogen-rich gas. *Fuel* **2014**, *128*, 442–450. [CrossRef]
27. Howaniec, N.; Smoliński, A. Influence of fuel blend ash components on steam co-gasification of coal and biomass—Chemometric study. *Energy* **2014**, *78*, 814–825. [CrossRef]
28. Smoliński, A. Coal char reactivity as a fuel selection criterion for coal-based hydrogen-rich gas production in the process of steam gasification. *Energy Convers. Manag.* **2011**, *52*, 37–45. [CrossRef]
29. Kaufman, L.; Rousseeuw, P.J. *Finding Groups in Data; an Introduction to Cluster Analysis*; Wiley: New York, NY, USA, 1990.
30. Romesburg, H.C. *Cluster Analysis for Researchers*; Lifetime Learning Publications: Belmont, CA, USA, 1984.
31. Howaniec, N.; Smoliński, A.; Cempa-Balewicz, M. Experimental study of nuclear high temperature reactor excess heat use in the coal and energy crops co-gasification process to hydrogen-rich gas. *Energy* **2015**, *84*, 455–461. [CrossRef]
32. Massart, D.L.; Kaufman, L. *The Interpretation of Analytical Data by the Use of Cluster Analysis*; Wiley: New York, NY, USA, 1983.

energies

MDPI

Article

Application of Rumen Microorganisms for Enhancing Biogas Production of Corn Straw and Livestock Manure in a Pilot-Scale Anaerobic Digestion System: Performance and Microbial Community Analysis

Wenyao Jin, Xiaochen Xu * and Fenglin Yang *

Key Laboratory of Industrial Ecology and Environmental Engineering (MOE), School of Environment Science and Technology, Dalian University of Technology, Linggong Road 2, Dalian 116024, China; jinwenyao@mail.dlut.edu.cn
* Correspondence: xxcepdut@sina.com (X.X.); yangfl@dlut.edu.cn (F.Y.); Tel.: +86-411-84706328 (X.X. & F.Y.)

Received: 20 March 2018; Accepted: 9 April 2018; Published: 13 April 2018

Abstract: This study aimed to assess the feasibility of rumen microorganisms inoculated in a modified pilot-scale system for enhancing biogas production of (1) solely corn straw (CS) and (2) CS with livestock manure under different solid contents and mixture ratios. The biogas liquid was proven to pretreat CS at this scale. The digestion system was started up within 32 days at a retention time of 20 days. The rumen culture was found to have a positive response to the impact on temperature and pH. The optimal solid content of CS was detected to be 3%, resulting in a stable biogas yield of 395 L kg^{-1}·total solid (TS)$^{-1}$. A higher biogas yield of 400 L kg^{-1}·TS^{-1} – 420 L kg^{-1}·TS^{-1} was achieved at a solid content of 10% organic loading rate (OLR, 4.42 kg volatile solid (VS) m^{-3}·d^{-1}) in co-digestion systems with CS and livestock manure. The methane content could be maintained at about 60%. Hydrogenotrophic methanogens were dominated by *Methanobacterium* in the solely CS digestion system, and two methanogenetic pathways, including hydrogenotrophic and acetoclastic methanogens by *Methanosarcina* and *Methanobacterium*, co-occurred for methane production during the co-digestion of CS with pig manure (PM). This study indicates that rumen microbes could be utilized in a pilot-scale digestion system and that they greatly promoted the biogas yield.

Keywords: rumen microorganism; anaerobic digestion; pilot-scale test; corn straw; livestock manure

1. Introduction

Agricultural solid organic wastes (ASOWs), which are produced as a result of agricultural activities performed by human beings, have been confirmed as one of the main factors influencing the sustainable development of China. Among ASOWs, lignocellulosic biomass and livestock manure have been given wider attention due to their large yield, resource waste, and serious environmental pollution [1]. According to statistics, an estimated 598 million tons of crop straw were produced in 2014, including 472 million tons of collectable straw [2]. The yield of livestock manure reached 2.121 billion tons in 2011, which will continue to increase rapidly to 2.875 billion tons by 2020 and 3.743 billion tons by 2030 [3]. Crop straw, accounting for 38.9% of total biomass resources, is considered an ideal candidate for clean energy production [4], whereas the annual pig manure (PM) accumulation of 1.036 million tons caused 2.5 times the emissions as open straw burning in China due to improper treatment [5].

Several methods are currently used to treat and manage these wastes, such as anaerobic digestion, straw incorporation, aerobic composting, biomass power generation, pyrolysis, etc. [6–8]. Anaerobic digestion is one of the most widely used methods for its clean and positive production. Moreover, the biogas produced from that amount of treated bio-waste could potentially solve the energy-crisis.

Fundamental conditions, such as temperature and pH, are crucial for anaerobic degradation because they play a more important role in microbial metabolism, which would further influence the biogas yield [9]. It will be important to learn to control these two factors in a larger digestion system and to investigate the impact of temperature and pH on degradation performance. Furthermore, numerous studies have attempted to develop a feasible system with nitrogen-rich and carbon-rich materials as co-substrates to satisfy the applicable C/N ratio range of 20–30 [10].

Lignocellulosic biomass, consisting of cellulose, hemicelluloses, and lignin, is a naturally carbon-rich material. Restricted by its structure, including the crystalline cellulose, etc., a variety of pretreatment methods have been proposed and explored to increase the biogas yield of lignocellulosic biomass [11]. In a previous study, biogas liquid, as a residuum of anaerobic digestion, was adopted to pretreat maize straw in a laboratory-scale test, which achieved a similarly successful performance with other conventional physical and chemical methods. Because it is more economical and operational [12], the lignocellulosic waste pretreated by biogas liquid would have great potential for a wide application in anaerobic digestion projects if it can be verified in a larger-scale test.

Hydrolysis is the key process for achieving biological conversion of cellulosic materials. Rumen microorganisms in ruminants are a natural cellulose-degrading system, which are known to be able to effectively digest lignocellulosic biomass for a long period of time [13]. Not only in vivo but also in vitro, several culture resources have been applied as the inoculum for anaerobic bioconversion systems, demonstrating that rumen microorganisms are superior to other microbes for the hydrolysis and degradation of cellulosic substrates [14,15]. High yields of volatile fatty acids (VFAs) greater than 0.7 g g^{-1}·VS^{-1} with high degradation efficiency greater than 75% have been obtained [16,17]. In addition to bioconversion of cellulosic substrates, other systems, such as palm oil mill co-digested with rumen fluid or anaerobic acidification of papermill sludge, have shown a 90.3% COD removal rate and 62% neutral detergent fiber (NDF) degradation efficiency, respectively [18,19]. However, it is commonly neglected that 23–27% of the methane produced by human activities comes from ruminants on account of the 10^8–10^{10} g^{-1} methanogens in the rumen microbial ecosystem [20,21]. This indicates that rumen cultures not only have an ability to degrade cellulosic materials but also great potential to further convert VFAs into biogas. In this regard, research with respect to using rumen microorganisms as a single inoculum to produce methane and diversities of archaea have not been reported in a one-stage anaerobic digestion system for the bioconversion of ASOWs.

Moreover, it has not been demonstrated in any field whether rumen microorganisms are adoptable in pilot-scale production. Compared with lab-scale tests, pilot studies could not only identify potential practical problems in the research procedure, but also develop and test the adequacy of research instruments [22]. Thus, anaerobic digestion of corn straw (CS) and livestock manure by rumen microorganisms should be conducted to fulfill a range of important functions and provide valuable insights for other researchers.

This study aims to develop a high-efficiency and sustainable method for the application of rumen microorganisms in a pilot-scale anaerobic digestion system for enhancing the biogas production of CS and livestock manure. First, pilot-scale pretreatment of CS with biogas liquid was explored. After that, a system with solely CS and a system with CS and livestock manure were adopted to investigate (1) the optimal solid content of solely CS and (2) the effect of different substrates on the anaerobic digestion system. In this process, temperature, pH, method of starting up, VFAs production, biogas yield, etc., were examined to reveal the inherent relevance of these important factors for a better understanding of the ruminal digestion processes. Archaeal diversities and communities were developed to further reveal the generated mechanism of biogas. Finally, mass balance was made to evaluate the effectiveness of the conversion treatment process. The results from this study are expected to be helpful for designing and operating modified systems for the high-efficiency anaerobic digestion of ASOWs.

2. Results and Discussion

2.1. Pretreatment of CS with Biogas Liquid

The properties of pilot-scale pretreatment of CS by biogas liquid are discussed in the following section. Unlike the laboratory-scale test, during the initial period of pretreatment, the temperature increased rapidly to about 60 °C, followed by a gradual fall to 38 °C at the end of pretreatment. The results revealed that calorigenesis had been induced by the abundant microbial activity, where the microbes from the biogas liquid were widely distributed on the pretreated CS. Furthermore, the pretreated CS smelled like vinasse, indicating that some amount of alcohol had been synthesized. This production of alcohol from sugars or easily degradable materials during pretreatment was similar to the process of lignocellulosic forage pretreatment and bioalcohol production [23].

The extent of the relative crystallinity of the pretreated CS in the pilot-scale test was verified by powder X-ray diffraction patterns (Figure 1a,b). There was a decrease in the interplanar spacing of the *d* value as 2θ equal to 18 or 20, where the relative crystallinity of $59.4 \pm 1.8\%$ of the pilot-scale pretreated CS was similar to the laboratory-scale test of 57.5% but obviously lower than the untreated samples at 80%. After pretreatment, many obvious distinctions were detected between the untreated CS and the pretreated CS by FTIR spectra (Figure 1c,d). Specifically, the most effective changes for the following hydrolysis of lignocellulosic biomass were the weakness of absorbance at 1735 cm^{-1} due to the C=O stretch of the ester carbonyl group and the great intensity of the vibrations at 1335 cm^{-1} assigned to the O–H in-plane bending of lyocell and hydrolyzed lyocell [24,25]. This indicated that the connection structure between lignin and hemicelluloses had been destroyed and crystalline cellulose was converted into amorphous cellulose. It could be noted that the pilot-scale pretreatment test met expectations for accelerating the disruption of the structure of CS.

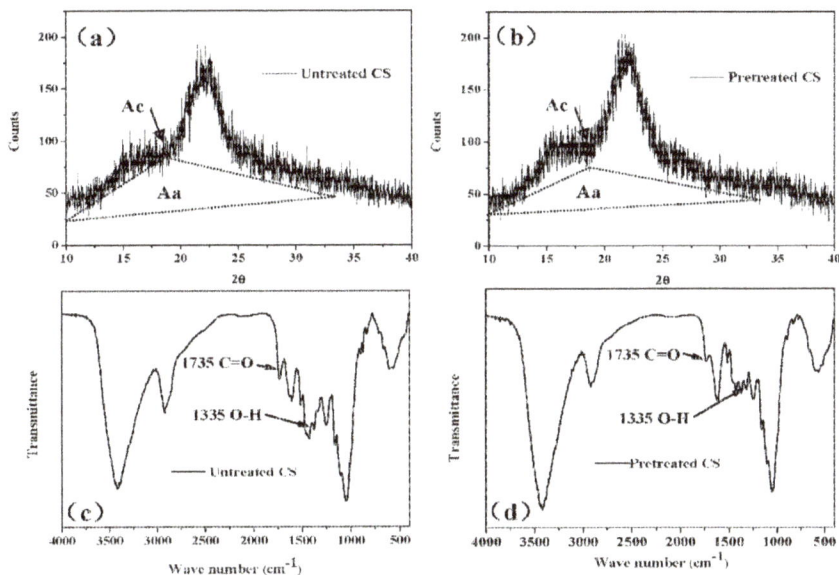

Figure 1. X-ray diffraction patterns and FTIR spectra of: (**a,c**) untreated corn straw (CS); and (**b,d**) pretreated CS with biogas liquid in the pilot-scale test. Ac and Aa represent the crystalline and amorphous portions of the X-ray diffractogram, respectively.

2.2. Temperature and pH Value

Figure S1 presents the temperature variation of the anaerobic digestion system. In the first three stages of the test, the electric tracing band was used to maintain the digestion temperature. However, the repetitive breakdown of this heating system resulted in a temperature fluctuation twice at 60 days and 85 days as a result of the weather turning cold suddenly. It was reported that use of biogas as a heat source is beneficial in terms of greenhouse gases and fossil fuel use [26]. In this case, the heating system was updated by using a biogas boiler for water heating together with rubber foam insulation material adhering to the external surface of the reactor. This modification was closer to the full-scale anaerobic digestion treatment plant [27], which could not only reduce the waste of energy but also lead to the stability of microbial metabolism.

The pH value of the biodegradation process was also explored (Figure S2). In stage I, 7.44 kg Na_2CO_3 and 19.7 kg $NaHCO_3$ were dissolved into the preheated substrate to avoid microbial acidosis, followed by the inoculation the rumen microorganisms. After adaption for a few days, the pH decreased to 6.1 within 7 days, during which the rumen microbial activity had recovered by achieving effective hydrolysis and acidogenesis. The pH value gradually stabilized at 6.8 and 6.9 in the end of stage I and stage II, respectively. It was demonstrated that the VFAs had been converted. Improving the solid content to 3%, the pH value decreased to 6.7 and stabilized at 6.5–6.6 in 80 days. Cellulose degradation increased with the pH from 6.0 to 7.5, whereas at a pH < 6.5, there was a negative effect on acidogenesis [28]. Therefore, the optimal solid content for the anaerobic digestion of carbon-rich materials was achieved at 3% without artificial factors. In consideration of the low C/N ratio of the CS, 2.6 kg of urea was added to the system to improve the nitrogen nutrient level of the digestion system. Nevertheless, unlike the fermentation cultures that could metabolize urea in rumen, the pH value increased up to 7.9 and still had to be adjusted with NaOH in stage IV at a solid content of 5% [29].

In stage V, livestock manure was co-digested with the CS. The pH value could not remain stable when 1% or 2% solid content of PM was mixed into the reactor. Specially, compared to the alkalescent cultures of the conventional anaerobic digestion systems, such as that in the literature by Zhou et al., at pH 7.01–7.25, a stable pH value was observed at 6.65 for the anaerobic digestion of both PM and CM by improving the solid content to 10% [30]. Regulating the ratio of substrates to 3:7, the pH value increased to 6.9 due to the high nitrogen content in CM. It was reported that a pH value of 6.5–8.2 was the optimal range for the growth of methanogens, and when the pH is below 6.6, the growth and methane productivity of methanogens is severely inhibited [31,32]. The lower pH value may be due to the fact that high-efficiency hydrolysis and acidogenesis occurred to degrade the high solid content CS with rumen microorganisms. It was demonstrated that these cultures were in accordance with the metabolic environment for cellulolytic microbes, acidogens, and methanogens.

2.3. VFAs Production

VFAs are a very useful index to reflect not only the microbial activity in the four phases of anaerobic digestion but also the comprehensive performance of degradation. Figure 2 presents the concentration of acetate, total volatile fatty acid (TVFA), the acetate/TVFA ratio, and the concentration of propionate, butyrate, and isobutyrate, respectively. During the initial days of starting up, acetate, propionate, and butyrate accumulation was observed up to 7560 mg·L^{-1}, 500 mg·L^{-1}, and 750 mg·L^{-1}, respectively, followed by a sharp decline to 3600 mg·L^{-1}, 220 mg·L^{-1}, and 250 mg·L^{-1}, respectively, at the end of stage I. The increase in VFAs indicated that the ruminal inoculum had adapted the environment to degrade the CS in vitro; meanwhile, the consumption for most VFAs proved that the methanogens had shown activity in utilization of VFAs.

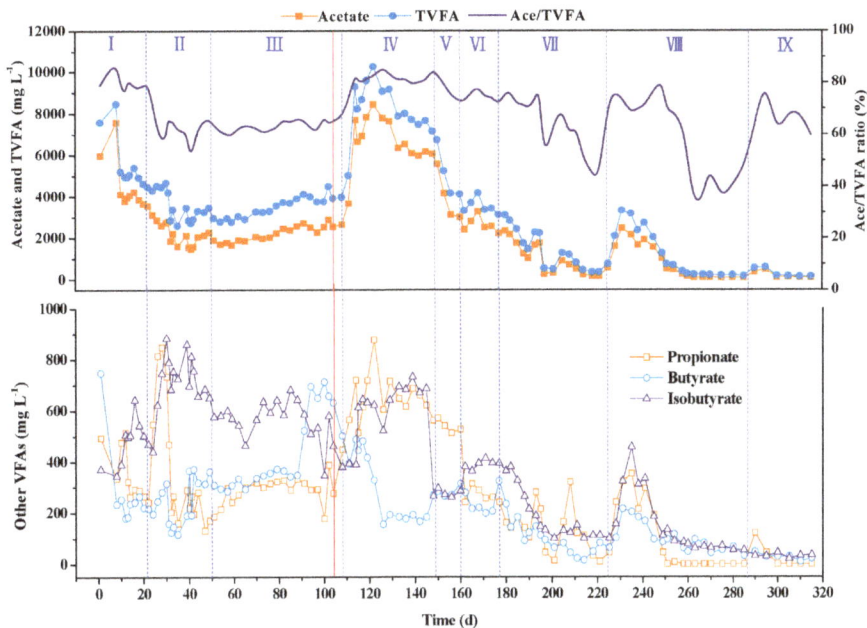

Figure 2. Concentrations of VFAs in the supernatant of anaerobic digestion. Ace = acetate.

The concentrations of VFAs, except isobutyrate, declined continuously in the process of stage II. Smooth growth of acetate, propionate, butyrate, and TVFA was obtained at 2000 mg·L^{-1}, 300 mg·L^{-1}, 350 mg·L^{-1}, and 3800 mg·L^{-1}, respectively by improving the solid content to 3%. The concentrations of VFAs were not affected by providing nitrogen to the digestion system. In order to investigate the performance on a higher solid content, 5% CS was added, which caused an obvious accumulation of VFAs, except butyrate. The concentrations of acetate and TVFA achieved at about 6200 mg·L^{-1} and 7500 mg·L^{-1}, respectively, at the end of stage IV were double and triple the concentrations of those with 3% solid content. Incorporating the tendency of the pH value, it could be verified that the VFAs could not be converted to methane completely in any metabolic pathway by methanogens, resulting in a maximum solid content of 3% for solely CS in the ruminal digestion system. Adequate branched-chain volatile fatty acids are required for the growth of most ruminal fiber-degrading microorganisms [33]. On the one hand, the opposite trend of isobutyrate during the first three stages may be due to the inhibition of microbial growth. On the other hand, high hydrogen partial pressure resulting from an accumulation of VFAs could decrease the conversion rate of long-chain fatty acids [34]. Overall, the activities of cellulolytic microbes and acidogens were not inhibited during any stage of part II, indicating that the micronutrients contained in the reactors and the C/N ratio could contend with such microbial survival conditions [35]. In other words, the accumulation of VFAs was mainly because of the low activity of the methanogens.

In stage V, part III, the livestock manure was added to the reactor, which formed a co-digestion system with the CS. A continuous decrease for all of the VFAs was detected with the addition of PM from 1% to 5% solid content. The concentrations of VFAs were similar to those with the 3% solid content CS digestion system in stage V and stage VI, indicating that the generated VFAs could not be entirely metabolized by the methanogens. The addition of PM was believed to provide buffer capacity and basicity to the anaerobic digestion system and improve the VFA consumption rate by methanogens [36]. In this context, the concentrations of VFAs, including acetate, propionate, butyrate, and isobutyrate, achieved at the end of stage VII were 150 mg·L^{-1}, 10 mg·L^{-1}, 150 mg·L^{-1}, and 100 mg·L^{-1}, respectively. This was in good agreement with the research by Kim et al. in which the reactor performance had achieved a steady state when the concentrations of VFAs were maintained at a lower level [37]. In conclusion, the solid content of this ruminal digestion system could be performed at 10% (i.e., a 1:1 mixing ratio of carbon-rich materials to nitrogen-rich materials).

CM was adopted to replace the PM to investigate the degradable properties of different manures in stage VIII. An apparent lack of adaptation was observed by the accumulation of VFAs. This process lasted for about 30 days, and then, the digestion system recovered to the previous status. Propionate was under the detection limit, and the concentrations of acetate, butyrate, and isobutyrate achieved were below 100 mg·L^{-1}, 60 mg·L^{-1}, and 50 mg·L^{-1}, respectively. Improving the solid content of CM to 7%, there was almost no fluctuation in the yield of VFAs. It could be deduced that this digestion system had an ability to resist the impact load.

Acetate is one of the main raw materials for synthesizing methane. Unlike fermentative culture in vivo, the acetate/TVFA ratio is more important than the acetate/propionate ration in a biogas production system [38]. It was demonstrated from the overall process of the pilot-scale test that when a phenomenon of inhibition of microbes or incomplete metabolism of intermediate products occurred during digestion, the acetate/TVFA ratio was above 0.6 in most cases. When the ruminal digestion system worked in good condition, the acetate/TVFA ratio decreased to nearly 0.4. This ratio may be considered an indirect index reflecting the properties of anaerobic digestion.

2.4. Biogas Yield

Figure 3 provides details of the total biogas yield and biogas per kilogram produced by the anaerobic digestion system. At the beginning of the inoculation and starting up, a relatively low yield of biogas was observed on account of insufficient methanogens in the original rumen microorganisms. The total biogas yield rapidly increased up to 410 L·d^{-1} on day 25 and then gradually stabilized at 500 L·d^{-1} on day 33. The biogas yield per kilogram was about 385 L kg^{-1}·TS^{-1}. Examining the biogas content at the beginning of degradation in Figure 4, it appeared that hydrogen was firstly detected in the reactor. The hydrogen liberated by hydrogenase enzymes arose from either phosphoroclastic reactions or nicotinamide adenine dinucleotide (NADH) during the drastic hydrolysis of CS by the rumen microorganisms [39]. Methane and carbon dioxide were detected almost at the same time on day 11. After a sharp increase from day 11 to day 25, the peak content of methane was achieved at 73.44%, followed by a gradual decline to about 60% on day 33. The variation tendency of methane was similar to the results of Zhao et al. [40], demonstrating that the acclimation of methanogens in the rumen microorganisms could be conducted within 25 days for biogas production. The content of carbon dioxide gradually rose up to about 40% on day 32. At this point, it could be identified from the stable biogas yield and biogas components that the ruminal pilot-scale anaerobic digestion system was operating successfully.

Remarkably, the production of biogas could be restored rapidly within 10 days after repairing the equipment in stage III, which implied that this improved system could resist the impact load of temperature. The final yields of total biogas and biogas per kilogram were 1550 L·d^{-1} and 395 L kg^{-1}·TS^{-1} at the end of stage III, respectively. A negligible impact on biogas production was realized after adding urea to the reactive system, whereas the biogas yield fell below 250 L·d^{-1} at a later improvement of the solid content of CS to 5%. Even though the biogas yield gradually recovered to 1494 L·d^{-1} and 230 L kg^{-1}·TS^{-1} on day 137, there was still a wide gap between the actual value and the predicted value. After that, no matter what feasible solutions were employed, such as adjusting the pH or supplying nitrogen nutrient, the biogas yield could not continue rising. It was found from the previous pH value that the ruminal digestion system had reached the ultimate solid content of CS at 3%. Further studies by Chandra et al. [41] proved that the high carbon content and low nitrogen content of CS led to a decrease in methanogenic activity, as also detected by the accumulation of VFAs at stage IV.

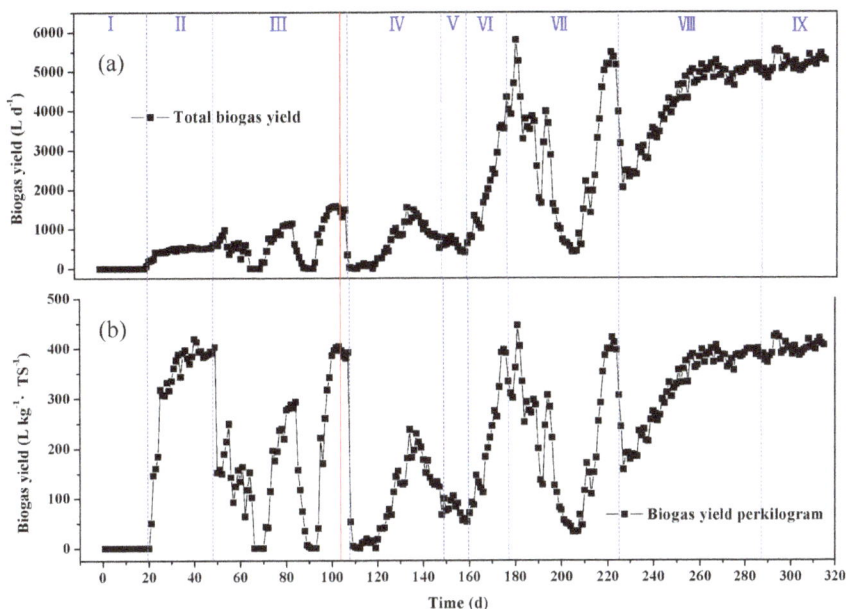

Figure 3. Biogas yield of the anaerobic digestion process in the pilot-scale test, (**a**) total biogas yield and (**b**) biogas yield per kilogram.

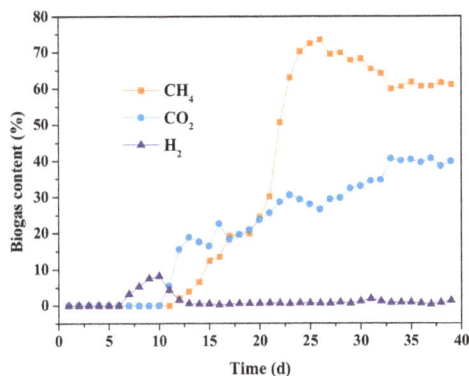

Figure 4. Biogas contents for the anaerobic digestion process in pilot-scale test.

In terms of a theory on ruminant nutrition, urea was used as the nitrogen source for improving the C/N ratio of this system [29]. Nonetheless, from the experimental results of gas production, the properties of anaerobic digestion were not significantly enhanced. The urea could be utilized by the rumen microorganisms, whereas most of the urea was converted into the alkalescent $(NH_4)_2CO_3$. Free ammonia nitrogen (FAN) was described as the primary cause of digester failure because of its direct inhibition of microbial activity [42]. However, according to an early study by Fernandes et al. [43], the highest FAN concentration was calculated at about 100 mg·L^{-1} in this period, which is far below the inhibitory levels. Additionally, the concentrations of accumulated VFAs, especially for propionate, had not reached the inhibitory concentrations, and the pH value was still at a suitable range for methanogenic metabolism [32,44]. Consequently, the inhibition of methanogens was not affected by the simple superposition of various factors. How to improve the sole carbon-rich solid content in these ruminal digestion systems should be further studied. Despite these facts, the biogas yield from the anaerobic digestion of CS by the rumen microorganisms was still higher than other studies, such as where Yu et al. [45] compared six different digestates as inoculums and achieved a maximum biogas yield of 325.3 mL $g^{-1} \cdot VS^{-1}$ from digested dairy manure.

A sustained inhibition of anaerobic digestion was observed at the beginning of part II with the addition of PM. Subsequently, when the solid content of the process was improved to 7% (i.e., 3% PM and 5% CS), the biogas yield was then sharply increased to 3500 $L \cdot d^{-1}$ and 390 L $kg^{-1} \cdot TS^{-1}$, and methanogenic utilization of VFAs was re-established. The frequency of adjusting the pH was extended but still unable to work on self-adaption. With the further increase of the solid content of PM to 5%, unfortunately, a malfunction of the screw conveyor caused fluctuations of the biogas yield from the maximum 410 L $kg^{-1} \cdot TS^{-1}$ to the minimum 30 L $kg^{-1} \cdot TS^{-1}$ at least three times. After attempting to restore repeatedly, the screw conveyor was repaired on day 207, and then, the biogas yield rapidly rose to 415 L $kg^{-1} \cdot TS^{-1}$ within 13 days corresponding to a total biogas yield of 5200 $L \cdot d^{-1}$. In stage VIII, CM was used instead of PM. An obvious adaptation problem of the methanogens resulted in a 50% reduction of the biogas yield. Afterwards, it took about 30 days to recover the methanogenic activity and to observe the sustained production of biogas at 5100 $L \cdot d^{-1}$ and 400 L $kg^{-1} \cdot TS^{-1}$. Adjusting the mixing ratio of CM and CS to 7:3, there was not a conspicuous impact on the biogas yield as in a previous stage. On the contrary, a minor increase was detected at 5300 $L \cdot d^{-1}$ and 420 L $kg^{-1} \cdot TS^{-1}$. The biogas content was perturbed by the fluctuation of the system. However, it could recover to stable levels soon after being repaired, and then, the components of biogas including CH_4, CO_2, and H_2 could be maintained at 60%, 40%, and 0%, respectively.

Compared with the sole CS digestion system, anaerobic co-digestion of livestock manure and CS provided great enhancement not only of the biogas production properties but also of the solid content. These improvements may be due to two main factors. First, the solid waste from farm animals has been extensively investigated as an alternate source of concentrates to feed ruminants, with these wastes potentially supplying sufficient crude protein, nitrogen, and mineral element nutrition for ruminal cellulolytic microorganisms [46]. The pathogenic microorganisms and heavy metals in livestock manure have a negative effect on ruminal organs rather than rumen microorganisms [47]. Second, when using PM or CM as co-substrates, extra nitrogen including other nutrients could be provided to meet the needs of methanogens [10]. Mata-Alvarez et al. also published a critical review on anaerobic co-digestion achievements in recent years [10]. In comparison with the performance summarized in that article, the ruminal anaerobic digestion system had many advantages. First, the organic loading rate (OLR) of this study could be converted into more than 4.42 kg VS $m^{-3} \cdot d^{-1}$, which is higher than all the mesophilic co-digestion systems and closer to the maximum value of thermophilic co-digestion systems using cow manure and PM as the main substrate. The higher OLR reflected that the ruminal digestion system had excellent degradation efficiency and a rational retention time. Second, the mixture ratio of CS could reach 50% of the total solid (TS), which was relatively higher than other operations. It was noted that the factors restricting the large-scale addition of carbon-rich substrates to anaerobic digestion systems had been solved by this improved system. It potentially provides a sustainable technology for solving environmental pollution problems caused by straw, especially for developing countries. Finally, the most notable of these features was the biogas yield. Xie et al. [48] conducted a similar pilot-scale test of a continuous stirred tank reactor (CSTR), and they detected 251 mL CH_4 $g^{-1} \cdot VS^{-1}$ at an organic load rate of 1.74 kg VS $m^{-3} \cdot d^{-1}$ with a hydraulic retention time of 30 days. The methane yields obtained from the PM and CM co-digestion systems of this study were 276 L $kg^{-1} \cdot TS^{-1}$ and 267 L $kg^{-1} \cdot TS^{-1}$, respectively, which are relatively higher. Better performance in biogas production would bring more potential economic benefits, thus making large-scale application more beneficial.

2.5. Microbial Community Analysis

Unlike the original rumen fermentation system, most of the VFAs in this new established digester were converted to CH_4 and CO_2 by archaea rather than absorbed by the stomach wall. In this case, archaeal diversities and communities might be crucial to reveal the metabolic and generated mechanism of biogas. The operational taxonomic unit (OTU) numbers and alpha diversity, including ACE estimation, Chao1, coverage, and Shannon index were summarized in Table 1. All the coverage values were above 0.99, suggesting that most of the archaea were detected. The archaeal communities showed more complicated patterns in the co-digestion system than that in the original rumen archaea and the solely CS period, suggesting that both the digestion condition in vitro and the co-substrates had positive effects on richness of archaea.

Table 1. Richness and diversity indexes of archaeal communities under different conditions.

Sample	OTU	ACE	Chao1	Coverage	Shannon
Original rumen archaea	44	46	45	0.999608	0.97
Days 100 (solely CS)	74	82	83	0.999356	1.38
Days 220 (CS with pig manure (PM))	100	104	102	0.999335	2.4

As shown in Figure 5, the most abundant archaea in the genus level was *Methanobrevibacter* in the original rumen microorganisms, the relative abundance of which was 93.42%. Genus *Methanobrevibacter* is known to synthesize CO_2 and H_2 to CH_4 in ruminal fermentation systems, indicating that hydrogenotrophic pathway was dominant [49].

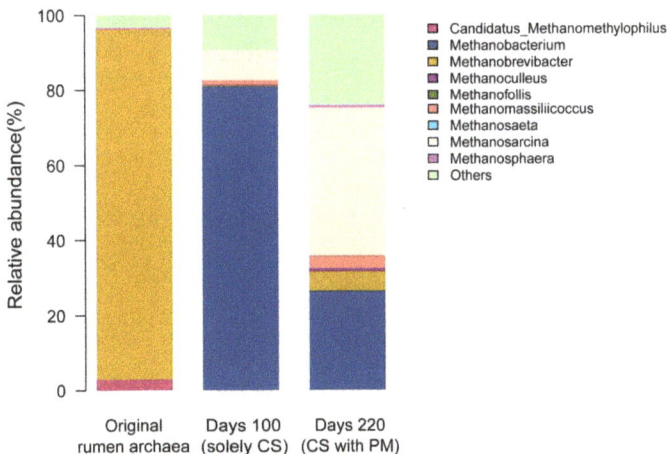

Figure 5. Composition of the archaeal community at the genus level.

Five genera of methanogens, *Methanobacterium* (80.91%), *Methanosarcina* (7.87%), and *Methanomassiliicoccus* (1.17%) were found in the solely CS digestion system. Compared with the original culture, the dominant hydrogenotrophic methanogenesis was shifted from *Methanobrevibacter* to *Methanobacterium*. This might be related to its higher VFAs content, in which VFAs concentrations of $50 \text{ mg} \cdot \text{L}^{-1}$ or higher would inhibit *Methanobrevibacter* [50]. This finding was surprising because the dominant genus *Methanobacterium* has been described to grow primarily at mesophilic temperatures (37–45 °C) [51]. A sharp increase in the relative abundance of the hydrogenotrophic methanogen *Methanobacterium* might be induced by the accumulation of VFAs presented in Figure 2, which would cause an increase in H_2 partial pressure [52]. This result was consistent with the findings by Yan et al. [53], who reported that hydrogenotrophic methanogens were dominant at the stabilization stage in the solid-state anaerobic digestion of rice straw. Furthermore, the relative abundance of *Methanosarcina* indicated the acetoclastic pathway existing in this system.

In a later stage, acetoclastic methanogen, *Methanosarcina*, accounted for 39.35%, of the populations, and hydrogenotrophic methanogens, *Methanobacterium* and *Methanobrevibacter*, accounted for 26.42% and 5.15%, respectively, suggesting that two methanogenetic pathways co-occurred for methane production during the co-digestion of CS with PM. It is known that hydrogenotrophic methanogens are most common in the digestion of agricultural waste [54]. In contrast, we found more abundant acetoclastic methanogens than hydrogenotrophic methanogens. This discrepancy might be explained by two aspects. First, the decreased concentrations of VFAs caused by the co-digestion system seemed to reduce H_2 partial pressure and resulted in a decrease of the relative abundance of *Methanobacterium*. Second, the dominance of *Methanosarcina* is, however, in line with previous studies of manure-based digesters [55–57]. High abundance of this methanogen has been suggested to be caused by its relatively high growth rate and ability to tolerate conditions inhibitory to other methanogens, such as presence of ammonia [58].

2.6. Physical Form and SEM Images of Substrates

From an appearance point of view, the digestate was brown rather than black, which was due to a high mixture ratio of CS. After anaerobic digestion, the particle size of CS in the digestate decreased obviously, and the pulverized CS was degraded from 0.5–1 cm to indiscernible minced material. The SEM images in Figure 6 demonstrated that the compact structure of the fibers had been swollen and destroyed by the amounts of bacteria in association with fungus [59,60], indicating that many ruminal microbes were still viable by these stages. These procedures of adherence and degradation were similar to the metabolic system in rumen, further confirming the positive effect of rumen microorganisms on the pilot-scale anaerobic digestion [13]. In addition, a similar smell was noted between the digestate and rumen fluid, which showed from a side view that it had remarkable homology in vivo to the newly conducted pilot-scale system.

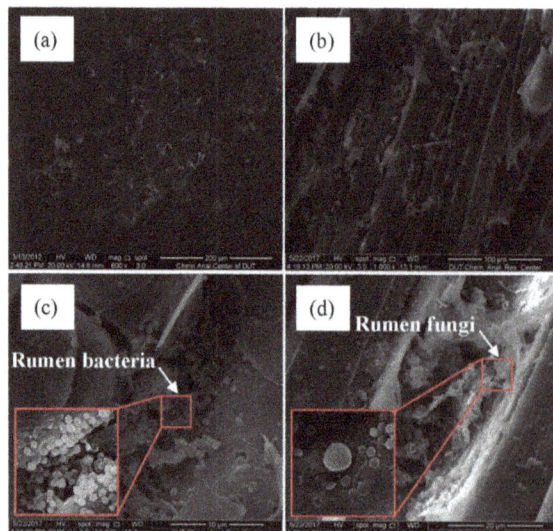

Figure 6. SEM images of: (**a**) undigested CS, (**b**) digested CS, (**c,d**) rumen microorganisms' adhesion on the surface of CS.

2.7. Mass Balance

Figure 7 shows the mass balances determined for the two typical and steady sections including stage III (solely CS at 3% solid content) and stage VII (1:1 mixture ratio of CS with PM at 10% solid content). As shown in Figure 7a, for the TS and VS mass balance, the input of solely CS was mostly converted into biogas, which achieved 72.56% and 77.35% removal rates, respectively. For the water, balance, the main pathway of the moisture in the feedstock included biogas liquid, biogas residue and gas–water separator. Biogas was used for the lamp and stove, resulting in 51.6% consumption. The mass balance and final fate of the product from the two sections were slightly different. The residual materials of the co-digestion system originated from two substrates, including the pretreated CS and PM, resulted in an improvement of the TS and VS removal rate to 76.38% and 79.33%. Due to the increase of biogas yield, the consumption rate declined to 46.88%.

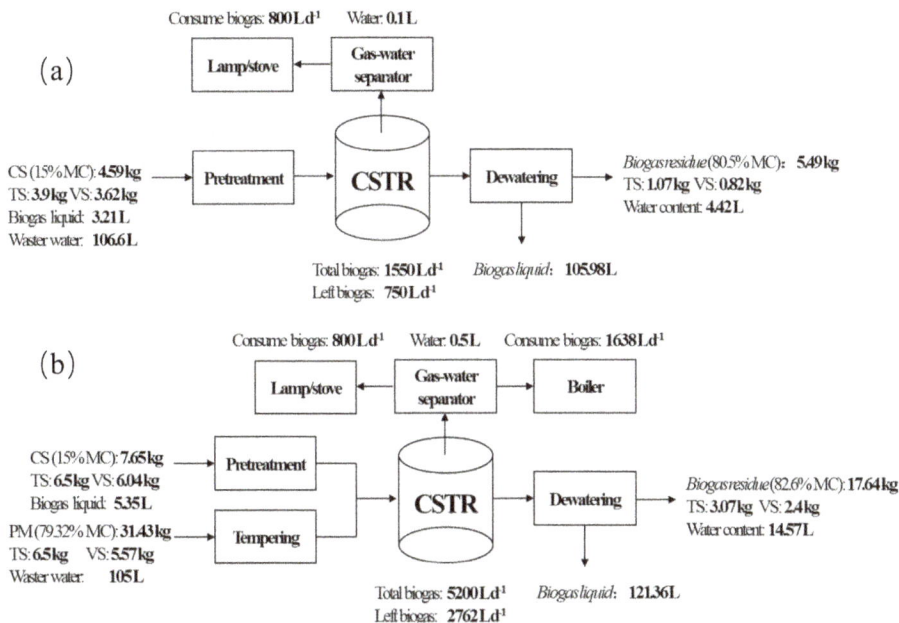

Figure 7. Mass balance of two sections, (**a**) stage III, solely CS at 3% solid content and (**b**) stage VII, 1:1 mixture ratio of CS with PM at 10% solid content (Moisture content, MC).

Furthermore, based on the calculation of removed VS, the biogas yield could be transformed into 553.57 L kg^{-1}·VS^{-1} removed and 564.6 L kg^{-1}·VS^{-1} removed. The similar conversion rate indicated that the rumen microorganisms had the same ability of degradation in both the solely CS digestion system and the co-digestion system. However, the accumulation of VFAs in stage III might imply that (1) the rumen microbes in co-digestion system needed more energy for self-growth or metabolism; and/or (2) a small quantity of intermediate products was not converted into biogas completed.

3. Materials and Methods

3.1. Inoculum and Substrates

Inoculum obtained from fresh bovine rumen in a slaughterhouse was poured into six pails (50 L, PVC) filled with N$_2$ for the anaerobic digestion system. This was conveyed to the pilot-scale test base immediately as the ruminal inoculum for starting up the anaerobic digestion reactor.

The naturally dried CS used as the carbon-rich material was collected from a farm near the test base. After being smashed to 0.5–1 cm pieces by a pulverizer, the CS was pretreated with biogas liquid coinciding with the method of Jin et al. in [12]. The nitrogen-rich materials were mainly composed of livestock manure, such as PM and CM, and they were obtained from a large-scale raising farm. The properties of the CS, PM, and CM are listed in Table 2.

Table 2. Characteristics of CS, PM, and chicken manure (CM).

Parameter	CS [b]	PM [b]	CM [b]
pH	-	7.89 ± 0.03	7.73 ± 0.01
Moisture content (%)	$15 \pm 2.2\%$	79.32 ± 0.05	73.36 ± 0.03
Volatile solid [a] (%)	92.90 ± 0.32	85.62 ± 0.08	80.67 ± 0.12
Total organic carbon [a] (%)	51.88 ± 0.82	43.57 ± 0.38	39.9 ± 0.29
NH_4^+–N (%)	-	0.85 ± 0.02	1.39 ± 0.05
Kjeldahlnitrogen [a] (%)	0.61 ± 0.13	3.28 ± 0.34	5.33 ± 0.42
C/N ratio	85.05 ± 0.48	13.28 ± 0.28	7.49 ± 0.36

[a] Dry weight basis. [b] Means $\pm$ S.E. (N = 3).

3.2. Experimental and Incubation Conditions

The pilot-scale experimental base had a total area of 2000 m^2. There was one mainstream and two sidestreams in this treatment system, including an anaerobic digestion unit, a pretreatment unit, and a biogas comprehensive utilization unit. The flow diagram and the schematic diagram of the technical system are shown in Figure 8.

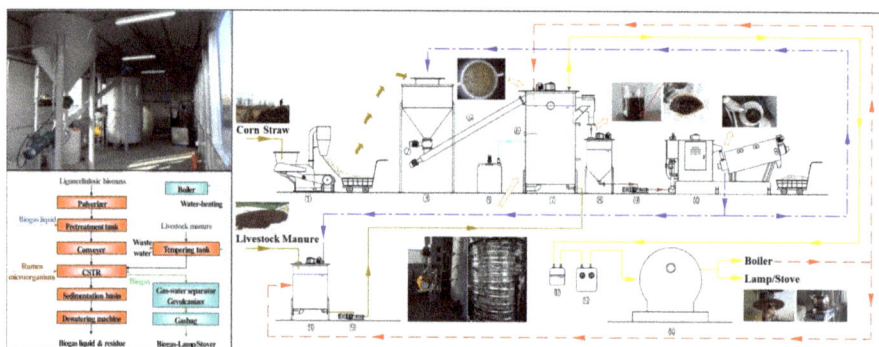

Figure 8. The flow and schematic diagram of the technical system including: (**1**) pulverizer, (**2**) unloader, (**3**) pretreatment tank, (**4**) conveyer, (**5**) pH adjustment bucket, (**6**) pH meter, (**7**) CSTR, (**8**) sedimentation basin, (**9**) screw pump, (**10**) screw type sludge dewatering machine, (**11**) tempering tank, (**12**) gas–water separator, (**13**) devulcanizer, and (**14**) gasbag. Pathway of substrates (————), digestate (————), pH adjustment (————), biogas (————), biogas liquid (—— · ——), and water-heating (—— ——).

Above all, the smashed CS was mixed with biogas liquid in a tank for improved digestibility. After pretreatment, the CS was accurately weighed, and these materials were used as the carbon-rich substrate for the anaerobic digestion system. Separately, feeding substrates with different densities could significantly lower the risk of pipeline block. A screw conveyor system with a hopper, an unloader, and a conveyer was installed on the top of the anaerobic reactor for adding the CS into the tank directly. After desilting and regulating the water content of the livestock manure, the mixture was pumped from the tempering tank into the bottom of the reactor by a screw pump.

Subsequently, a CSTR, equipped with an online monitoring and controlling system for pH and temperature, was constructed at a volume of 3 m^3 for anaerobic digestion in this test. The biogas overflowing from the top of the reactor under the ordinary pressure passed through a gas–water separator and a devulcanizer and then was temporarily stored in a 3 m^3 gasbag. Some of the facilities were adopted for comprehensive utilization of the biogas, such as a biogas lamp for illumination, a gas stove for cooking, and a boiler for heat addition. The screw type sludge dewatering machine was connected with a sedimentation basin to separate the digestate into biogas liquid and biogas residue.

The working volume of the CSTR was 2.6 m^3, and 0.26 m^3 of ruminal inoculum with 2.1 m^3 preheated livestock wastewater (10% v/v) was loaded into the reactor. After that, the batch stirred blender was turned on according to the index of working time at 15 min h^{-1} and a stirring speed at 60 rpm. The initial pH was adjusted to 7.0 ± 0.1, while based on the experimental requirements, the pH was maintained above 6.5 by an automatic controller with 10% NaOH. A water-heating pipe assisted by an electric tracing band was wrapped around the external wall of the reactor and tempering tank to maintain the temperature at 39 ± 1 °C. The retention time was 20 days. The anaerobic digestion test was divided into three parts including nine stages to comprehensively investigate the properties of inoculating and starting up, as well as CS as the sole substrate and mixed substrates by rumen cultures. The experimental design is listed in Table 3.

Table 3. Experimental design for the pilot-scale anaerobic digestion test.

Experimental Procedure	Experimental Stage	Solid Content			Remarks
		CS	PM	CM	
Part I	Stage I	1%	-	-	Inoculating and starting up
Part II	Stage II	1%	-	-	Anaerobic digestion of the CS
	Stage III	3%	-	-	CS associating with urea addition
	Stage IV	5%	-	-	Limiting solid content of solely carbon-rich substrate
Part III	Stage V	5%	1%	-	
	Stage VI	5%	2%	-	
	Stage VII	5%	5%	-	Anaerobic co-digestion of CM with livestock manures
	Stage VIII	5%	-	5%	
	Stage IX	3%	-	7%	

3.3. Analytical Methods

The temperature and pH during the digesting process were detected by automatic sensors (NB2-ICSS-18G-12, OMEGA; PP-100A, InPro 3250, Weihong, Mettler, Zurich, Switzerland). An anticorrosion wet gas flow meter (LMF-2, Kesion, Qingdao, China) was applied to monitor the biogas yield of the anaerobic digestion. Biogas yield expressed in L kg^{-1}·TS^{-1} was calculated as the volume of biogas produced per kg of the feedstock TS added to the digester. Biogas components such as CH_4, H_2, and CO_2 were analyzed by a portable gas detector (ZR-3110, Junray, Qingdao, China).

Powder X-ray diffraction diagrams (XRD, D/max-2400, Rigaku, Tokyo, Japan) were employed to estimate the relative crystallinity of the pretreated CS [61]. The functional groups of pretreated maize straw were characterized by a FTIR spectrometer (EQUINOX55, Bruker, Karlsruhe, Germany). The morphological characteristics of the digested substrates were observed using scanning electron microscopy (SEM, QUANTA 450, FEI, Hillsboro, FL, USA). After being washed several times with phosphate buffer, the samples were fixed with 3.0% glutaraldehyde for 24 h at 4 °C. Subsequently, the experimental samples were dehydrated through a graded series of tertiary ethanol solutions (from 50% to 100%). Finally, the samples were evacuated by vacuum drying and gold-coated by a sputter.

Total solids (TS), moisture content (MC), and volatile solids (VS) were analyzed according to standard methods [62]. The digested samples were transferred to the laboratory as soon as possible in a cold closet. After centrifuging the samples at 12,000 rpm for 10 min (CT14TD, Techcomp, Shanghai, China) and filtering with a 0.45-μm membrane, the concentrations of VFAs, including acetate, propionate, butyrate, and isobutyrate, in the samples were determined by a gas chromatograph (GC-2010, Shimadzu, Kyoto, Japan) equipped with a flame ionization detector (FID) and a 30 m × 0.53 mm × 0.1 μm fused-silica capillary column (DB-FFAP). The temperature of the injector and detector were maintained at 250 °C and 300 °C, respectively. The initial temperature of the oven was 70 °C for 3 min, followed by a ramp of 20 °C min⁻¹ for 5.5 min, and the final temperature was 180 °C for 6 min. High-purity nitrogen was used as the carrier gas with a flow rate of 32.8 mL min⁻¹. The VFA content was evaluated with calibration curves.

The digestate samples were collected on days 100 and 220 to determine the archaeal communities at both the solely CS and mixed substrates digestion phases, respectively. The samples, including original rumen archaea, were frozen at −80 °C immediately. Genomic DNA was extracted from each sample with an E.Z.N.A Soil DNA kit (OMEGA, Atlanta, GA, USA) according to the manufacturer's instructions. PCR amplification was performed using specifically synthesized primers with a barcode of Arch519F (5′-CAGCCGCCGCGGTAA-3′) and Arch915R (5′-CAGCCGCCGCGGTAA-3′) in a PCR thermal cycler Dice (BioRad Co. Ltd., Berkeley, CA, USA) using the following program: 2 min of denaturation at 95 °C, followed by 27 cycles of 30 s at 95 °C (denaturation), 30 s for annealing at 55 °C and 1 min at 72 °C (elongation), with a final extension at 72 °C for 10 min. After amplification, the PCR products were purified, quantified, and then pooled at equal concentrations. Finally, the PCR products were sent to Majorbio Bio-Pharm Technology Co., Ltd. (Shanghai, China) for high-throughput sequencing using paired-end sequencing with an Illumina MiSeq PE300 platform (San Diego, CA, USA).

For quality-control purpose, any sequences that contained mismatches and ambiguous reads (N) in primers were removed. After deleting unqualified sequences, total of 11,276, 18,623, and 12,027 high-quality 16S rRNA gene sequences were obtained for each sample. These clean, non-continuous sequences were then clustered by the complete linkage clustering method in the QIIME pipeline. OTUs were classified using 97% identity of the 16S rRNA gene sequence as a cutoff, and an OTUs table was generated for each sample and used for statistical analysis. Richness estimators of Chao1, ACE, and Shannon index were calculated by MOTHUR. Taxonomic classification at the genus level was performed using the Ribosome Database project (RDP) algorithm to classify the representative sequences of each OTU.

4. Conclusions

A pilot-scale anaerobic digestion system was successfully developed to demonstrate the biogas yield enhancement by rumen microorganisms in the bioconversion of solely pretreated CS and pretreated CS with livestock manures. Pretreatment of CS by biogas liquid could be adopted at this scale. The rumen culture had a positive response to the impact of temperature and pH. There was almost no promotion of the solely CS digestion system after adding urea. The ruminal pilot-scale digestion reactor was successfully started up within 32 days at a retention time of 20 days. The optimal solid content of solely CS was 3%, which resulted in a biogas yield of 395 L kg⁻¹·TS⁻¹. The addition of livestock manure was helpful for improving the comprehensive performance of the ruminal digestion system. The components of biogas including CH_4, CO_2, and H_2 could be maintained at about 60%, 40%, and 0%. Microbial community analysis demonstrated that hydrogenotrophic methanogens dominated in the solely CS digestion system. Hydrogenotrophic and acetoclastic methanogens co-occurred for methane production during the co-digestion of CS with PM. SEM images proved the effect of the rumen microorganisms. Mass balance revealed the internal relevance of the conversion treatment process. The experimental results demonstrated that the newly established ruminal anaerobic digestion system has great potential in the sustainable and comprehensive utilization of ASOWs.

Supplementary Materials: E-supplementary data of this work can be found in online version of the paper. The following are available online at http://www.mdpi.com/1996-1073/11/4/920/s1, Figure S1: Temperature variation of the anaerobic digestion system, Figure S2: pH value of the anaerobic digestion system.

Energies **2018**, *11*, 920

Acknowledgments: The authors wish to thank the Fund of Water Pollution Control and Management Projects in the Twelfth Five-Year Plan of the Chinese Government for the support of this study (2012ZX07202006-002).

Author Contributions: X.X. and F.Y. conceived and designed the experiments; W.J. performed the experiments; W.J. analyzed the data; X.X. and W.J. contributed reagents/materials/analysis tools; W.J. wrote the paper.

Conflicts of Interest: The authors declare no conflict of interest.

References

1. Ju, X.; Zhang, F.; Bao, X.; Römheld, V.; Roelcke, M. Utilization and management of organic wastes in Chinese agriculture: past, present and perspectives. *Sci. China Ser. C: Life Sci.* **2005**, *48*, 965–979.
2. Li, H.; Cao, Y.; Wang, X.; Ge, X.; Li, B.; Jin, C. Evaluation on the production of food crop straw in China from 2006 to 2014. *BioEnergy Res.* **2017**, *10*, 949–957. [CrossRef]
3. Ning, Z.; Ji, M. Changes and outlook about production amount of livestock and poultry manure in China. *Agric. Outlook* **2014**, *1*, 46–48.
4. Zhang, X.; Zhai, M.; Wang, Y.; Gao, Y.; Zhao, H.; Zhou, X.; Gao, J. Research on biomass energy utilization in Chinese rural area: Challenges and opportunities. *Emerald Group Publ. Ltd.* **2016**, 257–281. [CrossRef]
5. Zhang, L.; Liu, Y.; Hao, L. Contributions of open crop straw burning emissions to $PM_{2.5}$ concentrations in China. *Environ. Res. Lett.* **2016**, *11*. [CrossRef]
6. Prapinagsorn, W.; Sittijunda, S.; Reungsang, A.; Sciubba, E. Co-digestion of napier grass and its silage with cow dung for methane production. *Energies* **2017**, *10*, 1654. [CrossRef]
7. Powlson, D.S.; Riche, A.B.; Coleman, K.; Glendining, M.J.; Whitmore, A.P. Carbon sequestration in European soils through straw incorporation: Limitations and alternatives. *Waste Manag.* **2008**, *28*, 741–746. [CrossRef] [PubMed]
8. Yang, X.; Wang, H.; Strong, P.J.; Xu, S.; Liu, S.; Lu, K.; Sheng, K.; Guo, J.; Che, L.; He, Z.; et al. Thermal properties of biochars derived from waste biomass generated by agricultural and forestry sectors. *Energies* **2017**, *10*, 469. [CrossRef]
9. Mao, C.; Feng, Y.; Wang, X.; Ren, G. Review on research achievements of biogas from anaerobic digestion. *Renew. Sustain. Energy Rev.* **2015**, *45*, 540–555. [CrossRef]
10. Mata-Alvarez, J.; Dosta, J.; Romero-Güiza, M.S.; Fonoll, X.; Peces, M.; Astals, S. A critical review on anaerobic co-digestion achievements between 2010 and 2013. *Renew. Sustain. Energy Rev.* **2014**, *36*, 412–427. [CrossRef]
11. Brodeur, G.; Yau, E.; Badal, K.; Collier, J.; Ramachandran, K.B.; Ramakrishnan, S. Chemical and physicochemical pretreatment of lignocellulosic biomass: A review. *Enzym. Res.* **2011**, *4999*, 787532. [CrossRef] [PubMed]
12. Jin, W.Y.; Xu, X.C.; Gao, Y.; Yang, F.L.; Wang, G. Anaerobic fermentation of biogas liquid pretreated maize straw by rumen microorganisms in vitro. *Bioresour. Technol.* **2014**, *153*, 8–14. [CrossRef] [PubMed]
13. Hungate, R.E. *The Rumen and Its Microbes*; Academic Press Inc.: London, UK, 1966; pp. 466–525.
14. Li, N.; Yang, F.; Xiao, H.; Zhang, J.; Ping, Q. Effect of feedstock concentration on biogas production by inoculating rumen microorganisms in biomass solid waste. *Appl. Biochem. Biotech.* **2017**, 1–13. [CrossRef] [PubMed]
15. Yue, Z.; Wang, J.; Liu, X.; Yu, H. Comparison of rumen microorganism and digester sludge dominated anaerobic digestion processes for aquatic plants. *Renew. Energy* **2012**, *46*, 255–258. [CrossRef]
16. Yue, Z.; Yu, H.; Wang, Z. Anaerobic digestion of cattail with rumen culture in the presence of heavy metals. *Bioresour. Technol.* **2007**, *98*, 781–786. [CrossRef] [PubMed]
17. Hu, Z.H.; Yu, H.Q. Application of rumen microorganisms for enhanced anaerobic fermentation of corn stover. *Process Biochem.* **2005**, *40*, 2371–2377. [CrossRef]
18. Alrawi, R.A.; Ahmad, A.; Ismail, N.; Kadir, M.O.A. Anaerobic co-digestion of palm oil mill effluent with rumen fluid as a co-substrate. *Desalination* **2011**, *269*, 50–57. [CrossRef]
19. Gijzen, H.J.; Derikx, P.J.; Vogels, G.D. Application of rumen microorganisms for a high rate anaerobic digestion of papermill sludge. *Biol. Wastes* **1990**, *32*, 169–179. [CrossRef]
20. Change, O.C. Intergovernmental Panel on Climate Change. World Meteorological Organization, 2007. Available online: http://www.ipcc.ch/pdf/assessment-report/ar4/syr/ar4_syr_full_report.pdf (accessed on 12–17 November 2007).

21. Mosoni, P.; Martin, C.; Forano, E.; Morgavi, D.P. Long-term defaunation increases the abundance of cellulolytic ruminococci and methanogens but does not affect the bacterial and methanogen diversity in the rumen of sheep. *J. Anim. Sci.* **2011**, *89*, 783. [CrossRef] [PubMed]

22. Van, T.E.; Hundley, V. The importance of pilot studies. *Nurs. Stand.* **2010**, *16*, 33.

23. Chundawat, S.P.S.; Balan, V.; Sousa, L.D.C.; Dale, B.E. *Bioalcohol Production: Biochemical Conversion of Lignocellulosic Biomass*; Woodhead Publishing Limited: Cambridge, UK, 2010.

24. Carrillo, F.; Colom, X.; Sunol, J.J.; Saurina, J. Structural FTIR analysis and thermal characterisation of lyocell and viscose-type fibres. *Eur. Polym. J.* **2004**, *40*, 2229–2234. [CrossRef]

25. Phillipson, K.; Hay, J.N.; Jenkins, M.J. Thermal analysis FTIR spectroscopy of poly (ε-caprolactone). *Thermochim. Acta* **2014**, *595*, 74–82. [CrossRef]

26. Mezzullo, W.G.; Mcmanus, M.C.; Hammond, G.P. Life cycle assessment of a small-scale anaerobic digestion plant from cattle waste. *Appl. Energy* **2013**, *102*, 657–664. [CrossRef]

27. Lindorfer, H.; Braun, R.; Kirchmayr, R. Self-heating of anaerobic digesters using energy crops. *Water Sci. Technol.* **2006**, *53*, 159–166. [CrossRef] [PubMed]

28. Hu, Z.; Yu, H.; Zhu, R. Influence of particle size and pH on anaerobic degradation of cellulose by ruminal microbes. *Int. Biodeter. Biodeg.* **2005**, *55*, 233–238. [CrossRef]

29. Dixon, R.M.; Dixon, R.M. Effects of addition of urea to a low nitrogen diet on the rumen digestion of a range of roughages. *Aust. J. Agric. Res.* **1999**, *50*, 1091–1098. [CrossRef]

30. Zhou, J.; Yang, J.; Yu, Q.; Yong, X.; Xie, X.; Zhang, L.; Wei, P.; Jia, H. Different organic loading rates on the biogas production during the anaerobic digestion of rice straw: A pilot study. *Bioresour. Technol.* **2017**, *244*, 865–871. [CrossRef] [PubMed]

31. Lee, D.H.; Behera, S.K.; Kim, J.W.; Park, H.S. Methane production potential of leachate generated from Korean food waste recycling facilities: A lab-scale study. *Waste Manag.* **2009**, *29*, 876–882. [CrossRef] [PubMed]

32. Zhang, P.; Chen, Y.; Zhou, Q. Waste activated sludge hydrolysis and short-chain fatty acids accumulation under mesophilic and thermophilic conditions: Effect of pH. *Water Res.* **2009**, *43*, 3735–3742. [CrossRef] [PubMed]

33. Liu, Q.; Wang, C.; Pei, C.X.; Li, H.Y.; Wang, Y.X.; Zhang, S.L.; Zhang, Y.L.; He, J.P.; Wang, H.; Yang, W.Z. Effects of isovalerate supplementation on microbial status and rumen enzyme profile in steers fed on corn stover based diet. *Livest. Sci.* **2014**, *161*, 60–68. [CrossRef]

34. Schmidt, J.E.; Ahring, B.K. Effects of hydrogen and formate on the degradation of propionate and butyrate in thermophilic granules from an upflow anaerobic sludge blanket reactor. *Appl. Environ. Microbiol.* **1993**, *59*, 2546–2551.

35. Baek, G.; Kim, J.; Kim, J.; Lee, C. Role and potential of direct interspecies electron transfer in anaerobic digestion. *Energies* **2018**, *11*, 107. [CrossRef]

36. Yangin-Gomec, C.; Ozturk, I. Effect of maize silage addition on biomethane recovery from mesophilic co-digestion of chicken and cattle manure to suppress ammonia inhibition. *Energy Convers. Manag.* **2013**, *71*, 92–100. [CrossRef]

37. Kim, M.; Ahn, Y.H.; Speece, R.E. Comparative process stability and efficiency of anaerobic digestion: Mesophilic vs. Thermophilic. *Water Res.* **2002**, *36*, 4369. [CrossRef]

38. Russell, J.B. The importance of pH in the regulation of ruminal acetate to propionate ratio and methane production in vitro. *J. Dairy Sci.* **1998**, *81*, 3222–3230. [CrossRef]

39. Hegarty, R.S.; Gerdes, R. Hydrogen production and transfer in the rumen. *Recent Adv. Anim. Nutr. Aust.* **1999**, *12*, 37–44.

40. Zhao, B.; Yue, Z.; Ni, B.; Mu, Y.; Yu, H.; Harada, H. Modeling anaerobic digestion of aquatic plants by rumen cultures: Cattail as an example. *Water Res.* **2009**, *43*, 2047–2055. [CrossRef] [PubMed]

41. Chandra, R.; Takeuchi, H.; Hasegawa, T. Methane production from lignocellulosic agricultural crop wastes: A review in context to second generation of biofuel production. *Renew. Sustain. Energy Rev.* **2012**, *16*, 1462–1476. [CrossRef]

42. Moestedt, J.; Påledal, S.N.; Schnürer, A.; Nordell, E. Biogas production from thin stillage on an industrial scale—Experience and optimisation. *Energies* **2013**, *6*, 5642–5655. [CrossRef]

43. Fernandes, T.V.; Keesman, K.J.; Zeeman, G.; van Lier, J.B. Effect of ammonia on the anaerobic hydrolysis of cellulose and tributyrin. *Biomass Bioenergy* **2012**, *47*, 316–323. [CrossRef]

44. Wang, Y.; Zhang, Y.; Wang, J.; Meng, L. Effects of volatile fatty acid concentrations on methane yield and methanogenic bacteria. *Biomass Bioenergy* **2009**, *33*, 848–853. [CrossRef]

45. Yu, G.; Chen, X.; Liu, Z.; Zhou, X.; Zhang, Y. Effect of inoculum sources on the anaerobic digestion of rice straw. *Bioresour. Technol.* **2014**, *158*, 149–155.

46. Muller, Z.O. *Feed from Animal Wastes: Feeding Manual*; Food and Agriculture Organization of the United Nations: Rome, Italy, 1982.

47. Bhattacharya, A.N.; Taylor, J.C. Recycling animal waste as a feedstuff: A review. *J. Anim. Sci.* **1975**, *41*, 1438–1457. [CrossRef]

48. Xie, S.; Lawlor, P.G.; Frost, P.; Dennehy, C.D.; Hu, Z.; Zhan, X. A pilot scale study on synergistic effects of co-digestion of pig manure and grass silage. *Int. Biodeter. Biodeg.* **2017**, *123*, 244–250. [CrossRef]

49. Miller, T.L. *Methanobrevibacter*; William, B.W., Fred, R., Eds.; Wiley Online Library: Hoboken, NJ, USA, 2015.

50. Henderson, C. The effects of fatty acids on pure cultures of rumen bacteria. *J. Agr. Sci.* **1973**, *81*, 107–112. [CrossRef]

51. David, R.B.; Richard, W.C. *Bergey's Manual of Systematic Bacteriology: Volume One: Archaea and the Deeply Branching and Phototrophic Bacteria*; Springer Science & Business Media: Berlin, Germany, 2010.

52. Goberna, M.; Gadermaier, M.; García, C.; Wett, B.; Insam, H. Adaptation of methanogenic communities to the cofermentation of cattle excreta and olive mill wastes at 37 °C and 55 °C. *Appl. Environ. Microb.* **2010**, *76*, 6564–6571. [CrossRef] [PubMed]

53. Yan, Z.; Song, Z.; Li, D.; Yuan, Y.; Liu, X.; Zheng, T. The effects of initial substrate concentration, C/N ratio, and temperature on solid-state anaerobic digestion from composting rice straw. *Bioresour. Technol.* **2015**, *177*, 266–273. [CrossRef] [PubMed]

54. Ziganshin, A.M.; Liebetrau, J.; Pröter, J.; Kleinsteuber, S. Microbial community structure and dynamics during anaerobic digestion of various agricultural waste materials. *Appl. Microbiol. Biotechnol.* **2013**, *97*, 5161–5174. [CrossRef] [PubMed]

55. St-Pierre, B.; Wright, A.D.G. Comparative metagenomic analysis of bacterial populations in three full-scale mesophilic anaerobic manure digesters. *Appl. Microbiol. Biotechnol.* **2014**, *98*, 2709–2717. [CrossRef] [PubMed]

56. Demirel, B.; Scherer, P. The roles of acetotrophic and hydrogenotrophic methanogens during anaerobic conversion of biomass to methane: a review. *Rev. Environ. Sci. Biotechnol.* **2008**, *7*, 173–190. [CrossRef]

57. Karakashev, D.; Batstone, D.J.; Angelidaki, I. Influence of environmental conditions on methanogenic compositions in anaerobic biogas reactors. *Appl. Environ. Microbiol.* **2005**, *71*, 331–338. [CrossRef] [PubMed]

58. De Vrieze, J.; Hennebel, T.; Boon, N.; Verstraete, W. Methanosarcina: the rediscovered methanogen for heavy duty biomethanation. *Bioresour. Technol.* **2012**, *112*, 1–9. [CrossRef] [PubMed]

59. Akin, D.E. Ultrastructure of rumen bacterial attachment to forage cell walls. *Appl. Environ. Microbiol.* **1976**, *31*, 562–568. [PubMed]

60. Rezaeian, M.; Beakes, G.W.; Parker, D.S. Methods for the isolation, culture and assessment of the status of anaerobic rumen chytrids in both in vitro and in vivo systems. *Mycol. Res.* **2004**, *108*, 1215–1226. [CrossRef] [PubMed]

61. Segal, L.; Creely, J.J.; Martin, A.E., Jr.; Conrad, C.M. An empirical method for estimating the degree of crystallinity of native cellulose using the X-ray diffractometer. *Text. Res. J.* **1959**, *29*, 786–794. [CrossRef]

62. Apha, A.W. *Standard Methods for the Examination of Water and Wastewater*, 19th ed.; American Public Health Association: Washington, DC, USA, 1995.

![energies logo] *energies*   MDPI

Article

Digestion Performance and Microbial Metabolic Mechanism in Thermophilic and Mesophilic Anaerobic Digesters Exposed to Elevated Loadings of Organic Fraction of Municipal Solid Waste

Yiming Gao [1,2,3,4], Xiaoying Kong [2,3,4,*], Tao Xing [2,3,4], Yongming Sun [2,3,4], Yi Zhang [2,3,4,5], Xingjian Luo [2,3,4,5] and Yong Sun [1,*]

[1] College of Engineering, Northeast Agricultural University, Harbin 150030, China; gaoym@ms.giec.ac.cn
[2] Guangzhou Institute of Energy Conversion, Chinese Academy of Sciences, Guangzhou 510640, China; xingtao@ms.giec.ac.cn (T.X.); sunym@ms.giec.ac.cn (Y.S.); zhangyi@ms.giec.ac.cn (Y.Z.); luoxj@ms.giec.ac.cn (X.L.)
[3] CAS Key Laboratory of Renewable Energy, Guangzhou 510640, China
[4] Guangdong Provincial Key Laboratory of New and Renewable Energy Research and Development, Guangzhou 510640, China
[5] University of Chinese Academy of Sciences, Beijing 100049, China
* Correspondence: kongxy@ms.giec.ac.cn (X.K.), sunyong740731@163.com (Y.S.); Tel.:+86-020-87057783 (X.K.); Tel.:+86-0451-55191670 (Y.S.)

Received: 8 February 2018; Accepted: 20 March 2018; Published: 17 April 2018

Abstract: Mesophilic and thermophilic anaerobic digestion reactors (MR and TR) for the organic fraction of municipal solid waste (OFMSW) were tested to reveal the differential microbial responses to increasing organic loading rate (OLR). MR exhibited faster adaptation and better performance at an OLR range of 1.0–2.5 g $VS \cdot L^{-1} \cdot d^{-1}$, with average profiles of a biogas yield of 0.38 L/gVS_{added}*d at 0.5 g/L*d OLR and 0.69 L/gVS_{added}*d at 2.5 g/L*d OLR, whereas TR had a biogas yield of 0.07 L/gVS_{added}*d at 0.5 g/L*d OLR and 0.44 L/gVS_{added}*d at 2.5 g/L*d OLR. The pyrosequencing results of amplicons revealed the microbial mechanisms of OFMSW anaerobic digestion. Larger shifts in the bacteria composition were observed in the TR with OLR elevation. For methanogens in both reactors, *Methanothrix* dominated in the MR while *Methanosarcina* was favored in the TR. Moreover, analysis of the mode and efficiency of metabolism between the MR and TR demonstrated different performances with more efficiency related to the limiting hydrolytic acid step.

Keywords: anaerobic digestion; organic fraction of municipal solid waste; microbial mechanisms; metabolism analysis

1. Introduction

The treatment and disposal of the organic fraction of municipal solid waste (OFMSW) represent major challenges worldwide owing to growing production levels [1]. Anaerobic digestion (AD) is an efficient and eco-friendly treatment to transform OFMSW into energy.

Mesophilic (30–40 °C) and thermophilic (50–60 °C) AD systems are the most commonly used AD processes [2,3]. Several methods for increasing biogas production have been proposed, including raw material pretreatment and optimizing the fermenting digester [4,5]. According to the Arrhenius equation, chemical reaction rates could be doubled with a 10 °C temperature increase. Therefore, thermophilic AD is a possible strategy to improve the process efficiency owing to a higher reaction rates. Thermophilic AD shows several advantages such as higher loading rates of organic feedstock and a smaller pathogen degree [6]. Nevertheless, mesophilic AD would have a lower volatile fatty acid (VFA) concentration (especially propionate) and could usually support a higher organic loading rate (OLR) [7,8]. In general, OFMSW has a mutable complicated composition in the form of proteins, lipids, carbohydrates, and cellulose [9]. However, it is more uncertain whether the AD of OFMSW would lead to failure or low biogas production.

AD includes four major steps of microbial processes—hydrolysis, fermentation, acetogenesis, and methanogenesis—which are completed collectively by syntrophic interactions of different microorganisms [10]. Malfunctioning of the microbial population or activity at any one step can affect the overall microbial community, resulting in AD inefficiency or failure [11]. One common speculation for the lower efficiency of thermophilic AD at a high OLR is that higher VFA accumulation results in an imbalance between acidogenesis and methanogenesis in the microbial community [12].

To better understand the microbiome in AD reactors, molecular methods have been applied to replace traditional culture-dependent techniques [13]. With the development of next-generation sequencing technologies such as 454 pyrosequencing, a much larger amount of data can now be analyzed in a shorter period and has been successfully applied for examining the microbial community in different conditions [14], including investigations of the microbial diversity in AD [15]. Indeed, there is a great difference in the microbial community between mesophilic and thermophilic AD. For example, Guo reported that the methanogens *Methanosaeta* dominated the archaeal community in a mesophilic reactor (MR), while *Methanothermobacter* and *Methanoculleus* were favored in a thermophilic reactor (TR) [16]. Although various sequencing data have been obtained, information on the microbial community in AD is still very limited.

The aim of the study was to compare the dynamics and structure of the microbial communities using high-throughput sequencing, to compare metabolism between two lab-scale temperature-state ADs (MR and TR) that were operated at a gradually increasing OLR, and to determine their respective performances. Due to the inoculum, thermophilic AD took a longer time to adapt.

2. Results and Discussion

2.1. Performance of the Reactors

Initially, the OLR of the MR was gradually elevated with a gradient of 0.5 gVS· L^{-1}· d^{-1} from an initial level of 0.5 gVS·L^{-1}·d^{-1} over 7 days. The TR was acclimated with low concentrations of OFMSW for 43 days at an OLR of 0.5 gVS·L^{-1}·d^{-1}. Subsequently, the OLR of the MR was increased with a gradient of 0.5 gVS·L^{-1}·d^{-1} every week over the next two weeks, and was increased by 0.5 gVS·L^{-1}· d^{-1} every 25 days from Day 57 to Day 106.

Profiles of the biogas yield (SBY) of the added OLR, methane production and concentration, intermediate alkalinity to partial alkalinity ratio (IA:PA), and pH of the two reactors are shown in Figure 1. The two reactors showed different performances at the low OLR concentration, and the MR exhibited a much better ability for biogas and methane production, implying that a normal atmospheric temperature seed biomass allows for stronger adaptability of mesophilic AD. In the MR, the SBY increased in the first 7 days, peaking at 0.54 L/gVS$_{added}$*d, decreased over the next 25 days, and then showed an upward trend in the latter three stages. The SBY of the MR was 0.55 L/gVS$_{added}$*d at 2 g/L*d OLR and was 0.69 L/gVS$_{added}$*d at 2.5 g/L*d OLR. In the TR, the SBY remained steady in the first 43 days, increased to 0.26 L/gVS$_{added}$*d over the next week, and then showed an initial decrease, followed by an upward trend in the last 50 days. The SBY of the TR was 0.16 L/gVS$_{added}$*d at 2 g/L*d OLR and was 0.44 L/gVS$_{added}$*d at 2.5 g/L*d OLR. Thus, in this experiment, the MR showed greater biogas production, and had stronger adaptability and gas production efficiency than the TR for each OLR tested.

The pH of the two digesters ranged from 7.0 to 7.5 and remained in a stable state, as the suggested condition of anaerobic fermentation [17]. The IA:PA of alkalinity remained below 1 essentially the whole time for both reactors, implying the stability of the reactors [18]. However, at Day 35 of the TR, the IA:PA value was 1.16, suggesting that the microbial community showed preliminary adaptation to the production of organic acids. There was no accumulation of VFAs, as no VFAs could be detected. In addition, the NH₃–N concentration in both reactors was below 500 mg/L from start to finish, indicating that the effect of ammonia on the methane yield was negligible in both reactors [19,20]. Therefore, there were minimal inhibition effects in both reactors throughout the operation.

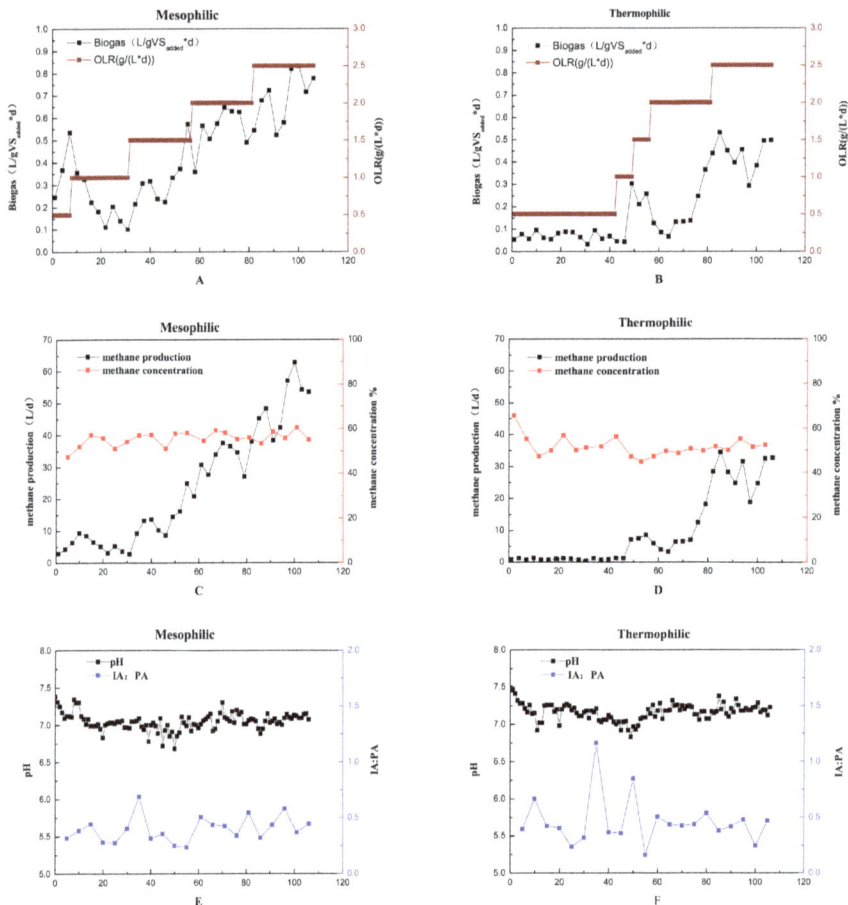

Figure 1. Evolution of biogas/methane yield (**A**,**B**), methane production and percentage (**C**,**D**), pH, and intermediate alkalinity to partial alkalinity (IA:PA) ratio (**E**,**F**) in both reactors.

2.2. Bacterial Communities Revealed by Pyrosequencing

Comparison of the sequencing data with the database and taxonomic analysis at the species level were carried out using a high-throughput pyrosequencing technique to investigate the microbial community structures and dynamics. The samples were collected at Days 0 (inoculum), 42, 81, and 106 from the MR and TR. Table 1 shows the sequence number (Seq num) and the alpha diversity estimators of three indices (Shannon, ACE, Chao1) for each sample. The Shannon index reflects the community diversity, and the ACE and Chao 1 indices are used to estimate the total number of species. The higher Shannon, ACE, and Chao 1 index values of the MR mutually implied the higher diversity of the bacterial community in this reactor. Nevertheless, Sample T42 had higher values of the diversity estimators than the inoculum, implying adaptation and an intense change of the bacterial community in the TR. For the TR, the values of the diversity estimators increased throughout most of the operation, and a subsequent decline in diversity indicated that the dominant bacterial community had a greater advantage in the end (Table 2). For the archaeal community, the diversity showed a different trend between the MR and TR, in which the diversity of the archaeal community increased from Day 0 to Day 106, indicating that more archaea were involved in the process of methane production than in the inoculum. The TR showed the highest diversity of the archaeal community at Day 42, and then decreased in the latter days. At Day 106, the TR had the lowest diversity of the archaeal community, implying that fewer archaea were involved in the process of methane production than in the MR. Overall, the MR and TR showed decreasing trends in the diversity of the bacterial community during the running time; the diversity of the archaeal community increased in the MR and decreased in the TR. The diversity of the MR was greater than that of the TR for both bacteria and archaea.

Table 1. Diversity indices of bacterial 16S rRNA gene libraries obtained from pyrosequencing. All values were calculated at the 0.03 distance limit.

Sample ID	Seq num	Shannon	ACE	Chao1
Inoculum	36,968	4.83	2085.72	2003.94
M42	50,721	4.35	1993.82	1993.82
M81	34,881	3.94	1482.31	1196.27
M106	43,495	3.87	1333.54	1290.58
T42	48,268	4.84	2107.76	2010.29
T81	24,640	3.53	819.58	740.92
T106	39,521	3.28	812.26	754.70

Table 2. Diversity indices of archaeal 16S rRNA gene libraries obtained from pyrosequencing. All values were calculated at the 0.03 distance limit.

Sample ID	Seq num	Shannon	ACE	Chao1
Inoculum	37,349	1.81	400.96	358.27
M42	35,726	1.91	479.86	395.59
M81	38,486	2.14	597.13	458.31
M106	38,703	2.33	622.17	497.01
T42	38,494	2.64	608.64	552.93
T81	38,095	2.12	476.57	330.21
T106	39,425	1.13	175.41	158

The phylum-level distribution of sequences is shown in Figure 2. For the bacteria, seven phyla existed in both digests with a relative abundance higher than 1%. *Chloroflexi* was the major phylum at the early stage for both digests. *Bacteroidetes* was the dominant phylum in the MR while *Thermotogae* was dominant in the TR for almost the entire process. *Firmicutes* appeared in both reactors throughout the whole period and its relative abundance was much higher in the TR than in the MR. *Synergistetes* and *Proteobacteria* existed throughout the operation period, indicating that these phyla might play important roles in both reactors even though their proportions were relatively low.

Moreover, the phyla distributions in the two digesters showed different patterns of change during the operation time. In the MR, the relative abundance of *Chloroflexi* decreased gradually after an initial increase. The proportion of the phylum *Bacteroidetes* increased obviously to 49.49% while those of *Synergistete* and *Proteobacteria* decreased to 7.11% and 2.47%, respectively. *Firmicutes* remained steady from Day 0 to Day 106. *Cloacimonetes* appeared at every stage in the MR, but only appeared initially in the TR. In the TR, the proportions of both *Chloroflexi* and *Bacteroidetes* showed a distinct downward trend and reached a very low level (0.25% and 3.41%, respectively) at Day 106. *Firmicutes* increased significantly and was the most abundant phylum (42.36%) at Day 106. Figure 2 shows that *Thermotogae* remained stable at the first and last stages (from Day 0 to Day 42 and from Day 81 to Day 106) but increased by up to 32.43% from Day 42 to 81, whereas the abundance of *Proteobacteria* declined slightly. After 106 days of operation, the two ADs showed a big difference in bacterial communities with *Bacteroidetes* dominant in the MR while *Thermotogae* and *Firmicutes* were dominant in the TR. Analysis of sequencing data provides a better approach to explaining the bacterial communities at the subdivision level. Hence, the relative abundance at the genus level was computed in every sample. In total, 39 genera were detected with a proportion higher than 0.1% in at least one sample screened as the abundant genera. The distributions at the genus level of each sample are shown in Figure 3.

Distribution Barplot

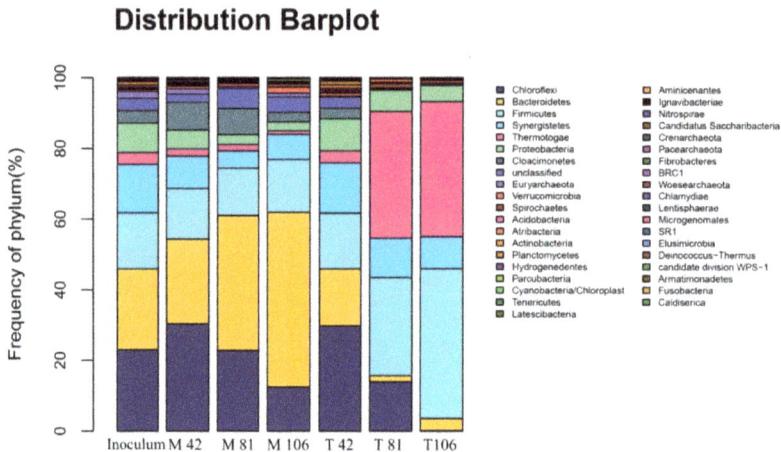

Figure 2. Taxonomic compositions of bacterial communities at the phylum level in each sample retrieved from pyrosequencing. The number in the sample names represents the day when sampling occurred. T: thermophilic reactor; M: mesophilic reactor.

As mentioned above, *Chloroflexi* was present in large proportions and then decreased in both reactors. The distributions also showed a downward trend in this phylum at the genus level. The major genus classified to the phylum *Chloroflexi* was *Levilinea*, which represented the greatest proportion in the inoculum as well as in samples M42, M81, T42, and then decreased dramatically from 19.27% to 8.87% in the MR and from 18.15% to 0 in the TR during operation. *Levilinea* can convert several carbohydrates to acetate and hydrogen under conditions of a high sodium concentration [21]. The genera *Longilinea* and *Ornatilinea* also decreased in both reactors, which might indicate their potential ability for organic matter degradation under hydrolysis acidification conditions [22].

For the MR, the genera in the dominant phylum *Bacteroidetes* showed obvious changes over the operation time. *Lutaonella* decreased initially and then increased significantly to 19.59% in M106, as this genus can use many organic acids and amino acids [23]. *Mariniphaga* and *Sunxiuqinia* showed an obvious decline in abundance. The former takes advantage of lipids and the latter ferments sugars with many kinds of acids as major fermentation products [24,25].

In the TR, the major genus assigned to the dominant phylum *Thermotogae* was *Fervidobacterium*, reaching very high levels of 35.63% at Day 81 and 38.07% at Day 106, which suggested its activities in acetate oxidation [26]. The relative abundance of *Clostridium_III* classified to the phylum *Firmicutes* also increased dramatically to 22.89% at Day 106; *Clostridium_III* converts complex macromolecules into alcohols, hydrogen, carbon dioxide, and volatile fatty acids [27].

%, Relative Abundance		Inoculum	M42	M81	M106	T42	T81	T106
Acidobacteria	Gp18	0.42	0	0	0	0.49	0	0
	Gp6	0	0	0	0	0	0.29	0
Atribacteria	Atribacteria_genera_incertae_sedis	0.52	0.39	0	0	0.54	0.93	0.22
		0	0	0	0	0	0	0
Bacteroidetes	Bacteroides	0	0	2.51	4.53	0	0	0
	Lutaonella	11.68	2.47	10.41	19.59	7.98	0	0
	Macellibacteroides	0	0	0.27	0.45	0	0	0
	Mangrovibacterium	0	0.44	0.23	0	0.52	0	0
	Mariniphaga	3.46	7.14	5.92	1.75	2.29	0	0
	Meniscus	1.26	2.36	2.97	1.3	0.76	0	0
	Paludibacter	0	0	0.67	0.97	0	0	0
	Petrimonas	0	0	0	0	0	0.22	2.88
	Prevotella	0	0	0	0.66	0	0	0
	Proteiniphilum	0.95	0	0	0	0	0	0
	Sunxiuqinia	0.7	7.38	5.18	3.6	0.78	0	0
	Tangfeifania	0	0	0.4	0	0	0	0
Chloroflexi	Bellilinea	0.53	0.84	0.46	0	0.63	0.37	0
	Leptolinea	0.48	0	0	0	0	0	0
	Levilinea	13.15	19.27	16.46	8.87	18.15	6.69	0
	Longilinea	2.36	4.29	1.46	0.86	3.37	0.83	0.14
	Ornatilinea	2.51	3.09	3.24	1.37	4.37	5.93	0
	Pelolinea	0.69	0	0	0	0.45	0	0
Cloacimonetes	Candidatus_Cloacamonas	3.61	7.87	7.39	2.78	3.05	0	0
Euryarchaeota	Methanothrix	1.83	1.36	0.51	0.99	0	0	0
Firmicutes	Acetanaerobacterium	0	0.4	0	0	0	0	0
	Anaeroarcus	0	0	0.27	0	0	0	0
	Anaerovorax	0.74	0.54	0.29	0.6	0.44	0	0
	Caloramator	0	0	0	0	0	0.17	0
	Catabacter	0.41	0	0	0	0	0	0
	Cellulosibacter	0	0	0	0	0	0.2	0.3
	Christensenella	0.44	0	0	0	0.53	0	0.15
	Clostridium_sensu_stricto	0	0	0	0	0	0.28	0.18
	Clostridium_III	0	0	0	0.44	0	12.46	22.89
	Clostridium_IV	0	0	0.4	1.2	0	0.62	0.44
	Clostridium_XIVa	0	0.54	0.4	0.66	0	1.15	4.57
	Coprothermobacter	0	0	0	0	0	0	0.16
	Defluviitalea	0	0	0	0	0	2.81	2.82
	Ercella	0	0.44	0.38	0	0	0	0
	Garciella	0	0	0	0	0	0	0.21
	Gracilibacter	0	0	0	0	0	0.24	0.25
	Hydrogenispora	0	0	0	0	0	0.71	1.54
	Intestinimonas	0	0	0	0	0	0	0.3
	Lactobacillus	0	0.4	0	0	0	0.48	0
	Lutispora	0	0	0	0	0	0.23	0.18
	Pelotomaculum	0	0	0	0	0	0.46	0
	Phascolarctobacterium	0	0.73	0.52	0.69	0	0	0
	Proteiniborus	0	0	0	0	1.05	0	0
	Pseudobacteroides	0	0	0	0	0	0	0.25
	Romboutsia	0.43	0	0	0	0.68	0	0
	Ruminococcus	0	0.66	0	0	0	0	0.32
	Saccharofermentans	2.89	1.26	1.26	1.15	1.83	0	0
	Syntrophomonas	0.47	0.67	0.44	0.52	0	3.32	1.77
	Tepidanaerobacter	0	0	0	0	0	0	0.17
	Tepidimicrobium	0	0	0	0	0	0	0.13
	Thermoflavimicrobium	0	0	3.16	3.34	0	0	0
	Thermoflavimicrobium	0	0.36	0	0	0	0	0
	Thermosyntropha	0	0	0	0	0	0.45	0
	Turicibacter	0	0	0	0.54	0	0	0
Hydrogenedentes	Candidatus_Hydrogenedens	0	0.56	0	0	0	0	0
Parcubacteria	Parcubacteria_genera_incertae_sedis	0	0	0	0.65	0	0	0
Planctomycetes	Thermogutta	0	0	0	0	0.92	0	0
Proteobacteria	Citrobacter	0	0	0	0	0	0	0.14
	Desulfatibacillum	1.08	0	0	0	1.28	2.64	0.22
	Smithella	2.08	2.52	1.12	1.22	2.32	0	0
	Syntrophobacter	0	0	0.26	0	0	0	0
	Syntrophorhabdus	2.06	0.73	0	0	1.87	0	0
	Syntrophus	0.86	0	0	0	0.54	0	0
	Tepidiphilus	0	0	0	0	0	2.89	3.71
Spirochaetes	Sphaerochaeta	0	0	0	0.43	0	0	0
	Treponema	0	0	0	0.53	0	1.34	1.02
Synergistetes	Aminivibrio	3.93	1.72	1.69	3.12	3.91	0	0
	Aminomonas	0	0	0	0	0	1.57	0.19
	Anaerobaculum	0	0	0	0	0	7.08	7.99
	Cloacibacillus	4.45	5.91	2.83	3.59	6.27	0.16	0
	Thermovirga	4.79	1.31	0	0	3.75	1.67	0.44
Thermotogae	Defluviitoga	2.69	1.26	0	0	2.58	0	0.19
	Fervidobacterium	0	0	0	0	0	35.63	38.07
	Mesotoga	0.71	0.81	1.71	0.95	0.88	0.26	0
Verrucomicrobia	Subdivision3_genera_incertae_sedis	0	0	0.36	0.97	0.52	0	0

Figure 3. Taxonomic compositions of bacterial communities at the genus level for the sequences retrieved from each sample. Grey, inoculum; red, mesophilic reactor (MR); purple, thermophilic reactor (TR) (higher relative abundances are shown in deeper colors).

2.3. Dynamics of Methanogen Communities

The archaeal community dynamics were also revealed by high-throughput pyrosequencing targeting 16S rRNA gene segments. The patterns of relative abundance at the genus level of archaea are shown in Figure 4. Overall, methanogens had lower diversity than the bacterial community partly because of the intrinsically low phylogenetic diversity of methanogens.

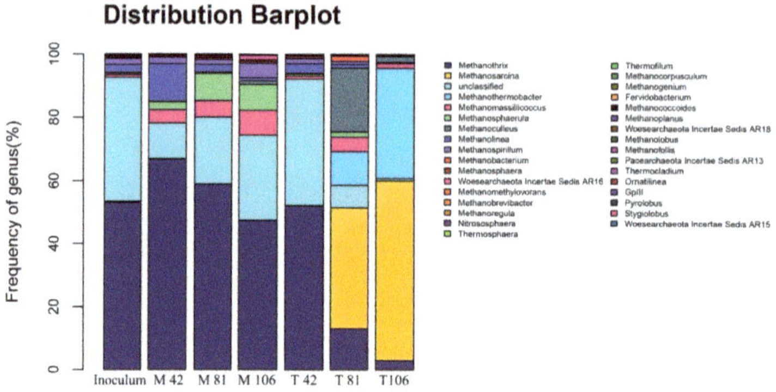

Distribution Barplot

Figure 4. Taxonomic compositions of methanogens at the genus level in each sample retrieved from pyrosequencing. The sample names are the same as those described in Figure 2.

Methanothrix was the most abundant methanogen in the MR throughout the operation and at Day 42 of the TR, and is considered to be the main methanogen involved in anaerobic fermentation [28]. Moreover, *Methanothrix* accounted for the majority of the sequence reads in the MR, while *Methanosarcina* was most abundant for the TR at Day 81 and Day 106 at the genus level, which is commonly observed in thermophilic digesters. *Methanothermobacter* ranked second in relative abundance in the TR at Days 81 and 106, which also implied a relation to hydrogenotrophic methanogens. In general, hydrogenotrophic methanogen species are more commonly detected in thermophilic digesters [28].

The genus *Methanothrix* was dominant in the MR throughout the running time with a relative abundance of 66.89% at Day 42, 58.94% at Day 81, and 47.39% at Day 106, representing an acetoclastic microorganism [29]. Comparatively, *Methanothrix* was only the most abundant genus (52.05%) in the TR at Day 42, and then *Methanosarcina* became the most abundant of the TR methanogens at 38.37% by Day 81 and at 57.07% by Day 106; *Methanosarcina* can produce methane by using H_2/CO_2, acetate, and methanol [30]. *Methanothermobacter* also played an important role in the TR, accounting for 10.62% at Day 81 and 34.86% at Day 106, indicating important effects of hydrogenotrophic methanogens [30]. Remarkably, *Methanoculleus* had a relative abundance of 20.20% with an ability for hydrogenotrophic activity [31]. In consideration of the rapid change in the OLR from Day 42 to Day 81, the microbial community might also go through a dramatic change in the TR. Overall, acetoclastic methanogens were dominant in the MR, while hydrogenotrophic methanogens were dominant in the TR, implying hydrogenotrophic and acetoclastic metabolism processes as the main pathways to methane production in the TR and MR, respectively.

2.4. Linkage between Metabolism and Reactor Performances

To further understand the impact of OLR and temperature on microbial communities, analysis of metabolic pathways was conducted. Based on the results of the two-level classification of the PICRUSt function, the differences of sample or intergroup abundance were compared, and the function of the significant differences between the samples or the abundance of the group was evaluated. Bacteria metabolism was compared between samples M106 and T106 to reveal the effect of the operation condition in both reactors. For bacteria, all of the metabolic functions are shown in Figure 5. Compared with that of the TR, the metabolism of the MR showed more active trends in amino acid metabolism, lipid metabolism, metabolism of cofactors and vitamins, carbohydrate metabolism, and energy metabolism, which contributed to the acid production step in anaerobic fermentation. Due to the complex substrate of the OFMSW and no accumulation of VFAs, the rate-limiting step was considered to be hydrolytic acid production. Amino acid metabolism, lipid metabolism, and carbohydrate metabolism refer to the degradation of amino acids, lipids, and carbohydrates, respectively. To a certain extent, the problem of a slow acid production rate has been solved because of the improvement of these metabolic pathways in the MR. By contrast, in the TR, cell motility, translation, and membrane transport were more active than in the MR, showing greater potential to adapt to environmental changes.

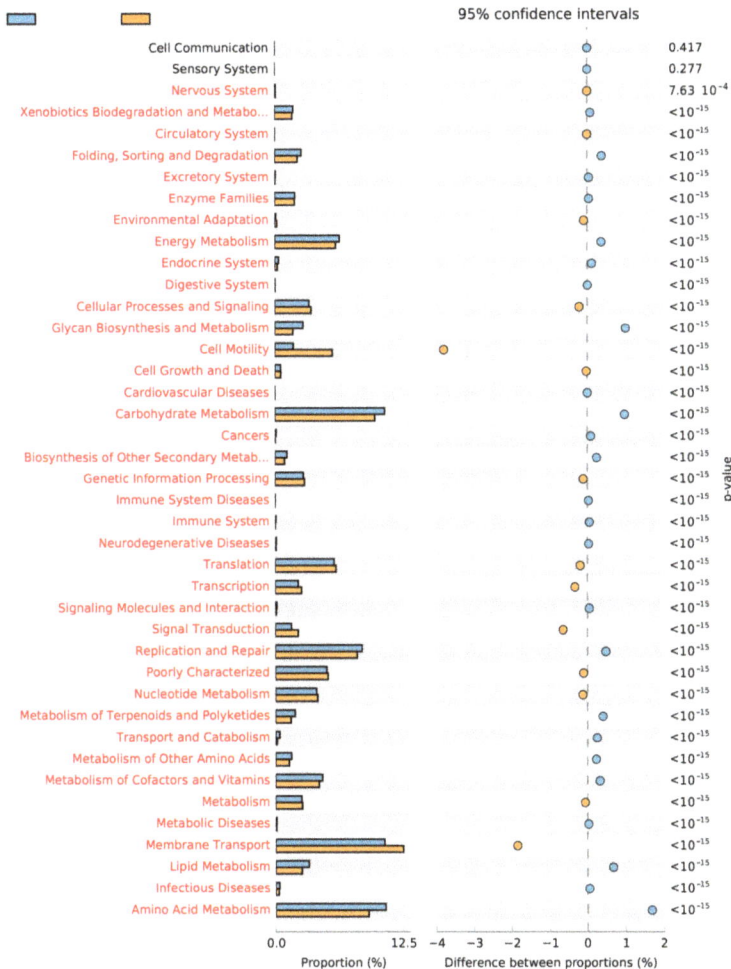

Figure 5. Different proportions of bacteria metabolic function at 95% confidence intervals between the MR and TR at Day 106.

For archaea, energy metabolism is considered as the primary methane production process in anaerobic fermentation [32]. The sequence patterns for metabolism are shown in Figure 6, demonstrating a proportion in the TR of 11.12%, which is much greater than the 8.39% in the MR. Thus, the TR might have more potential for methane production in the presence of a more acidic substrate in AD. Meanwhile, amino acid metabolism and lipid metabolism in the MR were more active than in the TR, which might be the main cause of the difference in AD performance.

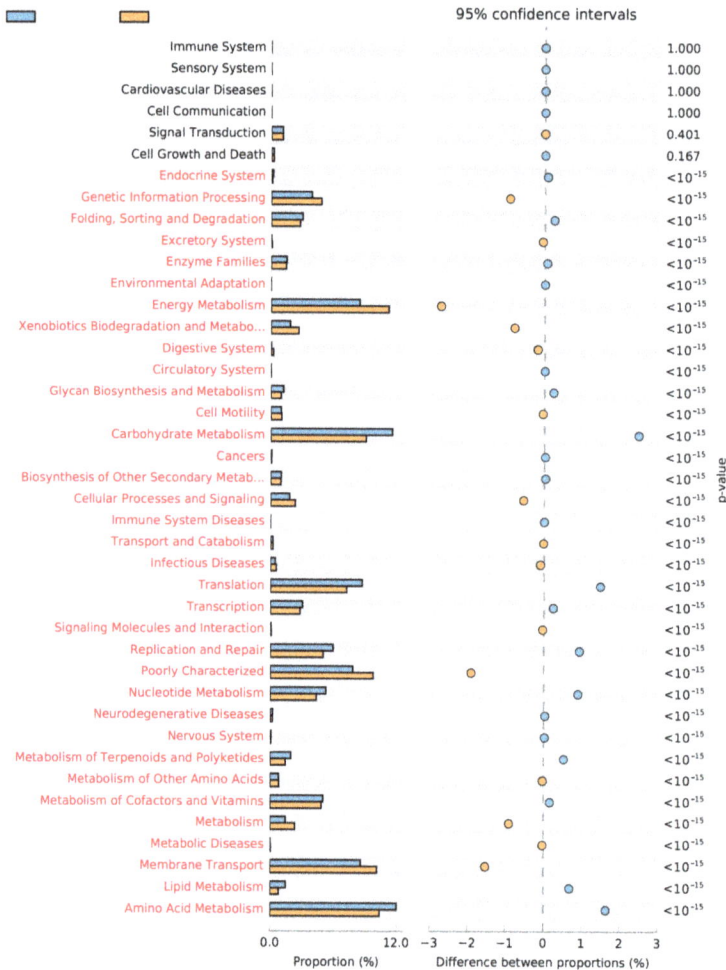

Figure 6. Different proportions of archaeal metabolic function at 95% confidence intervals between the MR and TR at Day 106.

3. Materials and Methods

3.1. Anaerobic Digesters

Two identical 60 L cylindrical anaerobic reactors with 50 L working volume and a heating apparatus were used to maintain internal temperatures at $55 \pm 1\ °C$ and $35 \pm 1\ °C$ for the TR and MR, respectively. The inoculum was taken from a normal atmospheric-temperature AD used for treating dairy farm wastewater (Kaiping, Guangzhou Province, China). Before feeding the reactors, the inoculum was filtered through a 1 mm sieve and starved for one week. The seed biomass was diluted with double-distilled water to 9.7 ± 0.7 gVS/L in reactors. The MSW was periodically obtained from a waste yard (Shaoguan, Guangdong province, China). The inorganic portions of the MSW such as glass and plastics were removed manually. The selected MSW was subpackaged in several sealing bags and stored at −18 °C after being ground and homogenized. The basic characteristics of the MSW are shown in Table S1 of the Supplementary Materials. The hydraulic retention time was 20 days, and the reactors were fed once a day. A stirrer was set at a speed of 60 rpm and ran constantly during the discharging and feeding time; at other time, it operated for 3 min every 15 min. The biogas was measured using a wet gas meter (Haide, Dalian, China).

3.2. Chemical Analysis

The total and volatile solid contents were measured by standard analytical methods for the examination of water and wastewater [33]. The C, H, and N contents were measured using a Vario EL element analyzer (VarioEL cube, Elementar, Langenselbold, Germany). The concentrations of NH_3-N and COD were analyzed using a commercially available kit (DR2800, Hach, Loveland, CO, USA). The pH was measured using a pH meter (pHS-3C, Rex, Shanghai, China). The contents of formate and VFAs—mainly acetate, propionate, and butyrate—were analyzed by high-performance liquid chromatography (e2695, Waters, Boston, MA, USA) using a refractive index detector and HPX-87 column. The mobile phase was 0.005 N H_2SO_4 with a flow rate of 0.5 mL/min, and the temperature of the column was 50 °C. Alkalinity was measured with 0.25 N H_2SO_4 to points of pH 5.7 and 4.3, to obtain data on IA, PA, and total alkalinity using a Titroline 5000 titrator (Julabo, Seelbach, Germany).

3.3. DNA Extraction and Amplification

DNA extraction was performed with an E.Z.N.A™ Mag-Bind Soil DNA Kit (Omega Bio-tek, Inc., Norcross, GA, USA) according to the manufacturer's specifications. The integrity of DNA was detected by agarose gel electrophoresis (UVP, Upland, CA, USA) and the Qubit2.0 DNA detection kit (Life Tech, Shanghai, China). Polymerase chain reaction (PCR) amplification was performed on a PCR instrument (T100™ Thermal Cycler, BIO-RAD, Hercules, CA, USA). The 16S rRNA genes were amplified by three rounds of PCR for archaea. The primers were 340F (5′-CCCTAYGGGGYGCASCAG-3′) and 1000R (5′-GGCCATGCACYWCYTCTC-3′) in the first round, and were 349F (5′-CCCTACACGACGCTCTTCCGATCTN(barcode)GYGCASCAGKCGMGAAW-3′) and 806R (5′-GACTGGAGTTCCTTGGCACCCGAGAATTCCAGGACTACVSGGGTATCTAAT-3′) in the second round. The genes were then PCR-amplified with Illumina nested primers in the third round. For bacteria, there were two rounds of PCR amplification, using the primers 341F (5′-CCCTACACGACGCTCTTCCGATCTG(barcode)CCTACGGGNGGCWGCAG-3′) and 805R (5′-GACTGGAGTTCCTTGGCACCCGAGAATTCCAGACTACHVGGGTATCTAATCC3′) in the first round, and then Illumina nested primers in the second round.

3.4. High-Throughput Pyrosequencing Analysis

The PCR products were sequenced through Illumina Miseq™, and the raw data were then transformed to sequenced reads through CASAVA Base Calling and deposited under accession number SRP131631 in the NCBI SRA database. The raw sequences were wiped off the primer sequences using cutadapt and clustered into operational taxonomic units using Usearch. A total of 295,925 and 282,765 effective 16S rRNA sequences were retrieved with average lengths of 452.91 bp and 417.18 bp for bacteria and archaea, respectively. Richness and diversity estimators (Shannon, ACE, and Chao1 indices) were calculated using MOTHUR. The microbial community structures were classified by blastn in Blast (coverage > 90%), and the comparison of metabolism was conducted using PICRUSt and STAMP.

4. Conclusions

The results of the present study suggest that that mesophilic AD of OFMSW showed better performance than thermophilic AD during both the adaptive phase and the OLR increasing phase. Larger shifts in the bacterial community were observed in the TR along with OLR elevation.

With respect to methanogens, *Methanothrix* dominated in the MR while *Methanosarcina* was favored in the TR. Variations in the mode and efficiency of metabolism between the MR and TR resulted in different performances with efficiency mainly related to the limiting hydrolytic acid step.

Supplementary Materials: The following are available online at http://www.mdpi.com/1996-1073/11/4/952/s1. Table S1: Basic characteristics of the food waste used in this study.

Acknowledgments: This work was funded by Guangdong Application Achievement Project (2017B020238005), Guangzhou science and technology project (201803030007) and Guangdong Technology Program of Industrial High Technology (2013B010204053).

Author Contributions: Xiaoying Kong, Tao Xing, Yi Zhang, and Yong Sun conceived and designed the experiments; Yiming Gao and Xingjian Luo performed the experiments; Yiming Gao analyzed the data; Yongming Sun contributed reagents/materials/analysis tools; Yiming Gao wrote the paper.

Conflicts of Interest: The authors declare no conflict of interest.

References

1. Murphy, J.D.; McKeogh, E. Technical, economic and environmental analysis of energy production from municipal solid waste. *Renew. Energy* **2004**, *29*, 1043–1057. [CrossRef]
2. Griffin, M.E.; McMahon, K.D.; Mackie, R.I.; Raskin, L. Methanogenic population dynamics during start-up of anaerobic digesters treating municipal solid waste and biosolids. *Biotechnol. Bioeng.* **1998**, *57*, 342–355. [CrossRef]
3. Song, Y.C.; Kwon, S.J.; Woo, J.H. Mesophilic and thermophilic temperature co-phase anaerobic digestion compared with single-stage mesophilic- and thermophilic digestion of sewage sludge. *Water Res.* **2004**, *38*, 1653–1662. [CrossRef] [PubMed]
4. Kang, X.; Sun, Y.; Li, L.; Kong, X.; Yuan, Z. Improving methane production from anaerobic digestion of Pennisetum Hybrid by alkaline pretreatment. *Bioresour. Technol.* **2018**, *255*, 205–212. [CrossRef] [PubMed]
5. Panepinto, D.; Genon, G. Analysis of the extrusion as a pretreatment for the anaerobic digestion process. *Ind. Crop. Prod.* **2016**, *83*, 206–212. [CrossRef]
6. Ahring, B.K. Perspectives for anaerobic digestion. In *Biomethanation I*; Ahring, B.K., Ed.; Springer: Berlin/Heidelberg, Germany, 2003; Volume 81, pp. 1–30.
7. Bayr, S.; Rantanen, M.; Kaparaju, P.; Rintala, J. Mesophilic and thermophilic anaerobic co-digestion of rendering plant and slaughterhouse wastes. *Bioresour. Technol.* **2012**, *104*, 28–36. [CrossRef] [PubMed]
8. Kim, M.; Ahn, Y.H.; Speece, R.E. Comparative process stability and efficiency of anaerobic digestion; mesophilic vs. thermophilic. *Water Res.* **2002**, *36*, 4369–4385. [CrossRef]
9. Ma, Q.; Shen, F.; Yuan, H.R.; Zou, D.X.; Liu, Y.P.; Zhu, B.N.; Li, X.J. Investigation on anaerobic digestion of the organic fraction of municipal solid waste (OFMSW) after pretreatment of fast aerobic fermentation. *Fresenius Environ. Bull.* **2015**, *24*, 1039–1046.
10. Muha, I.; Zielonka, S.; Lemmer, A.; Schonberg, M.; Linke, B.; Grillo, A.; Wittum, G. Do two-phase biogas plants separate anaerobic digestion phases?—A mathematical model for the distribution of anaerobic digestion phases among reactor stages. *Bioresour. Technol.* **2013**, *132*, 414–418. [CrossRef] [PubMed]
11. Gujer, W.; Zehnder, A.J.B. Conversion processes in anaerobic-digestion. *Water Sci. Technol.* **1983**, *15*, 127–167.
12. Schievano, A.; D'Imporzano, G.; Malagutti, L.; Fragali, E.; Ruboni, G.; Adani, F. Evaluating inhibition conditions in high-solids anaerobic digestion of organic fraction of municipal solid waste. *Bioresour. Technol.* **2010**, *101*, 5728–5732. [CrossRef] [PubMed]
13. Nelson, M.C.; Morrison, M.; Yu, Z.T. A meta-analysis of the microbial diversity observed in anaerobic digesters. *Bioresour. Technol.* **2011**, *102*, 3730–3739. [CrossRef] [PubMed]
14. Ye, L.; Zhang, T. Bacterial communities in different sections of a municipal wastewater treatment plant revealed by 16S rDNA 454 pyrosequencing. *Appl. Microbiol. Biotechnol.* **2013**, *97*, 2681–2690. [CrossRef] [PubMed]
15. Riviere, D.; Desvignes, V.; Pelletier, E.; Chaussonnerie, S.; Guermazi, S.; Weissenbach, J.; Li, T.; Camacho, P.; Sghir, A. Towards the definition of a core of microorganisms involved in anaerobic digestion of sludge. *ISME J.* **2009**, *3*, 700–714. [CrossRef] [PubMed]

16. Guo, X.H.; Wang, C.; Sun, F.Q.; Zhu, W.J.; Wu, W.X. A comparison of microbial characteristics between the thermophilic and mesophilic anaerobic digesters exposed to elevated food waste loadings. *Bioresour. Technol.* **2014**, *152*, 420–428. [CrossRef] [PubMed]
17. Oleszkiewicz, J.A.; Marstaller, T.; McCartney, D.M. Effects of ph on sulfide toxicity to anaerobic processes. *Environ. Technol. Lett.* **1989**, *10*, 815–822. [CrossRef]
18. Ripley, L.E.; Boyle, W.C.; Converse, J.C. Improved alkalimetric monitoring for anaerobic-digestion of high-strength wastes. *J. Water Pollut. Control Fed.* **1986**, *58*, 406–411.
19. Poggivaraldo, H.M.; Tingley, J.; Oleszkiewicz, J.A. Inhibition of growth and acetate uptake by ammonia in batch anaerobic-digestion. *J. Chem. Technol. Biotechnol.* **1991**, *52*, 135–143. [CrossRef]
20. Hansen, K.H.; Angelidaki, I.; Ahring, B.K. Anaerobic digestion of swine manure: Inhibition by ammonia. *Water Res.* **1998**, *32*, 5–12. [CrossRef]
21. Yamada, T.; Sekiguchi, Y.; Hanada, S.; Imachi, H.; Ohashi, A.; Harada, H.; Kamagata, Y. *Anaerolinea thermolimosa* sp. nov., *Levilinea saccharolytica* gen. nov., sp. nov and *Leptolinea tardivitalis* gen. nov., so. nov., novel filamentous anaerobes, and description of the new classes anaerolineae classis nov and Caldilineae classis nov in the bacterial phylum Chloroflexi. *Int. J. Syst. Evol. Microbiol.* **2006**, *56*, 1331–1340. [PubMed]
22. Yamada, T.; Imachi, H.; Ohashi, A.; Harada, H.; Hanada, S.; Kamagata, Y.; Sekiguchi, Y. *Bellilinea caldifistulae* gen. nov., sp. nov and *Longilinea arvoryzae* gen. nov., sp. nov., strictly anaerobic, filamentous bacteria of the phylum Chloroflexi isolated from methanogenic propionate-degrading consortia. *Int. J. Syst. Evol. Microbiol.* **2007**, *57*, 2299–2306. [CrossRef] [PubMed]
23. Arun, A.B.; Chen, W.M.; Lai, W.A.; Chou, J.H.; Shen, F.T.; Rekha, P.D.; Young, C.C. *Lutaonella thermophila* gen. nov., sp. nov., a moderately thermophilic member of the family Flavobacteriaceae isolated from a coastal hot spring. *Int. J. Syst. Evol. Microbiol.* **2009**, *59*, 2069–2073. [CrossRef] [PubMed]
24. Wang, F.Q.; Shen, Q.Y.; Chen, G.J.; Du, Z.J. *Mariniphaga sediminis* sp. nov., isolated from coastal sediment. *Int. J. Syst. Evol. Microbiol.* **2015**, *65*, 2908–2912. [CrossRef] [PubMed]
25. Irgens, R.L. Meniscus, a new genus of aerotolerant, gas-vacuolated bacteria. *Int. J. Syst. Bacteriol.* **1977**, *27*, 38–43. [CrossRef]
26. Patel, B.; Morgan, H. A Combination of Stable Isotope Probing, Illumina Sequencing, and Co-occurrence Network to Investigate Thermophilic Acetate- and Lactate-Utilizing Bacteria. *Microb. Ecol.* **2018**, *75*, 113–122.
27. Zhang, Y.; Alam, M.A.; Kong, X.; Wang, Z.; Li, L.; Sun, Y.; Yuan, Z. Effect of salinity on the microbial community and performance on anaerobic digestion of marine macroalgae. *J. Chem. Technol. Biotechnol.* **2017**, *92*, 2392–2399. [CrossRef]
28. Demirel, B.; Scherer, P. The roles of acetotrophic and hydrogenotrophic methanogens during anaerobic conversion of biomass to methane: A review. *Rev. Environ. Sci. Biotechnol.* **2008**, *7*, 173–190. [CrossRef]
29. Kendall, M.M.; Boone, D.R. The Order Methanosarcinales. In *The Prokaryotes: Volume 3: Archaea. Bacteria: Firmicutes, Actinomycetes*; Dworkin, M., Falkow, S., Rosenberg, E., Schleifer, K.-H., Stackebrandt, E., Eds.; Springer: New York, NY, USA, 2006; pp. 244–256.
30. Garcia, J.L.; Patel, B.K.C.; Ollivier, B. Taxonomic phylogenetic and ecological diversity of methanogenic Archaea. *Anaerobe* **2000**, *6*, 205–226. [CrossRef] [PubMed]
31. Garcia, J.-L.; Ollivier, B.; Whitman, W.B. The Order Methanomicrobiales. In *The Prokaryotes: Volume 3: Archaea. Bacteria: Firmicutes, Actinomycetes*; Dworkin, M., Falkow, S., Rosenberg, E., Schleifer, K.-H., Stackebrandt, E., Eds.; Springer: New York, NY, USA, 2006; pp. 208–230.
32. Zhang, J.X.; Mao, L.W.; Zhang, L.; Loh, K.C.; Dai, Y.J.; Tong, Y.W. Metagenomic insight into the microbial networks and metabolic mechanism in anaerobic digesters for food waste by incorporating activated carbon. *Sci. Rep.* **2017**, *7*, 10. [CrossRef] [PubMed]
33. Cleceri, L.S.; Greenberg, A.E.; Eaton, A.D. *Standard Methods for the Examination of Water and Waste Water*; American Public Health Association: Washington, DC, USA, 1998; p. 113.

Review

Opportunities and Barriers to Bioenergy Conversion Techniques and Their Potential Implementation on Swine Manure

Mahmoud A. Sharara [1] and Sammy S. Sadaka [2,*]

[1] Biological Systems Engineering Department, University of Wisconsin-Madison, Madison, WI 53706, USA; msharara@wisc.edu

[2] Biological and Agricultural Engineering Department, Division of Agriculture University of Arkansas, Little Rock, AR 72204, USA

* Correspondence: ssadaka@uaex.edu; Tel.: +1-501-671-2298

Received: 28 March 2018; Accepted: 13 April 2018; Published: 17 April 2018

Abstract: The objectives of this article are to offer a comprehensive evaluation of the opportunities and barriers for swine manure conversion technologies and to shed light on the gaps that might require further investigation to improve the applicability of these technologies. The challenges of manure management have been propagated alongside the global growth of swine production. Various technologies that target the production of energy, fuels, and bioproducts from swine manure have been reported. These technologies include pretreatments, i.e., drying, and solid separation; biological techniques, i.e., composting, anaerobic digestion, and biodrying; and thermochemical techniques, i.e., combustion, gasification, pyrolysis, liquefaction, and carbonization. The review highlights the yields and qualities of products, i.e., energy, gaseous fuel, liquid fuel, and solid fuel, of each technology. It exhibits that the choice of a conversion technology predominantly depends on the feedstock properties, the specifics of the conversion technique, the market values of the end products as well as the local regulations. The challenges associated with the presented techniques are discussed to ameliorate research and development in these areas. The notable finding of this paper is that there is a need for full-scale research in the area of thermochemical conversion of solid-separated swine manure.

Keywords: swine manure; biological conversion; anaerobic digestion; composting; biodrying; thermochemical conversion; combustion; gasification; pyrolysis; liquefaction; carbonization

1. Introduction

The United States Department of Agriculture (USDA) inventory reported that the number of pigs had reached 70.7 million heads in 2017 [1]. The growth in the pig inventory can be correlated to an increase in the swine manure production. The annual cumulative swine manure generated could reach 120 million metric tons (MMT) year^{-1} based on the assumption that the daily manure production per pig is about 4.67 kg·day^{-1} [2]. This vast amount of manure production has created a substantial challenge to the swine sector mainly due to the concentration of swine production [3]. It has been reported that about 85% of the national swine inventory produced in the USA is centered in the Midwestern states including Iowa, Minnesota, Illinois, Indiana, Nebraska and Kansas which are among the top ten producing states [4]. Research has also revealed that, out of the 69,000 swine operations, about 87% house 2000 heads or more [1].

The concentrated swine manure production, with the limitation of available land to apply manure as fertilizer has created a need to develop various technologies to convert this feedstock into bioenergy and/or value-added products. Bioenergy signifies the most essential renewable energy source [5].

It can add to fulfil, in a sustainable manner, the future energy requirements [6,7]. The conversion technologies could be classified as biological and thermochemical technologies. Biological techniques include composting, anaerobic digestion, and biodrying while thermochemical techniques include combustion, gasification, pyrolysis, liquefaction, and carbonization. Despite the success of developing various swine manure conversion technologies, some challenges still limit their large-scale applicability. The difficulties of swine manure conversion techniques include high moisture content, low calorific value, the scatter distribution that makes collection difficult and results in economic problems and deficiencies in transportation, handling, storage, and conversion. The present review focuses on the illustration of these technologies as well as the challenges facing them.

2. Swine Manure Composition and Characteristics

Swine manure composition data is highly varied, as shown by a survey of the available literature. This occurrence is occasioned by the fact that manure composition is incumbent upon various factors including, animal feed composition, animal age, and genetics. Additionally, the manure handling system alters the properties and composition of raw manure. Moreover, manure composition data are often reported using different matrices, i.e., as excreted, or as-removed basis. The latter option, as-removed basis, factors in the influence of the handling method on the composition and the properties of the manure. Table 1 below presents the various characteristics of swine manure for the animals at different stages. It is clear that, regardless of stage, the swine excrements are predominantly water, with the total solids making only 10% w/w of the manure.

Table 1. The weight of pigs at different stages and the corresponding characteristics of manure as-excreted [2].

Component	Units	Gestating Sow	Lactating Sow	Boar
Animal weight	kg	200	192	200
Total Manure	kg·day^{-1}animal^{-1}	5.00	12.00	3.80
Moisture Cont.	%, w.b.	90.00	90.00	90.00
TS	kg·day^{-1}animal^{-1}	0.50	1.20	0.38
VS	kg·day^{-1}animal^{-1}	0.45	1.00	0.34
BOD	kg·day^{-1}animal^{-1}	0.17	0.38	0.13
N	kg·day^{-1}animal^{-1}	0.03	0.09	0.03
P	kg·day^{-1}animal^{-1}	0.01	0.03	0.01
K	kg·day^{-1}animal^{-1}	0.02	0.05	0.02

3. Swine Manure Pretreatments

Manure is an essential source of nutrients, predominantly nitrogen (N), phosphorus (P), and potassium (K), to plants when applied in an ecologically sustainable manner [8,9]. Conventionally, land application of swine manure is considered the most recognized practical and economical utilization technique [10]. However, repeated manure application on the same field elevate the P concentration levels, which may harm the environment via surface runoff or leaching of P [11,12]. As mentioned earlier, the moisture content and total solids of fresh swine manure are approximately 90% w.b. and 10% w.b., respectively [2]. The high moisture content of manure tends to limit its long-distance transportation. Accordingly, research focused on reducing manure moisture content and concentrate the solid content via drying and solid separation.

3.1. Manure Drying

Manure could be dried to reduce its moisture content and to concentrate the solids via evaporation of water, thus reducing its volume. Animal manure used to be dried using old-fashioned techniques, i.e., oven-drying and hot air-drying. Typically, additional heat and/or forced air is desirable to enhance faster evaporation of moisture [13,14]. The advantage of dried manure is the reduction of microorganisms or insects growth. Additionally, dried manure has less odor as compared with fresh manure. Also, dry manure could be utilized as a soil fertilizer in the same manner that composted manure is utilized. The substantial disadvantages of drying manure are the high energy requirements and the costs associated with moisture removal. In a 100% efficiency dryer, the thermal energy required to evaporate 1 kg of moisture from manure was described to be 2.3 MJ [15]. This energy and the associated cost are incredibly high and cannot be fully recovered from the sale of the dried manure, which makes it economically unacceptable to producers. Drying systems also must be covered to keep manure away from rainfall.

3.2. Manure Solid Separation

Liquid and solid segments of swine manure could be separated and employed via various techniques to convert them to value-added products and to minimize their harmful effects [16]. Employing solid-liquid separation as a technique for manure treatment not only potentially enhances its management properties but also produces manure solids that might be exploited for either energy generation or compost [17]. Separation of solids from liquid manure could be achieved via sedimentation (gravity settling), centrifugation, filtration (using belt presses, screw presses, and screens), or chemical amendment [18,19]. The separated solid fraction of swine manure is rich in nutrients, particularly P, and can be transported for land application on soils with poor P concentrations. The liquid portion can be used to irrigate soils without considerably increasing their P levels [20]. Xiu et al. [21] stated that chemical separation of manure solids could be achieved using metal salts and organic flocculants that transport solubilized nutrients out of solution as fine particles. Following, it forms agglomerates of these salts, which can then rapidly precipitate. Normally used chemical amendments are metallic salts, i.e., iron (Fe), calcium (Ca), or aluminum (Al) and synthetic organic polymers, i.e., polyacrylamide (PAM) formulations [22]. Addition of coagulants, i.e., $FeCl_3$, $Fe_2(SO_4)_3$, $Al_2(SO_4)$, and $CaCO_3$, to manure results in coagulation of suspended particles by neutralizing the particles' negative surface charge and enhances P removal via coagulation of P by the cations constituting the coagulants [23]. Christensen et al. [24] and Sadaka and VanDevender [16] reported that treatments of animal manure with coagulants and flocculants enhanced solid-liquid separation. It has been reported that the overall energy balance presented a reduction of 0.596 MJ·kg^{-1} for the pyrolysis of the chemically pre-treated manure, whereas positive values of 0.352 MJ·kg^{-1} and 0.817 MJ·kg^{-1} were found for anaerobically digested sample and un-pretreated solid swine manure, respectively [25].

3.3. Challenges Associated with Swine Manure Pretreatments

As mentioned earlier, the energy consumption to dry manure is exceptionally high. Accordingly, an enhancement of the manure drying technology can considerably advance the overall impacts of swine manure management via thermochemical conversion. Manure solid separation using chemicals or mechanical methods add more costs to the manure management system which reduce the overall benefits to producers. Therefore, simple and affordable techniques to separate the solid portions from the liquid portion of manure need to be investigated. Swine manure drying using infrared technology or microwave drying systems could potentially be economic ways to reduce manure moisture content.

4. Biological Treatments of the Swine Manure

Biological processes target the elimination of pathogens, weed seeds, and parasites. Additionally, they amend the manure composition from a complex, malodorous effluent to an odor-free, plant-accessible one. Furthermore, biological treatments decrease the manure nutrient loading, thus minimizing the risk of over-application or nutrient runoff [26]. In general, most swine farms utilize at least one type of biological treatment. Biological treatments could take place in aerobic and anaerobic conditions as well as low temperature (mesophilic, 25–35 °C) or high temperature (thermophilic, above 50 °C) [27]. Manure matrix maintained within thermophilic conditions for a few days is usually sufficient to kill pathogens. However, manure composition modification is achieved through a sequence of decomposition stages that digest the complex organic species in the manure (proteins, fibers, fat, and organic acids) into uncomplex, short-chain compounds while releasing gaseous emissions of NH_3, CO_2, CH_4, and VFAs through aerobic and anaerobic digestion.

Manure management in farms above 2500 heads falls under the purview of Environmental Protection Agency's (EPA) concentrated animal feeding operation (CAFO) guidelines [28]. Soil application in such cases is regulated through a permitting process. Application permits are tied to manure quality only from a nutrient loading perspective, unlike with municipal sludge in which permitting tracks both composition and pathogen loading before permitting soil application. Permitting also ensures that the targeted agricultural land is capable of assimilating the intended manure volumes through soil nutrients analysis. In the following sections, we will explore swine manure biological treatments, i.e., composting, anaerobic digestion, and biodrying.

4.1. Swine Manure Composting

For centuries, animal farming has practiced aerobic composting of animal manure. It serves as a treatment that reduces odors and moisture, also, to eliminate pathogens and to improve the manure properties as a soil conditioner. The process is achieved through a series of oxidation and mineralization stages carried out by aerobic microorganisms (mesophilic and thermophilic), which convert the biomass matrix into a stabilized, humus-like substance [29]. The composting process requirements and outcomes are given by the following formula [30]:

$$\text{fresh organic waste} + O_2 \xrightarrow{\text{microbial metabolism}} \text{stabilized organic residue} + CO_2 + H_2O + \text{Heat} \quad (1)$$

The rate of carbon and nitrogen mineralization is triggered by the elevated temperatures achieved during activation of thermophilic bacteria. Incomplete aerobic digestion happens naturally in the bedding mixture (crop residue, and animal manure) under hoop structure swine feeding. By scraping it into piles or windrows, the bedding-manure mixture is often enabled to continue compositing between herds [31]. To facilitate aerobic conditions in wet-handling systems, where manure is flushed and collected in lagoons, a solid-separation step is necessary to increase the solid content of the matrix.

The levels of microbial activity and the phytotoxicity of the decomposed matrix remains the most critical determinants of the composting process [32]. From a temperature standpoint, the process can be separated into three consecutive stages: heating, thermophilic, and cooling phases. The heating phase, which starts immediately after mixing the biomass and setting the piles, is typically the shortest. The stage usually lasts 1–3 h, during which the compost pile temperatures increase rapidly from ambient to thermophilic levels, above 50 °C. The temperatures are sustained all across the thermophilic phase, the period of which depends on the nature of the mixture, and the aeration levels. Mineralization and oxidation rates are at their highest during this stage, resulting in volatilization of NH_3, CO_2, and evaporation of moisture. This phase is critical for killing pathogens, parasites, and weed seeds. Therefore, for the compost to be deemed safe, it should remain at this stage, at 55 °C, for at least three days [33]. The last stage is cooling and stabilization, which is typically the longest and it ends with the material thoroughly degraded and pathogen-free.

The composting process parameters (aeration, C/N ratio, moisture, pH, and temperature) should be appropriately managed to produce a stabilized, pathogen-free compost within a suitable timeframe. Most importantly, there are no sufficient C and N levels in swine manure (C/N ratio) to initiate and sustain composting by itself as it is evident in Table 2. Accordingly, introducing a carbon-rich source, i.e., wood chips, sawdust, or crop residue, to adjust the C/N ratio is the first step to initiate manure composting [34]. The recommended C/N range of 25–35 and the moisture content range of 50% and 60% w.b. are optimum conditions for the starting of the composting process. Excessively high temperatures (>60 °C), or rapid drying of the pile resulted in rapid decomposition in the initial phase but a limited activity of bacterial population, and eventual termination of the composting process [31].

Effects of aeration rates, both perpetual and intermittent, on maintaining a thermophilic environment for a blend of swine manure separated solids, and peat moss in an in-vessel composting setup was investigated [35]. Within 15 days of composting, the C/N ratio decreased from 15–18 to 10–14. Likewise, the pH of the mixture decreased from nearly neutral to acidic range (5.2–6.9). Intermittent aeration regime (between 0.04 and 0.08 $L \cdot min^{-1} \cdot Kg_{VM}^{-1}$) was proposed to attain pathogen-eradication (above 55 °C for three days) and odor control without volatilizing significant amount of the organic matter content in the mixture. Chemical composition changes during a 63-day composting experiment of swine manure blended with sawdust in a pile composting were investigated [36]. There was a rapid decline in organic-C from 45 to 36% within the first 14 days, followed by a slow decline in the remaining time, totaling a 10% drop in the process. Swine manure was also found to lose up to 72% of its organic C and 60% of the total N through emissions as part of the decomposition and mineralization stages [37]. Carbon dioxide and ammonia volatilization remain the primary causes of the losses. Various greenhouse gas (GHG) emissions from unturned and turned compost piles of manure and bedding material mixtures were monitored for over 80 days [38]. Higher GHG emissions were reported from turned piles (1.98 $kg_{CO2-eq} \cdot kg_{VSdegraded}^{-1}$), in comparison to unturned ones which reported to be 1.55 $kg_{CO2-eq} \cdot kg_{VSdegraded}^{-1}$. CO_2 emissions comprised the bulk of GHG emissions (75–80%), followed by CH_4, at 18–21%, then finally N_2O at 2–4%.

Table 2. Typical composition of liquid and solid swine manure [38].

Manure	Dry Matter $(g \cdot kg^{-1})$	Organic C $(g \cdot kg^{-1})$	Total N $(g \cdot kg^{-1})$	NH_4-N $(g \cdot kg^{-1})$	pH -
Liquid	4.9–152	1.0–65	0.6–7.8	0.3–6.6	6.7–8.9
Solid	150–330	42–132	3.5–11	0.5–6.0	8.1

4.2. Swine Manure Anaerobic Digestion

Anaerobic digestion of manure provides a method to treat animal waste by allowing microorganisms to digest the feedstock. This process produces biogas (with similarity to natural gas) suitable for energy production through burning, internal combustion engines, and gas turbines. The solids produced after the digestion process can be used as fertilizer or bedding for farm animals. About 200 anaerobic digesters were installed at dairy farms through the USA as a reaction to the energy deficient in the early 1970s. Several of these systems shut down or failed after just a few years of operation. In 2014, about 260 anaerobic digesters were operational in USA farms with only 39 digesters operational in swine farms according to data compiled by AgStar (AgStar Financial Services, Mankato, MN, USA), an EPA program to promote CH_4 mitigation projects in the livestock sector [39]. Anaerobic digestion process works efficiently with dairy and beef manures, although research continue to improve the process for swine and other animals. With enhancements in anaerobic digester equipment, as well as a better understanding of the method, there was a recovery of anaerobic digesters being installed on large dairy farms starting in the late 1990s and continuing to this day.

In mostly confined swine houses, anaerobic digestion occurs in the manure collection pits, and in the last storage lagoons. To activate specific bacterial and microbial communities that digest the organic matter, this process requires anaerobic conditions (oxygen-free). The composite manure substrate is converted to an effluent of degraded humic acids, and a mixture of gases (CH_4, CO_2, NH_3, and N_2O) by anaerobic bacteria. This process has been understood on USA swine farms in manure collection lagoons before the annual, or bi-annual, land application. However, most of these lagoons are not covered, meaning that the process only occurs within certain anaerobic sections of the lagoon.

Most farms do not actively flush or collect volatilized gaseous species resulting from digestion, making these gases a prominent source of greenhouse gas (GHG) emissions in agriculture. For instance, between 1990 and 2011 CH_4 and N_2O emissions from USA manure management were found to have increased from 31.5 to 52.0 Tg_{CO2-eq}, and from 14.4 Tg_{CO2-eq} to 18.0 Tg_{CO2-eq}, respectively [40].

These increases were credited, to the increase in the size of production units, and the spread of liquid handling and storage in dairy and swine production. Only a handful of USA farms have active digestion processes with covered anaerobic digesters where temperature, pH, and organic matter loading are controlled for purposes of biogas generation for energy or heating or both.

As shown in Figure 1, anaerobic digestion is a sequential process with different groups of bacteria carrying out different tasks to fully digest the biological substrate. After hydrolysis of soluble components, complex organic acids are transformed into volatile fatty acids (VFAs) by acidogenic bacteria. The VFAs are then transformed to acetate by acetogenic bacteria [41]. The VFDs were shown to decrease the pH of the mixture, thus damaging the pH-sensitive acetogenic and methanogenic bacteria. After inducing perturbation (changes in hydraulic retention time of substrate, temperature, and organic matter loading) to the anaerobic digestion of a cattle-swine manure mixture, changes in the VFAs concentration were reported [42]. Ammonia (NH_3) as a by-product of the acidogenesis was found to have an inhibitory effect on the digestion process. For instance, anaerobic digestion literature indicates that 1.7 to 14 $g \cdot L^{-1}$ total ammonium nitrogen (TAN), can lower CH_4 production by 50% [43]. Swine manure was proven to be challenging in anaerobic digestion studies, due to its high NH_4-N and sulfide contents [44,45]. Carbon-rich sources such as crop residue, glucose or glycerol are added to the swine manure to adjust the starting C/N ratio and to elucidate this challenge. Investigation of crop residue additive (wheat straw, corn stalks, and oat straw) and the C/N ratio (16, 20, and 25) on the gas yield from anaerobic digestion of swine waste was carried out [46]. There was a definite correlation between increasing the C/N ratio and the yields of both the CH_4 and the total biogas. Wheat straw blends yielded significantly lower CH_4 and biogas at all C/N levels when compared to corn stalks and oat straw blends. A better explanation for the occurrence is that wheat straw contained much higher lignin content, which is inaccessible to most digestive bacteria, in comparison to the other crop residues.

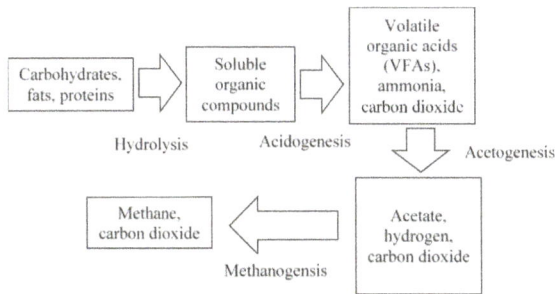

Figure 1. Biomass transformation with the various stages of anaerobic digestion [47].

As mentioned earlier, there are two temperature ranges for achieving anaerobic digestion process namely mesophilic and thermophilic. The removal of chemical oxygen demand (COD), biogas production, and CH_4 production disparity effected by mesophilic and thermophilic anaerobic digestion were studied [48]. Only minimal differences in COD reduction, 63% versus 67%, and volatile solids (VS) reduction of 64% versus 65%, were observed between mesophilic and thermophilic conditions, respectively. However, CH_4 productions were lower in thermophilic conditions compared to mesophilic conditions of 3.3 versus 3.5 $L_{reactor\ volume}^{-1}\ day^{-1}$, respectively. On the other hand, biogas yields and CH_4 concentrations in the biogas under thermophilic conditions were 494–611 $L \cdot kg_{VS}^{-1}$ (vs denotes to volatile solids) and 59.8–61.7%, which were higher than those under mesophilic conditions 315–419 $L \cdot kg_{VS}^{-1}$ and 56.9–57.7%, for the anaerobic digestion of three different maize varieties [49]. Thermophilic digestion increased the metabolism of bacterial communities but resulted in a buildup of VFA, in case of high-protein and urea substrates, which resulted in an accumulation of

NH_3 [50]. Increases in free NH_3 concentration (0.75–2.6 $gm_N \cdot L^{-1}$), VFA accumulation ($gm_{acetate} \, L^{-1}$) and a decrease in methane yield (188–220 $mL_{CH4} \, gm_{VS}^{-1}$) were observed with temperature increases in a study of swine manure anaerobic digestion [46]. There arises a critical need for continual monitoring and control of the swine manure anaerobic digestion due to the difficulty of the process and reluctance of stakeholders to adopt the technology. Furthermore, challenges arise due to the toxic effect of the liberated NH_3 in the digestion of nitrogen-rich sludge, such as swine manure [51].

4.3. Swine Manure Biodrying (Partial Composting)

The majority of the manure thermochemical conversion technologies or applications require manure drying to facilitate both transportation and conversion. Animal waste drying is a challenging procedure given its relatively low heating value of 19.7 $MJ \cdot kg^{-1}$ on a dry basis [52]. Table 3 shows the heating value of swine manure as well as both conventional fossil fuels, and different biomasses.

Table 3. Typical higher heating values (HHV, $MJ \cdot kg^{-1}$) for various feedstocks [53,54].

Feedstock	HHV ($MJ \cdot kg^{-1}$)-Dry Basis
Bituminous coal	31.60
Peat	21.22
Cellulose	17.30
Lignin	26.70
Poplar wood chips	20.75
Oil shale	12.44
Wheat straw	17.55
Corn stover	18.10
Rice straw	15.95
Poultry litter	17.14
Cattle manure	17.36
Swine manure	19.70

Manure drying is the leading obstacle to the broader application of thermochemical conversion due to the lack of economic manure drying. Biodrying has become a remedy to this obstacle. Biodrying is a biological process by which biodegradable waste is quickly heated through initial stages of biodegradability to remove excess moisture. The manure biodrying technique is a physical-biological treatment of organic feedstock generating heat from exothermic reactions, along with excess aeration, to volatilize most of the moisture yielding minimum carbon conversion. Except for the required aeration rates, biodrying and composting procedures are similar in their operating parameters. Biodrying operating parameters including pile temperature, reactor configuration, and aeration rate have been studied numerously. Adani et al. [55] carried out studies on the effects of temperature of biomass on the biodrying of urban solid waste. They adjusted aeration by setting three different temperature levels; 70, 60 and 45 °C. They attributed high temperatures with favorable biomass degradation, more stable high moisture material with low energy content. In previous studies [56,57] researchers indicated that elevated temperatures facilitated microbial activities until a threshold of 55–60 °C, beyond which microbial activity and heat generation diminished. Sadaka and Ahn [57] biodried swine manure-corn stover mixture. They reported that with no external heat provided to swine manure-corn stover mixture, about 50% reduction of the initial water content was achieved after biodrying for 2–4 weeks. A thermophilic temperature of 60 °C was attained after four days for swine manure. Biodrying did not significantly affect the biodegradability of the mixture as measured by volatile solids and heating value reductions.

The deliberation is presently continuing regarding the determination of suitable aeration levels for both composting and biodrying. Various composting studies have recommended various aeration rates. Li et al. [58] reported that several aeration rates on dairy manure composting indicate that 0.25 $L \cdot min^{-1} \cdot kg^{-1}_{volatile \, matter}$ (VM) achieved the highest temperature and kept it for the longest retention time, with lowest emissions of odorous gases. Hong et al. [59] also mentioned that the maximum degradation rate in dairy manure composting occurs with an air flow rate between 0.87 and 1.87 L·min^{-1}·kg^{-1}$_{\text{volatile matter}}$. Gao et al. [60] investigated the effects of three aeration levels; 0.3, 0.5 and 0.7 L·min^{-1}·kg^{-1}$_{\text{volatile matter}}$ on the composting of poultry manure and sawdust. They found that the best quality of composted manure (regarding maturity and stability) was under the air flow rate of 0.5 L·min^{-1}·kg^{-1}$_{\text{volatile matter}}$. Aeration levels for the biodrying process are recommended by few studies [61]. Consequently, there is a need to assess the effects of various aeration levels on the biodrying process of swine manure mixed with crop residues.

4.4. Challenges Associated with Swine Manure Biological Conversion

It is often quite challenging to operate biological conversion systems due to the complexity and interdependencies in the process. There are several attainment stories of biological conversion systems that have helped producers in meeting their goal of converting swine manure to value-added products. On the other hand, some biological conversion systems failed despite the enhancements of the biological conversion techniques. These failures can be attributed to the following causes:

- Poor design and equipment selection: Design and equipment selection for biological conversion systems can include decisions related to the manure pumping and conveyance systems, the gas cleaning and electrical generation equipment. Therefore, it is essential that the proper technology and equipment be selected. A failure of any one of these system components can result in operation failure, reduced revenue generation or added capital costs to replace faulty equipment.
- Lack of appropriate technical expertise: There is a lack of technical expertise in managing biological conversion systems in many cases due to the complexity of manure treatment systems. Although animal farmers routinely manage other complex systems, and training programs on biological conversion systems are available, there is a need to enhance the manure management skills to be able to sustain steady-state operations.
- Lack of system maintenance: Biological conversion systems need to be well maintained. Accordingly, producers need to dedicate some time and costs for maintaining these systems to avoid downtime.
- Lack of commitment by the operator: Given the seasonal nature of farming, there can be times of the year when the biological conversion systems may not receive the required attention and the careful maintenance. Again, this can be associated with the view that these systems are not an essential business function of the animal farm.

5. Thermochemical Conversion of Swine Manure

Thermochemical conversion, mainly combustion, remains the oldest known thermal technique to convert organic and biological residues to heat. Thermochemical conversion techniques include combustion, gasification, pyrolysis, liquefaction, and carbonization. Thermochemical processes utilize elevated temperatures, at different ranges for each process, aided by an oxidative or an inert agent. These processes target energy production or destruction of hazardous wastes.

5.1. Swine Manure Combustion

Combustion enabled people to utilize energy contained in organic matter for heating and cooking. Embedded in the hydrocarbon bonds, i.e., coal, petroleum and natural gas, and biomass, i.e., forest, crop or livestock residue, is fossilized energy which can be released under oxidative conditions using heat. A direct correlation can be drawn between the embedded energy and hydrogen (H), carbon (C), and oxygen (O), sulfur (S) and ash contents of the feedstock. Feedstock heating value is the primary characteristic that determines whether it could be logistically and economically used as an energy source. The heating value of common solid fuels is highly dependent on its characteristics and composition [62]. Various models were developed to forecast the heating value of biomass from its composition (C, H, N, O, S, and ash); proximate analysis (volatile matter, and fixed carbon), or summative analysis (cellulose, hemicellulose, and lignin) [63]. Ultimate analysis models: C, H, S, N, O, and Ash proved to be the most accurate in predicting higher heating values in biomass. Channiwala and Parikh [64] developed a correlation to estimate the HHV (MJ·kg^{-1}) of solid, liquid and gaseous fuels as follows:

$$HHV = 0.3491\,C + 1.1783\,H + 0.1050\,S - 0.1034\,O - 0.0151\,N - 0.0211\,Ash \qquad (2)$$

where

C, H, S, O, N, and Ash represent the elemental analysis of the feedstock.
HHV is the higher heating value, MJ·kg^{-1}.

It should be mentioned that HHV is the energy released upon oxidation of unit mass of the feedstock taking into consideration the enthalpy of vaporization for the generated water, whereas LHV accounts only the oxidation enthalpy. The correlation [65] below can be used to calculate one from the other, in units of MJ·kg^{-1}:

$$HHV = LHV + 21.978 \times H \qquad (3)$$

where:

H is the hydrogen weight fraction in the sample.
LHV is the lower heating value, MJ·kg^{-1}.

The combustion process is an accumulation of sequential steps of drying, pyrolysis (devolatilization), gasification, char combustion, and gas-phase oxidation. The biomass moisture evaporates at rates dependent on the particle size, and the vapor pressure in the surrounding space under atmospheric pressure (101.3 kPa) and at temperatures above 100 °C. Volatile organic species pyrolysis takes place at temperatures between 250 °C and 500 °C. Depending on the biomass type, and based on the size of biomass particles this range varies accordingly. Exothermic mixed reactions (gas-solid gasification, and char combustion) and exothermic homogenous reactions (gasification, and gas combustion) take place at higher temperatures between 600 °C and 1200 °C. Thermal energy and flue gas (CO_2, H_2O, NO_2, and SO_2), in addition to an inert ash residue, are released by these reactions. An organic molecule's embedded energy is the enthalpy of the complete oxidation of its hydrocarbons, into oxides and water.

Majority of incinerators used with biomass are grate, or fluidized-bed systems. They are more flexible to the fuel type, suspension burners which only allow for co-firing biomass at specific ratios (25% by energy share) with particular moisture, ash content and particle size prerequisites [66].

Combustion technology is effective in reducing animal manure volume and concentrating nutrients. Park et al. [67] evaluated the feasibility of using solids separated from swine wastewater treatment practice as a fuel source for heat production. Proximate analysis of the solid recovered from the swine wastewater after flocculation with organic polymer showed that the high heating value and the moisture content were 22.3 MJ·kg^{-1} and 15.38%, respectively. The combustion efficiency of the solids was found to be in the range of 95% and 98% with varied temperatures. Based on their results they recommended that solid separated from swine manure can be used as an alternate source of fuel.

Swine manure, coal, and their co-combustion (10% dry weight of manure) were studied by simultaneous TG/MS dynamic runs [68]. Furthermore, the non-isothermal kinetic analysis presented that the Arrhenius activation energy parallel to the combustion of the blend (125.8–138.9 kJ·mol^{-1}) was slightly higher than that of manure (106.4–114.4 kJ·mol^{-1}) or coal (107.0–119.6 kJ·mol^{-1}). The valorization of three different manure samples through pyrolysis and combustion processes was evaluated [69]. Dairy manure and swine manure were biologically pretreated by anaerobic digestion and biodrying processes, respectively. Thermal behavior of manure samples was studied using thermogravimetric analysis coupled with mass spectrometry. These researchers reported that the processes could be divided into four stages to include dehydration, devolatilization, char transformation and chemical matter decomposition. They attributed the differences among the samples to their different composition and pretreatment. Combustion characteristics of bio-oil derived from swine manure were examined using thermogravimetry techniques [70]. The results designated that the combustion processes of bio-oil happened in three stages, namely the water and the lighter compound evaporation, i.e., the release of the volatile compounds, ignition, and burning of the more massive compounds (central carbon), and finally decomposition of the carbonate compounds.

Various biomasses lean towards the slagging and fueling in the combustion reactors. Several indices were developed to correlate the composition and alkaline minerals to the slagging tendency. Determining the weight of alkali oxides (potassium and sodium oxides) per unit energy (heat) in the fuels used is among the standard measures to monitor slagging and fouling upon co-firing agricultural and biomass residue. The upper limit for alkali levels in fuel is 0.17 kg·GJ^{-1} to minimize the chance of slagging and fueling. Biomass S and Cl were influential in fouling and minerals deposition mostly because of the formation of alkali sulfates and chlorides that condensed on fly-ash, gas exits and downstream. On the other hand, the interaction of K and P with Si and Ca was responsible for the formation of agglomerates in fluidized bed combustors [71].

Combustion of animal manure produces ash that has concentrated nutrients. Komiyama et al. [71] tested the chemical characteristics of ashes from three manure types namely cattle, poultry, and swine manure. The concentration of P and K did not change in cattle and layer manure due to the existence of Si and Ca. However, the concentration of P increased by 3.6 and 4.6 folds for swine and broiler manure, respectively as compared to the original materials. In another study, after incineration at 700 °C, the resultant manure ashes were investigated [72]. High pH (>10), increased the concentration of P, K, and heavier metals like zinc (Zn), copper (Cu) and manganese (Mn) for all ashes. N was never detected in the manure. Swine manure ashes had the highest P concentration in ash residue, around 10–12%. These P levels are close to those in phosphate rocks, regularly used as soil amendment. Combustion temperature impact of solids on plant-available P in the swine manure ash was investigated [73]. An insoluble crystalline form of phosphorous (hydroxyapatite, $Ca_5(PO_4)_3(OH)$) was formed at temperatures of 700 °C and above. Therefore, to retain the functionality of ash-bound P, low-temperature combustion or gasification technologies (400–700 °C) were recommended.

5.2. Swine Manure Gasification

During gasification, that takes place in a starved oxygen environment, biomass particles undergo drying, devolatilization, solid-gas and gas-phase reactions that produce producer gas, char, and a small fraction of condensable [74]. Gasification primary objective is the production of an energy-rich blend of gases called "producer gas" that could be combusted in boilers, internal combustion (IC) engines or gas turbines. Producer gas contains carbon dioxide (CO_2) carbon monoxide (CO), hydrogen (H_2), methane (CH_4) and nitrogen (N_2) typically. Gasification takes place under temperature levels of 700 °C and 1000 °C, which is often lower than those of combustion. However, the type of feedstock is used to determine the exact gasification temperature. Coal, because of its low volatile matter content and low reactivity, has higher reaction temperatures than biomass. The gasification reactions commence after the biomass feedstock undergoes drying and devolatilization. Biomass has a much higher volatile matter content (around 80% of dry weight) and a more reactive char because of the catalytic effect of the ash minerals. The overall biomass gasification reaction in the air and steam environment can be represented by Equation (4), which proceeds with several gasification reactions as shown in the subsequent equations [75,76]:

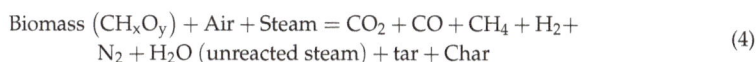

$$\text{Biomass } (CH_xO_y) + \text{Air} + \text{Steam} = CO_2 + CO + CH_4 + H_2 + \\ N_2 + H_2O \text{ (unreacted steam)} + \text{tar} + \text{Char} \tag{4}$$

Combustion reactions:

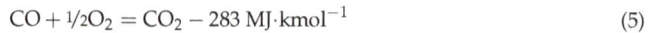

$$CO + \frac{1}{2}O_2 = CO_2 - 283 \text{ MJ·kmol}^{-1} \tag{5}$$

Fischer-Tropsch reaction:

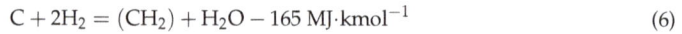

$$C + 2H_2 = (CH_2) + H_2O - 165 \text{ MJ·kmol}^{-1} \tag{6}$$

The Boudouard reaction:

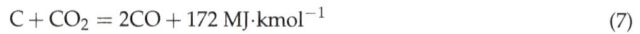

$$C + CO_2 = 2CO + 172 \text{ MJ·kmol}^{-1} \tag{7}$$

The water gas reaction:

$$C + H_2O = CO + H_2 + 131 \text{ MJ·kmol}^{-1} \tag{8}$$

The methanation reaction:

$$C + 2H_2 = CH_4 - 75 \text{ MJ·kmol}^{-1} \tag{9}$$

Methane steam reforming:

$$CH_4 + H_2O = CO + 3H_2 + 260 \text{ MJ·kmol}^{-1} \tag{10}$$

Additionally, some other elements, N and S, are liberated together with producer gas in the form of ammonia (NH_3), hydrogen sulfide (H_2S), hydrogen cyanide (HCN), and nitrous oxides (NO_x).The reaction temperature, heating mode (auto-thermal or externally heated), amount of oxidant present (typically oxygen) in the reaction volume per mole of biomass, use of a catalyst, physical and chemical characteristics of the biomass particles, and the type of gasification system are the key factors that influence gasification efficiency.

Pure oxygen, steam, carbon dioxide or blends of these oxidants can be used as the gasifying agent apart from the atmospheric air [77,78]. Energy-intensive oxidants such as pure oxygen or steam should only be used with high calorific value feedstock according to thermodynamic efficiency of the conversion. The use of air as the gasifying agent means the dilution of the product gas with nitrogen from the air (78.09 vol. %). It should be mentioned that the gasification process can be achieved without supplying an external oxidant given that most biomasses contain O_2 as part of their structures. There is a requirement for an external source of heat when no oxidizing agent is supplied because all devolatilization reactions: the water-gas reaction, and the Boudouard reaction are endothermic. On the other hand, supplying an oxidizing agent could eliminate the need for external heat beyond the startup (autothermal mode) because the exothermic reactions, full- or partial-oxidation reactions, can sustain the conversion.

Extensive research has gone into biomass gasification leading to commercial adoptions in different gasifier configurations. Gasification platforms resemble combustion units as relates to feeding mechanisms and reactor types. These systems are classified as a fixed bed, fluidized bed, and plasma gasifiers. In fixed bed gasifiers, biomass feed is fed into metered or gated entrance moving gradually, by gravity, across the various reaction zones. The feedstock is incrementally fed in, as part of the stack is reacted away into producer gas, char, and tars. The stages of the gasification process of drying, devolatilization or pyrolysis, char gasification (Reduction) and combustion proceed within distinct stratified zones in the biomass stack. Dependent on the thermodynamics of reaction taking place in each zone, temperatures inside a fixed-bed gasifier are also stratified. Downdraft, updraft, and cross draft systems represent fixed-bed gasifier's examples based on the relative movement of the producer gas concerning the feedstock. Downdraft gasifiers are the most common among fixed bed systems due to their ease of operation, and the superior quality of producer gas (low tar and condensable) [79].

Tar and condensable are deposited on low-temperature surfaces, typically downstream, causing blockages and considerable pressure drops during gasification process. Additionally, tar content corrodes the combustion chamber and forms undesirable deposition applications where the producer gas is directly fired. The majority of commercial gasification systems (75%) are downdraft gasifiers with updraft systems accounting for only 2.5% [80]. In a laboratory-scale gasifier, agglomeration (formation of an ash-layer as well as condensation in upper parts of a countercurrent/updraft) was reported with nutshells, and olive husks gasification [81].

Fluidized-bed gasification is a relatively newer model of conversion comparatively. Fluidized beds are rapid and efficient options that are also suitable for up-scaling. They consist of an inert, thermally stable media like silica, olivine, or alumina particles, blended at times with specialty catalysts for tar reforming. This inert media serves as the heat-transfer medium that facilitates conversion as the feedstock is fed into the reactor. The heat transfer is higher than in fixed bed reactors due to the fluidization of the biomass-bed material mixture increasing the contact between hot media and ambient biomass particles. Compared to fixed-bed reactors, the fluidized bed does not have thermal stratification or distinct conversion zones (isothermal conditions), helping it achieve steady-state operation faster. In fact, both the drying and the devolatilization stages are assumed to be instantaneous once the biomass enters the reactor bed. Circulating fluidized bed systems are also utilized in biomass conversion as well as a bubbling fluidized system described above. Unlike a bubbling fluidized bed unit, this system consists of two reactors connected in series to form a closed loop for the solids. Fluidization velocities in this system tend to be higher than in bubbling fluidization. By using gravity, char particles are entrained out of the first reactor and are allowed to react in the secondary-bed with the residues collected and circulated back to the main reactor bed.

Plasma gasification is an allothermal process occurring when organic matter is exposed to an extreme thermal practice using a plasma torch to produce syngas. Because feedstock varies in physical form, moisture content, ash content, and chemical composition, a plasma reactor must be designed to meet these differences. Therefore, a single plasma reactor design would not apply for all waste treatment cases [82]. Rajasekhar et al. [83] described plasma gasification as involving gases such as air, oxygen, nitrogen or noble gases, and this produces exceptionally high temperatures, which depolymerizes feedstock to individual atoms such as carbon, hydrogen, and oxygen. These atoms react with each other to form high-quality syngas. At these high temperatures, the inorganic part (metals, glass, silicates) extrudes from the bottom of the reactor and forms an inert, non-leachable vitrified slag that can be landfilled [84,85]. Campos et al. [85] reported that by utilizing plasma arc technology, it was possible to generate 816 kWh per ton of municipal solid waste (MSW); this is a higher energy producer than incineration, pyrolysis or conventional gasification. Plasma gasification has several advantages including better environmental benefits through lower emissions, vitrified slag has numerous usages compared to ash, metals, and slag from gasification, disposal of hazardous wastes (dioxins), higher gross energy recovery resulting from complete decomposition to the elemental level, and smaller installation size increasing process controllability [86,87]. There are some disadvantages related to plasma gasification including lack of reliability, high energy consumption, high consumption of electrical energy, which is economically costly, and high initial investment [88].

Biomass waste such as swine manure separated solids (SMSS), with the molecular formula $CH_{1.69}O_{0.5}$, is shown on the C-H-O triangular (ternary) diagram (see Figure 2) [89]. The area around dotted lines shows an equilibrium state between gaseous species and solid carbon (i.e., gasification region). Below the gasification, the region is fully oxidized carbon gaseous species while within the region above, only solid carbon exists.

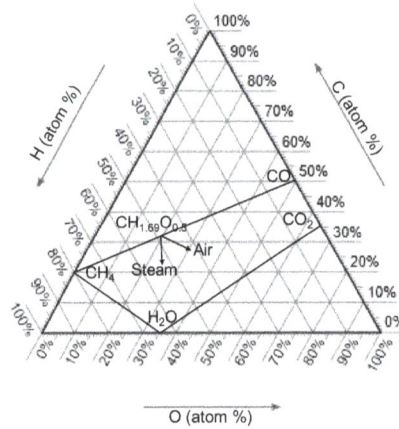

Figure 2. A ternary C-H-O plot is illustrating biomass waste gasification using either air or steam as the gasification medium.

The impact of gasification variables, including, equivalence ratio (ER), bed temperature, freeboard temperature, and the existence of a tar-reforming stimulus (calcined dolomites: MgO. CaO) on quality of producer gas was evaluated [90]. The ER remained the essential factor in determining the quality of the producer gas and the yield of tars generated. Increasing the ER from 0.18 to 0.41 and from 7.2 gm Nm^{-3} to 4.6 gm Nm^{-3} (Nm^3 is a cubic meter normalized to standard temperature and pressure conditions, 25 °C and 101.3 kPa) caused a drop in the tar-to-producer gas ratio [91]. Fixed bed gasifiers produce higher tar yields, representing between 12 wt. % and 20 wt. % of the carbon in the biomass, compared to tar in fluidized-bed systems which represent 4.3 wt. % of biomass carbon at 750 °C [92]. For optimal conversion, ER values between 0.25 and 0.30 were recommended. Temperature influence on gasification of feedlot cattle manure in a fluidized bed system was evaluated with the temperatures varied between 627 °C and 727 °C. Burner gas (generated from propane-air burning), was used as the fluidizing-gasifying agent. The producer gas yield reached 0.54 $Nm^3 \cdot kg^{-1}$ with 19.53 $MJ \cdot Nm^{-3}$ higher heating value. Burner gas as the gasifying agent has undoubtedly augmented the producer gas with combustion hydrocarbons. There was a 38.7%, 26.1% and 14.7% yield of H_2, CO, and CH_4 energy-rich gases within the dry produced gas composition.

The producer gas can be purified and catalytically upgraded through the Fischer-Tropsch (F-T) process to produce liquid hydrocarbons or used in direct combustion on gas-burning systems [93]. A catalytic reforming step to adjust the ratio of hydrogen to carbon monoxide above unity, $H_2/CO > 1$ and also to convert all CH_4 and higher hydrocarbons to CO and H_2 must be done to the producer gas [94]. As shown below in Equation (11), the Fischer-Tropsch reaction proceeds at a temperature between 200 °C and 250 °C, and pressures between 25 and 60 bars [95]. Liquid long hydrocarbon chains, C5+ are among the output products released along with gaseous fuel that can be suitable for power production in gas turbines. Adjusting the H_2/CO to numbers around two can be done through the water gas shift reaction as shown in Equation (12):

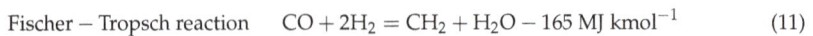

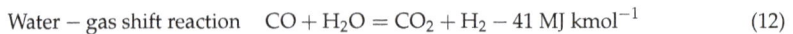

$$\text{Fischer} - \text{Tropsch reaction} \quad CO + 2H_2 = CH_2 + H_2O - 165 \text{ MJ kmol}^{-1} \quad (11)$$

$$\text{Water} - \text{gas shift reaction} \quad CO + H_2O = CO_2 + H_2 - 41 \text{ MJ kmol}^{-1} \quad (12)$$

Researchers assessed both dry and wet livestock manures gasification. Dry wastes alike poultry litter and feedlot manures can be treated directly via pyrolysis and air/steam gasification techniques. The solids produced from dairy and swine operations aqueous waste streams can be treated via wet gasification or direct liquefaction processes [96]. Current testing of the dry feedstocks, i.e., poultry litter and feedlot manure has been restricted to dry gasification systems using air as the oxidizing agent. Fixed-bed gasification yielded a low-heating value gas with an average HHV of 4.5 MJ·m^{-3} for poultry litter (TS) 92.5 wt. % and 4.1 MJ·m^{-3} for feedlot manure with total solids of 92.4 wt. %. The product gases contained a combustible portion consisting on average of 5.8% H_2, 27.6% CO, and 1.0% CH_4. Unfortunately, the product gases were rigorously diluted with nitrogen, thus lessening the potential HHV by approximately 60%. By blending the high-ash manure with a low-ash feedstock, as well as using acid washing, ash reduction is achievable. Evaluation of the ash minerals out of the biomass matrix can also be achieved efficiently by soaking biomass in diluted acid. In a swine and hen manures study, acid washing led to a drop in inorganic minerals (Fe, Ca, K, Zn, P, and S) and a tentative drop in the char reactivity [97]. The reduced char reactivity was strongly correlated with low Ca concentration in acid-washed manures. Washing (soaking) and fractionation (removing <1 mm particle size fraction), two approaches for ash reduction, were studied as pretreatment for peach stones gasification [98].

The critical challenge of manure gasification remains stratification of the reactive zones in fixed-bed systems because it results in the formation of hot spots and agglomerations. Investigations were done on the effects of mixing cow manure with sawdust, at different mixing ratios, on gasification efficiency in a downdraft gasifier [99]. Increasing the ratio of cow manure in the blends from 0 to 90% caused drops in the reduction zone temperatures, the producer gas heating value, and the conversion efficiency. The low conversion efficiency in cow manure was attributed to fixed carbon content within the manure being higher than sawdust translating to more endothermic char reduction reactions that cause a drop in the temperatures and the conversion efficiency. Pelletized poultry litter gasification was done in a commercial downdraft gasifier at temperatures between 825 °C and 925 °C [100]. Formation of clinkers (fused ash particles) in the reactor bed caused interruptions in gasification.

The presence of tar in the producer gas is a significant issue facing biomass gasification. Tar refers to mixtures of phenolic, and aromatic hydrocarbons formed during the devolatilization (pyrolysis) stage of gasification without sufficient decomposition inside the reactor to generate the targeted gases. Tar also refers to the subsidiary hydrocarbon compounds formed under high reaction rigor with higher molecular weights and higher aromaticity (polyaromatic). The volatilized organics formed during the pyrolysis stage are further digested by the increased gasification temperatures by either the gasifying agent or original oxygen and moisture (steam) in the feedstock. Persistence of tar in the produced gas can be attributed to the short residence time of vapor phase in high-temperature regions since most tar decomposition reactions are endothermic, as is the case with updraft gasifiers. In the absence of external oxidizing agents, volatiles undergoes subsidiary decomposition reactions at temperatures higher than 650 °C that increases permanent gases (H_2, CO, and CO_2) generation and decreases gravimetric tars [101]. By using mass spectrometry, tars were classified by their molecular masses into primary, secondary and tertiary tars [102]. With temperature increases, tars underwent both decomposition and repolymarization. This phenomenon explains the transformation of tars retrieved in biomass conversion from primary tars under common thermal conditions to secondary and tertiary tars under raised gasification temperatures, as shown in Table 4. Primary tars are comparatively more reactive and susceptible to thermal cracking in comparison to polycyclic aromatic hydrocarbons (PAH) and tertiary tars [103]. The existence of heteroatoms (O, N) and side-groups (OH, CH_3) in primary tars makes them more reactive than aromatic (ring) compounds. Free-radical reactions necessitated the primary thermal cracking mechanisms. The reactions take place through breakage of chemical bonds in a tar compound, the formation of free radicals that undergo propagation, isomerization and termination stages in which H_2 and CH_4 are liberated, and polyaromatic (tertiary) compounds are created.

Table 4. Classes of tars formed during biomass gasification [102].

Class	Tar Class			
	Primary	Secondary	Tertiary (Alkyl)	Tertiary (Condensed)
Compounds	Levoglucosan, hydroxyacetaldehyde, furfurals, and methoxyphenols	Phenolics, moreover, olefins	Methyl acenaphthylene, methylnaphthalene, toluene and indene	Benzene, naphthalene, acenaphthylene, pyrene
Temperature range	500–800 °C	500–1000 °C	700–1000 °C	700 °C ≥ 1000 °C

A study of biochar from manure gasification in a circulating fluidized bed (at 730 °C), and acid-treated ash produced from manure combustion was done comparatively to test the use of both as phosphorous (P) fertilizers [104]. The two types of thermal residue (ashes and chars) yielded no discernible differences on P availability in the soil. Gasification char can be used as a phosphate fertilizer to augment soil P levels but not as a starter P fertilizer as indicated by the results.

5.3. Swine Manure Pyrolysis

Pyrolysis, as mentioned earlier, is the essential stage to both the combustion and gasification of biomass. During the pyrolysis stage, the solid biomass matrix undergoes thermal depolymerization producing gaseous species, volatile organic compounds, and causing a rearrangement of both the volatilized and solid components. Biomass pyrolysis is optimized to generate condensable organic compounds resembling naturally-occurring crude oil (often referred to as bio-oil or bio-crude) in addition to char and gaseous products. Pyrolysis does not need any oxidizing agent to facilitate the conversion, unlike combustion or gasification. It is necessary to rapidly remove the volatilized species from the reactor to the cold condensation unit. Quick removal and cooling of pyrolysis vapors (quenching) are crucial in avoiding further thermal decomposition of volatiles into permanent gases which equally minimizes the solid-vapor reactions that facilitate char formation.

Pyrolysis is classified as either slow or fast (flash) pyrolysis based on the process duration. Slow pyrolysis tends to provide high yields because of low heat transfer rates and longer residence times for both the solids and vapors. On the other hand, fast or flash pyrolysis has a high heating rate (10^3–10^4 °C·s^{-1}), a residence time less than 2 s, and rapid cooling of the volatile species which increases the condensable (bio-oil) yield [105]. Fluidized bed (bubbling and circulating), ablative (rotating cone, and vortex), and vacuum reactors were investigated in the context of fast pyrolysis [106,107]. Lignocellulosic biomass (mainly wood) products of fast pyrolysis average yields 60–75 wt. % bio-oil, 15–25 wt. % biochar, and 10–20 wt. % non-condensable gases [108]. The main product in bio-oil is an acidic mixture (pH between 2.5 and 3.5) of water (15–30 wt. %) with a significant number of oxygenated hydrocarbons of different molecular weights. Acids, alcohols, aldehydes, ketones, aromatics, phenols and sugars are among more than 200 chemical species contained in bio-oils in varying concentrations depending on reaction conditions, and composition of original biomass [109,110]. Most woody and lignocellulosic feedstock pyrolysis reactions were grouped into cellulose, hemicellulose, and lignin pyrolysis reactions [111].

Phase separation of pyrolysis oil into water-rich and hydrocarbon-rich phases was achieved by adding surplus water to the bio-oil. Compared to liquid petroleum fuels, the oxygen content of bio-oil is quintessentially 45–50 wt. % which, in combination with the water content in the bio-oil, result in low heating values of about 18–26 MJ·kg^{-1}.

Biomass bio-oil is both thermally and temporally unstable because of its high water content, acidity, and corrosiveness. Repolymarization and phase separation reactions between various products including acids, alcohols, and aldehydes through various mechanisms were outlined [112]. Shortly after production, bio-oil repolymarization and phase separation complicate storage and transportation. Therefore, bio-oil must be upgraded first to overcome these qualities that deem the bio-oil unfit for most applications. In the presence of catalysts, through hydrodeoxygenation (HDO), steam bio-oil upgrading can be achieved. One or two high stage pressures of catalytic hydrotreatment are useful for bio-oil upgrading [113]. The process involves the injection of hydrogen at an elevated pressure (more than 100 bar) within temperatures between 250 °C and 400 °C in the presence of catalysts (sulfide CoMo or NiMo). Inhibition of repolymarization reactions is a crucial challenge in bio-oil upgrading mainly caused by increased reaction severity (pressure and temperature). Additionally, plugging and coking (deactivation) caused by the failure of oxygen molecules to dissociate on the catalyst is a setback of larger bio-oil compounds. Under atmospheric conditions, zeolite catalysts facilitate bio-oil upgrading without the need for further hydrogen injection [114].

Syngas products (CO, and H_2) have been produced through catalytic steam reforming of bio-oil products. As a means to expand the usefulness of the pyrolysis products, this technique has been investigated on bio-oil and model compounds such as phenol, acetone, and acetic acid [115]. The amount of energy spent in facilitating hydrogen production from bio-oil was equated to the energy necessary to reform natural gas into syngas products [116]. Steam reforming of bio-oil has challenges due to catalyst deactivation and coking [117].

Pyrolysis oil and gases have been investigated as fuel in compression (diesel) engines, turbines, and boilers. Hossain and Davies [118] reviewed the performance of pyrolysis products as fuel in internal compression engine. They reported various problems associated with pyrolysis oil as fuel, such as reduced ignition quality, corrosion, increased emissions and particulate matter in addition to low thermal efficiency. These challenges were attributed to the characteristics of the original feedstock, chemical and physical instability of the pyrolysis oil, and the high water and solids content. The change in bio-oil viscosity and composition with storage is a crucial barrier facing its utilization. Upgrading the pyrolysis oil, blending with diesel fuel, and modifying the engines showed improvement in the performance of pyrolysis oil as fuel. Indirect injection of pyrolysis oil in engines was recommended to ensure an even air-fuel mixture. Hossain et al. [119] evaluated pyrolysis oil blends (20% and 30%) with butanol, waste cooking oil, and diesel as fuel in compression engines. The tests showed the pyrolysis oil blends to reduce thermal brake efficiency by 3–7% and produce lower CO emissions when compared to diesel fuel. Vihar et al. [120] utilized a tire pyrolysis oil as a fuel in a turbo-charged compression engine without blending or engine modification. They reported stable performance that is comparable to diesel fuel at full load. Heating the intake manifold was highlighted as a critical requirement to achieve stable operation using pyrolysis oil. They reported increased NO_x and SO_2 emissions. These increases were attributed to the increased nitrogen and sulfur content in the bio-oil when compared to the fossil fuels. Pyrolysis oil from switchgrass, miscanthus, and eucalyptus were blended with ethanol and tested as fuel in a commercial boiler (40 kWth) [121]. A 20–80% blend of pyrolysis oil and ethanol resulted in a clean burn. Adjusting the boiler settings were necessary since the stoichiometric air: fuel ratios for ethanol and pyrolysis oil were lower than that for fuel oil. Utilizing low N feedstock to produce the pyrolysis oil was found to result in a reduction to the boiler NO_x emissions when compared to the type #2 fuel oil.

Pyrolysis gas has also been investigated as fuel in both compressions and spark ignition engines. Increased NO_x emissions under pyrolysis gas combustion were attributed to the high flame temperature which resulted from the increased H_2 amount in the gas. Lean conditions were found to enable the reduction of NO_x emissions and increased the engine thermal efficiency but resulted in a relative increase in CO emissions [122]. Shah et al. [123] found CO and NO_x concentrations in the exhaust from syngas-powered generators to be lower than under gasoline operation. Adjustment to engine design, such as increased compression ratio, optimizing injection timing and stratified charging can improve engine performance and reduce emissions under gasification and pyrolysis-derived gases [124].

Animal manure as feedstock for pyrolysis conversion exists in a few studies compared to the majority of available literature focusing on poultry litter pyrolysis. Pyrolysis of poultry litter yielded bio-oil 15–30 wt. % of the original feed, an HHV of 26–29 MJ·kg^{-1} and dynamic viscosity of 0.01–27.9 Pa, respectively [125]. The temperature remained the most instrumental parameter impacting conversion in a multi-parameter study of pyrolysis of poultry litter-wood shaving mixture within a fluidized-bed reactor; the other factors were biomass feed rate and N_2 flow rate respectively [126]. The highest bio-oil yield was 51 wt. %, at pH, 4.85 within temperatures of 475 °C. Investigations were carried out for swine compost, wood chips, and sewage sludge as a potential feedstock for bio-oil production in a fluidized-bed pyrolysis unit [127]. The swine compost bio-oil yielded H/C ratio of 1.63, a higher heating value (HHV) of 31.2 MJ·kg^{-1}, in comparison to 1.68 and 27.0 MJ·kg^{-1} for sewage sludge, and 1.51 and 23.9 MJ·kg^{-1} for wood shavings. Implementation of pyrolysis conversion and thermochemical conversion are impeded by the high moisture and ash contents of animal waste. Mineral and alkali salts control the pyrolysis conversion pathways, and the resultant oxygenates which are available in varying quantities in manure, further adding uncertainty to the bio-oil products generated. Production of char could be an added advantage of animal wastes pyrolysis that recently came to attention. Char or "charcoal", is the solid residuals of the devolatilization of volatile organics and the partial reaction of biomass fixed carbon.

Researchers pyrolyzed untreated separated solid swine manure as well as solids which had undergone chemical or biological pretreatment before separation [25]. The chemically pre-treated and anaerobically digested swine manure solids had comparable properties and showed similar performance during thermogravimetric analysis. Nonetheless, the energy content of the gas resulted from the pyrolysis process contain about 30% of the original energy of the swine manure solids in the case of the anaerobically digested materials and about 60% of the chemically pre-treated swine manure solids. Swine separated-solids; paved-feedlot manure; dairy manure; poultry litter; and turkey litter were pyrolyzed. Manure-based biochar physicochemical and thermochemical characteristics were evaluated [128]. It was found that dairy biochars contained the most significant volatile matter, C, and energy content and the lowest ash, N, and S contents. Swine biochars had the most significant P, N, and S contents alongside the lowest pH and EC values.

5.4. Swine Manure Hydrothermal Liquefaction

The liquefaction process is a promising technology for converting wastewater feedstock into bio-oil [129]. It has been utilized to convert a variety of wastewater feedstocks including swine manure, cattle manure, microalgae, macroalgae, and sludge [130,131]. During liquefaction, water serves as the reaction medium, lessening the need to dry the feedstock. As mentioned earlier, drying is a significant energy input for biofuel production. Liquefaction takes place under the temperature range of 200–350 °C and pressure of 5–15 MPa. Under these conditions feedstock macromolecules breakdown and reform biofuel [132]. Self-separation of the bio-oil from water is then enabled as the reaction solution returns to standard conditions. The produced bio-oil can be combusted or upgraded to be similar to petroleum oils [133]. Liquefaction bio-oil contains a wide range of chemical compounds such as straight and branched aliphatic compounds, aromatics and phenolic derivatives, carboxylic acids, esters, and nitrogenous ring structures [132,134]. Biller and Ross [130] reported that the class of compounds identified in the bio-oil is affected by the existence of protein and carbohydrate. The high oxygen/nitrogen content is the core factor differentiating bio-oils from petroleum oils [134] and results in detrimental biofuel qualities such as bio-oil acidity, polymerization, high viscosity, and high-boiling distribution. Furthermore, the varied chemical composition of bio-oil affects the combustion enactment, storage constancy, upgrading response, and economic value.

Liquefaction treatment system was developed to liquefy swine and dairy manures. The excess water content serves as the carrier fluid and reaction medium for direct liquefaction [96]. This process employs exceptional water properties that only occur in the vicinity of the critical region. The supercritical water denotes to the water at temperatures above its critical point, i.e., 374 °C and 22.1 MPa, with its density varying from 0.2 to 0.7 g·cm^{-3}. As the supercritical water density decreases substantially from that of ordinary liquid water, hydrogen bonds between water molecules profoundly weaken resulting in shallow values of dielectric constant, comparable to those polar organic solvents. Liquefaction hydrolyzes the lignocellulosic components in the feedstock and converts it into lighter organic oils (bio-oils). The metal salts naturally present in the waste assist in catalyzing the hydrolysis reactions. When compared to pyrolysis, liquefaction takes place at lower temperatures and higher pressure. Researchers developed batch and continuous liquefaction experiments on swine manure with total solids of 20–27 wt. % [135]. They reported that in the batch system under reactor temperatures ranging between 285 and 350 °C and CO atmosphere, the oil yield reached 76.2% of the volatile solid. The swine bio-oil had a higher heating value of 36.4 MJ·kg^{-1}. They also mentioned that additional processing of the oils is needed due to the existence of nitrogen and sulfur. Continuous operation under the reactor temperature of 305 °C and pressure of 10.3 MPa, resulted in slight decreases for both the maximum oil production to 70.4% and the bio-oil's heating value, to 31.1 MJ·kg^{-1} (a 23.4% decrease). As a waste treatment alternative, liquefaction reduced the initial swine waste stream's COD by 64.5%. The authors concluded that further treatment of the wastewater is necessary before discharging because no reductions in the N, P, and K of the post-treated stream were observed. Supercritical liquefaction processing was used to convert swine manure into a liquid fuel [136]. Swine manure ethanol was used as a solvent in an autoclave in the temperature range of 240–360 °C. The oil yield was evaluated by Fourier Transform Infrared spectroscopy (FTIR), elemental analyses, heating values, water content analyses, ash content and solids content. The results showed that the yield of the liquefaction products was significantly influenced by the reaction temperature. The maximum oil yield of 26.7% (of dry matter) with low oxygen content (11.48%) and heating value of 33.98 MJ·kg^{-1} were achieved at a reaction temperature of 300 °C.

5.5. Swine Manure Carbonization

Biochar is the main solid byproduct of the carbonization process. The characteristics of this char are a function of the composition of the first biomass and the conversion severity. On the other hand, ash (mineral oxides with no carbon) is produced from complete combustion. Contrastingly, carbonization, pyrolysis, and gasification produce a carbon-rich solid constituent (char) that also contains the ash minerals. In a process less severe than pyrolysis or gasification, char can be produced in a dedicated procedure at temperatures between 250 °C and 400 °C in the absence of air or oxygen. This process is referred to as slow-pyrolysis, carbonization or torrefaction. Torrefaction is a mild thermal treatment taking place at lower temperatures (200–300 °C) in which the biomass matrix remains unmodified mainly except for the easily devolatilized fraction (hemicellulose). The process improves friability and energy density of the biomass, facilitating co-firing with coal, or stand-alone conversion [137]. Char is utilized for soil quality improvement, including filtration and adsorption media [138], incineration as a blend-in with fossil coal or as a stand-alone solid fuel. Biochar is used to indicate the specific end-use of the produced char as a soil amendment [139]. Compared with other sophisticated thermochemical conversion technologies, such as fluidized bed gasification and fast pyrolysis, carbonization (or slow pyrolysis) necessitates relatively low technical resources, making the process suitable for farm-scale [140]. The biochar produced from animal wastes can easily be transported and stored without irritation odor and deterioration. It can be eagerly used as a cooking fuel and feedstock for existing coal power plants. Bio-char is higher in quality to charcoal produced from coal due to its low sulfur content and high reactivity.

Numerous advantages of incorporation of biochar in the soil have been learned through studying these soils. Biochar contains aromatic carbon structures that provide a stable carbon form facilitating nutrients retention. Furthermore, porous biochar structures ameliorate the soil qualities facilitating the growth of microbial microorganisms. Soil cation exchange capacity (CEC) is also improved due to the increased alkalinity of the soil. A study was carried out on impacts of biochar origins (cotton trash, grass clippings, and prunings) verses biochar application rates (10, 50 and 100 t·ha^{-1}) on a radish crop in Alfisol soil [141]. Even with the highest rate of biochar application, nitrogen fertilization was shown to be necessary despite the fact that C/N for biochar is quite high about 200. Radish dry matter production improved with increasing biochar application rates as well as the interaction between N fertilization and biochar. Additionally, there were general improvements in soil quality; soil pH, exchangeable alkali ions (Na, K and Ca) and a decrease in the soil tensile strength, with the incorporation of biochar. Recent studies revealed that incorporation of biochar in the soil not only improves soil properties and immobilizes heavy metals, it also mitigate carbon emissions by sequestering carbon in a stable form.

With temperature increase, from 100 to 700 °C, biomass underwent progressive structural transformation classified into four stages: transition char, amorphous char, composite char, and lastly, turbostratic char [142]. There was an inverse correlation between the biochar degradation in the soil and pyrolysis temperatures [143]. Under low-severity conversion, residual cellulose and hemicellulose in the biochar matrix are quickly degraded and lost in the soil compared to aromatic biochar produced under increased temperatures (525–575 °C). However, higher pyrolysis/biochar production temperatures reduce the recoverable biochar mass. During corn stover pyrolysis, the influence of air injection on the quality of biochar as a soil amendment was investigated [144]. Higher organic carbon content was exhibited for chars produced at 0% and 10% air injection until week 6 of the study where differences became insignificant. There was more biologically available carbon in the 0% and 10% chars utilized during the first few weeks by the soil microorganisms. On the other hand, extractable P started to increase in week four due to increased microbial activity that facilitated char decomposition and P demineralization.

The great macro- and micro-nutrients content of manure solids has created interest in using them as biochar feedstock. Characteristics of biochar produced at 380 °C from different biomass feedstocks were studied [145]. Highest electrical conductivity, 2.90 mS·cm^{-1}, water retention capacity, 294%, organic N, 0.25 mg·kg^{-1}, and phosphate, 0.76 mg·kg^{-1} was exhibited by cattle sludge biochar. Soil incorporated swine manure and woodchip biochars were monitored for carbon emissions [146]. Soils treated with biochar maintained organic C with CO_2 emissions equal to those from control soil. When compared with biochar-free soils, biochar application facilitated a reduction in carbon emissions with the application of manure digestate. Soils amended with both manure digestate and manure biochar reported the largest drop in Olsen P levels, but they still maintained more than 50 mg·kg^{-1}. Temperature effect on the quality of swine manure solids biochar was studied [147]. Increasing the charring temperature from 350 to 700 °C resulted in a drop in char yields from 62.3 to 36.4% of the original mass. A corresponding increase in aromatic and carbonyl carbon from the loss of alcoholic, paraffinic, and carboxylate carbon in the char carbon was reported. Physical characteristics of swine manure solids biochar generated at different temperatures, 400–800 °C, through slow pyrolysis conditions (no air) were investigated [148]. The biochar yields varied from 39 wt. % at 400 °C to 34 wt. % at 800 °C. Increasing the pyrolysis temperature from 400 to 800 °C, increased the pH of biochar solution (5 g biochar in 10 mL deionized water) from 7.5 to 11.4. Furthermore, it increased biochar P content, porosity, and surface area to 7.7 wt. %, 0.13, and 63 m^2·g^{-1}, respectively. A series of biochars were produced from dried swine manure waste via slow pyrolysis at different temperatures, i.e., 400–800 °C [149]. The produced biochar was examined for its prospective use as a soil amendment. It was observed that increasing the reactor temperature improved the pore characteristics, ash contents and pH values of all swine-manure-derived biochars, whereas it decreased the yield and N/O contents.

Most biochar literature reported adoption of torrefaction, slow pyrolysis (carbonization), or flash pyrolysis as the conversion technologies [149]. Although pyrolysis-derived biochar contains more char carbon in principle, only a fraction of this carbon is raw unconverted cellulosic carbon readily degraded by microorganisms upon soil application as revealed by studies. Consequently, a rapid carbon loss from the biochar reduces its sequestration potential. Bio-oil studies point out some challenges facing pyrolysis including; storage, transportation, pretreatment and upgrading, the energetics and economics of biochar production via pyrolysis would, therefore, be further constrained.

5.6. Challenges Associated with Swine Manure Thermochemical Conversion

In addition to the manure thermochemical conversion challenges mentioned earlier, some other issues might hinder the advancements of these conversion technologies. This section will cover issues associated with generic biomass types due to the nonexistence of full-scale research on thermochemical conversion of swine manure. The following items are a collection of previously cited issues which are listed here to highlight the reasons for failures. Accordingly, they might initiate research strategies to overcome these issues. It should be revealed that this list is not so conclusive. The failures of biomass thermochemical conversion systems have resulted from design problems, i.e., poorly designed conveyor systems, wrong selection of vessels walls thickness, mistakes in selecting the steam and expansion joints. Additionally, operational and maintenance problems, i.e., lack of maintenance of material handling systems, insufficient control and monitoring, poor bio-oil conversion to gasoline, and cracks in concrete chambers due to heat and corrosion. Moreover, downstream issues, i.e., excess of the limits of dioxins emission, ignition of produced hot char, pipe burst, and low energy recovery.

6. Conclusions

- Increases in scale and aggregation of swine production farms have resulted in manure accumulation problems in high production regions.
- Various manure management technologies, i.e., biological, and thermochemical, could be utilized to convert swine manure to value-added products while mitigating its negative impacts on surrounding ecosystems.
- Thermochemical conversion technologies are mature, stable and modular but, so far, underutilized in swine manure management.
- Gasification of swine manure solids, although under-investigated, can overcome the challenges associated with high-ash feedstock, and also generate a biochar stream.
- There is a need for integrating the swine manure biological and thermochemical conversion technologies to maximize the benefit of such a feedstock.
- There is a lack of research studies that investigate the kinetics of swine manure solids decomposition. These solids could be produced from biological conversion or solid separation technologies.
- It is crucial to develop comprehensive assessments of environmental impacts of thermochemical conversion as a manure management strategy.

Acknowledgments: Acknowledgment is due to the Agriculture and Food Research Initiative for their support of the Competitive Grant No. 2011-68002-30208 from the USDA National Institute of Food and Agriculture. Their financial support is much appreciated.

Conflicts of Interest: The authors declare no conflict of interest.

References

1. USDA-NASS. Quarterly Hogs and Pigs December 2017. ISSN 1949-1921. Available online: http://usda. mannlib.cornell.edu/usda/current/HogsPigs/HogsPigs-12-22-2017.pdf (accessed on 26 February 2018).
2. American Society of Agricultural and Biological Engineers (ASABE). Standard D3843.2. In *Manure Production and Characteristics 2005*; ASABE: St. Joseph, MI, USA, 2005.
3. Tilman, D.; Cassman, K.G.; Matson, P.A.; Naylor, R.; Polasky, S. Agricultural sustainability and intensive production practices. *Nature* **2002**, *418*, 671–677. [CrossRef] [PubMed]
4. USDA-NASS. *Overview of the U.S. Hog Industry*; NASS: Washington, DC, USA, 2009.
5. Fantozzi, F.; Bartocci, P.; D'Alessandro, B.; Arampatzis, S.; Manos, B. Public–private partnerships value in bioenergy projects: Economic feasibility analysis based on two case studies. *Biomass Bioenergy* **2014**, *66*, 387–397. [CrossRef]
6. Manos, B.; Partalidou, M.; Fantozzi, F.; Arampatzis, S.; Papadopoulou, O. Agro-energy districts contributing to environmental and social sustainability in rural areas: Evaluation of a local public–private partnership scheme in Greece. *Renew. Sustain. Energy Rev.* **2014**, *29*, 85–95. [CrossRef]
7. Manos, B.; Bartocci, P.; Partalidou, M.; Fantozzi, F.; Arampatzis, S. Review of public–private partnerships in agro-energy districts in Southern Europe: The cases of Greece and Italy. *Renew. Sustain. Energy Rev.* **2014**, *39*, 667–678. [CrossRef]
8. Hodgkinson, R.A.; Chambers, B.J.; Withers, P.J.A.; Cross, R. Phosphorus losses to surface waters following organic manure applications to a drained clay soil. *Agric. Water Manag.* **2002**, *57*, 155–173. [CrossRef]

9. Novak, J.M.; Watts, D.W.; Hunt, P.G.; Stone, K.C. Phosphorus movement through a coastal plain soil after a decade of intensive swine manure application. *J. Environ. Qual.* **2000**, *29*, 1310–1315. [CrossRef]

10. Aguirre-Villegas, H.A.; Larson, R.; Reinemann, D.J. From waste-to-worth: Energy, emissions, and nutrient implications of manure processing pathways. *Biofuels Bioprod. Biorefin.* **2014**, *8*, 770–793. [CrossRef]

11. Miller, J.J.; Chanasyk, D.S.; Curtis, T.W.; Olson, B.M. Phosphorus and nitrogen in runoff after phosphorus- or nitrogen-based manure applications. *J. Environ. Qual.* **2011**, *40*, 949–958. [CrossRef] [PubMed]

12. Salvano, E.; Flaten, D.N.; Rousseau, A.N.; Quilbe, R. Are current phosphorus risk indicators useful to predict the quality of surface waters in southern Manitoba, Canada? *J. Environ. Qual.* **2009**, *38*, 2096–2105. [CrossRef] [PubMed]

13. OSU. *Ohio Livestock Manure and Wastewater Management Guide*; Bulletin 604; Ohio State University Extension: Columbus, OH, USA, 2000; Available online: https://agcrops.osu.edu/sites/agcrops/files/imce/fertility/bulletin_604.pdf (accessed on 10 January 2018).

14. Prince Edward Island. *Guidelines for Manure Management for Prince Edward Island*; Government of Prince Edward Island: Charlottetown, PE, Canada, 2000. Available online: http://www.gov.pe.ca/agric/index.php3?number=70588&lang=F (accessed on 20 December 2017).

15. Day, D.; Funk, T.; Hatfield, J.; Stewart, B. *Processing Manure: Physical, Chemical and Biological Treatment, in Animal Waste Utilization: Effective Use of Manure as a Soil Resource*; CRC Press: Boca Raton, FL, USA, 1998.

16. Sadaka, S.; Van Devender, K. Evaluation of Chemically Coagulated Swine Manure Solids as Value-Added Products. *J. Sustain. Bioenergy Syst.* **2015**, *5*, 136–150. [CrossRef]

17. Zhang, R.H.; Westerman, P.W. Solid-liquid separation of animal manure for odor control and nutrient management. *Appl. Eng. Agric.* **1997**, *13*, 657–664. [CrossRef]

18. Hjorth, M.; Christensen, K.V.; Christensen, M.L.; Sommer, S.G. Solid-liquid separation of animal slurry in theory and practice—A review. *Agron. Sustain. Dev.* **2009**, *30*, 153–180. [CrossRef]

19. Ford, M.; Fleming, R. *Mechanical Solid-Liquid Separation of Livestock Manure—Literature Review*; Prepared for Ontario Pork; Ridgetown College, University of Guelph: Ridgetown, ON, Canada, 2002; Available online: http://mie.esab.upc.es/ms/informacio/residus_ramaders/Separator%20manure.pdf (accessed on 17 March 2015).

20. Rodriguez, M.D.E.; del Puerto, A.M.G.; Montealegre, M.L.M.; Adamsen, A.P.S.; Gullov, P.; Sommer, S.G. Separation of phosphorus from pig slurry using chemical additives. *Appl. Eng. Agric.* **2005**, *21*, 739–742. [CrossRef]

21. Xiu, S.; Zhang, Y.; Shahbazi, A. Swine manure solids separation and thermochemical conversion to heavy oil. *BioResources* **2009**, *4*, 458–470.

22. Hjorth, M.; Christensen, M.L.; Christensen, P.V. Flocculation, coagulation, and coagulation of manure affecting three separation techniques. *Bioresour. Technol.* **2008**, *99*, 8598–8604. [CrossRef] [PubMed]

23. Powers, W.J.; Flatow, L.A. Flocculation of swine manure: Influence of flocculant, rate of addition, and diet. *Appl. Eng. Agric.* **2002**, *18*, 609–614. [CrossRef]

24. Christensen, M.L.; Hjorth, M.; Keidling, K. Characterization of pig slurry with reference to flocculation and separation. *Water Res.* **2009**, *43*, 773–783. [CrossRef] [PubMed]

25. Wnetrzak, R.; Kwapinski, W.; Peters, K.; Sommer, S.G.; Jensen, L.S.; Leahy, J.J. The influence of the pig manure separation system on the energy production potentials. *Bioresour. Technol.* **2013**, *136*, 502–508. [CrossRef] [PubMed]

26. Mohammed, M.; Egyir, I.S.; Donkor, A.K.; Amoah, P.; Nyarko, S.; Boateng, K.K.; Ziwu, C. Feasibility study for biogas integration into waste treatment plants in Ghana. *Egypt. J. Pet.* **2017**, *26*, 695–703. [CrossRef]

27. Labatut, R.A.; Angenent, L.T.; Scott, N.R. Conventional Mesophilic vs. Thermophilic anaerobic digestion: A trade-off between performance and stability? *Water Res.* **2014**, *53*, 249–258. [CrossRef] [PubMed]

28. Sweeten, J.M.; Korenberg, J.; LePori, W.A.; Annamalai, K.; Parnell, C.B. Combustion of cattle feedlot manure for energy production. *Energy Agric.* **1986**, *5*, 55–72. [CrossRef]

29. Bernal, M.P.; Sánchez-Monedero, M.A.; Paredes, C.; Roig, A. Carbon mineralization from organic wastes at different composting stages during their incubation with soil. *Agric. Ecosyst. Environ.* **1998**, *69*, 175–189. [CrossRef]

30. Finestein, M.S.; Miller, F.C.; Strom, P.F. Monitoring and evaluating composting process performance. *J. WPCF* **1986**, *58*, 272–278.

31. Tiquia, S.; Richard, T.; Honeyman, M. Effect of windrow turning and seasonal temperatures on composting of hog manure from hoop structures. *Environ. Technol.* **2000**, *21*, 1037–1046. [CrossRef]

32. Sadaka, S.S.; Richard, T.L.; Loecke, T.D.; Liebman, M. Determination of Compost Respiration Rates Using Pressure Sensors. *Compos. Sci. Util.* **2006**, *14*, 124–131. [CrossRef]

33. Gamroth, M.J. *Composting: An Alternative for Livestock Manure Management and Disposal of Dead Animals*; The Ohio State University: Columbus, OH, USA, 2012.

34. Gajalakshmi, S.; Abbasi, S.A. Solid Waste Management by Composting: State of the Art. *Crit. Rev. Environ. Sci. Technol.* **2008**, *38*, 311–400. [CrossRef]

35. Lau, A.K.; Lo, K.V.; Liao, P.H.; Yu, J.C. Aeration experiments for swine waste composting. *Bioresour. Technol.* **1992**, *41*, 145–152. [CrossRef]

36. Huang, G.; Wu, Q.; Wong, J.; Nagar, B. Transformation of organic matter during co-composting of pig manure with sawdust. *Bioresour. Technol.* **2006**, *97*, 1834–1842. [CrossRef] [PubMed]

37. Bernal, M.; Alburquerque, J.; Moral, R. Composting of animal manures and chemical criteria for compost maturity assessment. A review. *Bioresour. Technol.* **2009**, *100*, 5444–5453. [CrossRef] [PubMed]

38. Ahn, H.K.; Mulbry, W.; White, J.; Kondrad, S. Pile mixing increases greenhouse gas emissions during composting of dairy manure. *Bioresour. Technol.* **2011**, *102*, 2904–2909. [CrossRef] [PubMed]

39. U.S. EPA. *Operating Anaerobic Digester Projects December 2014*; U.S. EPA: Washington, DC, USA, 2014.

40. EPA. *Inventory of U.S. Greenhouse Gas Emissions and Sinks: 1990–2011*; EPA 430-R-13-001; EPA: Washington, DC, USA, 2013.

41. Wang, Q.; Kuninobu, M.; Ogawa, H.I.; Kato, Y. Degradation of volatile fatty acids in highly efficient anaerobic digestion. *Biomass Bioenergy* **1999**, *16*, 407–416. [CrossRef]

42. Ahring, B.K.; Sandberg, M.; Angelidaki, I. Volatile fatty acids as indicators of process imbalance in anaerobic digestors. *Appl. Microbiol. Biotechnol.* **1995**, *43*, 559–565. [CrossRef]

43. Chen, H.; Li, B.; Yang, H.; Yang, G.; Zhang, S. Experimental investigation of biomass gasification in a fluidized bed reactor. *Energy Fuels* **2008**, *22*, 3493–3498.

44. Angelidaki, I.; Ahring, B. Thermophilic anaerobic digestion of livestock waste: The effect of ammonia. *Appl. Microbiol. Biotechnol.* **1993**, *38*, 560–564. [CrossRef]

45. Hansen, K.H.; Angelidaki, I.; Ahring, B.K. Anaerobic digestion of swine manure: Inhibition by ammonia. *Water Res.* **1998**, *32*, 5–12. [CrossRef]

46. Wu, X.; Yao, W.; Zhu, J.; Miller, C. Biogas and CH$_4$ productivity by co-digesting swine manure with three crop residues as an external carbon source. *Bioresour. Technol.* **2010**, *101*, 4042–4047. [CrossRef] [PubMed]

47. Lyberatos, G.; Skiadas, I. Modelling of anaerobic digestion—A review. *Glob. Nest Int. J.* **1999**, *1*, 63–76.

48. Gallert, C.; Winter, J. Mesophilic and thermophilic anaerobic digestion of source-sorted organic wastes: Effect of ammonia on glucose degradation and methane production. *Appl. Microbiol. Biotechnol.* **1997**, *48*, 405–410. [CrossRef]

49. Vindis, P.; Mursec, B.; Janzekovic, M.; Cus, F. The impact of mesophilic and thermophilic anaerobic digestion on biogas production. *J. Achiev. Mater. Manuf. Eng.* **2009**, *36*, 192–198.

50. Chen, Y.; Cheng, J.J.; Creamer, K.S. Inhibition of anaerobic digestion process: A review. *Bioresour. Technol.* **2008**, *99*, 4044–4064. [CrossRef] [PubMed]

51. Rajagopal, R.; Massé, D.I.; Singh, G. A critical review on inhibition of anaerobic digestion process by excess ammonia. *Bioresour. Technol.* **2013**, *143*, 632–641. [CrossRef] [PubMed]

52. Mukhtar, S.; Goodrich, B. *Dairy Biomass as a Renewable Fuel Source*; Texas Agri-Life Extension Service Publication No. L-5494; Texas A&M University: College Station, TX, USA, 2008.

53. Jenkins, B.; Baxter, L.; Miles, T., Jr.; Miles, T. Combustion properties of biomass. *Fuel Process. Technol.* **1998**, *54*, 17–46. [CrossRef]

54. Energie Centrum Nederland (ECN). *Phyllis, the Composition of Biomass and Waste*; ECN: Sint Maartensvlotbrug, The Netherlands, 2007; Available online: www.ecn.nl/phyllis2 (accessed on 12 October 2014).

55. Adani, F.; Baido, D.; Calcaterra, E.; Genevini, P. The influence of biomass temperature on biostabilization—Biodrying of municipal solid waste. *Bioresour. Technol.* **2002**, *83*, 173–179. [CrossRef]

56. Suler, D.J.; Finstein, M.S. Effect of temperature, aeration, and moisture on CO$_2$ formation in bench-scale, continuously Thermophilic composting of solid waste. *Appl. Environ. Microbiol.* **1977**, *33*, 345–350. [PubMed]

57. Sadaka, S.; Ahn, H. Evaluation of a biodrying process for beef, swine, and poultry manures mixed separately with corn stover. *Appl. Eng. Agric.* **2012**, *28*, 457–463. [CrossRef]

58. Li, X.; Zhang, R.; Pang, Y. Characteristics of dairy manure composting with rice straw. *Bioresour. Technol.* **2008**, *99*, 359–367. [CrossRef] [PubMed]

59. Hong, J.H.; Matsuda, J.; Ikeuchi, Y. High rapid composting of dairy cattle manure with crop and forest residues. *Trans. ASAE* **1983**, *26*, 533–541. [CrossRef]

60. Gao, M.; Li, B.; Yu, A.; Liang, F.; Yang, L.; Sun, Y. The effect of aeration rate on forced-aeration composting of chicken manure and sawdust. *Bioresour. Technol.* **2010**, *101*, 1899–1903. [CrossRef] [PubMed]

61. Sharara, M.A.; Sadaka, S.; Costello, T.A.; VanDevender, K. Influence of aeration rate on the physio-chemical characteristics of biodried dairy manure-wheat straw mixture. *Appl. Eng. Agric.* **2012**, *28*, 407–415. [CrossRef]

62. Mason, D.M.; Gandhi, K.N. Formulas for calculating the calorific value of coal and coal chars: Development, tests, and uses. *Fuel Process. Technol.* **1983**, *7*, 11–22. [CrossRef]

63. Cordero, T.; Marquez, F.; Rodriguez-Mirasol, J.; Rodriguez, J. Predicting heating values of lignocellulosics and carbonaceous materials from proximate analysis. *Fuel* **2001**, *80*, 1567–1571. [CrossRef]

64. Channiwala, S.A.; Parikh, P.P. A unified correlation for estimating HHV of solid, liquid and gaseous fuels. *Fuel* **2002**, *81*, 1051–1063. [CrossRef]

65. Zhang, Y.; Li, B.; Li, H.; Liu, H. Thermodynamic evaluation of biomass gasification with air in autothermal gasifiers. *Thermochim. Acta* **2011**, *519*, 65–71. [CrossRef]

66. Thygesen, A.; Wernberg, O.; Skou, E.; Sommer, S.G. Effect of incineration temperature on phosphorus availability in bio-ash from manure. *Environ. Technol.* **2011**, *32*, 633–638. [CrossRef] [PubMed]

67. Park, M.H.; Kumar, S.; Ra, C. Solid Waste from Swine Wastewater as a Fuel Source for Heat Production. *Asian-Aust. J. Anim. Sci.* **2012**, *25*, 1627–1633. [CrossRef] [PubMed]

68. Otero, M.; Sánchez, M.E.; Gómez, X. Co-firing of coal and manure biomass: A TG–MS approach. *Bioresour. Technol.* **2011**, *102*, 8304–8309. [CrossRef] [PubMed]

69. Fernandez-Lopez, M.; Puig-Gamero, M.; Lopez-Gonzalez, D.; Avalos-Ramirez, A.; Valverde, J.; Sanchez-Silva, L. Life cycle assessment of swine and dairy manure: Pyrolysis and combustion processes. *Bioresour. Technol.* **2015**, *182*, 184–192. [CrossRef] [PubMed]

70. Xiu, S.; Rojanala, H.K.; Shahbazi, A.; Fini, E.H.; Wang, L. Pyrolysis and combustion characteristics of Bio-oil from swine manure. *J. Therm. Anal. Calorim.* **2011**, *107*, 823–829. [CrossRef]

71. Komiyama, T.; Kobayashi, A.; Yahagi, M. The chemical characteristics of ashes from cattle, swine and poultry manure. *J. Mater. Cycles Waste Manag.* **2013**, *15*, 106–110. [CrossRef]

72. Petrus, L.; Noordermeer, M.A. Biomass to biofuels, a chemical perspective. *Green Chem.* **2006**, *8*, 861–867. [CrossRef]

73. Yung, M.M.; Jablonski, W.S.; Magrini-Bair, K.A. Review of catalytic conditioning of biomass-derived syngas. *Energy Fuels* **2009**, *23*, 1874–1887. [CrossRef]

74. Sadaka, S.S.; Ghaly, A.E.; Sabbah, M.A. Two-phase biomass air-steam gasification model for fluidized bed reactors: Part III—Model validation. *Biomass Bioenergy* **2002**, *22*, 479–487. [CrossRef]

75. Lv, P.; Xiong, Z.; Chang, J.; Wu, C.; Chen, Y.; Zhu, J. An experimental study on biomass air–steam gasification in a fluidized bed. *Bioresour Technol.* **2004**, *95*, 95–101. [CrossRef] [PubMed]

76. Kumar, A.; Wang, L.; Dzenis, Y.A.; Jones, D.D.; Hanna, M.A. Thermogravimetric characterization of corn stover as gasification and pyrolysis feedstock. *Biomass Bioenergy* **2008**, *32*, 460–467. [CrossRef]

77. González, J.; Román, S.; Bragado, D.; Calderón, M. Investigation on the reactions influencing biomass air and air/steam gasification for hydrogen production. *Fuel Process. Technol.* **2008**, *89*, 764–772. [CrossRef]

78. Prins, M.J.; Ptasinski, K.J.; Janssen, F.J.J.G. Thermodynamics of gas-char reactions: First and second law analysis. *Chem. Eng. Sci.* **2003**, *58*, 1003–1011. [CrossRef]

79. Reed, T.; Reed, T.B.; Das, A.; Das, A. *Handbook of Biomass Downdraft Gasifier Engine Systems: Biomass Energy Foundation*; National Technical Information Service, U.S. Department of Commerce: Alexandria, VA, USA, 1988.

80. Balat, M.; Balat, M.; Kırtay, E.; Balat, H. Main routes for the thermo-conversion of biomass into fuels and chemicals. Part 2: Gasification systems. *Energy Convers. Manag.* **2009**, *50*, 3158–3168. [CrossRef]

81. Di Blasi, C.; Signorelli, G.; Portoricco, G. Countercurrent fixed-bed gasification of biomass at laboratory scale. *Ind. Eng. Chem. Res.* **1999**, *38*, 2571–2581. [CrossRef]

82. Van der Walt, I.J.; Nel, J.T.; Glasser, D.; Hildebrandt, D.; Ngubevana, L. An Economic Evaluation for Small Scale Thermal Plasma Waste-to-Energy Systems. 2012, pp. 1–4. Available online: https://www.yumpu.com/en/document/view/24349230/an-economic-evaluation-for-small-scale-plasma-waste-to-energy (accessed on 14 February 2018).

83. Rajasekhar, M.; Rao, N.V.; Rao, G.C.; Priyadarshini, G.; Kumar, N.J. Energy generation from municipal solid waste by innoviative technologies—Plasma gasification. *Procedia Mater. Sci.* **2015**, *10*, 513–518. [CrossRef]

84. Bosmans, A.; Wasan, S.; Helsen, L. Waste to clean syngas: Avoiding tar problems. In Proceedings of the 2nd International Academic Symposium on Enhanced Landfill Mining, Houthalen-Helchteren, Belgium, 15–16 October 2013.

85. Campos, U.; Zamenian, H.; Koo, D.D.; Goodman, D.W. Waste-to-energy technology applications for municipal solid waste (MSW) treatment in the urban environment. *Int. J. Emerg. Technol. Adv. Eng.* **2015**, *5*, 504–508.

86. Gray, L. Plasma gasification as a viable waste-to-energy treatment of municipal solid waste. In *MANE 6960-Solid and Hazardous Waste Prevention and Control Engineering*; Rensselaer Hartford: Hartford, CT, USA, 2014; pp. 1–15.

87. Heberlein, J.; Murphy, A.B. Thermal plasma waste treatment. *J. Phys. D Appl. Phys.* **2008**, *41*, 1–20. [CrossRef]

88. Sirillova, I.L. Waste treatment by plasma technology. *Transf. Inovacii* **2015**, *31*, 40–42.

89. Cairns, E.; Tevebaugh, A. CHO Gas Phase Compositions in Equilibrium with Carbon, and Carbon Deposition Boundaries at One Atmosphere. *J. Chem. Eng. Data* **1964**, *9*, 453–462. [CrossRef]

90. Narvaez, I.; Orio, A.; Aznar, M.P.; Corella, J. Biomass gasification with air in an atmospheric bubbling fluidized bed. *Effect of six operational variables on the quality of the produced raw gas. Ind. Eng. Chem. Res.* **1996**, *35*, 2110–2120.

91. Gai, C.; Dong, Y. Experimental study on non-woody biomass gasification in a downdraft gasifier. *Int. J. Hydrog. Energy* **2012**, *37*, 4935–4944. [CrossRef]

92. Baker, E.; Brown, M.; Elliott, D.; Mudge, L. Characterization and treatment of tars and biomass gasifiers. In Proceedings of the American Institute of Chemical Engineers Summer National Meeting, Denver, CO, USA, 21–24 August 1988.

93. Boerrigter, H.; den Uil, H.; Calis, H. Green diesel from biomass via Fischer-Tropsch synthesis: New insights in gas cleaning and process design. In Proceedings of the Pyrolysis and Gasification of Biomass and Waste Expert Meeting, Strasbourg, France, 30 September–1 October 2002; pp. 371–383.

94. Tijmensen, M.J.; Faaij, A.P.; Hamelinck, C.N.; van Hardeveld, M.R. Exploration of the possibilities for production of Fischer Tropsch liquids and power via biomass gasification. *Biomass Bioenergy* **2002**, *23*, 129–152. [CrossRef]

95. Higman, C.; Van der Burgt, M. *Gasification*; Gulf Professional Publishing: Houston, TX, USA, 2003.

96. Cantrell, K.; Ro, K.; Mahajan, D.; Anjom, M.; Hunt, P.G. Role of thermochemical conversion in livestock waste-to-energy treatments: Obstacles and opportunities. *Ind. Eng. Chem. Res.* **2007**, *46*, 8918–8927. [CrossRef]

97. Zhang, S.; Cao, J.; Takarada, T. Effect of pretreatment with different washing methods on the reactivity of manure char. *Bioresour. Technol.* **2010**, *101*, 6130–6135. [CrossRef] [PubMed]

98. Arvelakis, S.; Gehrmann, H.; Beckmann, M.; Koukios, E.G. Preliminary results on the ash behavior of peach stones during fluidized bed gasification: Evaluation of fractionation and leaching as pre-treatments. *Biomass Bioenergy* **2005**, *28*, 331–338. [CrossRef]

99. Roy, P.C.; Datta, A.; Chakraborty, N. Assessment of cow dung as a supplementary fuel in a downdraft biomass gasifier. *Renew. Energy* **2010**, *35*, 379–386. [CrossRef]

100. Gautam, G.; Adhikari, S.; Brodbeck, C.; Bhavnani, S.; Fasina, O.; Taylor, S. Gasification of wood chips, agricultural residues, and waste in a commercial downdraft gasifier. *Trans. ASABE* **2011**, *54*, 1801–1807. [CrossRef]

101. Morf, P.; Hasler, P.; Nussbaumer, T. Mechanisms and kinetics of homogeneous secondary reactions of tar from continuous pyrolysis of wood chips. *Fuel* **2002**, *81*, 843–853. [CrossRef]

102. Evans, R.J.; Milne, T.A. Molecular characterization of the pyrolysis of biomass. *Energy Fuels* **1987**, *1*, 123–137. [CrossRef]

103. Vreugdenhil, B.; Zwart, R.; Neeft, J.P.A. *Tar Formation in Pyrolysis and Gasification*; ECN: Sint Maartensvlotbrug, The Netherlands, 2009.

104. Kuligowski, K.; Poulsen, T.G.; Rubæk, G.H.; Sørensen, P. Plant-availability to barley of phosphorus in ash from thermally treated animal manure in comparison to other manure based materials and commercial fertilizer. *Eur. J. Agron.* **2010**, *33*, 293–303. [CrossRef]

105. Bridgwater, A.; Meier, D.; Radlein, D. An overview of fast pyrolysis of biomass. *Org. Geochem.* **1999**, *30*, 1479–1493. [CrossRef]

106. Meier, D.; Faix, O. State of the art of applied fast pyrolysis of lignocellulosic materials—A review. *Bioresour. Technol.* **1999**, *68*, 71–77. [CrossRef]

107. He, B.J.; Zhang, Y.; Yin, Y.; Funk, T.L.; Riskowski, G. Preliminary characterization of raw oil products from the thermochemical conversion of swine manure. *Trans. ASAE* **2001**, *44*, 1865–1871.

108. Mohan, D.; Pittman, C.U.; Steele, P.H. Pyrolysis of wood/biomass for bio-oil: A critical review. *Energy Fuels* **2006**, *20*, 848–889. [CrossRef]

109. Mullen, C.A.; Boateng, A.A. Chemical Composition of Bio-oils Produced by Fast Pyrolysis of Two Energy Crops. *Energy Fuels* **2008**, *22*, 2104–2109. [CrossRef]

110. Maggi, R.; Delmon, B. Comparison between 'slow'and 'flash' pyrolysis oils from biomass. *Fuel* **1994**, *73*, 671–677. [CrossRef]

111. Demirbaş, A. Mechanisms of liquefaction and pyrolysis reactions of biomass. *Energy Convers. Manag.* **2000**, *41*, 633–646. [CrossRef]

112. Diebold, J.P. *A Review of the Chemical and Physical Mechanisms of the Storage Stability of Fast Pyrolysis Bio-Oils*; National Renewable Energy Laboratory Golden: Golden, CO, USA, 2000.

113. Elliott, D.C. Historical developments in hydroprocessing bio-oils. *Energy Fuels* **2007**, *21*, 1792–1815. [CrossRef]

114. Huber, G.W.; Iborra, S.; Corma, A. Synthesis of transportation fuels from biomass: Chemistry, catalysts, and engineering. *Chem. Rev.* **2006**, *106*, 4044–4098. [CrossRef] [PubMed]

115. Rioche, C.; Kulkarni, S.; Meunier, F.C.; Breen, J.P.; Burch, R. Steam reforming of model compounds and fast pyrolysis bio-oil on supported noble metal catalysts. *Appl. Catal. B Environ.* **2005**, *61*, 130–139. [CrossRef]

116. Vagia, E.C.; Lemonidou, A.A. Thermodynamic analysis of hydrogen production via steam reforming of selected components of aqueous bio-oil fraction. *Int. J. Hydrog. Energy* **2007**, *32*, 212–223. [CrossRef]

117. Takanabe, K.; Aika, K.; Seshan, K.; Lefferts, L. Sustainable hydrogen from bio-oil—Steam reforming of acetic acid as a model oxygenate. *J. Catal.* **2004**, *227*, 101–108. [CrossRef]

118. Hossain, A.K.; Davies, P.A. Pyrolysis liquids and gases as alternative fuels in internal combustion engines–A review. *Renew. Sustain. Energy Rev.* **2013**, *21*, 165–189. [CrossRef]

119. Hossain, A.K.; Serrano, C.; Brammer, J.B.; Omran, A.; Ahmed, F.; Smith, D.I.; Davies, P.A. Combustion of fuel blends containing digestate pyrolysis oil in a multi-cylinder compression ignition engine. *Fuel* **2016**, *171*, 18–28. [CrossRef]

120. Vihar, R.; Seljak, T.; Oprešnik, S.R.; Katrašnik, T. Combustion characteristics of tire pyrolysis oil in turbo charged compression ignition engine. *Fuel* **2015**, *150*, 226–235. [CrossRef]

121. Martin, J.A.; Boateng, A.A. Combustion performance of pyrolysis oil/ethanol blends in a residential-scale oil-fired boiler. *Fuel* **2014**, *133*, 34–44. [CrossRef]

122. Shudo, T.; Nagano, T.; Kobayashi, M. Combustion characteristics of waste-pyrolysis gases in an internal combustion engine. *Int. J. Automot. Technol.* **2003**, *4*, 1–8.

123. Shah, A.; Srinivasan, R.; To, S.D.F.; Columbus, E.P. Performance and emissions of a spark-ignited engine driven generator on biomass based syngas. *Bioresour. Technol.* **2010**, *101*, 4656–4661. [CrossRef] [PubMed]

124. Hagos, F.Y.; Aziz, A.R.A.; Sulaiman, S.A. Trends of syngas as a fuel in internal combustion engines. *Adv. Mech. Eng.* **2014**, *6*, 401587. [CrossRef]

125. Kim, S.; Agblevor, F.A.; Lim, J. Fast pyrolysis of chicken litter and turkey litter in a fluidized bed reactor. *J. Ind. Eng. Chem.* **2009**, *15*, 247–252. [CrossRef]

126. Mante, O.D.; Agblevor, F.A. Parametric study on the pyrolysis of manure and wood shavings. *Biomass Bioenergy* **2011**, *35*, 4417–4425. [CrossRef]

127. Cao, J.; Xiao, X.; Zhang, S.; Zhao, X.; Sato, K.; Ogawa, Y.; Wei, X.Y.; Takarada, T. Preparation and characterization of bio-oils from internally circulating fluidized-bed pyrolyses of municipal, livestock, and wood waste. *Bioresour. Technol.* **2011**, *102*, 2009–2015. [CrossRef] [PubMed]

128. Cantrell, K.B.; Hunt, P.G.; Uchimiya, M.; Novak, J.M.; Ro, K.S. Impact of pyrolysis temperature and manure source on physicochemical characteristics of biochar. *Bioresour. Technol.* **2012**, *107*, 419–428. [CrossRef] [PubMed]

129. Yin, S.; Dolan, R.; Harris, M.; Tan, Z. Subcritical hydrothermal liquefaction of cattle manure to bio-oil: Effects of conversion parameters on bio-oil yield and characterization of bio-oil. *Bioresour. Technol.* **2010**, *101*, 3657–3664. [CrossRef] [PubMed]

130. Biller, P.; Ross, A.B. Potential yields and properties of oil from the hydrothermal liquefaction of microalgae with different biochemical content. *Bioresour. Technol.* **2011**, *102*, 215–225. [CrossRef] [PubMed]

131. Peterson, A.A.; Vogel, F.; Lachance, R.P.; Fröling, M.; Antal, M.J.; Tester, J.W. Thermochemical biofuel production in hydrothermal media: A review of suband supercritical water technologies. *Energy Environ. Sci.* **2008**, *1*, 32–65. [CrossRef]

132. Duan, P.; Savage, P.E. Upgrading of crude algal bio-oil in supercritical water. *Bioresour. Technol.* **2011**, *102*, 1899–1906. [CrossRef] [PubMed]

133. Zhou, D.; Zhang, L.; Zhang, S.; Fu, H.; Chen, J. Hydrothermal liquefaction of macroalgae Enteromorpha prolifera to bio-oil. *Energy Fuels* **2010**, *24*, 4054–4061. [CrossRef]

134. Demirbas, F. Biorefineries for biofuel upgrading: A critical review. *Appl. Energy* **2009**, *86*, S151–S161. [CrossRef]

135. He, B.J.; Zhang, Y.; Funk, T.L.; Riskowski, G.L.; Yin, Y. Thermochemical conversion of swine manure: An alternative process for waste treatment and renewable energy production. *Trans. ASAE* **2000**, *43*, 1827–1833.

136. Xiu, S.; Shahbazi, A.; Wang, L.; Wallace, C.W. Supercritical ethanol liquefaction of swine manure for bio-oils production. *Am. J. Eng. Appl. Sci.* **2010**, *3*, 494–500. [CrossRef]

137. Bergman, P.C.; Kiel, J.H. Torrefaction for biomass upgrading. In Proceedings of the 14th European Biomass Conference & Exhibition, Paris, France, 17–21 October 2005.

138. Qiu, Y.; Zheng, Z.; Zhou, Z.; Sheng, G.D. Effectiveness and mechanisms of dye adsorption on a straw-based biochar. *Bioresour. Technol.* **2009**, *100*, 5348–5351. [CrossRef] [PubMed]

139. Sohi, S.; Krull, E.; Lopez-Capel, E.; Bol, R. A review of biochar and its use and function in soil. *Adv. Agron.* **2010**, *105*, 47–82.

140. Ro, K.S.; Cantrell, K.B.; Hunt, P.G.; Ducey, T.F.; Vanotti, M.B.; Szogi, A.A. Thermochemical conversion of livestock wastes: Carbonization of swine solids. *Bioresour. Technol.* **2009**, *100*, 5466–5471. [CrossRef] [PubMed]

141. Chan, K.; Van Zwieten, L.; Meszaros, I.; Downie, A.; Joseph, S. Agronomic values of greenwaste biochar as a soil amendment. *Soil Res.* **2008**, *45*, 629–634. [CrossRef]

142. Keiluweit, M.; Nico, P.S.; Johnson, M.G.; Kleber, M. Dynamic molecular structure of plant biomass-derived black carbon (biochar). *Environ. Sci. Technol.* **2010**, *44*, 1247–1253. [CrossRef] [PubMed]

143. Bruun, E.W.; Hauggaard-Nielsen, H.; Ibrahim, N.; Egsgaard, H.; Ambus, P.; Jensen, P.A.; Dam-Johansen, K. Influence of fast pyrolysis temperature on biochar labile fraction and short-term carbon loss in a loamy soil. *Biomass Bioenergy* **2011**, *35*, 1182–1189. [CrossRef]

144. Unger, R.; Killorn, R. Effect of Three Different Qualities of Biochar on Selected Soil Properties. *Commun. Soil Sci. Plant Anal.* **2011**, *42*, 2274–2283. [CrossRef]

145. Shinogi, Y.; Yoshida, H.; Koizumi, T.; Yamaoka, M.; Saito, T. Basic characteristics of low-temperature carbon products from waste sludge. *Adv. Environ. Res.* **2003**, *7*, 661–665. [CrossRef]

146. Marchetti, R.; Castelli, F.; Orsi, A.; Sghedoni, L.; Bochicchio, D. Biochar from swine manure solids: Influence on carbon sequestration and Olsen phosphorus and mineral nitrogen dynamics in soil with and without digestate incorporation. *Ital. J. Agron.* **2012**, *7*, e26. [CrossRef]

147. Cantrell, K.B.; Martin, J.H. Stochastic state-space temperature regulation of biochar production. *Part II: Application to manure processing via pyrolysis. J. Sci. Food Agric.* **2012**, *92*, 490–495. [PubMed]

148. Tsai, W.T.; Liu, S.C.; Chen, H.R.; Chang, Y.M.; Tsai, Y.L. Textural and chemical properties of swine-manure-derived biochar pertinent to its potential use as a soil amendment. *Chemosphere* **2012**, *89*, 198–203. [CrossRef] [PubMed]

149. Meyer, S.; Glaser, B.; Quicker, P. Technical, economical, and climate-related aspects of biochar production technologies: A literature review. *Environ. Sci. Technol.* **2011**, *45*, 9473–9483. [CrossRef] [PubMed]

energies

MDPI

Article

The Effect of Temperature on the Methanogenic Activity in Relation to Micronutrient Availability

Kessara Seneesrisakul [1], Twarath Sutabutr [2] and Sumaeth Chavadej [1,3,*]

1   The Petroleum and Petrochemical College, Chulalongkorn University, Soi Chula 12, Phayathai Road, Pathumwan, Bangkok 10330, Thailand; seneesrisakul.k@gmail.com
2   Energy Policy and Planning Office, Ministry of Energy, 121/1-2 Phetchaburi Road, Ratchathewi, Bangkok 10400, Thailand; twarath@eppo.go.th
3   Center of Excellence on Petrochemical and Materials Technology, Chulalongkorn University, Soi Chula 12, Phyathai Road, Pathumwan, Bangkok 10330, Thailand
*   Correspondence: sumaeth.c@chula.ac.th; Tel./Fax: +66-2-218-4139

Received: 6 April 2018; Accepted: 23 April 2018; Published: 25 April 2018

Abstract: In the view of microbial community, thermophilic microorganisms were reported to have faster biochemical reaction rates, which are reflected by a higher methane production rate. However, there has no research to discuss the effect of temperature on methanogenic activity in relation to micronutrient transport and availability. The objective of this study was to investigate the effect of temperature on methanogenic activity in relation to nutrient uptakes, micronutrient transports, and mass balance using anaerobic sequencing batch reactors (ASBR) with recycled biogas for treating ethanol wastewater at mesophilic (37 °C) and thermophilic (55 °C) temperatures. The increase in temperature from 37 to 55 °C increased in both of the optimum chemical oxygen demand (COD) loading rate and methanogenic activity, corresponding to the results of N and P uptakes, energy balance, and mass balance. The higher temperature of the thermophilic operation as compared to the mesophilic one caused a lower water solubility of the produced H_2S, leading to lowering the reduction of divalent cation micronutrients. The thermophilic operation could prevent the deficit of micronutrients, thus causing a higher methanogenic activity, while the mesophilic operation still had the deficit of most micronutrients, leading to the lower activity.

Keywords: anaerobic digestion; energy balance; ethanol wastewater; mass balance; micronutrients; nutrient uptakes

1. Introduction

Sustainable and renewable energy resources are of great interest in research and development in order to replace the limited and essentially non-renewable fossil fuels. Biogas is one potential renewable resource that is produced via anaerobic digestion (AD) of a variety of waste materials under ambient temperature and pressure, and is typically composed of 60–70% methane (CH_4), 30–40% carbon dioxide (CO_2), and a trace amount of hydrogen sulfide (H_2S) [1–4]. An AD is economically applied for industrial wastewaters, because it provides the dual benefits of the energy gain from the produced biogas and the reduction in wastewater treatment cost [5–7].

Anaerobic sequencing batch reactors (ASBRs), which are classified as high rate anaerobic systems, have been known to be able to handle wastewater containing a high level of suspended solids because they can maintain a high microbial concentration in the system [8,9]. A conventional ASBR system for CH_4 production from wastewater uses a mechanical stirrer for mixing, which causes high power consumption for operation. As reported previously, the use of the produced biogas for mixing can enhance the CH_4 transport from the aqueous to gaseous phases by entraining CH_4, mostly adhered to the biomass and bacterial cell membranes, to exit the aqueous phase more easily, and so it results in a higher CH_4 yield [10]. The recirculation of the produced biogas was reported to enhance the CH_4 production level under a mesophilic temperature by approximately 12–26% [11,12].

The biogas production from organic compounds via an anaerobic degradation process consists of the four sequential stages of hydrolysis, acidogenesis, acetogenesis, and methanogenesis, as summarized in Table 1 [2,13–17]. Basically, the methanogenesis is the most vulnerable step of AD, since its reaction rate is much slower than those of the first three steps. The overall process performance depends on several environmental factors, including temperature, solution pH, the presence of toxic compounds, and nutrient levels [1,18].

Table 1. Biochemical pathways of the anaerobic digestion of polysaccharides for methane (CH_4) production by different groups of microorganisms [2,13–17].

Chemical Reaction	Equation
Hydrolysis:	
$(C_6H_{10}O_5)_n + nH_2O \rightarrow nC_6H_{12}O_6$	(1)
Acidogenesis (favorable pH of ~4.5–5.5):	
$C_6H_{12}O_6 \rightarrow 3CH_3COOH$	(2)
$C_6H_{12}O_6 + 2H_2 \leftrightarrow 2CH_3CH_2COOH + 2H_2O$	(3)
$C_6H_{12}O_6 \rightarrow CH_3(CH_2)_2COOH + 2CO_2 + 2H_2$	(4)
$C_6H_{12}O_6 \rightarrow 2CH_3CH(OH)COOH$	(5)
$C_6H_{12}O_6 \rightarrow 2CH_3CH_2OH + 2CO_2$	(6)
$CH_3CH_2COOH + CH_3(CH_2)_2COOH \leftrightarrow CH_3(CH_2)_3COOH + CH_3COOH$	(7)
Acetogenesis (favorable pH of ~6):	
$CH_3CH_2COOH + 2H_2O \leftrightarrow CH_3COOH + CO_2 + 3H_2$	(8)
$CH_3(CH_2)_2COOH + 2H_2O \leftrightarrow 2CH_3COOH + 2H_2$	(9)
$CH_3(CH_2)_3COOH + 4H_2O \leftrightarrow CH_3CH_2COOH + 2CO_2 + H_2$	(10)
$CH_3(CH_2)_3COOH + 2H_2O \leftrightarrow CH_3CH_2COOH + CH_3COOH + 2H_2$	(11)
$C_6H_{12}O_6 + 2H_2O \leftrightarrow 2CH_3COOH + 2CO_2 + 4H_2$	(12)
$2CH_3CH_2OH + 2H_2O \leftrightarrow 2CH_3COOH + 4H_2$	(13)
Methanogenesis (favorable pH of ~7–8):	
Hydrogenotrophic methanogenesis:	
$4H_2 + CO_2 \leftrightarrow CH_4 + 2H_2O$	(14)
Acetotrophic methanogenesis:	
$CH_3COOH \rightarrow CH_4 + CO_2$	(15)
$C_6H_{12}O_6 \rightarrow 3CH_4 + 3CO_2$	(16)

Trace metals, which are exerted as micronutrients, such as iron (Fe^{2+}), copper (Cu^{2+}), zinc (Zn^{2+}), nickel (Ni^{2+}), cobalt (Co^{2+}), manganese (Mn^{2+}), and molybdenum (Mo^{2+}) play important role on the process performance and the stability of AD. Their deficiency is usually a primary reason of poor process efficiency of AD, in spite of proper management and operational control [19]. The major reason of micronutrient deficiency in AD results from the chemical precipitation of all the divalent cations (micronutrients) with sulfide ions (S^{2-}) being produced from the reduction of sulfate and the decomposition of sulfur-containing organic compounds [20].

Mesophilic (37 °C) and thermophilic (55 °C) AD systems are the most commonly used AD processes. Several studies reported the temperature effect on the process performance of AD [21–24]. The thermophilic AD has been claimed advantages over mesophilic AD. Firstly, the capability to produce pathogen free streams with no restrictions on crop type, harvesting, or site access for land application [24]. Secondly, thermophiles give a faster biochemical reaction rate, as compared with mesophiles [22]. Concurrently, the thermophilic AD was reported to be more sensitive to operational conditions than mesophilic AD, while some studies claim that it is no any problem with the long time adaptation of biomass in the thermophilic AD [23]. In the view of the microbial community, thermophilic microorganisms were reported to have higher metabolic activity and substrate conversion rates reflected by a higher methane production rate than mesophilic microorganisms [25]. Up to now, there has no research to discuss why thermophiles have higher methanogenic activity than mesophiles, and the effect of temperature on methanogenic activity in relation to micronutrient transport and availability. In this investigation, it was, for the first time, hypothesized that the thermophilic AD had a lower precipitation of the metal sulfides, leading to a higher availability of most micronutrients for methanogenic activity in thermophilic AD, as compared with the mesophilic AD. This hypothesis is based on the fundamental knowledge that the solubility of H_2S in water decreases with an increasing temperature, and so the produced H_2S in AD is present in the gaseous phase at the thermophilic temperature higher than that in the mesophilic AD [26].

The aim of this study was, for the first time of its kind, to investigate the effect of temperatures (37 and 55 °C) on methanogenic activity in relation to micronutrient availability using anaerobic sequencing batch reactor (ASBR) with recycled biogas. The COD loading rates were firstly varied to determine the optimum values under the different two temperatures. In addition, the macro- and micro- nutrient transport and the overall mass and energy balances were taken to explain why AD operated at 55 °C has a higher methanogenic activity than that at 37 °C.

2. Materials and Methods

2.1. Ethanol Wastewater

The ethanol wastewater used in this study was obtained from a cassava root fermentation plant, and was kindly supplied by Sapthip Co., Ltd., Lopburi, Thailand. The wastewater was collected from a centrifuge, where a large quantity of unfermented cassava roots in the discharge from the bottom of the distillation columns was removed. The wastewater still contained a small quantity of solid particles of a small particle size (<425 μm). The collected wastewater was kept at 4 °C until use, and was used as received.

2.2. Seed Sludge Preparation

A seed sludge sample, collected from the upflow anaerobic sludge blanket reactor (UASB) unit of a biogas production plant at the same factory that provided the ethanol wastewater (Section 2.1), was firstly concentrated by sedimentation. The concentrated sludge was then added to each ASBR to obtain an initial mixed liquid volatile suspended solids (MLVSS) value of approximately 20,000 mg/L.

2.3. ASBR Operation

Each ASBR unit was made from a glass column with an inner diameter of 8 cm and a height of 70 cm (Figure 1). The bioreactors, with a total volume of 3.5 L, were operated at a working liquid volume of 2 L. Each reactor was entirely covered with a black foam sheet in order to prevent the photosynthetic activities from both algae and bacteria. The temperature in the ASBR unit was controlled by a water circulating bath via a water jacket on each bioreactor. The ASBR operation consisted of the four sequential steps of feed, react, settle, and decant, all of which were controlled by a set of digital timers. For the feed step (2 min), the ethanol wastewater was pumped into the top of the bioreactor using a peristaltic pump with a level probe. For the react step (116 min), the feed pump was turned off and another peristaltic pump was turned on to recycle the produced biogas at the top of the ASBR to the bottom of the ASBR column to achieve a homogeneous mixing of the bioreactor's liquor, using a biogas flow rate of 0.25 L/min. For the settle step (120 min), the biogas recirculation was stopped so as to allow for the microbial cells to settle in the bioreactor. For the final decant step (2 min), the clarified effluent was drained out of the bioreactor by a peristaltic pump that was equipped with a level probe. The ASBR was operated at 6 cycles/d, as previously reported [27]. Each ASBR unit was operated at the given COD loading rate at either 37 or 55 °C until reaching a steady state, which was around four weeks. The steady state was determined as the onset of relatively constant values (standard deviation < 5%) of both the gas production rate and the effluent COD value over time. Then, the biogas and effluent samples were collected for analysis and measurement.

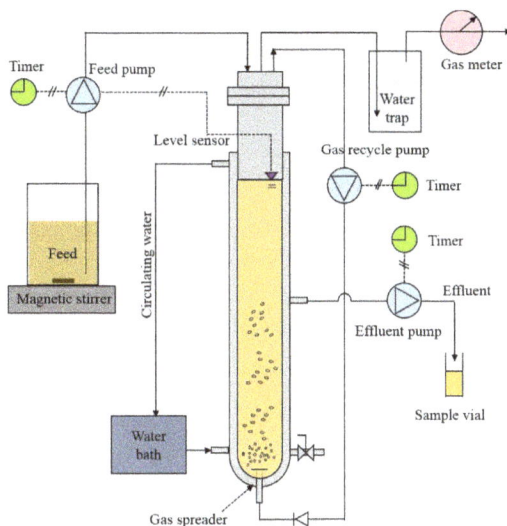

Figure 1. Schematic diagram of the anaerobic sequencing batch reactors (ASBR) used in this study with recycled biogas.

2.4. Measurements and Analytical Methods

The produced gas from the ASBR unit was passed through a water trap flask and measured by a wet gas meter to determine the biogas production rate. The gas composition was analyzed by gas chromatography (GC; Perkin-Elmer (Waltham, MA, USA), AutoSystem), equipped with a thermal conductivity detector (TCD) and a series of two packed columns (HayeSep D 100/120 mesh and Molecular sieve, Altech (Flemington, NJ, USA)). The injector, column, and detector temperatures were kept at 150, 60, and 200 °C, respectively. Argon was used as the carrier gas. The concentration of H_2S in the produced gas was analyzed by another GC (Shimudsu (Kyoto, Japan), GC-2014), equipped with a flame photometric detector (FPD) and a capillary column (Agilent (Santa Clara, CA, USA), DB-1). The temperatures of injector, column and detector were kept at 100, 80, and 250 °C, respectively.

All volatile fatty acids (VFAs) in the influent and effluent samples were determined qualitatively and quantitatively by high performance liquid chromatography (HPLC, Shimudzu, LC-20A), equipped with a refractive index detector (RID, Shimudzu, RID-20A) and an Aminex HPX-87H column (Bio-Rad Laboratories, Hercules, CA, USA). The mobile phase (5 mM sulfuric acid) was kept at a flow rate of 0.6 mL/min and the column temperature was controlled at 60 °C.

The samples that were taken from the ASBR unit during the react step for any COD loading rate under the steady state condition were analyzed for the MLVSS content, which was used to represent the microbial concentration in the system. The effluent samples during the decant step were analyzed for the VSS to represent the microbial washout from the system. Both the MLVSS and VSS were determined according to the standard methods [28].

The organic contents in the ethanol wastewater and the effluent samples, in terms of the COD, were determined by the dichromate oxidation method using a COD digester (HACH, DRB 200) and measurement of the absorbance by a spectrophotometer (HACH, DR3800). The N concentration, as total-N, NH_4^+-N, NO_3-N, and NO_2^--N, in the influent and effluent samples were measured using the persulfate digestion, salicylate, cadmium reduction, and diazolization methods, respectively. The organic nitrogen (org-N) concentration was calculated by subtracting the sum of the NH_4^+-N,

NO_3^--N, and NO_2^--N from the total-N [29]. The total P concentrations of the influent and effluent samples were determined by the molybdovanadate/acid persulfate digestion method (Hach Company, Loveland, CO, USA) [30]. The sulfur (S) concentrations, as sulfate and soluble sulfide, in the influent and effluent samples were measured by the SulfaVer 4 method and methylene blue method, respectively, (Hach Company) [30]. The effluent samples from all of the ASBR units were also measured for pH and total alkalinity. The total alkalinity was measured by the titration method with a standard acid solution (0.020 N H_2SO_4), following the method 8221 (HACH Company). The concentrations of some micronutrients (Fe^{2+}, Co^{2+}, Ni^{2+}, Cu^{2+}, Zn^{2+}, Mn^{2+}, and Mo^{2+}) in the influent and effluent samples were measured using atomic absorption spectrophotometry (AAS; Varian (Palo Alto, CA, USA), SpectrAA 300).

The samples of influent and effluent were dried in an oven overnight at 105 °C before being analyzed for their calorific value and elemental content. The calorific value of each dried sample was determined using a bomb calorimeter (Leco (St Joseph, MI, USA), AC-500). The elemental composition of the dried sample was determined by an elemental analyzer (Leco, TruSpec-CHN). Oxygen, helium, and air were used as carrier gases with a combustion and burner temperatures of 950 and 850 °C, respectively.

3. Results and Discussion

3.1. Characteristics of the As-Used Ethanol Wastewater

The characteristics of the ethanol wastewater used in the present work are summarized in Table 2. The chemical oxygen demand (COD) values were in the range of 63,000 ± 900 mg/L, with a COD:Nitrogen (N):Phosphorous (P) ratio of 100:1:0.4. When compared to the theoretically required COD:N:P ratio of 100:0.5:0.1 for AD for biogas production [30], the studied ethanol wastewater likely contained sufficient amounts of both N and P, and so they were not supplemented in this study. The N in the ethanol wastewater was mostly in the form of organic nitrogen (org-N) with a significant amount of ammonium nitrogen (NH_4^+-N), but only a very small amount of nitrate nitrogen (NO_3^--N) and no detectable nitrite nitrogen (NO_2^--N). The high concentration of volatile fatty acids (VFAs, 11,100 ± 90 mg/L as HAc), with a very high concentration of HLa, resulted in the low pH value (3.5 ± 0.2) of the ethanol wastewater [31]. The self-degradation of wastewater under an anaerobic condition was the main reason for the high content of VFAs in the ethanol wastewater [30].

Table 2. Characteristics of the cassava-based ferment alcohol wastewater used in this study.

Parameter	Unit	Value
pH	-	3.5 ± 0.2
COD	mg/L	63,000 ± 900
Total VFA:	mg/L as HAc	11,100 ± 90
Lactic acid (HLa)	mg/L	10,500 ± 50
Acetic acid (HAc)	mg/L	3660 ± 2
Propionic acid (HPr)	mg/L	300 ± 1
Butylric acid (HBu)	mg/L	30 ± 1
Valeric acid (HVa)	mg/L	10 ± 2
Ethanol	mg/L	540 ± 5
Total solid (TS)	mg/L	38,200 ± 1800
Total volatile solid (TVS)	mg/L	34,400 ± 1600
Total carbon (C)	mg/L	19,000 ± 300
Total nitrogen (N)	mg/L	640 ± 7
NH_4^+-N	mg/L	100 ± 1
NO_3^--N	mg/L	2.0 ± 0.1
NO_2^--N	mg/L	0
Org-N	mg/L	538
Total P	mg/L	230 ± 28
Total SO_4^{2-}	mg/L	27 ± 1.4
Total S	mg/L	0.11
COD:N:P:S	-	100:1:0.4:0.002

3.2. Process Performance Results

3.2.1. Organic Removal

For the mesophilic operation (37 °C) of the ASBR, the COD removal efficiency was slightly decreased from 96 to 93% while increasing the COD loading rate from 2 to 6 kg/m^{3}d, and then more strongly decreased to 83% when increasing the COD loading rate to 8 kg/m^{3}d (Figure 2a). The slightly decrease in the COD removal level with an increased COD loading rate to 6 kg/m^{3}d resulted from the increased level of organic compounds that are available to support the growth of the anaerobes. The COD removal efficiency then significantly decreased at COD loading rates above 6 kg/m^{3}d was potentially because the VFA production rate then exceeded the VFA consumption rate, causing the VFA accumulation in the system to exceed the toxic level to the methanogens (Section 3.3).

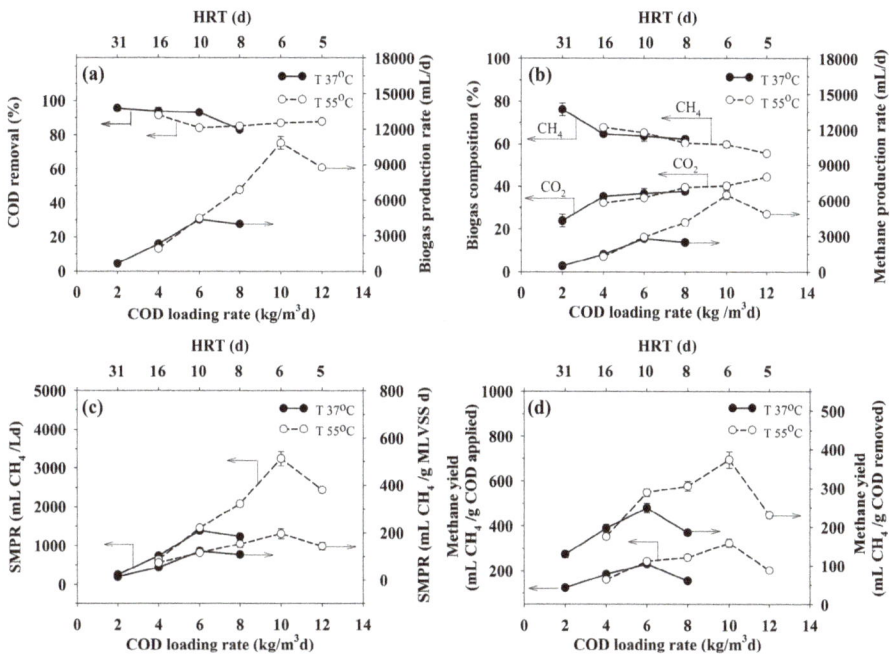

Figure 2. The effects of the chemical oxygen demand (COD) loading rate and hydraulic retention time (HRT) on the steady-state parameters; (**a**) COD removal and gas production rate; (**b**) gas composition and CH$_4$ production rate; (**c**) specific methane production rates (SMPRs); and, (**d**) CH$_4$ yields in the ASBR operated under a mesophilic (37 °C) or thermophilic (55 °C) temperature. Data are shown as the mean ± 1SD, derived from five independent repeats.

For the thermophilic operation (55 °C) of the ASBR, the COD removal level decreased from 92 to 84% when the COD loading rate was increased from 4 to 6 kg/m^{3}d, and it then remained almost unchanged at higher COD loading rates from 6 to 12 kg/m^{3}d. The highest COD removal efficiency (92%) was found at the lowest COD loading rate of 4 kg/m^{3}d. The COD removal profiles of the mesophilic and thermophilic ASBR systems were not very different, with COD removal efficiencies in the range of 83–96%. These results suggest that the cassava-based ethanol wastewater contained a high content of easily biodegradable organic compounds. The COD removal efficiency that was obtained in this study was superior to those of other studies of AD of different wastewaters, such as 58% from cassava stillage [32] and 64% for potato waste in a two-stage CSTR [33], 78% for ethanol stillage in a mesophilic continuous stirred tank reactor (CSTR) [34], 65 and 72% COD removal for cassava wastewater in mesophilic and thermophilic UASB units, respectively [15,35], and 80% for petrochemical wastewater in a mesophilic CSTR with recirculated biogas [12]. These comparisons could lead to the conclusion that this ASBR with recycled biogas is the most efficient bioreactor in terms of COD removal because it can provide a better settleability for both the bacterial sludge and suspended solids that are present in the wastewater during the settle step of the process [36].

3.2.2. Biogas and CH$_4$ Production Rates

As shown in Figure 2a,b, the biogas and the CH$_4$ production rates increased with increasing COD loading rates up to a maximum at a COD loading rate of 6 or 10 kg/m^{3}d for the mesophilic or thermophilic temperature, respectively. Beyond these optimum COD loading rates, the biogas and CH$_4$ production rates decreased markedly with increasing COD loading rates. Interestingly, when the ASBR system was operated at a COD loading rate that was lower than 6 kg/m^{3}d, the process performance, in terms of both the COD removal efficiency and biogas production rates, were more or less the same at both of the temperatures, because the systems were under a low range of organic loads. The increased COD loading rate caused an increased level of organic substrates that are available for conversion to biogas, resulting in increased biogas and CH$_4$ production rates. However, for either operational temperature, increasing the COD loading rate above the respective optimum value led to a reduction in both the biogas and CH$_4$ production rates because of the increased toxicity of the accumulating VFAs (Section 3.3), as mentioned before. The thermophilic operation provided a higher optimum organic loading rate with a much higher biogas and CH$_4$ production rates than the mesophilic operation, which is in good agreement with a previous investigation [23].

The biogas produced under the studied conditions consisted mainly of CH$_4$ and CO$_2$ without any H$_2$, due to high self-retained pH (7–8), resulting in all of the produced H$_2$ being completely converted anaerobically to CH$_4$ by hydrogenotrophic methanogens. At both of the temperatures, the CH$_4$ concentration gradually decreased and the CO$_2$ concentration conversely increased with increasing COD loading rates. The increased COD loading rate increased the amount of organic compounds available for both hydrolytic and acidogenetic microbes, which grow some 10-fold faster than the methanogens [2], and so led to the higher production of CO$_2$ (Table 1). In addition, H$_2$ production was limited by unfavorable pH values for acetogens (Equations (8)–(13) in Table 1, resulting in a low reduction rate of CO$_2$ for CH$_4$ production (Equation (14)).

3.2.3. SMPRs and CH$_4$ Yields

As shown in Figure 2c,d, both SMPRs (mL CH$_4$/L d and mL CH$_4$/g mixed liquor volatile suspended solids (MLVSS) d) and CH$_4$ yields (mL CH$_4$/g COD applied and mL CH$_4$/g COD removed) showed similar trends, where they increased to maximum values, and then decreased with COD loading rates that were above their optimum COD loading rates. The optimum COD loading rates, 6 and 10 kg/m^{3}d at 37 and 55 °C, respectively, provided the maximum values of both the SMPRs and CH$_4$ yields for the ASBR systems. For the ASBR system that was operated at the mesophilic temperature, the highest SMPR values (1390 mL CH$_4$/L d or 122 mL CH$_4$/g MLVSS d) were much higher than those obtained from previous works (910 mL CH$_4$/L d for the digestion of cassava wastewater in a two-stage UASB [15], 682 mL CH$_4$/L d from digestion of potato waste in a two-stage CSTR [33], and 673 mL CH$_4$/L d from the digestion of biodiesel wastewater with added glycerin in a two-stage ASBR [37]). Similarly, the highest SMPR value of 3240 mL CH$_4$/L d or 196 mL CH$_4$/g MLVSS d in the thermophilic ASBR system was much higher than that of previous work, which was 650 mL CH$_4$/l d for the second tank of a two-stage UASB for cassava wastewater at a higher COD loading rate of 15 kg/m^{3}d [35]. However, the maximum SMPR value, which was based on the microbial concentration of the present study, was lower than those of the previous studies, indicating that the mesophilic ASBR of this study could maintain a high concentration of microbial cells with a high methanogenic activity. In comparison, the thermophilic operation of the ASBR greatly outperformed the mesophilic operation, permitting a higher optimum COD loading rate and a higher CH_4 production rate and yield.

3.3. VFA Concentration and Composition

Although VFAs are precursors for both steps of the acetogenesis and the acetotrophic methanogenesis to finally produce CH_4 and CO_2, the high concentration of VFAs that resulted from the higher rates of acidogenesis and acetogenesis than that of methanogenesis that is directly inhibited the methanogens [38]. The VFA concentration and composition profiles in the mesophilic and thermophilic ASBRs as a function of the COD loading rate are shown in Figure 3. For the mesophilic operation, the total VFA concentration was very low (around 5 mg/L as HAc), and then decreased slightly with an increasing COD loading rate that is up to the optimum value of 6 kg/m^{3}d. Above the optimum COD loading rate, the total VFA concentration sharply increased up to about 2500 mg/L as HAc at a COD loading rate of 8 kg/m^{3}d. The total VFA concentration profile showed the opposite trend to the SMPRs and CH_4 yields (Figure 2), supporting that the reduction in CH_4 productivity resulted from the VFA accumulation, which exceeded the inhibiting level (about 400–800 mg/L as HAc) to microbes, especially methanogens [35,37].

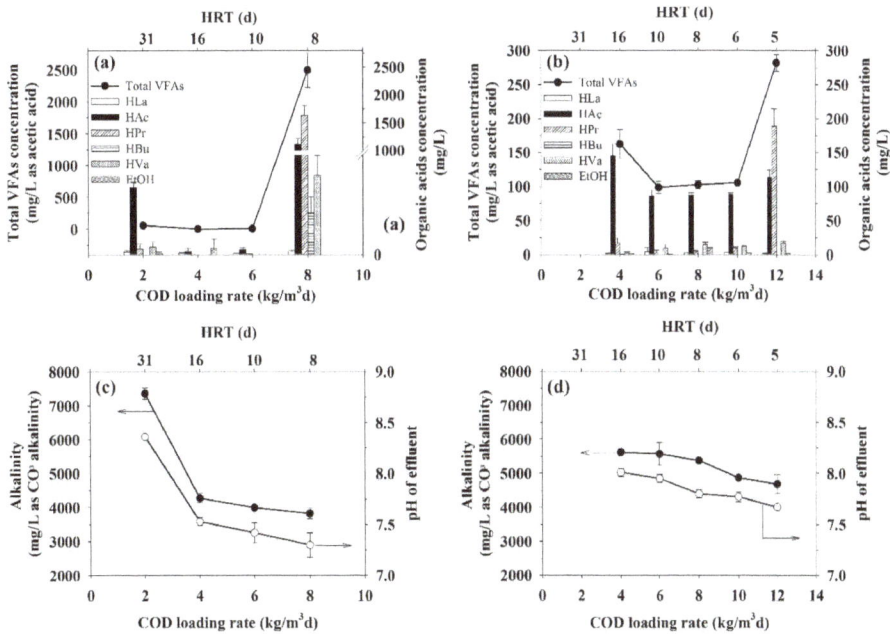

Figure 3. The (**a,b**) total volatile fatty acids (VFA), VFA composition and ethanol concentration and the (**c,d**) alkalinity and effluent pH in the ASBR operated at different COD loading rates under a (**a,c**) mesophilic and (**b,d**) thermophilic temperature. Data are shown as the mean ± 1SD, derived from five independent repeats.

For the thermophilic operation, the ASBR system had a low total VFA concentration (100 mg/L as HAc) in the COD loading rate range of 6–10 kg/m^{3}d, but it then increased to about 280 mg/L as HAc at a COD loading rate of 12 kg/m^{3}d, corresponding to the reduction in CH_4 productivity (Figure 2). From the result, the latter VFA concentration did not meet the inhibition level, suggesting that the reduction in CH_4 productivity might be influenced by micronutrient availability, as discussed later.

At both operational temperatures, the produced VFAs contained mainly HAc with small amounts of HLa, HPr, HBu, HVa, and ethanol at any COD loading rate that was lower than the respective optimum COD loading rate (6 kg/m^{3}d for the mesophilic temperature or 10 kg/m^{3}d for the thermophilic temperature). At a COD loading rate that is greater than the optimum COD loading rate, the produced biogas contained significant amounts of the organic acids, ranked in the order: HPr > HAc > HVa > HBu >> HLa for the mesophilic temperature and HPr > HAc >> HVa >> HLa >> HBu for the thermophilic temperature. The results suggested that under a low COD loading rate at 37 °C, all of the produced organic acids were completely converted to HAc and the produced HAc was mostly consumed by the methanogens. When the COD loading rate exceeded its optimum value, the methanogenic rate was lower than the acetogenic and acidogenic rates, as indicated by the high total VFA concentrations.

It is worthwhile to point out that the COD loading rates of 2 and 4 kg/m^{3}d for the mesophilic and thermophilic operations, respectively, had total VFA concentrations that were much higher than those of the higher COD loading rates of 4–6 and 6–10 kg/m^{3}d, respectively. This is because the methanogenic bacterial growth rate was much lower (10-fold) than the acidogenic bacterial growth rate [2], and so at the early start-up period with a low COD loading rate, the microorganisms contained a lower quantity of methanogens, resulting in the higher VFA concentration, as compared to those with a higher COD loading rate.

3.4. Alkalinity and pH

Alkalinity in an AD unit is the capability of the solution to resist the pH drop that is caused by the production of organic acids in the system, and it is referred to as the system buffer capacity [38]. The higher the alkalinity value (buffer capacity), the higher the ability of the methanogens to withstand a higher VFA concentration. Thus, the alkalinity and pH are basically used as process stability indicators for a CH$_4$ production process [37]. Figure 3c,d show the alkalinity and the pH profiles at various COD loading rates under the mesophilic and thermophilic operation of the ASBR. At both temperatures, they tended to decrease with increasing COD loading rate due to the increased total VFA concentration. In addition, the decreased pH values were consistent with the increased CO$_2$ levels in the biogas produced at increased COD loading rates. Interestingly, the level of reduction in both the alkalinity and pH in the ASBR at 55 °C were much lower than those at 37 °C, which is because the thermophiles had a much higher activity to convert the VFAs to CH$_4$ than the mesophiles.

3.5. Microbial Concentration and Washout

Figure 4 shows the profiles of the mixed liquor volatile suspended solids (MLVSS), representing the microbial community concentrations and the volatile suspended solids (VSS), representing the amount of microbial washout, from both ASBR units at different COD loading rates. Fundamentally, the solids retention time (SRT) has to be longer than the HRT for a successful operation of an AD system. All of the studied conditions showed much higher SRT than HRT values, indicating that the ASBR operation under the studied conditions did not have a hydraulic washout problem [9]. At both temperatures, the microbial washout, in terms of the VSS, increased with increasing COD loading rates, as is consistent with the decreased sludge settleability, resulting from the increased toxicity to the microbes from the increasing VFA concentrations. The microbial concentration in the mesophilic ASBR system gradually decreased with increasing COD loading rates, but the thermophilic ASBR system showed the opposite trend. Thus, the thermophiles, with a higher methanogenic activity, could withstand a higher COD loading rate due to the lower VFA concentration, leading to a better sludge settleability, as compared with the mesophiles. Hence, the microbial concentration (MLVSS) of the thermophilic ASBR system increased with increasing COD loading rates.

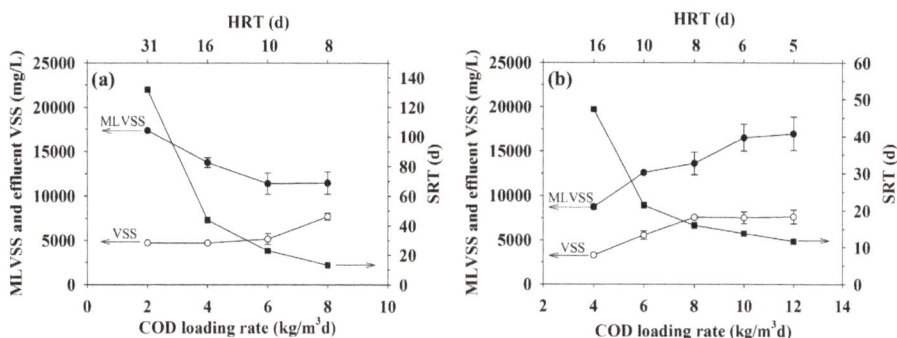

Figure 4. The microbial concentration (MLVSS), effluent volatile suspended solids (VSS) and solids retention time (SRT) at different COD loading rates and HRT in the ASBR operated under a (**a**) mesophilic (37 °C) or (**b**) thermophilic (55 °C) temperature. Data are shown as the mean ± 1SD, derived from 5 independent repeats.

3.6. Macronutrient Transport

The macronutrients (N and P) are essential for microbial growth in the process of biogas production [39]. They have also been considered as a major factor in eutrophication, and so the removal of both N and P has become an additional objective in wastewater treatment [40]. Hence, a study of N and P transport is of great interest for obtaining a better understanding about the process performance.

Nitrogen is basically found in both organic and inorganic forms, including NH_4^+-N, NO_3^--N, and NO_2^--N, which can be taken up by the bacteria during AD [39]. As shown in Table 2, most N in the ethanol wastewater was in the form of org-N (80–90%), with a N:P ratio of about 4:1. As shown in Figure 5a, the uptakes of both N and P in the mesophilic ASBR system increased with increasing COD loading rates to maximum values (44% for N uptake and 75% for P uptake) at the optimum COD loading rate (6 kg/m^{3}d). Above this optimum COD loading rate, the N, and especially the P uptake level, decreased markedly. The N and P uptake profiles corresponded well to those of methanogenic activity in terms of the CH_4 production rate, SMPRs, and CH_4 yields. As shown in Figure 5b, for the thermophilic ASBR system, the N uptake remained very low (27%) at a COD loading rate of 4–8 kg/m^{3}d and rose abruptly to 55% at the optimum COD loading rate of 10 kg/m^{3}d, while the P uptake did not vary significantly and it was relatively high (about 95%) throughout the studied range of COD loading rates. Above the optimum COD loading rate, both the N and P uptake levels remained almost constant. At each respective optimal COD loading rate, the N and P uptake levels of the thermophilic ASBR were significantly higher than those of the mesophilic ASBR, which corresponded well to the methanogenic activity results. Interestingly, the uptake ratios of N to P at the optimum COD loading rates for the mesophilic and thermophilic operation were not significantly different.

Figure 5c,d show the transformation of N compounds as a function of the COD loading rate in the mesophilic and thermophilic ASBR systems, respectively. Both NH_4^+-N and org-N can be directly utilized by anaerobes. The org-N in the ethanol wastewater was mostly metabolized to release NH_4^+-N as byproducts [41], resulting in a higher NH_4^+-N in the effluent, when compared to the influent (Figure 5c). For any COD loading rate, the levels of NH_4^+-N in the mesophilic and thermophilic ASBR units (190–270 and 260–360 mg NH_4^+-N/L, respectively) were much lower than the inhibition level to methanogens (1100–6000 mg NH_4^+-N/L), suggesting that NH_4^+-N inhibition could be ruled out in this study [42]. Under the studied conditions, the NO_3^--N and NO_2^--N levels were close to zero because of the anaerobic environment. The org-N was the main nitrogen source for the anaerobes, since the N present in the ethanol wastewater was mainly in organic forms.

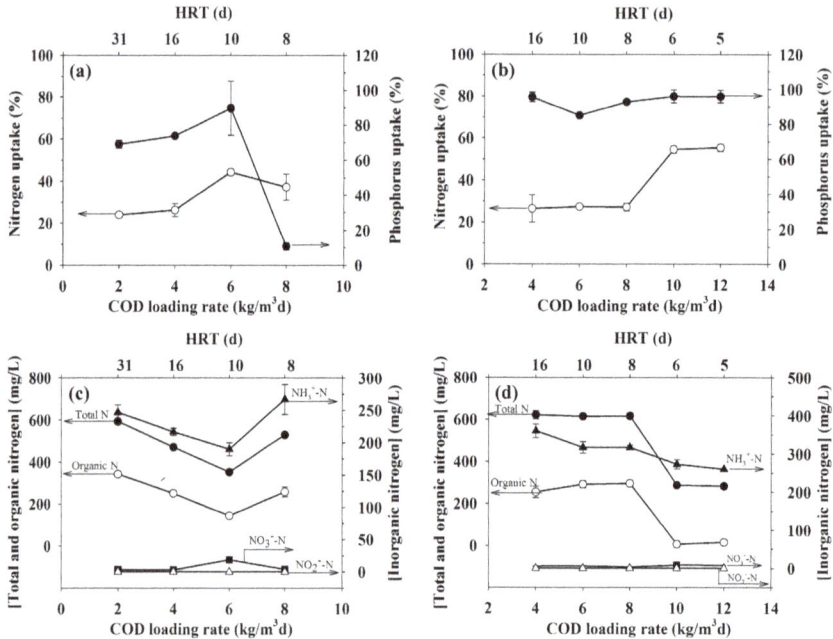

Figure 5. The (**a,b**) macronutrient (N and P) uptake and (**c,d**) total-N, org-N and inorganic-N levels in the ASBR operated at different COD loading rates under a (**a,b**) mesophilic (37 °C) and (**c,d**) thermophilic (55 °C) temperature. Data are shown as the mean ± 1SD, derived from 5 independent repeats.

3.7. Micronutrient Transport

In this study, the concentrations of Fe^{2+}, Zn^{2+}, Cu^{2+}, Ni^{2+}, Co^{2+}, Mn^{2+}, and Mo^{2+} in the feed and the effluent samples at various COD loading rates in both the mesophilic and thermophilic ASBR units were compared with the required concentrations for anaerobic decomposition [43–47] in order to reveal any micronutrient deficit. The measured concentrations of all studied micronutrients in the ethanol wastewater were found to be higher than the minimum stimulatory concentrations for AD (Table 3), but they tended to decrease with increasing COD loading rates at both of the temperatures. This reduction in the micronutrients in both ASBR systems likely resulted from the precipitation of the divalent cations (Fe^{2+}, Ni^{2+}, Co^{2+}, Zn^{2+}, Cu^{2+}, Mn^{2+}, and Mo^{2+}) with the sulfide ions (S^{2-}), which are the dissociated form of H_2S produced anaerobically from the sulfate and sulfur-containing organic compounds that are present in the ethanol wastewater. Under the studied conditions, the mesophilic ASBR system had a deficit of some micronutrients, especially Fe^{2+}, Zn^{2+}, Mo^{2+}, Co^{2+}, and Ni^{2+} at a COD loading rate of 8 kg/m^3d, corresponding to the decrease in the gas production rate of mesophilic ASBR. For the thermophilic ASBR system operated at the highest COD loading rate of 12 kg/m^3d showed no micronutrient deficit except Mo^{2+}. The addition to Fe^{2+}, Co^{2+}, and Ni^{2+}, and potentially Mo^{2+}, to the ASBR systems to improve the biogas productivity will be further investigated. Interestingly, the reduction of most micronutrients of the thermophilic ASBR system was lower, causing higher quantities of micronutrients available for methanogenic activity, as compared to that of the mesophilic ASBR system, as discussed later.

Table 3. Concentration of micronutrients in the ASBR system operated at a mesophilic or thermophilic temperature at different COD loading rates, as compared with the recommended values for AD (with S.D. less than 10%).

Parameters	Micronutrients (ppb)													
	Fe^{2+}		Mn^{2+}		Zn^{2+}		Cu^{2+}		Ni^{2+}		Mo^{2+}		Co^{2+}	
Recommended concentration	1000–10,000 [a] 100–400 [e]		5–50,000 [c] 10–50 [e]		1000–3000 [b] 100–1000 [e]		60–64,000 [c] 10–50 [e]		5–500 [d] 50–300 [e]		3–50 [a]		3–60 [a]	
Feed	17,500		2900		1260		870		150		250		80	
Effluent at different COD loading rate (kg/m³d)	37 °C	55 °C	37 °C	55 °C	37 °C	55 °C	37 °C	55 °C	37 °C	55 °C	37 °C	55 °C	37 °C	55 °C
4	1400	1000	180	20	12	30	70	30	20	20	0	100	20	20
6	280	2250	123	140	15	70	70	150	20	120	40	40	20	80
8	150	2300	120	170	10	60	25	120	6	90	0	40	20	40
10	-	350	-	25	-	30	-	110	-	30	-	40	-	20
12	-	270	-	10	-	30	-	70	-	40	-	0	-	20

[a] [43,44], [b] [45], [c] [46], [d] [47], [e] [19].

3.8. The Mass Balance

The mass balance results for both the mesophilic and thermophilic ASBR systems are shown in Table 4. The mass balance results, calculated from on the COD and carbon (C) content, were quite similar, while those that are based on the total solid (TS) were significantly lower. This was because the TS values included inorganic matter as well as organic compounds. The carbon content in the ethanol wastewater was mostly organic carbon, while the COD analysis is derived from chemical oxidation, and so the mass balance results based on either COD or C did not differ much. The mass balance results suggest that the ASBR could uptake more than 85% of the organic fraction of the ethanol wastewater and most of it was converted to biogas. In comparison between the two operational temperatures, the thermophiles showed a higher organic removal efficiency and higher biogas productivity (methanogenic activity) than the mesophiles at their respective optimal COD loading rates.

Table 4. The mass balance for the ASBR operated under a mesophilic or thermophilic temperature under steady state conditions at the optimum COD loading rates of 6 and 10 kg/m^{3}d, respectively (with S.D. less than 5%).

Sample	Mass Balance (% (w/w) of Feed)			
	TS	COD	C	S
Mesophilic ASBR system:				
Effluent	35	15	29	0
Biogas	50 (77%) *	69 (81%) *	62 (87%) *	27
Sludge	15 (23%) *	16 (19%) *	10 (14%) *	73
Thermophilic ASBR system:				
Effluent	13	11	10	0
Biogas	72 (83%) *	79 (89%) *	82 (91%) *	74
Sludge	15 (17%) *	10 (11%) *	8 (9%) *	26

* Percentage mass balance (shown in parenthesis) is calculated based on the mass removed.

For the sulfur (S) mass balance results, all of the S-containing compounds that were present in the ethanol wastewater were completely removed by the ASBR system when being operated at either temperature. As shown in Table 4, a significant portion of S was formed and was found in the sludge (73% w/w) for the mesophilic ASBR system, but found in the biogas 74% for the thermophilic ASBR system. The results are consistent with the higher micronutrient depletion in the mesophilic ASBR than in the thermophilic ASBR, implying that the higher precipitation level following the formation of sulfide ions to form metal sulfides occurred in the mesophilic ASBR systems.

3.9. The Energy Balance

The energy balance was evaluated for the two ASBR systems at their respective optimum COD loading rates using the calorific values of the ethanol wastewater, and the effluents and the energy gain from the produced biogas, calculated from the heating value of CH$_4$ of 35.8 kJ/L [48]. It should be mentioned here that the energy consumption for the ASBR operation, including the feeding, mixing, and maintaining the bioreactor temperature, was not considered in the energy balance. The energy content of the ethanol wastewater was found in the range of 13.8–14.1 kJ/g COD, as consistent with the reported energy values of most wastewaters (13–15 kJ/g COD) [49]. The energy extraction efficiencies for the biogas production of the mesophilic and thermophilic ASBR systems were 85 and 92%, respectively, (Table 5), which corresponded well to the mass balance results based on both the COD and C content. These values were superior to the reported 47% of energy extraction efficiency of the anaerobic digester treating the excess sludge from municipal wastewater [50]. The energy loses of 15 and 8% of the mesophilic and thermophilic ASBR systems, respectively, were assumed to contribute to bacterial metabolism. The specific energy values that were produced by the mesophilic and thermophilic ASBR systems were 8.6 and 10.9 kJ/g COD applied (3.6 and 5.2 kJ/g MLVSS d), respectively, which were comparable to the reported energy yield values (10.1–10.6 kJ/g COD applied) of a single digester [51]. Thus, the thermophilic ASBR system potentially had higher extraction efficiency with a lower energy requirement for bacterial metabolic activities, when compared to the mesophilic ASBR system. This can be explained by the fact that the thermophiles had a higher methanogenic activity with a higher optimum organic loading rate. Additionally, the energy recovery from the studied ethanol wastewater was in the range of 9.9–12.3 kJ/g TS, which was significantly higher than those of other wastes of cow and pig manure (6.2–7 kJ/kg TS), slaughterhouse waste (9.4 kJ/kg TS), and straw waste (7.17 kJ/kg TS) [52].

Table 5. The energy balance of the ABR system operated at a mesophilic or thermophilic temperature under steady state conditions at the optimum COD loading rates of 6 and 10 kg/m^{3}d, respectively (with S.D. less than 5%).

Value	Mesophilic ASBR	Thermophilic ASBR
Energy extraction efficiency (%)	85	92
Energy for bacterial metabolism (%)	15	8
Specific energy production rate (kJ/L d)	51.6	108.7
Specific energy production rate (kJ/gMLVSS d)	3.6	5.2
Energy yield (kJ/g COD applied)	8.6	10.9
Energy yield (kJ/g COD removed)	9.2	12.5
Energy yield (kJ/g TS applied)	9.9	12.3

3.10. New Explanation of the Methanogenic Activity of Mesophiles and Thermophiles

Up to now, the explanation for the higher methanogenic activity of thermophiles than that of mesophiles lies on the difference in microbial activity, and no concrete scientific evidences are given. In this study, new evidences of micronutrient transport and sulfur mass balance were used to explain why thermophiles have a higher methanogenic activity than mesophiles. The solubility of produced H_2S in pure water as mole fraction is 1.4469×10^{-3} and 1.0609×10^{-3} at 37 and 55 °C, respectively [26], suggesting that the produced H_2S in the ASBR was likely present in the biogas phase at 55 °C higher than that at 37 °C, as confirmed experimentally by the sulfur mass balance results (Table 4). As a consequence, the lower dissolved H_2S in the thermophilic ASBR caused lower precipitation of micronutrients in the form of metal sulfides, as compared with the mesophilic ASBR, as described before (Table 3). The present results can lead to a conclusion that the lower water solubility of produced H_2S in thermophilic AD, as compared with the mesophilic one, plays a crucial role to make thermophiles having higher methanogenic activity because the micronutrient deficit condition that was generally occurring under the mesophilic temperature can be eliminated under the thermophilic temperature.

4. Conclusions

Methane production from ethanol wastewater using an ASBR with recycled biogas under mesophilic and thermophilic operation without controlled pH was investigated to relate the process performance to macro- and micro-nutrient transport, and overall mass and energy balance. The thermophilic ASBR was superior to the mesophilic ASBR in terms of a higher optimum COD loading rate, (10 to 6 kg/m^{3}d) and a higher CH$_4$ yield (324 to 232 mL CH$_4$/g COD applied). The CH$_4$ production in this study demonstrated that the thermophiles had higher macronutrient uptakes and lower tolerance levels to VFAs than the mesophiles. A deficiency of most micronutrients was found in the mesophilic ASBR system, while the thermophilic ASBR system still had sufficient amounts of all micronutrients except for Mo^{2+}. The energy and mass balance results indicated that the studied ASBR systems could effectively extract energy from the ethanol wastewater to biogas (85% and 92% for mesophilic and thermophilic temperatures, respectively). Additionally, the mass balance that was based on COD and C showed good agreement with the energy balance. The higher methanogenic activity of thermophiles, for the first time of its kind, was found to result from lower precipitation of all micronutrients, which was caused by lower water solubility of produced H_2S (36%), as compared to that of mesophiles.

Author Contributions: Conceptualization, K.S., T.S. and S.C.; Methodology, K.S.; Software, K.S.; Validation, K.S., T.S. and S.C.; Formal Analysis, K.S.; Investigation, K.S.; Resources, T.S. and S.C.; Data Curation, K.S.; Writing-Original Draft Preparation, K.S.; Writing-Review & Editing, K.S. and S.C.; Visualization, K.S.; Supervision, S.C.; Project Administration, S.C.; Funding Acquisition, T.S. and S.C.

Funding: This research was funded by the Thailand Research Fund (TRF) through the Royal Golden Jubilee Ph.D. Program Grant (No. PHD/0244/2552) to the first author. TRF Senior Scholar Research Grant (No. RTA5780008) and Industrial Research Grant (No. RDG6050068) to the corresponding author, and the Thai Oil Group Company are acknowledged and greatly appreciated. The National Science and Technological Development Agency (NSTDA) and The Ministry of Energy also provided research grants (FDA-CO-2559-2569-TH and 459042-AE1) to support this study.

Acknowledgments: The authors thank the Center of Excellence on Petrochemical and Materials Technology, Chulalongkorn University, for providing some of the equipment for this research. Bangchak Bioethanol (Chachoensao) Co. Ltd. (Chachoengsao, Thailand) also provided a partial support for this project. Additionally, the authors thank Sapthip Lopburi Co., Ltd., Lopburi, Thailand, for kindly providing the sludge and ethanol wastewater used in this study.

Conflicts of Interest: The authors declare no conflict of interest.

References

1. Khan, M.A.; Ngo, H.H.; Guo, W.S.; Liu, Y.; Nghiem, L.D.; Hai, F.I.; Deng, L.J.; Wang, J.; Wu, Y. Optimization of process parameters for production of volatile fatty acid, biohydrogen and methane from anaerobic digestion. *Bioresour. Technol.* **2016**, *219*, 738–748. [CrossRef] [PubMed]
2. Christy, P.M.; Gopinath, L.; Divya, D. A review on anaerobic decomposition and enhancement of biogas production through enzymes and microorganisms. *Renew. Sustain. Energy Rev.* **2014**, *34*, 167–173. [CrossRef]
3. Weiland, P. Biogas production: Current state and perspectives. *Appl. Microbiol. Biotechnol.* **2010**, *85*, 849–860. [CrossRef] [PubMed]
4. Karve, A.D.; Karve, P.; Kulkarni, G. A new compact biogas system based on sugary/starchy feedstock. *Energy Sustain. Dev.* **2005**, *9*, 63–65. [CrossRef]
5. Searmsirimongkol, P.; Rangsunvigit, P.; Leethochawalit, M.; Chavadej, S. Hydrogen production from alcohol distillery wastewater containing high potassium and sulfate using an anaerobic sequencing batch reactor. *Int. J. Hydrogen Energy* **2011**, *36*, 12810–12821. [CrossRef]
6. Grimsby, L.K.; Fjørtoft, K.; Aune, J.B. Nitrogen mineralization and energy from anaerobic digestion of jatropha press cake. *Energy Sustain. Dev.* **2013**, *17*, 35–39. [CrossRef]
7. Reungsang, A.; Pattra, S.; Sittijunda, S. Optimization of key factors affecting methane production from acidic effluent coming from the sugarcane juice hydrogen fermentation process. *Energies* **2012**, *5*, 4746–4757. [CrossRef]
8. Ziganshin, A.M.; Schmidt, T.; Lv, Z.; Liebetrau, J.; Richnow, H.H.; Kleinsteuber, S.; Nikolausz, M. Reduction of the hydraulic retention time at constant high organic loading rate to reach the microbial limits of anaerobic digestion in various reactor systems. *Bioresour. Technol.* **2016**, *217*, 62–71. [CrossRef] [PubMed]
9. Angenent, L.T.; Karim, K.; Al-Dahhan, M.H.; Wrenn, B.A.; Domíguez-Espinosa, R. Production of bioenergy and biochemicals from industrial and agricultural wastewater. *TRENDS Biotechnol.* **2004**, *22*, 477–485. [CrossRef] [PubMed]
10. Gary, R.K. The concentration dependence of the δs term in the gibbs free energy function: Application to reversible reactions in biochemistry. *J. Chem. Educ.* **2004**, *81*, 1599. [CrossRef]
11. Al-mashhadani, M.K.H.; Wilkinson, S.J.; Zimmerman, W.B. Carbon dioxide rich microbubble acceleration of biogas production in anaerobic digestion. *Chem. Eng. Sci.* **2016**, *156*, 24–35. [CrossRef]
12. Siddique, N.I.; Munaim, M.S.A.; Wahid, Z.A. Role of biogas recirculation in enhancing petrochemical wastewater treatment efficiency of continuous stirred tank reactor. *J. Clean. Prod.* **2015**, *91*, 229–234. [CrossRef]

13. Venkata Mohan, S.; Mohanakrishna, G.; Sarma, P.N. Integration of acidogenic and methanogenic processes for simultaneous production of biohydrogen and methane from wastewater treatment. *Int. J. Hydrogen Energy* **2008**, *33*, 2156–2166. [CrossRef]

14. Martinez, F.A.C.; Balciunas, E.M.; Salgado, J.M.; González, J.M.D.; Converti, A.; de Souza Oliveira, R.P. Lactic acid properties, applications and production: A review. *Trends Food Sci. Technol.* **2013**, *30*, 70–83. [CrossRef]

15. Intanoo, P.; Chaimongkol, P.; Chavadej, S. Hydrogen and methane production from cassava wastewater using two-stage upflow anaerobic sludge blanket reactors (UASB) with an emphasis on maximum hydrogen production. *Int. J. Hydrogen Energy* **2016**, *41*, 6107–6114. [CrossRef]

16. Conrad, R. Contribution of hydrogen to methane production and control of hydrogen concentrations in methanogenic soils and sediments. *FEMS Microbiol. Ecol.* **1999**, *28*, 193–202. [CrossRef]

17. Prapinagsorn, W.; Sittijunda, S.; Reungsang, A. Co-digestion of napier grass and its silage with cow dung for methane production. *Energies* **2017**, *10*, 1654. [CrossRef]

18. Mao, C.; Feng, Y.; Wang, X.; Ren, G. Review on research achievements of biogas from anaerobic digestion. *Renew. Sustain. Energy Rev.* **2015**, *45*, 540–555. [CrossRef]

19. Demirel, B.; Scherer, P. Trace element requirements of agricultural biogas digesters during biological conversion of renewable biomass to methane. *Biomass Bioenergy* **2011**, *35*, 992–998. [CrossRef]

20. Dąbrowska, L. Speciation of heavy metals in sewage sludge after mesophilic and thermophilic anaerobic digestion. *Chem. Pap.* **2012**, *66*, 598–606. [CrossRef]

21. Gao, Y.; Kong, X.; Xing, T.; Sun, Y.; Zhang, Y.; Luo, X.; Sun, Y. Digestion performance and microbial metabolic mechanism in thermophilic and mesophilic anaerobic digesters exposed to elevated loadings of organic fraction of municipal solid waste. *Energies* **2018**, *11*, 952. [CrossRef]

22. Labatut, R.A.; Angenent, L.T.; Scott, N.R. Conventional mesophilic vs. Thermophilic anaerobic digestion: A trade-off between performance and stability? *Water Res.* **2014**, *53*, 249–258. [CrossRef] [PubMed]

23. Gebreeyessus, G.D.; Jenicek, P. Thermophilic versus mesophilic anaerobic digestion of sewage sludge: A comparative review. *Bioengineering* **2016**, *3*, 15. [CrossRef] [PubMed]

24. Kim, M.; Ahn, Y.-H.; Speece, R.E. Comparative process stability and efficiency of anaerobic digestion; mesophilic vs. Thermophilic. *Water Res.* **2002**, *36*, 4369–4385. [CrossRef]

25. Lu, J.; Gavala, H.N.; Skiadas, I.V.; Mladenovska, Z.; Ahring, B.K. Improving anaerobic sewage sludge digestion by implementation of a hyper-thermophilic prehydrolysis step. *J. Environ. Manag.* **2008**, *88*, 881–889. [CrossRef] [PubMed]

26. Gevantman, L.H. The Solubility of Selected Gases in Water. In *CRC Handbook of Chemistry and Physics*, 72nd ed.; Lide, D.R., Ed.; CRC Press: Boca Raton, FL, USA, 1992; pp. 82–83.

27. Sreethawong, T.; Chatsiriwatana, S.; Rangsunvigit, P.; Chavadej, S. Hydrogen production from cassava wastewater using an anaerobic sequencing batch reactor: Effects of operational parameters, COD: N ratio, and organic acid composition. *Int. J. Hydrogen Energy* **2010**, *35*, 4092–4102. [CrossRef]

28. Eaton, A.; Clesceri, L.; Rice, E.; Greenberg, A. *Standard Methods for the Examination of Water and Wastewater*, 23rd ed.; American Public Health Association (APHA); American Water Works Association (AWWA); Water Enviroment Federation: Washington, DC, USA, 2017.

29. Pehlivanoglu, E.; Sedlak, D.L. Bioavailability of wastewater-derived organic nitrogen to the alga *Selenastrum capricornutum*. *Water Res.* **2004**, *38*, 3189–3196. [CrossRef] [PubMed]

30. Intanoo, P.; Rangsunvigit, P.; Namprohm, W.; Thamprajamchit, B.; Chavadej, J.; Chavadej, S. Hydrogen production from alcohol wastewater by an anaerobic sequencing batch reactor under thermophilic operation: Nitrogen and phosphorous uptakes and transformation. *Int. J. Hydrogen Energy* **2012**, *37*, 11104–11112. [CrossRef]

31. Lin, Y.; Tanaka, S. Ethanol fermentation from biomass resources: Current state and prospects. *Appl. Microbiol. Biotechnol.* **2006**, *69*, 627–642. [CrossRef] [PubMed]

32. Luo, G.; Xie, L.; Zou, Z.; Wang, W.; Zhou, Q.; Shim, H. Anaerobic treatment of cassava stillage for hydrogen and methane production in continuously stirred tank reactor (CSTR) under high organic loading rate (OLR). *Int. J. Hydrogen Energy* **2010**, *35*, 11733–11737. [CrossRef]

33. Zhu, H.; Stadnyk, A.; Béland, M.; Seto, P. Co-production of hydrogen and methane from potato waste using a two-stage anaerobic digestion process. *Bioresour. Technol.* **2008**, *99*, 5078–5084. [CrossRef] [PubMed]

34. Luo, G.; Xie, L.; Zhou, Q.; Angelidaki, I. Enhancement of bioenergy production from organic wastes by two-stage anaerobic hydrogen and methane production process. *Bioresour. Technol.* **2011**, *102*, 8700–8706. [CrossRef] [PubMed]

35. Intanoo, P.; Rangsanvigit, P.; Malakul, P.; Chavadej, S. Optimization of separate hydrogen and methane production from cassava wastewater using two-stage upflow anaerobic sludge blanket reactor (UASB) system under thermophilic operation. *Bioresour. Technol.* **2014**, *173*, 256–265. [CrossRef] [PubMed]

36. Shao, X.; Peng, D.; Teng, Z.; Ju, X. Treatment of brewery wastewater using anaerobic sequencing batch reactor (ASBR). *Bioresour. Technol.* **2008**, *99*, 3182–3186. [CrossRef] [PubMed]

37. Tangkathitipong, P.; Intanoo, P.; Butpan, J.; Chavadej, S. Separate production of hydrogen and methane from biodiesel wastewater with added glycerin by two-stage anaerobic sequencing batch reactors (ASBR). *Renew. Energy* **2017**, *113*, 1077–1085. [CrossRef]

38. Sakar, S.; Yetilmezsoy, K.; Kocak, E. Anaerobic digestion technology in poultry and livestock waste treatment—A literature review. *Waste Manag. Res.* **2009**, *27*, 3–18. [CrossRef] [PubMed]

39. McCarty, P.L.; Rittmann, B.E. *Environmental Biotechnology: Principles and Applications*; McGraw-Hill: New York, NY, USA, 2001; pp. 19–55.

40. Li, B.; Wu, G. Effects of sludge retention times on nutrient removal and nitrous oxide emission in biological nutrient removal processes. *Int. J. Environ. Res. Public Health* **2014**, *11*, 3553–3569. [CrossRef] [PubMed]

41. Schimel, J.P.; Bennett, J. Nitrogen mineralization: Challenges of a changing paradigm. *Ecology* **2004**, *85*, 591–602. [CrossRef]

42. Kinyua, M.N. Effect of Solids Retention Time on the Denitrification Potential of Anaerobically Digested Swine Waste. Master's Thesis, University of South Florida, Tampa, FL, USA, 22 March 2013.

43. Schattauer, A.; Abdoun, E.; Weiland, P.; Plöchl, M.; Heiermann, M. Abundance of trace elements in demonstration biogas plants. *Biosyst. Eng.* **2011**, *108*, 57–65. [CrossRef]

44. Weiland, P. Anforderungen an pflanzen seitens des biogasanlagenbetreibers, tll-jena, eigenverlag, 12. *Thüringer Bioenergietag* **2006**, 26–32.

45. Bożym, M.; Florczak, I.; Zdanowska, P.; Wojdalski, J.; Klimkiewicz, M. An analysis of metal concentrations in food wastes for biogas production. *Renew. Energy* **2015**, *77*, 467–472. [CrossRef]

46. Sahm, H. Biologie der methan-bildung. *Chemie Ingenieur Technik* **1981**, *53*, 854–863. [CrossRef]

47. Pobeheim, H.; Munk, B.; Lindorfer, H.; Guebitz, G.M. Impact of nickel and cobalt on biogas production and process stability during semi-continuous anaerobic fermentation of a model substrate for maize silage. *Water Res.* **2011**, *45*, 781–787. [CrossRef] [PubMed]

48. Liu, X.; Li, R.; Ji, M.; Han, L. Hydrogen and methane production by co-digestion of waste activated sludge and food waste in the two-stage fermentation process: Substrate conversion and energy yield. *Bioresour. Technol.* **2013**, *146*, 317–323. [CrossRef] [PubMed]

49. Olsson, G. *Water and Energy: Threats and Opportunities*; IWA Publishing: London, UK, 2015. [CrossRef]

50. Shizas, I.; Bagley, D.M. Experimental determination of energy content of unknown organics in municipal wastewater streams. *J. Energy Eng.* **2004**, *130*, 45–53. [CrossRef]

51. DiStefano, T.D.; Palomar, A. Effect of anaerobic reactor process configuration on useful energy production. *Water Res.* **2010**, *44*, 2583–2591. [CrossRef] [PubMed]

52. Berglund, M.; Börjesson, P. Assessment of energy performance in the life-cycle of biogas production. *Biomass Bioenergy* **2006**, *30*, 254–266. [CrossRef]

energies

MDPI

Article

Economic Analysis of Pellet Production in Co-Digestion Biogas Plants

Dávid Nagy, Péter Balogh *, Zoltán Gabnai, József Popp, Judit Oláh and Attila Bai

Faculty of Economics and Business, University of Debrecen, Böszörményi Street 138, H-4032 Debrecen, Hungary; nagydavid4220@gmail.com (D.N.); gabnai.zoltan@econ.unideb.hu (Z.G.); popp.jozsef@econ.unideb.hu (J.P.); olah.judit@econ.unideb.hu (J.O.); bai.attila@econ.unideb.hu (A.B.)
* Correspondence: balogh.peter@econ.unideb.hu; Tel.: +36-52-508-444

Received: 30 March 2018; Accepted: 26 April 2018; Published: 3 May 2018

Abstract: In our paper we examine the economics of a technological process which utilizes the separated biogas plant digestate as the primary material and, as auxiliary material, the waste heat produced by the cogeneration process, to produce a marketable pellet which can be used in two ways (to supply soil nutrients and heat energy). Using multivariate linear regression model we developed an equation for the biogas yield from the modelled production recipe and expected nutrient pellet prices, and sensitivity analysis were also performed for the substrate dry matter content. We found that pellets can be produced at a cost of 88–90 EUR/ton with a 6 to 10% dry matter substrate content and that, primarily, sales of pellets for heating justify pelleting; producer's own use and use for nutrient purposes can only be justified in exceptional cases. In the case of dry solid content above 5%, the process does not require the total amount of waste heat; some of this can be used to cover other heat requirements.

Keywords: biogas; manure; economics; pellet; waste heat

1. Introduction

The planned targets of the European Union (EU) for 2020 are aimed at reducing emissions of harmful gases and, in addition to increasing efficiency, promoting the production of renewable energy sources. Among renewables, the number of biogas plants in the EU has already exceeded 17,000 and their number has grown dynamically over the past few years [1]. The longer-term goals of the EU are even more ambitious: by 2030, a 40% reduction of greenhouse gases (GHGs) is projected compared to 1990, and a 27% reduction compared to 2021 [2]. This objective is unattainable without a significant increase in renewable energy consumption, of which biomass accounts for 45% in the EU and almost 90% in Hungary [3]. The most dynamically growing segment of biomass-based heat (possibly electricity) production is the use of heating pellets.

At present, most biogas plants are equipped with combined heat and power (CHP), but the use of heat generated by gas engines (especially in the summer) is problematic, and its improvement is of decisive importance in environmental and economic terms. In addition, the 35–40% efficiency gains from the production of green electricity can be increased up to 40–50% with the use of waste heat, which can result in a 85% overall efficiency of CHP units [4,5]. This is more doubtful in larger plants, but if it is achieved, it will increase the corresponding plant size and is closely linked to the improvement of the break-even point [6]. Typical heat utilization areas in the biogas plant include: fermentation heating, hygiene, and biogas upgrading and, according to Lindkvist et al. (2017), the typical proportions for these areas are 62%, 37% and 1% respectively [7]. Upgrading biogas is the only technological option for eliminating waste heat generation which also helps to improve the calorific value and marketability of biogas [8].

In agricultural biogas plants, where much of the feedstock is derived from liquid manure from animal breeding (with typically less than 10% dry matter content), wet processes and mesophilic fermentation are used most often in Europe [9], while fermentation technology based on psychrophilic conditions has a decisive role among agricultural digesters globally, mainly in developing countries [10–12]. The latter (psychrophilic) technology typically operates without artificial heating and at a low cost level, but our present research focuses on those plants which are dominant in the developed countries under moderate climatic conditions, so we analyze later only these types of biogas factories. Hydraulic retention time (HRT) is typically less than 30 days in mesophilic conditions, when using a high proportion of manure and few energy crops [13,14].

This paper examines the economics of a technological process using the separated biogas plant digestate and as auxiliary material the waste heat from the cogeneration process to produce a marketable pellet, which can also be used in two ways (soil nutrient supply and thermal energy utilization). As a basic case, we consider the actual technological parameters of a biogas plant in Hungary, which can be considered typical in the literature, but our results can be used in other operating conditions with our sensitivity analyses.

2. Literature Review

2.1. The Characteristic Technological Parameters of Cogeneration Biogas Plants and the Resulting End Products

While the economical aspect of biogas production can be most easily improved by increasing the organic load [15] or by optimizing the recipe [16], the use of waste heat also significantly influences operational results.

Onthong–Juntarachat (2017) investigated the biogas yield of five different plant-derived by-products in mesophilic conditions with 15–35 days of hydraulic residual time, resulting in three of the five feedstocks having the highest biogas yield at a 25-day residual time, and two at a time of 15 and 30 days [17]. In Aramrueang et al.'s (2016) examination algae substrate (Spirulina (Arthrospira) platensis) received the highest yield of biogas and biomethane at 25 days of HRT and 1.0 (g VS) L^{-1} d^{-1} OLR (0.342 and 0.502 L (g VS)$^{-1}$, respectively) [18]. At the same time, due to economic considerations it may be advisable to increase the organic loading rate (OLR), the economic optimums in the same experiments of 2.0 (g VS) L^{-1} d^{-1} OLR and 25 d HRT were recommended for achieving biogas and methane yields of 0.313 and 0.490 L (g VS)$^{-1}$, respectively. It should be noted, however, that the heat demand and investment requirement of biogas production from algae is significantly higher than that of traditional biogas feedstocks [19] and can be limited by the low C/N ratio of high protein content algae species which may lead to ammonia inhibition [18].

Typically, 90% of the energy produced in CHP units can be utilized, 35–40% of which is electricity and 50–55% heat energy. On average, however, 25 to 30% of the total heat produced is used to heat fermenters in biogas plants which use mesophilic technology (37–43 °C fermentation temperature) [20–22]. It is advisable to divide the heating of the fermenters into summer and winter periods, as in the summer 25 kWh of heat is needed to heat up a ton of incoming substrate, while in winter this figure is 36 kWh. On the basis of other results, a 1 MW capacity biogas plant generates an average of 1000 kWh of heat per day, of which 250 kWh in winter and 100 kWh of heat in summer is needed to heat the fermenters at 38 °C, i.e., using 25% and 10% of the generated heat energy [23–25]. The seasonal fluctuations in its own heat consumption play a significant role in the fact that a total heat utilization solution is quite problematic, since in the case of heat utilization the heat generated during the summer period is lost, and so the technological (daily evenly uniform) use of waste heat available in different amounts in winter and summer makes the purchase of external heat energy necessary. In theory, the pelleting we examine can (with the right quantity of raw materials) be used to solve total heat utilization, because the raw material to be pelletized (the solid phase of the separated organic fertilizer) and the final product (the pellet) can both be stored well.

The use of waste heat is one of the most effective solutions for sustainable heat utilization. Among the many alternative uses for biogas plants—such as heating of buildings, greenhouses, aquaculture installations, absorption cooling or crop drying—separated fermentation drying is a process which can be carried out throughout the year, so that a substantial part of the generated waste heat can also be used for agricultural purposes [20].

Biochars (including digestate pellets) can be considered a promising material for the continuously developing biogas purification industry, especially when CO_2 and H_2S are successfully captured [26].

One of the major challenges for biogas plants is the utilization of circulated fermentation liquid for agricultural use, since nutrient supply in the EU is restricted to a limited amount (typically 170 kg N/ha/year) and can only be carried out over a specified period [27]. At the EU level the relevant regulatory framework is provided by the Nitrate Directive (91/676/EEC). In addition, transport to nutrient-poor areas and the typical half year storage period can have a very negative impact on the profitability of bio-fertilization [28,29].

The organic content of the organic matter introduced, however, is significantly reduced during anaerobic fermentation. Digestate can contain 50–80% less TS compared to the incoming substrate [30], and the organic part of the TS content of digestate can be up to 70%. Oenema and Tamminga (2005) found that 55–95% of N (nitrogen) and large proportions of phosphorus (P) and potassium (K) in animal diets were excreted [31]. Animal herbs and slurries are accordingly rich in plant nutrients.

2.2. Potential Markets for Pellet Production

For the production of a marketable product, biogas plants need to conduct further stages of processing (such as separation, drying, pelleting), following which fermentation sludge can be directly used for nutrient supply, bedding material or even burning [28,32].

All these workflows, combined with the costs of storage, shipping, or delivery, result in significant investment and operating costs due to the large amount of material and low nutrient content. A study conducted by the Association for Technology and Structures in Agriculture (KTBL) (Darmstadt, Germany) [33] considered a model biogas plant (50% manure and 50% corn silage) with an electrical capacity of 500 kWe and an annual digestate production of 30,000 m^3 and investigated 6 differential digestate management models. The results showed that the specific net cost varied from 1.94 €/m^3 (direct land application) to 6.80 €/m^3 (separation using a screw press before dehydrating the solids with a belt dryer) [34]. Due to the multiple cost increases, further processing of organic fertilizer can be recommended mainly for sales purposes and when there is a significant quantity of the producer's own waste heat.

2.2.1. A Possible Market for Heating Pellets

A heating pellet is pressed at high pressures, and mostly held together by its own material. Due to the strict standards, its heat value depends mainly on the raw material used: wood pellets average 18.5 MJ/kg, while agripellets typically have a calorific value of 16.5 MJ/kg. In terms of the ash content, the difference is even greater, with wood pellets having 0.5–1%, and agripellets 7.5–11% [35]. The bio-fertilizer pellet we tested was similar to the latter value.

In 2016, of all pellets worldwide, the EU produced half (14.8 MMT (million metric tons)), but used 75% (22.2 MMT), of which about one third (6.4 million tons) were imported [36]. Both home produced and imported pellets must comply with sustainability requirements [2]. The total utilization of production capacities is 76%, with Germany the largest producer accounting for 57% of the total. In Hungary, pellets are produced in a small quantity (about 5000 t/year). The largest consumers are the United Kingdom (2016: 7.3 MMT) and Italy (2016: 3.4 MMT). The latter country also has the highest proportion of imports.

Households and small public institutions (e.g., hospitals, swimming pools) currently dominant pellet use, but in the near future, with the tightening of environmental regulations, the role of heating pellets in power plant use will also increase in co-firing, mainly for coal substitution [37].

Power plant utilization is not expected to require new pellet plants, but can be achieved by better utilizing existing capacities.

By 2020, a further dynamic growth in use in the EU (mainly due to power plant use) is expected, with use predicted to reach up to 28 million tons. The widespread use of organic fertilizer pellets could also help reduce import expenditures, which in 2016 accounted for expenditure of 920 million euros just to the USA. Prices in 2015 in the EU are typically around 228–240 €/t, while the price of imported pellets is considerably lower (2016: 188 €/t, [38]). The most important competitor to the pellet is natural gas (12–20 €/GJ) in the domestic sector, wood waste in smaller public institutions (19 €/t) and wood pulp (25 €/t) in the power plant sector [35]. By 2020, the EU's current 60 ML of cellulose-based bioethanol production capacity is expected to grow to 200 ML, which may also bring significant demand to the European pellet and wood chip markets [39].

In summary, the spread of fermentation sludge pelleting technology and the use of the end-product for heating fuels can, in the long run, contribute to meeting the renewable energy, environmental, or new generation fuel targets of the EU's Renewable Energy Directive (2009/28/EC).

2.2.2. The Market for Upgraded Organic Manures

Managing the biogas origin of digestate products and especially their marketization is one of the major challenges for the biogas sector because of the large quantity involved and the variable fertilizer value.

Almost all (95%) of the digestate produced in the EU is used for soil management purposes, as an organic fertilizer on cropland [40]. Direct application on the producer's own land is typically the best option [41], because of savings on nutrient purchases and improvements in soil quality [42]. Because of the upgrading technologies and the developing markets, less than 3% of the digestate is currently being upgraded to different products.

Solid digestate (compost and pellets) has wider possibilities for the non-agricultural market, unlike fluid digestate. 17 percent of the solid digestate was sold to private gardeners and soil manufacturers, while in the case of liquid digestate, the proportion sold to conventional and organic farmers was 99% [43].

However if the density of livestock farming and biogas plants is huge, this can result in a scarcity of land for digestate and lead to high land rental prices.

Currently pellet prices vary between 0–200 €/t, while digestate pellets are usually sold at an average price of 20 €/t, but can have a fertilizer value of 90 €/t. Prices depend mainly on the packaged amount, the content and the season. In the case of plant soil for domestic use, a 9 €/L shelf price can also be reached, but increasing prices do not automatically result in higher margins, due to other costs emerging related not only to manufacturing, but to packaging and marketization, too.

Digestate as raw material, with mid-level Nitrogen-Phosphorus-Potassium (NPK) values could be an optimal solution for most home gardeners who over-fertilize [44].

All in all, the factors which can be considered an advantage when selling organic fertilizer pellets are; bionutrients, upgradability (organic farming) and new applications. The disadvantages are, above all, marketing barriers including transport costs and negative perceptions by customers [32].

2.3. Technological Processes of Biogas Fuel Pellet Production

The first step in the processing of the fermentation sludge is separation, during which the solid and liquid fractions are separated. Various separation methods can be used with mechanical means such as decanting centrifuges or screw separators [45].

According to Bauer et al.'s (2009) data, the mass of the starting material after separation was 79.2% liquid and 20.8% solid, but the total dry matter was 61.8% in the solid fraction [46]. In addition, according to Möller et al. (2010), the amount of solid separated mass makes up 22.5% of the starting material after separation [47].

During the separating operation, a solid fraction of 0.5–2% liquid matter and 25–35% dry matter is formed [48]. Similar results have been presented by Möller et al. (2002) [49] and Tambone et al. (2017) [50], since after separating the 4–6% dry matter content fermentation liquid, a solid separation of 30–32% dry matter is produced, with a volume of 0.5 t/m³ [51].

Tambone et al. (2017) investigated pig and cow liquid manure and fermentation sludge using different separators, on the basis of which it can be stated that the dry matter content of the feedstocks was between 3.74% and 6.37%, and during separation the share of the dry matter of the solid fraction increased to between 24.97% and 36.47%, increasing by an average of 25.4% relative to the starting material when using a screw press separator [50].

After separation, the solid phase is typically dried on a belt dryer at 70–80 °C during which the heat energy from the CHP plant passes through the belt and the material on it from bottom to top in the form of hot air. During the drying, the specific heat energy required to extract one ton of water is 1–1.1 MWh [25,52].

2.4. Ingredients and Fertilization Effects of Separated Biodiesel

According to Awiszus et al. (2018) [53], after drying the separated fermentation liquid with its 24% dry matter content at 70–80 °C, it can provide as much as 91–93% dry matter content. However, during the process, total nitrogen content is reduced by 47–48%, ammonium nitrogen by 82–87%, phosphorus by 6–8%, and sulphur content by 6–12% in comparison with the separated material. After drying, the total nitrogen content in the dry matter is 1.5%, ammonium nitrogen is 0.2%, phosphorus is 1.0–1.1%, and potassium is 2.0–2.1%. Other results showed a higher value in the pelletized final product since the ratio of nitrogen in the dry matter was 1.5–2.9%, the phosphorus ratio was 1.1–1.3%, while the ratio of potassium was 1.4–1.6% [54].

Vanden Nest et al. (2015) [55] investigated the use of different fertilizer types in comparison with four different inorganic fertilizer cultures. Two dried fermentation residues were also used among the manure types, derived from organic waste, energy crops and fermented liquid animal manure. On the basis of the results, after drying the fermentation residue is suitable for use as nutrient replenishment, and without loss of yield to replace fertilizers and liquid cow manure.

Thomas et al. (2017) investigated not only organic fertilizer, matured fermentation liquid, and solid separated matter, but also supplemented this with the already pelletized fermentation residue, during which the effects of the individual fertilizer pulp on autumn barley were measured for four years. On the basis of their results, during the fermentation and processing, the material undergoes substantial internal changes in content until it becomes pellet. The dry matter content increased by an average of 50% in relation to the mature fermented liquid, and the total amount of carbon in the dry matter by 4%. By contrast, the total nitrogen content is reduced by 42%, the proportion of ammonium nitrogen by 90% and the total phosphorous content by 44%. Based on the yields of autumn barley, the yield of the plant treated with fermented liquid exceeded the yield of other barley treated with manure by 30–50%. In the case of pellets, the yield was only 4.5% lower than that of organic fertilizers. Nitrogen absorption was the lowest in the case of pellets, at only 2%, while it was highest in the matured fermentation liquid, at 41%. In addition, it was argued that pellet from fermented sludge could be an effective, but slow-release nutrient source, while also providing carbon replenishment [56].

According to Ross et al. (2018), the pellet or granule derived from fermented sludge can improve the transportability and storage of the fermented liquid. They found that products derived from fermented sludge treated in different ways (composted, pelletized and granulated) only slowly break down during nutrient supply and the release of nutrients is low, especially for nitrogen. The effect of fertilizer on plant growth was observed only after the first week after application. On this basis, the nutritional utility of the processed fermented sludge should be considered limited, irrespective of the actual nutrient content. However, the previously mentioned fertilizers, as a result of increasing carbon content, are well suited to improving the humus content of soil poor in organic matter [57].

Ball et al. (2004) also found that pelletizing is a useful method for preserving the nutrient content of organic materials and that nitrogen is slowly and permanently released during application of the organic fertilizer pellet [58].

In relation to the fattening of growing broiler chickens, Hammac et al. (2007) [59] formulated the problems of the disposal and utilization of manure, the solution to which could be pelleting, given that while the application of bedding fertilizer is economical only to a certain extent, pelleting increases bulk density, particle size uniformity and economical delivery distance. This can reduce pressure in areas where large amounts of fertilizer can cause nutrient and heavy metal contamination, similar to fermented sludge. In the investigation of pellets from poultry manure, it was found that the nitrogen and phosphorus uptake of different plants is lower than with inorganic nitrogen and phosphorus fertilizer. Dry biomass production from pellets was lower than from ammonium nitrate, but showed similarity to calcium phosphate. Finally, it can be stated that the pelletized broiler litter can serve primarily as nitrogen and phosphorus fertilizer, but its nutrient content and supply do not match that of inorganic fertilizers.

Table 1 shows pellets derived from different fermented organic fertilizers, which show that they typically have a lower proportion of dry matter, nitrogen and phosphorus content, and a higher proportion of ammonium-N than poultry fertilizers.

Table 1. Characteristics of different types of digestate and poultry pellets.

Pellet Types	Dry Matter (%)	Total N (%/DM)	NH_4-N (%/DM)	Total P (%/DM)	Total K (%/DM)	Total C (%/DM)
Digestate from cattle manure	72.0	2.0	0.07	0.6	n.d. *	41.6
Digestate from household bio-waste	80.0	2.0	0.17	0.4	1.7	n.d. *
Digestate from maize silage, grass and grass silage, potatoes	90.8	2.9	n.d. *	1.3	1.4	45.3
Digestate from maize silage, sugar sorghum, poultry manure, corn cob mix	90.1	1.5	n.d. *	1.1	1.6	43.2
Interval from different digestates	**72.0–90.8**	**1.5–2.9**	**0.07–0.17**	**0.4–1.3**	**1.4–1.7**	**41.6–45.3**
Poultry manure	90	4.5	0.46	1.8	n.d. *	n.d. *

* No data. Source: Thomas et al. (2017) [56]; Ross et al. (2018) [57]; Kratzeisen et al. (2010) [54]; Haggard et al. (2005) [60].

2.5. Properties of Heating Pellets Made from Separated Organic Fertilizer

Shirani and Evans (2012) carried out studies on the drying and pelleting of the fermentation residue, which found that a combustible product from cattle liquid manure could be produced. The moisture content of the pellet produced was 9.9% and the amount of organic dry material was 74.6%. The ash content of the entire combustion process was 5.4%, while the gross calorific value was 17.6 MJ/kg. During the combustion, there was some smoke, unpleasant smell or slag, at different intensities of boiler operation. Their investigations were supplemented with others into additional fuels, in which the fermented separated solid matter was mixed into different biomass combinations, with the results for combustion of the solid separated matter, bovine liquid manure and sawdust shown in Table 2. It was found that the separated matter from anaerobic fermentation could serve as a raw material for high quality pellets. Their calorific values and mechanical durability are similar to softwood pellets [61].

On the basis of Kratzeisen et al.'s (2010) [54] investigation, it can be said that pellets from biogas plant fermentation sludge can be used as heating fuel, as despite their high ash content, their physical and chemical properties were found to be adequate; these are also summarized in Table 2. In addition, the energy demand for pellet production (3.24 kWh/kg) did not exceed the value of the energy output (4.56–4.81 kWh/kg), so the pellet production energy balance is also positive.

Table 2. Characteristics of different pellet types.

Parameter	Unit	Digestate [1]	Digestate *[2]	Straw [3]	Wood [3]
Diameter	mm	5.82–9.97	-	6.0	6.0
Moisture content	%	9.2–9.9	12.3	9.40	7.14
Ash content	%	14.6–18.3	5.7	9.10	0.65
Sulphur content	%	0.33–0.86	0.2	0.13	0.01
Fine fraction	%	2.2–3.1	-	1.6	0.34
Nitrogen content	%	1.54–2.86	1.03	0.91	0.40
Gross calorific value	MJ/kg	16.4–17.3	16.9	18.25	20.31
Net calorific value	MJ/kg	15.0–15.8	15.6	15.15	17.38

* Pellet raw materials: fermented solid separated matter, beef cattle manure and sawdust. Source: [1] Kratzeisen et al. (2010) [54]; [2] Shirani-Evans (2012) [61]; [3] Verma et al. (2012) [62].

3. Material and Methods

3.1. Introduction of the Model Biogas Plant and of the Utilized Feedstocks

The most important characteristic of the Hungarian cogeneration (637 kWe capacity) biogas plant we examined is that the majority of the substrate components are produced within the farm, thus ensuring continuous operation in the long run. Plant cultivation is currently carried out on 4000 hectares of arable land, and the basic objective of this activity is to provide the livestock and forage feed needs of the livestock sector, although it also participates in the fermentation process by providing a certain amount of maize silage and grass silage.

The livestock sector includes the dairy cattle sector with an average cow herd size of nearly 2000, while the pig breeding operation has 1200 breeding sows, and approximately 20,000 porkers are released annually. The types of fertilizer produced at the livestock farm provide more than 90% of the pellet ingredient, and there is also whey and dairy sludge from the nearby cheesery ready for disposal. The primary objective of the operation of the biogas plant is to manage these continuously generated, unmarketable and environmentally problematic by-products. In order to improve the fermentation processes, its own raw materials with a significant dry matter content (maize silage, grass silage, litter fertilizer) will also be used. The amount of raw materials used daily in the biogas plant recipe over the last 3 years is shown in Table 3. The HRT used corresponds to the 25 days considered typical of the mesophilic technique in the literature.

Due to the available fermentation capacity and the applied organic matter load, the amount of substrate fed daily was relatively stable during the period examined (2015–2017), averaging between 174 and 183 m³. Its composition can be characterized by minor seasonal and annual changes (Table 3).

Table 3. Daily feedstocks of biogas plant in different years.

Used Feedstocks	2015		2016		2017	
	Average	±SD	Average	±SD	Average	±SD
Cattle liquid manure (m³/d)	98.7	±8.9	96.3	±8.7	103.8	±12.0
Pig liquid manure (m³/d)	48.1	±7.5	45.7	±7.2	46.3	±7.3
Silo maize (m³/d)	3.4	±2.4	2.6	±2.7	2.1	±2.1
Grass haylage (m³/d)	3.5	±2.5	1.6	±1.5	1.2	±1.6
Solid fraction of cattle liquid manure (m³/d)	8.0	±3.7	11.3	±4.9	11.0	±5.2
Solid fraction of pig liquid manure (m³/d)	0.0	±0.0	1.1	±1.8	1.0	±0.9
Whey (m³/d)	9.1	±1.1	9.2	±0.9	7.8	±10.6
Dairy sludge (m³/d)	11.0	±2.2	12.7	±2.1	0.6	±2.8
Cattle manure (m³/d)	1.0	±1.5	0.4	±1.6	0.0	±0.0
Feed substrate (m³/d)	182.9	±12.1	181.0	±8.2	173.9	±12.2

Source: processing plant data (Hajdúböszörmény, Hungary). SD: standard derivation.

There is currently no built-in technology for the use of heat, which results in a low overall efficiency of the cogeneration unit. The utilization of the fermented liquid for agricultural purposes entails high costs (mainly during storage and transport), but further processing (separation, drying and pelleting) is also required for the production of a marketable product. The foregoing factors largely justify the introduction of pelleting for the efficient use of significant waste heat and the marketing of organic fertilizers.

3.2. Material Flow and Formulas Used in the Model

The process of treating the electricity and heat energy produced in the biogas plant and its related fermentation sludge is based on the quantities of the various substrate components ($SQ^{1\dots n}$) and their specific dry matter ($DM\%^{1\dots n}$) and the organic content ($OM\%^{1\dots n}$), which are summarized in Table 4.

Table 4. The quality of different raw materials.

Parameter	Cattle Liquid Manure (SQ^1)	Pig Liquid Manure (SQ^2)	Silo Maize (SQ^3)	Grass Haylage (SQ^3)	Solid Fraction of Cattle Liquid Manure (SQ^4)	Solid Fraction of Pig Liquid Manure (SQ^5)	Whey (SQ^6)	Dairy Sludge (SQ^7)	Cattle Manure (SQ^8)
Dry matter ($DM\%^{1\dots n}$)	3.6% [2]	4% [1]	26.1% [2]	24.8% [2]	21.6% [3]	21.9% [3]	3.5% [2]	2% [2]	21.2% [2]
Organic matter ($OM\%^{1\dots n}$)	75% [4]	75% [1]	72% [4]	85% [4]	85% [4]	81.1% [5]	80% [4]	85% [4]	85% [4]

Source: [1] Al Seadi (2001) [63]; [2] Mézes (2011) [51]; [3] Möller et al. (2000) [64]; [4] Al Seadi (2001) [63] and Mézes (2011) [51]; [5] Sáez et al. (2017) [65].

The amount of daily substrate (SS^d) and outflow fermented liquid (AD^d), as well as the daily dry matter (DM^d), the organic load (OM^d), and the amount of organic matter contained in the raw material ($SO^{1\dots n}$) used can be determined for all substrate components with the following formulas:

$$SS^d = \sum SQ^1 + SQ^2, \dots, SQ^n \text{ and } AD^d = SS^d \tag{1}$$

$$DM^d = \sum (SQ^1 \times DM\%^1) + (SQ^2 \times DM\%^2), \dots, (SQ^n \times DM\%^n) \tag{2}$$

$$OM^d = \sum (SQ^1 \times DM\%^1 \times OM\%^1) + (SQ^2 \times DM\%^2 \times OM\%^2), \dots, (SQ^n \times DM\%^2 \times OM\%^n) \tag{3}$$

$$SO^1 = SQ^1 \times DM\%^1 \times OM\%^1 \tag{4}$$

In the course of our analysis we used multi-variable linear regression calculations to determine the coefficients of the amount of raw material used in the various organic matter models. Using the database available, we investigated the most suitable model, based on the composition fed at different times (1–25 days). For this we applied R^2 values of the functions and we examined the residue from the models. After comparing models with different dates, the 25-day feed model was selected ($R^2 = 0.412$; $F = 27.429$; $p < 0.001$). For the parameters of this model, the absolute values of *t*-tests are 2.034–8.157; the *p* values are 0.001–0.043; the variance inflation factor values varied from 1.127 to 2.100. Residues showed a normal distribution, as the one-sample Kolmogorov-Smirnov test was not significant ($p = 0.215$). In our opinion, it was appropriate to use the values of this model in the simulation Excel model, and it suits the optimal HRT as measured by Onthong–Juntarachat (2017) [17] and Aramrueang et al. (2016) [18].

From the daily organic load, the biogas yield (BY^d) can be determined and, using the two-variable regression equation ($R^2 = 0.938$, $F = 5257.247$, $p < 0.001$), we estimated electricity production (EP^d). The small difference between the biogas and the electricity volume may be due to the slight differences in biogas quality (methane content) at different periods.

The production of heat (WH^d) was derived from all electricity used, which, on the basis of Rutz et al. (2015) [20], is 157% of the electricity produced (35% of the electricity produced, and

55%—the previous figure multiplied by 1.57—of the waste heat). The amount of waste heat available for pellet drying is calculated separately for winter and summer, as the amount of heat required to heat the fermenters is significantly different. In the winter (DH^W) this was 36 kWh/tonne, and in the summer (DH^S) 25 kWh/tonne, calculated with substrate specific values, according to Pullen's (2015) data [24]. Based on the above, the equation used for the amount of biogas produced is:

$$BY^d = 10879 - 1037 \times SO^1 - 1020 \times SO^2 + 135 \times SO^3 - 260 \times SO^4 - 168 \times SO^4 + 83 \times SO^5 + 1874 \times SO^6 - 3514 \times SO^7 - 718 \times SO^8 \quad (5)$$

For the equation for the amount of green electricity produced, the *t*-values were 2.301 and 72.507 and the *p* values were <0.001 and 0.022. Residues showed a normal distribution, as the one-sample Kolmogorov-Smirnov test was not significant ($p = 0.302$):

$$EP^d = 363 + 1.858 \times BY^d \quad (6)$$

The equation for the quantity of waste heat produced:

$$WH^d = EP^d \times 1.57 \quad (7)$$

Finally, the equation for waste heat that can be used for pelleting (drying) during winter and summer:

$$DH^W = 36 \times SS^d \text{ and } DH^S = 25 \times SS^d \quad (8)$$

The dry matter content ($AD^{DM\%}$) of the matured fermentation liquid (AD^d) can be determined as the ratio of dry matter load to daily substrate volume. After the separation, the amount of matured fermented liquid decreases by an average of 77.5% and the volume of the solid fraction is 0.5 t/m³, which can be used as a permanent value to determine the amount of daily solid fraction (SF^d) expressed in kilograms [47,51].

According to Tambone et al.'s (2017) results, the dry matter content of the solid fraction ($SF^{DM\%}$) for screw press separators increased by an average of 25.4% compared to the starting material [50].

Based on the above,

- the dry matter content of the circulated fermentation liquid:

$$D^{DM\%} = (DM^d \div SS^d) \quad (9)$$

- daily amount of solid fraction:

$$SF^d = AD^d \times 0.225 \times 0.5 \times 1000 \quad (10)$$

- solid fraction of dry content:

$$SF^{DM\%} = AD^{DM\%} + 0.254 \quad (11)$$

Factors taken into account in the quantities of winter dried solid fraction (DD^W) and the summer dried solid fraction (DD^S):

- moisture which can be removed in addition to the heating performed by the fermenters (1.1 kWh/kg of water),
- the dry matter content of unit separation ($SF^{DM\%}$),
- the dry matter content ($DD^{DM\%}$) required for pelleting, which is in this case is constant at 90.0%.

To calculate, we applied the IF function of the Microsoft Excel program, which allowed us to ensure that the material flow rates cannot be negative. The IF function performs a certain logic test which, after the test is completed or not completed, solves the predetermined formulas, or even issues a fixed value. Based on the IF function, the partial results of the applied simulation method can modify the changes of the following other calculated model parameters. Accordingly, the calculations are as follows:

DD^W = using the IF function, where:

$$DD^W = \begin{cases} SF^d \times (1 - (DD^{DM\%} - SF^{DM\%})); & \frac{WH^d - DH^W}{1,1} / (DD^{DM\%} - SF^{DM\%}) \geq SF^{DM\%} \\ (SF^d - NDSF^W) \times (1 - (DD^{DM\%} - SF^{DM\%})); & \frac{(WH^d - DH^W)}{1,1} / (DD^{DM\%} - SF^{DM\%}) < SF^{DM\%} \end{cases}$$

- the logic test: $(WH^d - DH^W) \div 1.1 \div (DD^{DM\%} - SF^{DM\%}) \geq SF^d$; i.e., the quantity of available material dried by heat energy is greater than the amount of solid separated matter
- if true: $DD^W = SF^d \times (1 - (DD^{DM\%} - SF^{DM\%}))$; i.e., the mass of the dried solid fraction is equal to the product of the weight of the solid separated matter and the percentage of the extracted moisture content,
- if false: $DD^W = (SF^d - NDSF^W) \times (1 - (DD^{DM\%} - SF^{DM\%}))$; the mass of the dried solid fraction is equal to the product of the entire available separated matter and the percentage of the extracted moisture content subtracted from the non-dryable separated matter.

Accordingly, the amount of DD^S was calculated, during which the logical test is $(WH^d - DH^S) \div 1.1 \div (DD^{DM\%} - SF^{DM\%}) \geq SF^d$. If the logical test is true, it computes the value of $SF^d \times (1 - (DD^{DM\%} - SF^{DM\%}))$, otherwise the value of $(SF^d - NDSF^S) \times (1 - (DD^{DM\%} - SF^{DM\%}))$.

For the purpose of carrying out the logical tests, we determined the difference between the potentially dryable quantity of separated matter with waste heat and the actual (moist) separated matter, i.e., the non-dryable separated matter, which may differ in winter ($NDSF^W$) and summer ($NDSF^S$) periods depending on the thermal energy used for the fermenter heat exchanger.

$NDSF^W$ = using the IF function, where:

$$NDSF^W = \begin{cases} 0, & \frac{WH^d - DH^W}{1,1} / (DD^{DM\%} - SF^{DM\%}) \geq SF^d \\ SF^d - \frac{\frac{(WH^d - DH^W)}{1,1}}{(DD^{DM\%} - SF^{DM\%})}; & \frac{(WH^d - DH^W)}{1,1} / (DD^{DM\%} - SF^{DM\%}) < SF^d \end{cases}$$

- the logic test: $(WH^d - DH^W) \div 1.1 \div (DD^{DM\%} - SF^{DM\%}) \geq SF^d$, i.e., the amount of decompressed separated matter is greater than or equal to the available quantity
- if true: $NDSF^W = 0$, i.e., the total separated matter has been dried by the heat energy
- if false: $NDSF^W = SF^d - ((WH^d - DH^W) \div 1.1 \div (DD^{DM\%} - SF^{DM\%}))$, i.e., the amount of undried separated matter is equal to the difference between the available separated matter and the potentially dryable separated matter.

Accordingly, the amount of $NDSF^S$ was calculated, with the logical test $(WH^d - DH^S) \div 1.1 \div (DD^{DM\%} - SF^{DM\%}) \geq SF^d$. If the logical test is true, the $NDSF^S$ value is zero, and if false, then it calculates the value of $SF^d - ((WH^d - DH^S) \div 1.1 \div (DD^{DM\%} - SF^{DM\%}))$.

The amount of waste heat not used in winter (UWH^W) and summer (UWH^S) was also calculated using the IF function, whereby the function is as follows:

UWH^W = using the IF function, where:

$$UWH^W = \begin{cases} \left(\frac{\frac{WH^d - DH^W}{1,1}}{DD^{DM\%} - SF^{DM\%}} - SF^d \right) \times 1,1; & \frac{WH^d - DH^W}{1,1} / (DD^{DM\%} - SF^{DM\%}) \geq SF^d \\ 0; & \frac{(WH^d - DH^W)}{1,1} / (DD^{DM\%} - SF^{DM\%}) < SF^d \end{cases}$$

- logic test: $(WH^d - DH^W) \div 1.1 \div (DD^{DM\%} - SF^{DM\%}) \geq SF^d$
- if true: $UWH^W = ((WH^d - DH^W) \div 1.1 \div (DD^{DM\%} - SF^{DM\%}) - SF^d) \times 1.1$, i.e., the difference between the potentially dryable and the actually available separated matter multiplied by the 1.1 kWh/kg constant value.

- **if false**: $UWH^W = 0$, i.e., in this case the total waste heat is used.

Accordingly, the UWH^S quantity was also calculated, with the logical test $(WH^d - DH^S) \div 1.1 \div (DD^{DM\%} - SF^{DM\%}) \geq SF^d$. If the logic test is true then the UWH^S can be calculated using the $((WH^d - DH^S) \div 1.1 \div (DD^{DM\%} - SF^{DM\%}) - SF^d) \times 1.1$ formula, while if the logical test is false, UWH^S is zero.

The total amount of separated matter dried during the winter and summer months was calculated and expressed in pellets (DP) per tonne per year, calculated by using the formula $(DD^W \times 182 + DD^S \times 183)/1000$, where 182 days is the period between the first of October and the end of March (winter), and 183 days the period between the first of April and the end of September (summer).

The calculations were carried out for the year 2016 because we found the closest correlation relationship between the biogas yield calculated by our regression equation and the actual biogas yield $(R^2_{2015} = 0.830; R^2_{2016} = 0.938; R^2_{2017} = 0.571)$.

Figure 1 shows the daily values of the material flow and energy demand of the biogas plant, from the arrival of the raw material to pelleting.

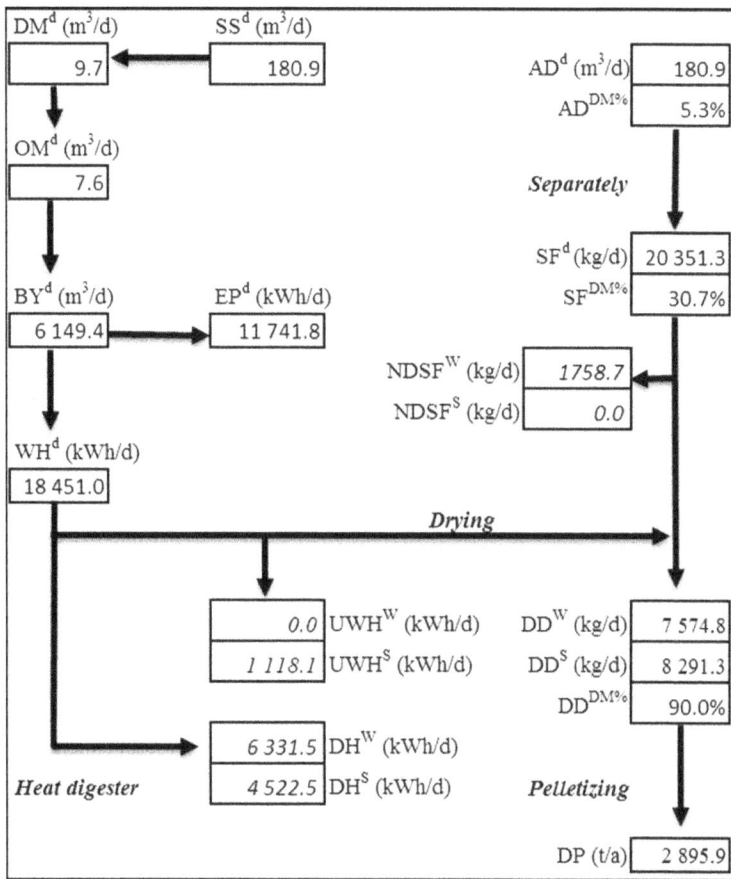

Figure 1. Biogas plant material flow based on the average raw material consumption of the year 2016 (Source: authors' own calculations).

3.3. Substrate Dry Matter Content Optimization

Using the material flow model, we investigated how the increase in the dry matter content of the substrate influences biogas yield and hence thermal energy production. It is expected that an increase in the dry matter content of the substrate will result in an increase in the dry matter and organic content of the separated solid fraction, thus increasing the amount of waste heat generated and at the same time reducing the amount of heat required for pellet drying.

To optimize the dry matter content of the substrate, a linear programming (LP) model was used, similarly to our previous study [16]. Additionally, LP has been used as a model by several authors in biogas plant testing [66,67].

The values of our target function in the LP model are fixed, i.e., sensitivity analysis of the amount of pellet produced from raw material with a 6%, 7%, 8%, 9% and 10% dry matter content. The range is justified by the 5.3% in our basic model and the 10% suggested as an ideal model in Stolze et al. (2015) [9]. The variable cell of the model is the quantity of feedstocks ($SQ^{1...n}$) determined as a limiting condition so that the SS^d value corresponds to the actual data of the plant (180.9 m^3 substrate per day). In the composition of the feedstocks, we have further limitations in that the amount of liquid manure can be changed by ±50% and other feedstocks by −90% and +50% compared to the data from 2016. In the case of maize silage and grass silage which play a significant role in increasing dry matter content, the upper limit is 10 times the 2016 average. In establishing the values of the limiting factors, due to the constant livestock numbers, almost the same internal and highly variable external base raw material was taken into account, and as regards the growth of energy crops at the farm, the maximum coverage area of was also taken into account.

3.4. Methodology of Cost and Income Calculations of Pellet Production

The economic evaluation of the processes used in the material flow model was carried out using cost and income calculations based on our calculated and secondary basic data. Table 5 shows the special invested assets for the processing of mature fermented liquid and their investment costs, amortization (permanent) and operating costs for 3000 to 3500 t pellet production capacity per year. The total investment cost of the equipment is €381,380, the vast majority of which (79%) is the cost of the belt dryer. The amortization cost of the assets was calculated using a 10-year (10% annual) linear depreciation. Within the variable costs, the cost of personnel was 3.4 € per hour, as well as the associated costs which can be considered typical in the agricultural sector in Hungary.

Table 5. Equipment in pellet production and their operating costs.

Name of the Equipment	Investment Cost (€)	Fixed Cost (€/Annual)	Variable Cost (€/t)
Screw press separator [1]	11,370	1137	0.09
Belt dryer [2]	300,000	30,000	1.65
Pelletizer [3]	63,000	6300	37.41
Bagger [3]	7010	701	33.70 *
Total	381,380	38,138	-

* Labour hourly rate: €3.4 + 28.5% personal contributions. Source: [1] Möller et al. (2000) [64]; [2] Bolzonella et al. (2017) [25]; [3] Kröger et al. (2016) [68].

In the case of cost and income calculations, separate calculations were made for the mode of utilization (nutrient or heat energy target), marketability (own use or sales), and—for the purpose of demonstrating the significance of waste heat utilization—calculations were made relating to the use of waste heat or natural gas as auxiliary material during the drying.

The composition of organic fertilization pellet was calculated on the data of Thomas et al. (2017) [56], Ross et al. (2018) [57], Kratzeisen et al. (2010) [54] and Haggard et al. (2005) [60]. Since the nitrogen content significantly reduces during drying, we considered only its phosphorus- (0.8%) and potassium-content (1.5%) for pricing.

Regarding calculation of the value of heating pellet, we considered the prices of agripellets with similar composition. Heating value of digestate pellets: 15 MJ/kg, ash content: 17%, based on Kratzeisen et al. (2010) [54].

In the case of organic fertilizer pellets, two values were taken into account for economic evaluation:

- 22.4 €/t, calculated on the Hungarian and German market prices of different quality pellets, with multivariation linear regression, depending on PK content. ($R^2 = 0.870$; $F = 20.054$; $p = 0.02$)

Equation of the price:

$$-9.54 + 43.345 \times P - 12.08 \times K \tag{12}$$

- We used 90 €/t as a proportional value for accounting the price, which, according to Dahlin et al. (2015) [32] reflects the true economic value when it is used by the producer. In this case, this value can be determined as the product of the nutrient content of the pellets (NPK) and the fertilizer prices.

For heating pellets two prices were also considered:

- 143.8 €/t as the lowest and the most affordable domestic market price of Hungarian agri-food pellets.
- Although we did not take other market prices into account in our calculations, bearing the future in mind, we used a price comparison of 250–318 €/t as the comparison price for straw pellets prices in the UK, which is relevant for export sales.

4. Results and Discussion

Calculations made in this study were based on the results of the case study we discussed in Section 3.1; however, due to the sensitivity analysis, our results can be considered as typical for medium-sized cogeneration plants in general.

4.1. Cost and Income Calculation of Organic Fertilizer and Heating Pellets

Based on the actual data of the biogas plant under investigation (181 m^3, 5.3% dry matter content of fermented liquid per day), according to our calculations, after separation and pelleting, 2896 tons of marketable final products can be produced annually. The sales price of the organic fertilizer pellet according to our statistical tests is based on the internal parameters of 22.4 €/t, which is close to Dahlin et al.'s (2015) [32] calculated market value (20 €/t), which can be expected to reflect the expected price well.

The annual turnover is very different in the case of use of pellets for nutrient or for burning for heat, as is the case with pellets produced for the producer's own use or for sale. Since the annual cost of production and the cost of the final product are independent of the previous four cases, the expected income is also primarily a function of utilization and marketability. At the same time, it would significantly increase the final cost of the final product (by about 46%) if natural gas were used instead of waste heat in the technological process. According to Fürstaller et al. (2010), the use of natural gas can increase the cost of pellets by up to 25% compared to drying with free waste heat [28]. The difference is probably attributable to the price and technological changes which have occurred in the meantime, but we agree with the significance of the decisive importance of waste recovery.

The proportion of fixed costs is relatively modest (10–15% of the cost of production), therefore—while of course efforts should be made for continuous operation—pelleting is less sensitive to capacity utilization than the biogas plant (Table 6).

Table 6. Cost and income calculation of waste heat and natural gas pellet production.

Type	Drying with Waste Heat		Drying with Natural Gas	
	Organic Manure Pellet	Heating Pellet	Organic Manure Pellet	Heating Pellet
Yield (t)	2896	2896	2896	2896
Sales price (€/t)	22.4	143.8	22.4	143.8
Revenue (€)	64,915	416,361	64,915	416,361
Fixed costs (€)	40,152	40,152	40,152	40,152
-Heat energy used (MWh)	4550	4550	4550	4550
-Heat energy used (MWh/t pellet)	1.6	1.6	1.6	1.6
-Heat energy used (€)	-	-	123,402	123,402
Variable costs (€)	223,736	223,736	347,138	347,138
Production costs (€)	263,888	263,888	387 290	387,290
Producer's costs (€/t)	91.1	91.1	133.7	133.7
Income (€)	−198,972	152,473	−322,374	29,071
Winter waste heat utilization (%)	100.0%	100.0%	-	-
Summer waste heat utilization (%)	93.9%	93.9%	-	-

Source: authors' own calculation.

In the basic scenario presented, profitable operation is expected only in the case of heating pellet sales:

- using waste heat: 52.7 EUR/t income
- using natural gas: 10.1 EUR/t income

In the United Kingdom, the use of fire pellets for their producers' own use is largely supported, and savings on natural gas, together with subsidies can reach a combined total of 621 €/t [61]. Within this, the natural gas savings from the burning of one ton of pellets are 167.5 €/t, which is lower than the sales prices (250–318 €/t) in the UK, but far exceeds the cost of pellet production (88–90 €/t), which means that thanks to the support available it is more economical when used by the producer. In Hungary, in the absence of a subsidy for the same purpose, use by the producer can only be recommended in the following cases:

- if the application of leftover fermented liquid to arable land presents problems (the biogas plant has little arable land) or
- if the winter energy demand of the agricultural enterprise exceeds the amount of waste heat available in this period, i.e., in this case the waste heat from the summer can be converted into winter fuel.

In the case of producers using their own pellets produced by waste heat, the producer's cost (91.1 EUR/t) and the value-based price calculated in the literature (EUR 90/t) [32] are nearly the same. The production of nutrient pellets with natural gas and the sale of nutrient pellets will inevitably lead to a loss (-69 + 111 EUR/t).

Taking account of the use of waste heat, the sale of heating pellets can lead to a profit of €152,473 per annum for the biogas plant, and in addition to the sale of green electricity under guaranteed purchasing, the by-products (waste heat, bio-fertilizer) can further improve the profitability of the plant. In the case of nutrient pellets, however, the price differential of more than six times results in a loss of almost €200,000 per annum.

With respect to the organic fertilizer and heating pellets, it was found that 4550 MWh waste heat per year can be used to dry the solid fraction. In winter, waste heat consumption may reach 100%, while in summer the figure is 93.9%, which could be further increased by improving stock management (the undried amount during the winter period is 1.7 t/day). All in all, in a basic case pelleting is suitable for virtually all waste heat.

When comparing drying by waste heat and by natural gas, it can be stated that the use of natural gas is recommended in plants where:

- waste heat from the combustion of biogas is used throughout the whole year (e.g., as technological heat) or is fed to gas pipelines after purification (in these cases there is no usable heat energy generated), and
- when the product is intended to be sold on the fuel heating fuel market.

4.2. The Effect of the Dry Matter Content of the Substrate on Economic Indicators

Further calculations were made to determine the potential for increasing substrate dry matter content (using the LP programme), the results of which are shown in Table 7.

Based on our results, the increase in the dry matter content of the substrate by one percent (between the 6% and 10% interval) increases the amount of pellets which can be produced by 2.5% (74–75 t/year), i.e., the spread fertilizer is able to produce ever greater numbers of pellets.

The reduction in the producer's own cost, although only in the case of a unit increase in dry matter content, is 0.4%, but due to the higher number of pellets, the annual income growth is higher.

Increasing dry matter production does not only affect the demand for drying energy, but also increases the production of heat and electricity in cogeneration units. Based on our calculations, the increase in unit solid content increases thermal energy production on average by 1.05% and reduces the specific heat consumption of pellet production by 0.1 MWh/t. With the reduction of the specific heat consumption, the winter and summer quantities of waste heat that can be used for other purposes also increase significantly. While in the basic case pelleting (or the drying which occurs beforehand) uses practically the total amount of generated waste heat, a one percent increase in dry matter content decreases the available amount of waste heat by 5.0–5.5% in both the winter and the summer. With 10% dry matter, half of the waste heat in the summer and one third in the winter is available for other heat energy utilization, which can increase the biogas plant's profit (mainly in winter).

Table 7. Revenue from pellet sales depending on the different sales directions and substrate dry matter content.

Dry Matter of Substrate (%)	6.0		7.0		8.0		9.0		10.0	
Utilization	Fertilizer	Fuel	Fertilizer	Fuel	Fertilizer	Fuel	Fertilizer	Fuel	Fertilizer	Fuel
DP (t/a)	3075	3075	3150	3150	3224	3224	3298	3298	3372	3372
Income (thousand €)	277	442	283	453	290	464	297	474	304	485
Production cost (th. €)	277	277	282	282	288	288	293	293	298	298
Unit cost (€/t)	**90.1**	**90.1**	**89.7**	**89.7**	**89.2**	**89.2**	**88.8**	**88.8**	**88.4**	**88.4**
Profit (th. €)	**−0.0**	**165**	**1**	**170**	**2**	**176**	**4**	**181**	**5**	**187**
Thermal energy production (MWh/annual)	7 396	7396	7477	7477	7557	7557	7638	7638	7710	7710
Waste heat use in winter (%)	90.0	90.0	84.6	84.6	79.1	79.1	73.5	73.5	68.2	68.2
Waste heat use in summer (%)	71.6	71.6	66.6	66.6	61.7	61.7	56.6	56.6	51.7	51.7
Specific heat utilization (MWh/t pellet)	1.41	1.41	1.31	1.31	1.21	1.21	1.11	1.11	1.01	1.01

Source: authors' own calculations.

According to the results of Fürstaller et al. (2010) [28], the pellet production costs will typically be in the range of 70–90 €/t, which was also confirmed by Shirani-Evans's (2012) [61] estimate of 89.9 €/t. The results of our calculations do not differ significantly from the aforementioned author, as the cost we calculate is between 88.4 and 90.1 €/t in the case of 6–10% substrate dry matter content.

5. Conclusions

The cost of pellet production can be reduced by approximately 46% through the use of waste heat, but even natural gas drying is profitable for heat-energy sales, which is why this is the primary reason for pelleting bio-fertilizer. It should be noted that Western European firewood prices (e.g., UK: 250–318 €/t) are considerably higher than in Hungary, so it is advisable for Hungarian producers to even consider sales in the EU. The economics of producer's own use can be doubtful, since waste heat (without pelleting) is a much cheaper way to meet local heat demands.

The investment cost of about €381 thousand (Table 5) is expected to be recovered in the case of heating pellets produced by waste heat over a period of 3 years if the total quantity is sold on the market.

Pelleting is not only suitable for the full or near recovery of the waste heat in the winter period, but is also capable of utilizing the bulk of the waste heat in the summer period, and thus helps to produce marketable products from bio fertilizer that would otherwise only be used locally. 10% of the substrate dry matter content already delivers a significant amount of waste heat savings, which can be used for biogas and linked farm processes and can be a source of considerable savings.

In the case of sales for nutrition, the substitution value for the producer's own use instead of sales is much higher (about 90 €/t) and is approximately the same as the producer's own costs, but this does not justify its use compared to when it is used without pelleting (as spread fertilizer). In addition to pelleting it might be used for precision cultivation technology and for better storage. All in all, utilization for nutrition can only be considered as a kind of "necessary evil" in the absence of thermal energy utilization/sales or land available for fertilization.

Author Contributions: Dávid Nagy participated in data collection, in the literature review, in the economic model and in the economic calculations. Péter Balogh performed the statistical analyses. Zoltán Gabnai participated in literature review and formatted the manuscript. József Popp revised and finalized the manuscript and participated in the development of the research concept. Judit Oláh revised and finalized the manuscript. Attila Bai participated in economic calculations and analysis, in literature review, and performed the conclusions.

Acknowledgments: This work was supported by the ÚNKP-17-2 New National Excellence Program of the Ministry of Human Capacities (Hungary).

Conflicts of Interest: The authors declare no conflict of interest.

Abbreviations

ADd	daily quantity of anaerobic digestate (m^3/day)
ADDM%	dry matter of anaerobic digestate (%)
BYd	daily biogas yield (m^3/day)
CHP	combined heat and power
DDDM%	dry matter of dried digestate (constant value: 90%)
DDS	quantity of dried digestate per day in summer period (kg/day)
DDW	quantity of dried digestate per day in winter period (kg/day)
DHS	daily heating demand of digester in summer period (kWh/day)
DHW	daily heating demand of digester in winter period (kWh/day)
DM%	dry matter (%)
DMd	daily quantity of dry matter loaded into the digester (m^3/day)
DP	quantity of digestate pellet per year (tons/year)
EPd	daily electricity production (kWh/day)
GHG	greenhouse gases
HRT	hydraulic retention time (day)

MMT	million metric tons
NDSFS	quantity of non-dried solid digestate per day in summer period (kg/day)
NDSFW	quantity of non-dried solid digestate per day in winter period (kg/day)
NPK	Nitrogen-Phosphorus-Potassium
OLR	organic loading rate
OM%	organic matter (%)
OMd	daily quantity of organic matter loaded into the digester (m^3/day)
SFd	quantity of solid fraction of digestate per day (kg/day)
SFDM%	dry matter of solid separated digestate (%)
SQ	daily quantity of substrate component (m^3/day)
SSd	substrate fed into the digester (m^3/day)
UWHS	quantity of unused waste heat per day in summer period (kWh/day)
UWHW	quantity of unused waste heat per day in winter period (kWh/day)
WHd	daily thermal energy production (kWh/day)

References

1. Pflüger, S. European biogas industry now and beyond 2020. In Proceedings of the EBA Workshop: Contribution of Biogas towards European Renewable Energy Policy beyond 2020, Brussels, Belgium, 8 February 2017.
2. European Commission (EC). *Proposal for a DIRECTIVE OF THE EUROPEAN PARLIAMENT AND OF THE COUNCIL on the Promotion of the Use of Energy from Renewable Sources*; Proposal, R.I., Ed.; European Commission: Brussels, Belgium; Luxembourg, 2016.
3. Eurostat. *Primary Production—All Products—Annual Date*; Wood Pellets; Eurostat: Luxembourg, 2017.
4. Mergner, R.; Rutz, D.; Wagner, I.; Amann, S.; Amann, C.; Kulišić, B.; Abramović, J.M.; Pozar, H.; Vorisek, T.; Bailón Allegue, I.; et al. *European Strategy Paper on Heat Use from Biogas Plants*; WIP Renewable Energies: Munich, Germany, 2013; p. 22.
5. Ramanauskaite, R.; Rutz, D.; Amann, S.; Amann, C.; Abramovic, J.M.; Vorisek, T.; Bailón Allegue, L.; Hinge, J.; Dzene, I.; De Filippi, F. *Biogas Markets and the Use of Heat of Biogas Plants in Austria, Croatia, Czech Republic, Denmark, Germany, Italy, Latvia, Poland and Romania*; BiogasHeat WP; WIP Renewable Energies: Munich, Germany, 2012.
6. Carlini, M.; Mosconi, E.M.; Castellucci, S.; Villarini, M.; Colantoni, A. An Economical Evaluation of Anaerobic Digestion Plants Fed with Organic Agro-Industrial Waste. *Energies* **2017**, *10*, 1165. [CrossRef]
7. Lindkvist, E.; Johansson, M.T.; Rosenqvist, J. Methodology for Analysing Energy Demand in Biogas Production Plants—A Comparative Study of Two Biogas Plants. *Energies* **2017**, *10*, 1822. [CrossRef]
8. Kokkoli, A.; Zhang, Y.; Angelidaki, I. Microbial electrochemical separation of CO_2 for biogas upgrading. *Bioresour. Technol.* **2018**, *247*, 380–386. [CrossRef] [PubMed]
9. Stolze, Y.; Zakrzewski, M.; Maus, I.; Eikmeyer, F.; Jaenicke, S.; Rottmann, N.; Siebner, C.; Pühler, A.; Schlüter, A. Comparative metagenomics of biogas-producing microbial communities from production-scale biogas plants operating under wet or dry fermentation conditions. *Biotechnol. Biofuels* **2015**, *8*, 14. [CrossRef] [PubMed]
10. Ghimire, P.C. SNV supported domestic biogas programmes in Asia and Africa. *Renew. Energy* **2013**, *49*, 90–94. [CrossRef]
11. Bond, T.; Templeton, M.R. History and future of domestic biogas plants in the developing world. *Energy Sustain. Dev.* **2011**, *15*, 347–354. [CrossRef]
12. Martí-Herrero, J.; Chipana, M.; Cuevas, C.; Paco, G.; Serrano, V.; Zymla, B.; Heising, K.; Sologuren, J.; Gamarra, A. Low cost tubular digesters as appropriate technology for widespread application: Results and lessons learned from Bolivia. *Renew. Energy* **2014**, *71*, 156–165. [CrossRef]
13. Bayerisches Landesamt für Umwelt. Biogashandbuch Bayern–Materialienband, Augsburg 2007. Available online: http://www.lfu.bayern.de/abfall/biogashandbuch (accessed on 11 December 2017).
14. Weiland, P. Biomass digestion in agriculture: A successful pathway for the energy production and waste treatment in Germany. *Eng. Life Sci.* **2006**, *6*, 302–309. [CrossRef]

15. Khanal, S.K. Bioenergy generation from residues of biofuel industries. In *Anaerobic Biotechnology for Bioenergy Production: Principles and Applications*; Wiley-Blackwell: Oxford, UK; John Wiley & Sons: New York, NY, USA, 2008; pp. 161–188.
16. Mézes, L.; Bai, A.; Nagy, D.; Cinka, I.; Gabnai, Z. Optimization of Raw Material Composition in an Agricultural Biogas Plant. *Trends Renew. Energy* **2017**, *3*, 61–75. [CrossRef]
17. Onthong, U.; Juntarachat, N. Evaluation of Biogas Production Potential from Raw and Processed Agricultural Wastes. *Energy Procedia* **2017**, *138*, 205–210. [CrossRef]
18. Aramrueang, N.; Rapport, J.; Zhang, R. Effects of hydraulic retention time and organic loading rate on performance and stability of anaerobic digestion of Spirulina platensis. *Biosyst. Eng.* **2016**, *147*, 174–182. [CrossRef]
19. Kumar, K.; Ghosh, S.; Angelidaki, I.; Holdt, S.L.; Karakashev, D.B.; Morales, M.A.; Das, D. Recent developments on biofuels production from microalgae and macroalgae. *Renew. Sustain. Energy Rev.* **2016**, *65*, 235–249. [CrossRef]
20. Rutz, D.; Mergner, R.; Janssen, R. *Sustainable Heat Use of Biogas Plants*; Euroheat & Power: Brussels, Belgium, 2015; ISBN 978-3-936338-35-5.
21. Elsenbruch, T. *Jenbacher Gas Engines a Variety of Efficient Applications*; GE Power: Atlanta, GA, USA, 2010; p. 73.
22. Deutsches Biomasseforschungszentrum (DBFZ). *Guide to Biogas from Production to Use*; Deutsches Biomasse Forschungs Zentrum, Fachagentur Nachwachsende Rohstoffe e.V: Gülzow, Germany, 2012; p. 229.
23. Kirchmeyr, F.; Anzengruber, G. Leitfaden zur Wärmenutzung bei Biogasanlagen. *ARGE Kompost und Biogas Österreich.* 2008. Available online: https://www.klimaaktiv.at/dam/jcr:75e19d28-6058-4d52-a4fe-4e9ef91ead7e/Leitfaden_Waermenutzung%20bei%20Biogasanlagen.pdf (accessed on 11 January 2018).
24. Pullen, T. *Anaerobic Digestion–Making Biogas–Making Energy: The Earthscan Expert Guide*; Routledge: Abingdon, UK, 2015; ISBN 1317673409.
25. Bolzonella, D.; Fatone, F.; Gottardo, M.; Frison, N. Nutrients recovery from anaerobic digestate of agro-waste: Techno-economic assessment of full scale applications. *J. Environ. Manag.* **2018**, *216*, 111–119. [CrossRef] [PubMed]
26. Sethupathi, S.; Zhang, M.; Rajapaksha, A.U.; Lee, S.R.; Mohamad Nor, N.; Mohamed, A.R.; Al-Wabel, M.; Lee, S.S.; Ok, Y.S. Biochars as Potential Adsorbers of CH_4, CO_2 and H_2S. *Sustainability* **2017**, *9*, 121.
27. Huttunen, S.; Manninen, K.; Leskinen, P. Combining biogas LCA reviews with stakeholder interviews to analyse life cycle impacts at a practical level. *J. Clean. Prod.* **2014**, *80*, 5–16. [CrossRef]
28. Fürstaller, A.; Huber, M.; Krueger, J.; Pfleger, M. Processing of digestate to pellets for usage as alternative solid fuel. In Proceedings of the 18th European Conference and Exhibition, Lyon, France, 3–7 May 2010; pp. 3–7.
29. Lukehurst, C.T.; Frost, P.; Al Seadi, T. *Utilisation of Digestate from Biogas Plants as Biofertiliser*; IEA Bioenergy: Paris, France, 2010; pp. 1–36.
30. ARBOR. Inventory: Techniques for Nutrient Recovery from Digestate. 2013. Available online: http://arbornwe.eu/downloads (accessed on 24 October 2017).
31. Oenema, O.; Tamminga, S. Nitrogen in global animal production and management options for improving nitrogen use efficiency. *Sci. China Ser. C Life Sci.* **2005**, *48*, 871–887.
32. Dahlin, J.; Herbes, C.; Nelles, M. Biogas digestate marketing: Qualitative insights into the supply side. *Resour. Conserv. Recycl.* **2015**, *104*, 152–161. [CrossRef]
33. KTBL. *Umweltgerechte, Innovative Verfahren zur Abtrennung von Nährstoffen aus Gülle und Gärrückständen—Technologischer Stand, Perspektiven und Entwicklungsmöglichkeiten*; KTBL: Darmstadt, Germany, 2008.
34. Drosg, B.; Fuchs, W.; Al Seadi, T.; Madsen, M.; Linke, B. *Nutrient Recovery by Biogas Digestate Processing*; IEA Bioenergy: Paris, France, 2015; pp. 7–11.
35. Agrárgazdasági Kutató Intézet (AKI). *Agrárpiaci Jelentések, Gabona és Ipari Növények (Agrarian Market Reports, Grain and Industrial Plants)*; AKI: Budapest, Hungary, 2015; p. 31.
36. AEBIOM. Statistical Report 2017. *European Bioenergy Outlook*. 2017. Available online: http://www.aebiom.org/wp-content/uploads/2017/10/KF17-v2.pdf (accessed on 12 January 2018).
37. Flach, B.; Lieberz, S.; Rossetti, A. *EU Biofuels Annual 2017*; GAIN Report; USDA Foreign Agricultural Service, Global Agricultural Information Network: Washington, DC, USA, 2017; p. 44.

38. GTIS, World Trade Atlas Database. Global Trade International Service. 2017. Available online: http://www.worldtradestatistics.com/english/gtis.html (accessed on 21 February 2018).
39. SFCCI. *Sustainable Forest Management Criteria & Indicators*; Standing Forestry Committee: 2015. Available online: https://ec.europa.eu/agriculture/sites/agriculture/files/forest/publications/pdf/sfcci-report_en.pdf (accessed on 10 February 2018).
40. Saveyn, H.; Eder, P. *End-of-Waste Criteria for Biodegradable Waste Subjected to Biological Treatment (Compost & Digestate): Technical Proposals*; European Commission, Joint Research Centre, Institute for Prospective Technological Studies: Sevilla, Spain, 2014.
41. Vaneeckhaute, C.; Meers, E.; Michels, E.; Buysse, J.; Tack, F. Ecological and economic benefits of the application of bio-based mineral fertilizers in modern agriculture. *Biomass Bioenergy* **2013**, *49*, 239–248. [CrossRef]
42. Jones, P.; Salter, A. Modelling the economics of farm-based anaerobic digestion in a UK whole-farm context. *Energy Policy* **2013**, *62*, 215–225. [CrossRef]
43. Luyten-Naujoks, K. BGK *Rechtliche Aspekte des Inverkehrbringens von Gärprodukten zu Düngezwecken*; Berlin, Germany, 2015. Available online: https://veranstaltungen.fnr.de/fileadmin/allgemein/pdf/veranstaltungen/gaerreste2015/16_Luyten-Naujoks.pdf (accessed on 3 February 2018).
44. Dahlin, J.; Halbherr, V.; Kurz, P.; Nelles, M.; Herbes, C. Marketing green fertilizers: Insights into consumer preferences. *Sustainability* **2016**, *8*, 1169. [CrossRef]
45. Al Seadi, T.; Lukehurst, C. *Quality Management of Digestate from Biogas Plants Used as Fertiliser*; IEA Bioenergy: Paris, France, 2012; Volume 37, p. 40.
46. Bauer, A.; Mayr, H.; Hopfner-Sixt, K.; Amon, T. Detailed monitoring of two biogas plants and mechanical solid–liquid separation of fermentation residues. *J. Biotechnol.* **2009**, *142*, 56–63. [CrossRef] [PubMed]
47. Möller, K.; Schulz, R.; Müller, T. Substrate inputs, nutrient flows and nitrogen loss of two centralized biogas plants in southern Germany. *Nutr. Cycl. Agroecosystems* **2010**, *87*, 307–325. [CrossRef]
48. Fechter, M.; Kraume, M. *Digestate Treatment Techniques*; Technical Transactions Mechanics; 2016, Volume 1-M, pp. 95–106. Available online: http://www.ejournals.eu/pliki/art/7547/ (accessed on 12 February 2018).
49. Møller, H.B.; Sommer, S.G.; Ahring, B.K. Separation efficiency and particle size distribution in relation to manure type and storage conditions. *Bioresour. Technol.* **2002**, *85*, 189–196. [CrossRef]
50. Tambone, F.; Orzi, V.; D'Imporzano, G.; Adani, F. Solid and liquid fractionation of digestate: Mass balance, chemical characterization, and agronomic and environmental value. *Bioresour. Technol.* **2017**, *243*, 1251–1256. [CrossRef] [PubMed]
51. Mézes, L. *Mezőgazdasági és élelmiszeripari biogáz-termelés optimalizálása. (Optimizing the Agricultural and Food Industrial Biogas Production)*; University of Debrecen: Debrecen, Hungary, 2011.
52. Turley, D.; Hopwood, L.; Burns, C.; Di Maio, D. *Assessment of Digestate Drying as an Eligible Heat Use in the Renewable Heat Incentive*; NNFCC: York, UK, 2016; p. 28.
53. Awiszus, S.; Meissner, K.; Reyer, S.; Müller, J. Ammonia and methane emissions during drying of dewatered biogas digestate in a two-belt conveyor dryer. *Bioresour. Technol.* **2018**, *247*, 419–425. [CrossRef] [PubMed]
54. Kratzeisen, M.; Starcevic, N.; Martinov, M.; Maurer, C.; Müller, J. Applicability of biogas digestate as solid fuel. *Fuel* **2010**, *89*, 2544–2548. [CrossRef]
55. Nest, T.V.; Ruysschaert, G.; Vandecasteele, B.; Cougnon, M.; Merckx, R.; Reheul, D. P availability and P leaching after reducing the mineral P fertilization and the use of digestate products as new organic fertilizers in a 4-year field trial with high P status. *Agric. Ecosyst. Environ.* **2015**, *202*, 56–67. [CrossRef]
56. Thomas, B.W.; Li, X.; Nelson, V.; Hao, X. Anaerobically digested cattle manure supplied more nitrogen with less phosphorus accumulation than undigested manure. *Agron. J.* **2017**, *109*, 836–844. [CrossRef]
57. Ross, C.-L.; Mundschenk, E.; Wilken, V.; Sensel-Gunke, K.; Ellmer, F. Biowaste Digestates: Influence of Pelletization on Nutrient Release and Early Plant Development of Oats. *Waste Biomass Valorization* **2018**, *9*, 335–341. [CrossRef]
58. Ball, B.; McTaggart, I.; Scott, A. Mitigation of greenhouse gas emissions from soil under silage production by use of organic manures or slow-release fertilizer. *Soil Use Manag.* **2004**, *20*, 287–295. [CrossRef]
59. Hammac II, W.; Wood, C.; Wood, B.; Fasina, O.; Feng, Y.; Shaw, J. Determination of bioavailable nitrogen and phosphorus from pelletized broiler litter. *Sci. Res. Essays* **2007**, *2*, 89–94.
60. Haggard, B.E.; DeLaune, P.B.; Smith, D.R.; Moore, P.A. Nutrient and β17-estradiol loss in runoff water from poultry litters. *JAWRA J. Am. Water Resour. Assoc.* **2005**, *41*, 245–256. [CrossRef]

61. Shirani, A.; Evans, M. *Driving Innovation in AD Optimisation—Uses for Digestates*; WRAP: Banbury, UK, 2012.
62. Verma, V.; Bram, S.; Delattin, F.; Laha, P.; Vandendael, I.; Hubin, A.; De Ruyck, J. Agro-pellets for domestic heating boilers: Standard laboratory and real life performance. *Appl. Energy* **2012**, *90*, 17–23. [CrossRef]
63. Al Seadi, T. *Good Practice in Quality Management of AD Residues from Biogas Production*, Report Made for the International Energy Agency, Task 24-Energy from Biological Conversion of Organic Waste; IEA Bioenergy; AEA Technology Environment: Oxfordshire, UK, 2001.
64. Møller, H.B.; Lund, I.; Sommer, S.G. Solid–liquid separation of livestock slurry: Efficiency and cost. *Bioresour. Technol.* **2000**, *74*, 223–229. [CrossRef]
65. Sáez, J.A.; Clemente, R.; Bustamante, M.Á.; Yañez, D.; Bernal, M.P. Evaluation of the slurry management strategy and the integration of the composting technology in a pig farm–Agronomical and environmental implications. *J. Environ. Manag.* **2017**, *192*, 57–67. [CrossRef] [PubMed]
66. García-Gen, S.; Rodríguez, J.; Lema, J.M. Optimisation of substrate blends in anaerobic co-digestion using adaptive linear programming. *Bioresour. Technol.* **2014**, *173*, 159–167. [CrossRef] [PubMed]
67. Gebrezgabher, S.A.; Meuwissen, M.P.; Prins, B.A.; Lansink, A.G.O. Economic analysis of anaerobic digestion—A case of Green power biogas plant in The Netherlands. *NJAS Wagening. J. Life Sci.* **2010**, *57*, 109–115. [CrossRef]
68. Kröger, R.; Reckermann, M.; Schaper, C.; Theuvsen, L. Gärreste als Gartendünger vermarkten? *Berichte über Landwirtschaft-Zeitschrift für Agrarpolitik und Landwirtschaft* **2016**, *94*. Available online: http://buel.bmel.de/index.php/buel/article/view/99/Kröger_Gärreste.html (accessed on 22 February 2018).

Article

Gasification of Agroresidues for Syngas Production

Nadia Cerone and Francesco Zimbardi *

ENEA, Energy Technologies Department, ss Jonica 106, 75026 Rotondella, Italy; nadia.cerone@enea.it
* Correspondence: francesco.zimbardi@enea.it; Tel.: +39-083-597-4486

Received: 13 April 2018; Accepted: 14 May 2018; Published: 17 May 2018

Abstract: Biomass residues from agriculture and agroindustry are suitable sources for the production of energy because they don't compete with the food chain and they are produced cheaply. Their transformation into heat and power or energy vectors depends on morphology and composition. Shells of almonds and hazelnuts can be easily gasified in fixed beds because of their low fines content and high gas permeation. In this work we investigated the overall process performances and syngas composition, especially the H_2/CO ratio, by changing the air and steam supply. The tests were carried out in a pilot updraft gasifier having a capacity of treating up to 20–30 kg/h of biomass. Experimental data were worked out by surface response analysis as function of the equivalence ratios (ER) in relation to the complete combustion and water reaction. By using only air at $ER(O_2)$ 0.24 the ratio H_2/CO in the syngas was 0.33 while adding steam at $ER(H_2O)$ 0.28 the ratio reached a value of 1.0. The energy conversion efficiency from solid to gas and oils reached maximum values of 76% and 28%, respectively. As anticipated by TGA, hazelnut shells produced less organic volatiles and gas efficiency was generally higher than for almond shells.

Keywords: gasification; agroresidues; syngas; biorefinery; steam gasification; oxygasification; biofuel

---

1. Introduction

Syngas, or synthetic gas, is a flexible feedstock for the chemical and energy industry. Syngas is a mixture of carbon monoxide, hydrogen, carbon dioxide and lower percentages of incondensable hydrocarbons, like methane. Liquid fuels can be produced from syngas trough the Fischer–Tropsch synthesis (FT) a catalytic reaction discovered over ninety years ago but still very relevant because of the demand for clean fuels and the emerging affordability of converting on-site natural gas [1]. The high content of hydrogen and carbon monoxide makes syngas suitable for feeding fuel cells and Integrated Gasification Combined Cycles (IGCCs) that efficiently generate power [2]. Not only is syngas a flexible energetic and chemical vector, but also the process that generates it, called gasification, is flexible regarding the type of feedstock. Virtually any kind of material containing carbon can be gasified: coal, biomass, urban wastes. The search for renewable and sustainable sources has made lignocellulosics from forestry, agriculture and waste, an interesting source of bioenergy. According to the International Energy Association (IEA) in 2015 bioenergy accounted for 11.2% of the world's total fuel consumption, i.e., 1051 Mtoe, which corresponds to an absolute increase of 72% compared to the 1975 data [3]. Several processes for thermal conversion of biomass, which provide directly heat, power or even energy vectors, like pyrolytic oil, torrefied biomass, and syngas, are available. Also, in the process of biorefining to produce liquid fuels or other chemicals with the fermentation route, the residual stream of lignin and other unconverted fibers is thermally converted [4,5]. Coal gasification technologies cannot be simply adapted to biomass gasification because of the different characteristic of the feedstocks, such as particle density and pyrolysis behavior [6,7]. Moreover, the business model is quite different because biomass resources are territorially diffused, seasonal, and available from fragmented ownership [8,9]. Fixed bed gasification is one of the most suitable process to exploit lignocellulosics because it can be carried out at small scale (<1 MW$_{th}$) with a relatively simple and robust technology; for this reason plants based on this technology can either be included in smart energy networks or be utilized by isolated small communities [10]. In fixed bed gasifiers, the syngas can be draft from the bottom (downdraft type) or from the top (updraft type). The latter process is characterized by higher energy efficiency conversion and enhanced flexibility of feedstock, in terms of moisture content, size, composition [11,12]. Recent advancements in feeding low bulk density biomass, tar reduction and tar reforming, provide even more chances for updraft gasifiers to be used with larger pool of residual feedstock and to achieve higher conversion in syngas [13–16]. One of the most important characteristics that defines the quality of the syngas is the ratio H_2/CO, which largely affects its further use in FT processes. Moreover, in the early 90s it was discovered that syngas could be used to obtain ethanol by fermentation. Recently, this combined thermochemical-biochemical route gained new interest because of innovative processes developed to obtain biofuels such as ethanol or butanol at a commercial scale [17]. The mechanisms of such bioreactions are still under investigation; however, initial observations suggest that the use of microorganisms allow some flexibility of gas composition without influencing the final product [18]. Nevertheless, the optimal syngas composition is not clear as the process requires transfer of soluble gas into the liquid phase. Tuning syngas composition is a desirable feature of biomass gasification; it can be achieved in several ways, including by introducing steam in the system, by using oxygen or enriched air, by a special reactor design (indirect gasification), by adding a catalytic step of Water Gas Shift (WGS) or a CO_2 adsorption step. The introduction of steam is the simplest way of tuning syngas composition, in particular the H_2/CO ratio. Moreover, the use of steam improves the thermal stability of the fixed bed and contributes to avoiding ash melting [4,19]. The effect of steam addition in the fixed bed gasifier has not been systematically investigated. Generally, in reporting steam gasification experiments the ratio steam/biomass is provided; this method does not take into account the difference in chemical composition of different feedstocks that require specific amount of water for a stoichiometric conversion. Indeed, while the effect of oxygen is generally discussed in terms of Equivalence Ratio, $ER(O_2)$ i.e., the ratio of used oxygen in relation to the stoichiometric for full combustion of the biomass, the analogous parameter $ER(H_2O)$ (i.e., the ratio of used water in relation to the stoichiometric value for full conversion of biomass into H_2 and CO_2) has only recently been introduced [19]. Nut shells are a residue of particular economic interest for small scale gasification because it they are produced at agro industries in high amounts; for example, the average shell:kernel weight yield is 1:2 for almond shells and 1:0.8 for hazelnut shells (referred to in this paper as M and N, respectively) [20,21]. Though with low bulk density, these shells are stiff and when broken to extract the kernels the resulting pieces still have the size and shape to allow good gas permeation through the fixed bed. For this reason, M and N are suitable for testing and modeling fixed bed and fluidized bed gasifiers. Gasification of N with air was successfully carried out in downdraft mode, without sign of bridging or ash fusion in the bed; moreover, the quality of syngas was good in terms of calorific value [22]. Fluidized bed reactors were used to gasify crumbled M; these experiments highlighted the importance of particle size in this kind of gasifier because of thermal resistance [23] and showed that with the same substrate naturally occurred catalytic substances enhanced the yield of syngas and reduced its tar content by reforming and cracking reactions [24].The aim of this paper is to investigate the gasification of M and N, which have optimal gas permeation but different chemical compositions, in order to highlight the effects of both the ERs on process yields, energy efficiency and syngas quality. In particular, from a literature survey, the updraft gasification of these residues has not been previously reported in detail and at pilot scale.

2. Materials and Methods

2.1. Feedstock

The feedstock M and N were purchased by local agroindustry. The average size of the M was 2.5–3.0 cm in length, 2 cm in width and 2–3 mm in thickness. The N shells showed a more homogeneous size than the almond shells and retained the initial shape after the particle size reduction process.

The average dimensions were 1.5 cm in length, 1.5 cm in width with a thickness of 1 mm. The chemical and physical characterization is reported in Table 1, along with the methods used.

Table 1. Characteristics and properties of the agroresidues.

	Almond Shells (M)	Standard Deviation	Hazelnut SHELLS (N)	Standard Deviation	Method
Bulk density, kg/m³	417		299		ASTM E873
HHV MJ/kg	19.5	0.1	19.4	0.1	ISO 1928
LHV [a], MJ/kg	18.1	0.1	17.8	0.1	
Moisture [b], %	11.8		5		
Fix carbon, %	18.2	0.2	20.9	0.4	ASTM D 3172
Volatile, %	80.6	0.2	78	0.4	ASTM D 3175
Ash, %	1.2	0.01	1.1	0.01	ASTM D 1102 (600 °C)
Hexosans %	31.2	0.8	22.2	0.7	NREL/TP-510-42623
Pentosans %	28.0	0.8	12.2	0.1	NREL/TP-510-42623
Lignin (Klason)%	30.2	0.3	40.9	0.5	TAPPI 13M-54
Lignin (Klason) ac. sol. %	1.98	0.1	1.3	0.1	TAPPI T250
C %	47.9	0.1	50.5	0.1	UNI EN 15104
H %	6.3	0.2	6.64	0.05	UNI EN 15104
N %	0.36	0.05	1.7	0.4	UNI EN 15104
O [c] %	45.4		40.0		
Cl %	nd		0.025	0.005	UNI EN 15289
S (ppm)	145	4	250	40	UNI EN 15289
Si (ppm)	484	40	106	6	CEN/TC 343
Al (ppm)	67.5	3.7	54	4	CEN/TC 343
Fe (ppm)	1136	60	380	30	CEN/TC 343
Ca (ppm)	941	26	3019	95	CEN/TC 343
K (ppm)	3513	200	2560	120	CEN/TC 343
Mg (ppm)	164	24	229	10	CEN/TC 343
Na (ppm)	nd		<53		CEN/TC 343
P (ppm)	nd		162	10	CEN/TC 343
Ni (ppm)	63.3	0.7	46	5	CEN/TC 345
Cr (ppm)	nd		78	9	CEN/TC 345
Mn (ppm)	13.4	0.6	23	2	CEN/TC 345
Zn (ppm)	26.1	0.6	4.5	0.3	CEN/TC 345
Pb (ppm)	nd		<2.2		CEN/TC 345
Cu (ppm)	9.4	0.1	7.4	0.3	CEN/TC 345
Mo (ppm)	8.6	0.1	nd		CEN/TC 345
Co (ppm)	nd		<2.19		CEN/TC 345
Cd (ppm)	nd		<2.19		CEN/TC 345
H/C, mol/mol	1.58		1.58		
O/C, mol/mol	0.71		0.59		
H₂O for oxidation, kg/kg	0.93		1.06		
O₂ for combustion, kg/kg	1.33		1.40		
Air for combustion, kg/kg	5.75		6.08		

[a] Calculated from HHV and H content; [b] As arrived; [c] By difference: $100 - (ash + C\% + H\% + N\%)$.

2.2. The Updraf Gasifier

Gasification tests were carried out in a pilot plant called PRAGA (for uP dRAft GAsification) designed and built at the ENEA Research Center of Trisaia (Rotondella, MT, Italy). The heart of the plant is the autothermic reactor operating slightly above atmospheric pressure (Figure 1). The typical gasification capacity is 20–30 kg/h of lignocellulosic biomass.

The gasifier is made of a 2.4 m cylindrical steel tube having external diameter of 0.5 m. The inner wall of the reactor is coated with a layer of insulating material of 0.1 m, the internal diameter of the reactor is 0.3 m. The lower part is shaped like a cone to collect and discharge the ash through a steel grid fixed at 0.70 m from the bottom.

Air, oxygen or mixtures of these with steam are introduced from the lower part of the reactor, under the steel grid that supports the biomass bed. Three infrared lamps right above the grid are used to ignite the biomass during the start up of the process. The steam is supplied by an external boiler that produces superheated steam at 160 °C and 1.2 bar.

The temperature profile along the reactor axis is measured by 11 thermocouples positioned in a steel probe at 147 and 303, 459, 615, 771, 979, 1187, 1395, 1603, 1759 and 1815 mm from the grid.

Average temperature in the bed and freeboard were calculated from the corresponding values given by the first seven and last four thermocouples respectively.

Through an active/passive control and supervision system (DCS) it is possible to remotely monitor the entire process by continuously recording the instrumental output (temperatures, pressures, mass flows) that can then be analyzed offline.

The gas leaves the reactor at the top and is conveyed to a cleaning and cooling section consisting of a scrubber containing biodiesel and two coalescing filters which remove the drops of liquid. The plant is equipped with sampling lines to convey a small flow of gaseous streams to the chromatograph (GC) for on-line analysis of non-condensable gases.

In addition to the gasification section and the purification section, the plant includes the upgrading section and the hydrogen separation section [4].

2.3. TGA of Feedstock and Data Processing

The thermal analysis of the M and N was carried out using an apparatus for the TGA micro (model TGA7, Perkin Elmer, Waltham, MA, USA) loading in each test 2–5 mg of ground sample. As a standard procedure, the heating rate was 10 °C/min from 60 °C to 900 °C, in 20 mL/min of gas flow. The temperature calibration of the device is based on the Curie temperature of nickel (354 °C) and iron (970 °C) and was repeated when the ferromagnetic transition of the nickel was detected in a temperature range that deviated more than 1 °C from the expected (theoretical) value. Under valid test conditions, weight loss in two tests performed with the same temperature profile, gas flow and initial weight, was reproduced with an average displacement of 0.5% by weight. The data were worked out with calculation sheets [25]. The 3D plots were obtained with the free program Essential Regression and Experimental Design for Chemists and Engineers, which works as a macro in the Excel ver. 2010 (Microsoft Redmond, WA, USA. The reported surfaces were the quadratic curves interpolating experimental data.

2.4. Gasification Test Procedure

Prior to the execution of the gasification tests, the feeding system was calibrated to determine the feed ratio, kg/h of dry substance, at a fixed engine speed for each type of biomass tested. The start of each test took place with the biomass gasifier filled up to an internal height of 1.3 m from the grate. The biomass was ignited by infrared lamps. After about an hour and a half, the air flow is reduced to the desired value and the process reaches a steady state at the conditions set for the test.

During the experiments, the fuel is supplied to the gasifier by means of the augers in semi-continuous filling the intermediate stub pipe at regular intervals and completed in a few seconds (typically 4.2–4.5 kg/h are loaded every 12 min), in a nitrogen atmosphere to avoid gas leaks produced in the fuel system. This process caused a variation of 5% of the height of the biomass bed. About 24 h after the end of the test, the gasifier was discharged from the bottom, ash and char were collected, weighed, and analyzed separately. More details on the procedures are reported elsewhere [4,19,25].

2.5. Sampling and Chemical Analyses

During the gasificaton tests, the composition of the conditioned syngas was measured by sampling the gaseous stream at the scrubber outlet and analyzed onsite with GC apparatus; the standard deviation for the incondensable gas analyzed by GC on three measurements was 2%. The condensable volatile organic products and water, sampled at the outlet of the gasifier and scrubber, were absorbed in a series of bottles containing isopropanol at room temperature and at −20 °C, according to the method CEN/TS15439). The tar and the water were determined in a solution filtered with the gravimetric method and Karl Fisher titration, respectively; the standard deviation of these determinations was 5%.

The reported gas volume refers to the standard temperature and pressure conditions (STP, 273.15 K and 105 Pa). The flows of air, steam, oxygen and synthesis gas were measured with calibrated instruments accredited for a maximum error of 1% on the measured mass. More details about sampling and chemical analyses are reported elsewhere [4,19]

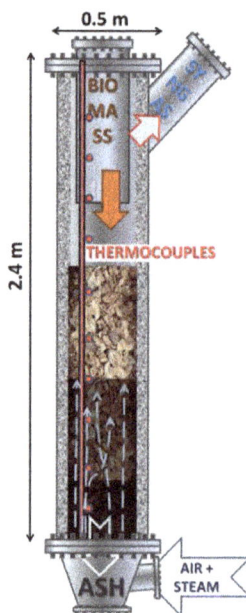

Figure 1. Updraft gasifier scheme.

3. Results and Discussion

3.1. Thermogravimetric Analysis

Gasification is the result of several chemical reactions, some taking place in the solid phase, others in the gas and others at the interface. The reference reactions that are currently used to describe the process are shown in Table 2.

Table 2. Main reactions of gasification.

Reaction	Stoichiometry	Enthalpy
Combustion	$C_{(grafite)} + O_2 \leftrightarrow CO_2$	$\Delta H = -393.5 \text{ kJ/mol}$
Partial combustion	$C_{(grafite)} + \frac{1}{2} O_2 \leftrightarrow CO$	$\Delta H = -110.5 \text{ kJ/mol}$
Bouduard reaction	$C_{(graphite)} + CO_2 \leftrightarrow 2CO$	$\Delta H = 172.6 \text{ kJ/mol}$
Water gas reaction	$C_{(grafite)} + H_2O \leftrightarrow CO + H_2$	$\Delta H = 131.4 \text{ kJ/mol}$
Thermal cracking	$C_nH_x \leftrightarrow nC + (x/2)H_2$	for n = 1, $\Delta H = 74.9 \text{ kJ/mol}$
Steam reforming	$CH_4 + H_2O \leftrightarrow CO_2 + 3H_2$	$\Delta H = 206 \text{ kJ/mol}$
Water Gas Shift	$CO + H_2O \leftrightarrow CO_2 + H_2$	$\Delta H = -41.2 \text{ kJ/mol}$
C-Methanation	$C + 2H_2 \leftrightarrow CH_4$	$\Delta H = -78.84 \text{ kJ/mol}$
CO-Methanation	$CO + 3H_2 \leftrightarrow CH_4 + H_2O$	$\Delta H = -206 \text{ kJ/mol}$
CO_2-Methanation	$CO_2 + 4H_2 \leftrightarrow CH_4 + 2H_2O$	$\Delta H = -165 \text{ kJ/mol}$
Reforming	$C_nH_x + mH_2O \leftrightarrow nCO + (m + x/2)H_2$	

The list is based mostly on the carbon chemistry and does not include the fate of heteroatoms and minerals contained in the feedstock which can largely affect conversion rates and gas quality [26,27]. At the typical process conditions, i.e., without catalyst and with a gas residence time of few seconds, the system does not reach the thermodynamic equilibrium. Definitively, the yield of syngas compared to byproducts (pyrolytic oil and char) and its composition depends on the kinetics. Thermogravimetric analysis (TGA) is one of the most used techniques to characterize the reactivity of the biomass as function of the temperature in a given atmosphere. The sensitivity of modern instruments and their coupling with computers provide large data sets that can be used to obtain empirical relationships or to speculate on the molecular mechanisms. Weight loss as a function of T and oxidant (or inert) gas encompasses a number of factors, such as topology, chemical composition, heat and mass diffusion [28,29]. The TGA was used in this work to compare thermal behavior of milled M and N linking it to their composition and to the results obtained at macroscale level.

TGA tests were carried out in air, pure oxygen (99%) and pure nitrogen (99.999%) with a heating rate of 10 °C/min from 60 °C to 900 °C. The atmospheres reproduced the pyrolysis and oxidation steps. The heating rate was set to simulate the average thermal profile inside the gasifier through which biomass particles passed to achieve full conversion after a typical residence time of 1.5 h.

The data were used to obtain the shell degree of conversion, defined as:

$$X, \% = \frac{W - ash}{W_0 - ash} \cdot 100 \tag{1}$$

where W_0 e W are the actual and starting weight and the specific conversion rate is given by:

$$r = -\frac{dX}{X \cdot dt} \, [min^{-1}] \tag{2}$$

From 60 °C to 150 °C, the mass loss was not significant and reached the maximum value of 2% in the case of N (Figure 2). This mass loss was due not only to the removal of moisture from the samples, but also to biomass decomposition processes. In fact, TGA-GCMS studies at low temperatures showed that starting from 60 °C various organic compounds can be released from lignocellulosic materials, including extractives and sugars degradation products. The bound water, on the other hand, cannot be completely removed even after 3 h at 120 °C. Thus, using TGA alone it is impossible to distinguish between the two contributions in this temperature range [30].

Between 150 °C and 210 °C there was practically no loss of mass; however, from this temperature onwards the loss became noticeable due to the decomposition of hemicellulose and cellulose. From 210 °C the differences between M and N became evident. Indeed, M are made up of 28% of hemicellulose and 30% of cellulose and therefore, between 210 °C and 350 °C lose more volatile substances than N, which contain 12% of hemicellulose and 22% of cellulose. Another observation was that up to about 350 °C, pyrolysis dominated on combustion. Indeed, in pure nitrogen at 300 °C N reached a conversion of 25%; in air and at the same temperature, the conversion was 30% and in pure oxygen it was 39% (Figure 2a,b).

The combined effect of pyrolysis and combustion caused the shift of the conversion rate to peak at lower temperature when the oxygen concentration was higher (Figure 2c). For the pyrolysis rate (in N_2) a greater intensity and a shift at lower temperatures was observed for M compared to N (335 °C and 0.0546 min^{-1} versus 342 °C and 0.0508 min^{-1}, Figure 2d). These findings could be explained by the higher content of carbohydrates and ash in M. A higher yield of char obtained from N compared to M at high temperatures, was likely due to the higher content of lignin [31] that in our cases was 42% versus 32%. In fact, lignin, in addition to being more resistant to thermal degradation, is transformed into char preserving part of the three-dimensional aromatic structure.

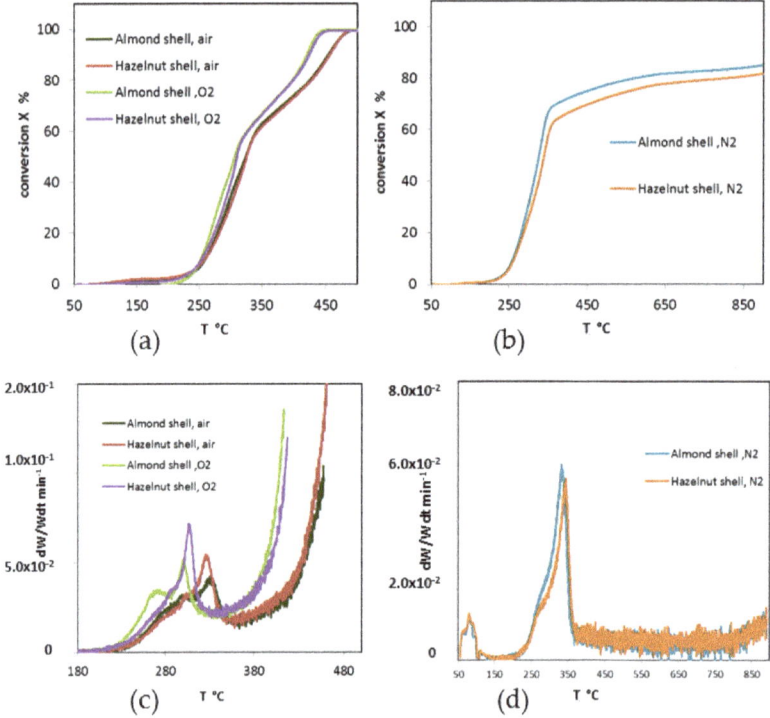

Figure 2. Thermogravimetric analysis of almond shells (M) and hazelnut shells (N): (**a**) Degree of conversion (X) vs. Temperature in air and O$_2$; (**b**) Degree of conversion (X) vs. Temperature in N$_2$; (**c**) Specific conversion rate in air and O$_2$; (**d**) Specific conversion (pyrolysis) rate in N$_2$.

3.2. Updraft Gasification

In this work the gasification process was investigated by varying the oxygen and steam flows, that affect the equivalence ratios ER(O$_2$) and ER(H$_2$O) as main operating parameters. The variation of flow rates affects a number of chemical and physical processes, such as the degree of combustion, the fluid dynamics, the thermal profile inside the gasifier, the syngas composition, and the production of condensable organic products. In the auto-thermal gasification mode these parameters are interlinked more than in the allothermal processes where the temperature is kept constant by external sources of power [32,33].

The equivalence ratio of combustion is the ratio between the available oxidant and the stoichiometric amount required for the complete reaction [34]. It would have a value of 1 for complete combustion and 0 for pyrolysis, while suitable values fall within the range 0.19–0.43 for gasification [35], up to 0.5 for fluid bed gasification of wet feedstock [36], resulting in a higher quality of the producer gas. It is defined as follows:

$$ER(O_2) = \frac{\text{feed } O_2 \left[\frac{kg}{h}\right]}{\text{flow of } O_2 \text{ for complete combustion } \left[\frac{kg}{h}\right]} \tag{3}$$

The availability of oxygen, both as free molecule and as atom in the water molecule, is a key factor in gasification; so similarly an equivalence ratio of water-reactions can be introduced:

$$ER(H_2O) = \frac{\text{feed } H_2O \text{ as steam } [\frac{kg}{h}]}{\text{flow of } H_2O \text{ for complete gasification } [\frac{kg}{h}]} \tag{4}$$

These ER are respectively linked to the stoichiometry of biomass oxidation by oxygen and water:

$$C_xH_yO_z + \left(X + \frac{Y}{4} - \frac{Z}{2}\right)O_2 = XCO_2 + \frac{Y}{2}H_2 \tag{5}$$

$$C_xH_yO_z + (2X - Z)H_2O = XCO_2 + \left(\frac{Y}{2} + 2X - Z\right)H_2 \tag{6}$$

Reaction (6) is obtained summing up the water gas reaction and WGS reaction of Table 2; it also takes into account that biomass itself contains H an O. In the tests presented here for gasification of M the $ER(O_2)$ fell in the range 0.22–0.25 and $ER(H_2O)$ from 0 to 0.30 while for gasification of N $ER(O_2)$ varied from 0.19 to 0.28 and the $ER(H_2O)$ from 0.18 to 0.28. In the auto-thermal process, the partial combustion reactions of the biomass generate the heat required to dry the feedstock, to conduct the endothermic reactions and to compensate for the heat loss through the reactor wall. Therefore, this kind of process can be tested in a narrower range of operating conditions than allothermal process.

The countercurrent gasification of M and N at pilot scale was carried out successfully without major problems and resulted in a very regular and reproducible process. All tests were carried out with an almost constant bed height, loading the biomass in a semi continuous mode, i.e., 4–5 kg of biomass were introduced into the gasification chamber at intervals of about 12 min and the process was completed in a few seconds. Air, oxygen and combinations of these with steam were used as oxidizing agents and introduced at the bottom of the grate that sustained the bed of biomass. The stabilization of the thermal profile inside the gasifier was used as indicator that steady conditions were achieved; while minor fluctuations in the freeboard reflected the cyclic feeding of the biomass. In Table 3 the experimental conditions and process parameters are reported as well the code assigned to each test: air and steam (A coded tests), oxygen and steam (O coded tests) and only air (AS coded tests). The numeric code of the test is referred to the corresponding percent value of $ER(O_2)$ and $ER(H_2O)$.

Table 3. Test coding of shells gasification: experimental parameters; calculated Equivalence Ratios, average temperature of the fixed bed, calculate residence time of the gas in the bed.

Experiment Code [a]	Oxidant	Feeding Rate kg dry /h	Particle Residence Time, h	Air kg/h	O_2 kg/h	ER (O_2) kg/kg	Steam kg/h	ER (H_2O) kg/kg	Average T in Bed, °C	Gas Residence Time [b], s
MAS1(24)	Air	12.4	2.81	16.7	0	0.24	0	0.00	767	6.2
MAS2(24)	Air	21.2	1.64	28.8	0	0.24	0	0.00	761	3.6
MA1(22/24)	Air and steam	22.4	1.55	28.9	0	0.22	4	0.24	701	3.1
MA2(24/28)	Air and steam	21.2	1.64	29.8	0	0.24	5.5	0.28	741	2.8
MA3(22/19)	Air and steam	22.4	1.55	29.0	0	0.22	4	0.19	715	3.2
MA4(24/25)	Air and steam	22.1	1.57	31.0	0	0.24	5.2	0.25	758	2.7
MA5(25/30)	Air and steam	21.6	1.60	31.5	0	0.25	6	0.30	739	2.7
MO1(23/28)	Oxigen and steam	21.2	1.64	0.0	6.5	0.23	5.5	0.28	748	6.5
NA1(19/28)	Air and steam	16.3	1.52	18.9	0	0.19	4.8	0.28	703	4.7
NA2(24/22)	Air and steam	20.4	1.22	29.3	0	0.24	4.8	0.22	760	3.3
NA3(22/18)	Air and steam	20.4	1.22	27.1	0	0.22	4	0.18	713	3.6
NO1(28/23)	Oxigen and steam	20.4	1.22	0.0	8	0.28	5	0.23	768	6.9
NO2(28/28)	Oxigen and steam	20.4	1.22	0.0	8	0.28	6	0.28	714	6.7
MAO(27/23)	Enric. air and steam	21.2	1.64	9.1	5.5	0.27	4.5	0.23	806	4.6

[a] Example for reading of the Table 3: MA1(24/28) means that test 1 of almond shells gasification was carried out with air at $ER(O_2)$ 0.24 and steam at $ER(H_2O)$ 0.28; [b] The average residence time of the gas inside the bed was calculated considering the average temperature of the bed, its void fraction as calculated from the bulk density (Table 1), a true density of lignocellulosics of 1530 kg/m³ [37] and the average molar flow of reactants and syngas (Table 4).

Figures 3 and 4 report the temperature profile at steady operating conditions recorded along the vertical reactor axis by the set of 11 thermocouples. For greater clarity only a few thermal profiles are shown for M and N. From the thermal point of view, the behavior of the two residues was very similar. The temperature at each height of the bed depended on the equilibrium between several endothermic and exothermic reactions that occurred at the solid-gas interface and in the gas phase as well as on heat and mass transfer and interactions. When the ER(O_2) increased, the temperatures increased above 1000 °C, but the introduction of the steam as co-gasification agent resulted in the temperature decreasing of about 200 °C in those bed regions (test MA4(24/25) versus test MAS(24)). The 3D plot of the maximum reactive bed temperatures for M gasification under the nine tested operating conditions (Figure 5) shows that with ER(H_2O) > 0.2 the temperature was kept below 950 °C. This ensured a smooth operation of the process avoiding hot spots within the ash layer that formed at the bottom of the fixed bed gasifier. Indeed, one of the most common problems in thermal conversion of biomass at pilot scale is unwanted ash sintering caused by high temperatures in the reactor. When a mix of oxygen and steam was used as oxidizing agents, the hottest zone was located close to the grate, where oxygen was in contact with hot char (tests NO1 and MO1) [38]. Instead, the use of air with steam caused the highest temperature at about 26 cm from the bottom where the ash was still dispersed in the char and the risk of its melting was lower.

Table 4. Process yields and plant performances.

Experiment Code	H_2g/kg	COg/kg	CO_2 g/kg	C_nH_m g/kg	Syngas [a] STP m³/kg	LHW MJ/m³	Density kg/STPm³	CGE %	Net CGE %	CLE %	Net CLE %	Plant Power [b] kW
MAS1(24)	15	635	451	40	1.78	5.76	0.98	57	57	11	11	35
MAS2(24)	18	615	360	34	1.79	5.62	1.11	56	56	13	13	59
MA1(22/24)	28	508	306	22	1.71	5.59	1.19	53	51	27	27	59
MA2(24/28)	39	548	213	29	2.09	5.56	1.05	64	62	13	13	68
MA3(22/19)	26	578	371	24	1.61	6.32	1.18	56	55	28	27	63
MA4(24/25)	36	630	643	34	2.03	6.10	1.11	69	66	15	14	76
MA5(25/30)	40	564	273	26	2.10	5.62	1.09	65	63	11	11	71
MO1(23/28)	35	580	479	37	1.15	10.4	1.10	66	64	20	19	70
MAO(27/23)	26	704	595	31	1.39	8.46	1.10	65	63	19	18	69
NA1(19/28)	25	564	403	55	1.77	6.43	1.14	64	61	19	18	52
NA2(24/22)	26	673	342	39	1.88	6.37	0.95	66	64	25	24	68
NA3(22/18)	29	699	265	24	2.01	5.90	1.09	66	64	23	22	67
NO1(28/23)	36	744	348	33	1.23	11.0	0.91	76	73	17	17	77
NO2(28/28)	40	691	259	23	1.22	10.6	1.10	73	70	19	18	74

[a] Conditioned and dry; [b] As thermal output in clean syngas.

The flow of steam at 160 °C introduced in the hot zones contributed to the cooling of the reactor promoting also endothermic reactions of water gas which increased the production of H_2 e CO.

In the updraft reactor, where the feed is introduced from the top and the particles of fuel move downwards, the heating rate was calculated from the space derivative of the temperature as follows:

$$\frac{dT}{dt}(°C\,s^{-1}) = \frac{dT}{dz}(°C\,m^{-1})\frac{dz}{dt}(ms^{-1}) \tag{7}$$

At 1.5 m, near the top of the bed where fresh biomass arrived, the thermal gradient showed a peak due mainly to exothermic reactions of pyrolysis of hemicellulose (Figure 3b tests NO1 and NA2).

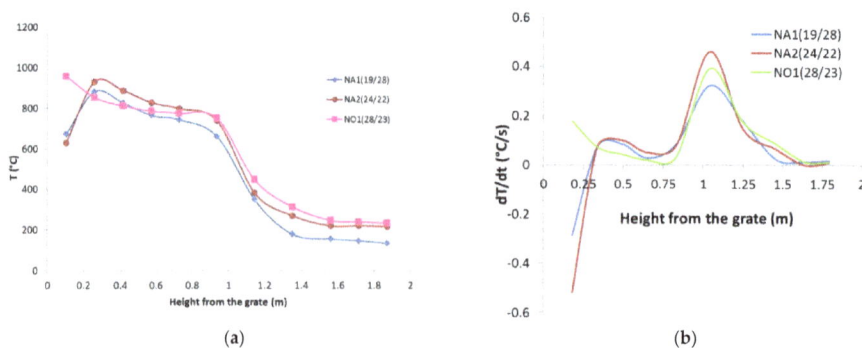

(a)

(b)

Figure 3. Thermal profile inside the gasifier in the gasification of hazelnut shells (N): (**a**) Temperatures measured in selected tests along the vertical axis (z); (**b**) Heating rate of the particles moving downward.

The extension of this region was larger at higher temperatures, as detected for NO1(28/23) and NA2(24/22) compared to NA1(19/28) whose average values were 220 °C, 250 °C and 160 °C respectively. A pronounced peak was found for all tests at 1 m and corresponded to the WGS reaction. This peak was lower for NA1(19/28) because of the longer residence time that improved heat transport between different regions reducing thermal gradients. The height where the heating rate approached its minimum was 0.7 m. It could be associated with gasification and cracking reactions which, due to their high endothermicity, used the enthalpy provided by the combustion in the zone below at about 0.5 m. The comparison between the gasification of M with absolute air and with the addition of steam pointed out the exothermic contribute of WGS reaction in the bed zone between 1.25 and 1.5 m corresponding to temperatures in the range of 400–600 °C (Figures 4b and 5a). In these zones differences were evident in peaks when steam is introduced both with air and oxygen. In the gasification with air only the exothermic contribute of the pyrolysis was observed coincident as regard the height in the bed, but lower temperatures were measured (about 200 °C).

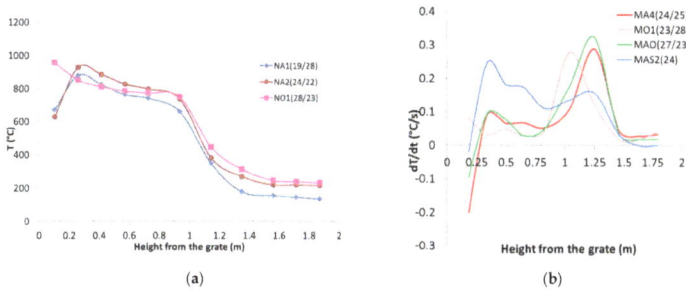

Figure 4. Thermal profile inside the gasifier in the gasification of almond shells (M): (**a**) Temperatures measured in selected tests along the vertical axis (z); (**b**) Heating rate of the particles moving downward.

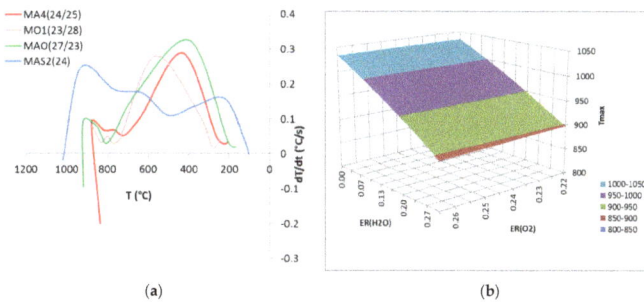

Figure 5. Thermal profile inside the gasifier in the gasification of almond shells (M): (**a**) Thermal gradients; (**b**) 3D plot of the maximum temperature in the bed.

The composition of the conditioned gas at steady conditions is reported in Figure 6. Supplying steam as a gasifying agent increased the partial pressure of H_2O inside the reactor promoting the endothermic water gas reactions in the regions with high temperatures. This led to increased H_2 production in the syngas as in the test NO1(28/23) versus NO2(28/28) for N and in series MAS1(24) and MAS2(24) versus MA2(24/28) and MA4(24/25) for M. The H_2/CO molar ratio significantly increased from 0.34 to 0.99 (MAS1(24) and MAS5(25/30)) by injecting of steam which favored the WGS. The H_2 and CO yields also showed the same trend (Table 4). However, it should be noted that there is a threshold of ER(H_2O) above which any further effects on the syngas composition could not be observed as shown by the direct comparison between tests MA1(22/24) and MA3(22/19). This is due to the temperature of the steam supplied to the reactor being lower than the gasification temperature. Moreover, the increase in the molar flow resulted in a decreasing of residence time and these two factors highly affect the syngas quality and composition.

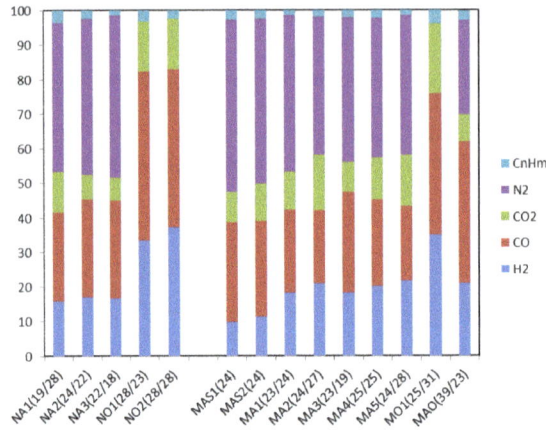

Figure 6. Composition of clean syngas (C_nH_m is the sum of CH_4, C_2H_6, C_3H_8).

From a qualitative point of view, the effects of the ERs on the gasification process were the same for the two feedstocks. In the oxy-gasification the producer gas had a higher H_2 content as shown in the series NO versus NA and MO1 versus MAS (see also Table 4). Moreover, higher corresponding gas heating values were obtained under these conditions than under those using air as gasification agent, because its nitrogen content dilute the products decreased proportionally the heating value of this stream. The gas heating value was comparable for M and N (Table 4) and reached the maximum value of 11.0 MJ/m^3 using oxygen (test NO1(28/23); at these conditions the highest value of the plant thermal power output of 77 kW was achieved. In the case of M gasification with enriched air and steam, the gas heating values were between those obtained with air and with oxygen. The yields of the dry and clean gas (g per kg of dry feedstocks) as functions of ER(O_2) and ER(H_2O) are shown in the 3D plot of Figures 7 and 8. The curves were drawn interpolating the data of Table 4 and were calculated starting from the gas composition of each component and the mass flow. The production of H_2 showed a positive correlation with both parameters in the examined range and reached a maximum at ER(O_2) and ER(H_2O) of 0.28 (see Figure 7 for N). Similarly, the production of CO was linked to the ERs but in a more complex way: it increased with ER(O_2) and decreased with ER(H_2O), with a minimum value for ER(O_2) of 0.19 and ER(H_2O) of 0.28 and a maximum value at 0.28 and 0.23 respectively. It is worth to point out how the surfaces of CO and H_2 appeared complementary to each other according to WGS reaction.

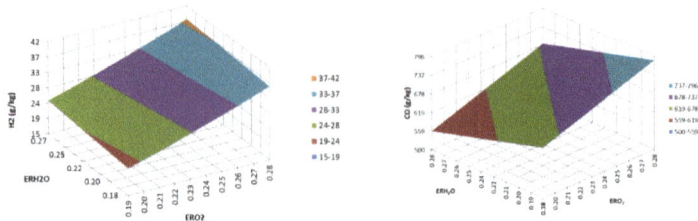

Figure 7. Yield of H_2 and CO as function of the equivalence ratio ER(O_2) and ER(H_2O) for hazelnut shells.

The equivalence ratio $ER(H_2O)$ had a strong influence on the H_2 production as reported in Figure 8 for M gasification and reached a maximum at $ER(H_2O)$ 0.30. The CO yield showed a correlation with both ERs: it increased with $ER(O_2)$ while had a generally negative trend with $ER(H_2O)$. In the examined range, the highest values of molar ratio H_2/CO corresponded to higher values of $ER(H_2O)$ parameter, while the dependence on $ER(O_2)$ appeared weak. This effect was observed comparing MAS1 and MA2 tests carried out keeping the same $ER(O_2)$ and tuning the $ER(H_2O)$ from 0 to 0.98. The ratio H_2/CO increased from 0.34 in air gasification to 0.97 in air-steam gasification so demonstrating the effect of steam addiction.

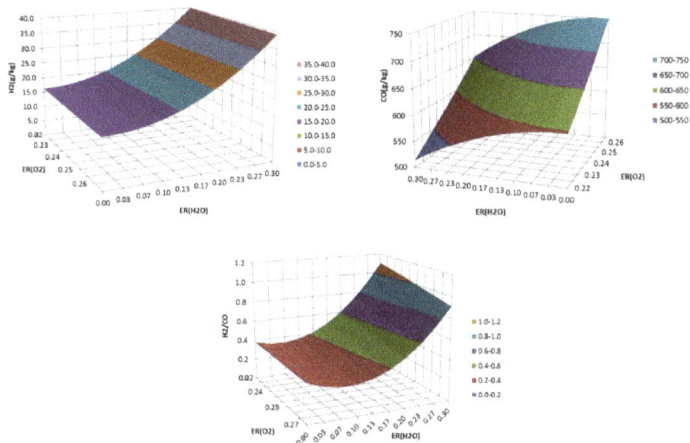

Figure 8. Yield of H_2, CO and H_2/CO ratio as function of the equivalence ratio $ER(O_2)$ and $ER(H_2O)$ for almond shells.

For both residues, the tar content showed a strong correlation with $ER(O_2)$. As shown in Figure 9a, the tar content increased with $ER(O_2)$, while for $ER(H_2O)$ there was a weaker dependence even if, at the same $ER(O_2)$, the increase of steam caused a rapid raise of tar production (NO2(28/28) versus NO1(28/23)). The tar yields (Figure 9b) confirmed these trends in the air-steam gasification, although with a significant lower production of tar, at higher $ER(O_2)$ when oxygen and steam were used as gasification medium. To explain the behavior of organic condensable species, it was considered its correlation with the fluid dynamic of the system and more specifically with the residence time of the syngas in the bed as shown in Figure 10a. The use of air as gasification agent resulted in higher total syngas flow with a corresponding decrease in the residence time leading to larger tar content in the gas. In fact, tar decrease was linked to the longer residence time that allowed volatile organic molecules to undergo thermal cracking cycles into incondensable hydrocarbons and hydrogen according to consecutives reactions scheme in (8):

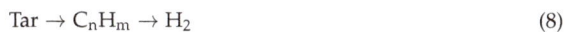

$$Tar \rightarrow C_nH_m \rightarrow H_2 \tag{8}$$

Figure 10a,b show the opposite trends of tar and C_nH_m and H_2 as a function of residence time for N. The addition of steam, which caused an increase of gas flow, produced the same effect on the residence time reduction (NO1(28/23) versus NO2(28/28)). Moreover, the use of steam led to a reduction of the temperature in the reactor and this condition increased tar production, as observed in tests NO1(28/23) and NO2(28/28) to which corresponded average bed temperatures of 768 °C and 714 °C respectively.

For M the tar content in the syngas decreased up to $ER(O_2)$ 0.25, then the trend reverses as shown in Figure 11a. The analysis of the gas yield with $ER(O_2)$ (Figure 11b) showed a positive correlation, with a maximum value for $ER(O_2)$ of 0.25. The complementary trends between syngas and tar yield were inherent in the chemistry of the process that provided only these two types of products and confirmed the correctness of the mass balance.

The increase of $ER(O_2)$ led to an increase of bed temperatures, (Table 3) thus favoring the attainment of higher syngas yields associated with progressively lower tar production. However, higher gas flow rate inside the reactor caused a decrease in the residence time and an increase in the risk of ash melting. The comparison of the graphs in Figure 11a,c shows the opposite correlation between the tar production and the production of incondensable hydrocarbons.

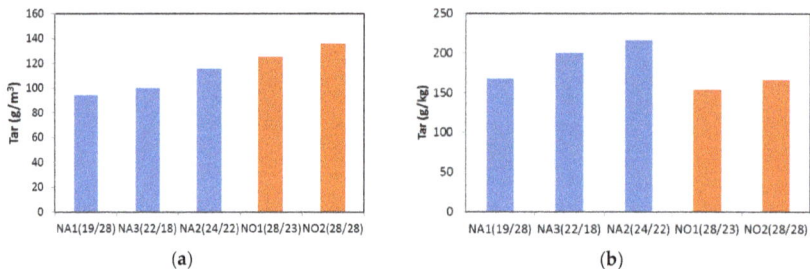

Figure 9. Organic volatiles at the exit of the updraft reactor in the gasification of almond shells (M) and hazelnut shells (N) with air (blue bars) and oxygen (orange bars); (a) Gravimetric tar contents per unit of conditioned syngas volume (method CEN TS 15439); (b) Tar yields referred to the dry feedstock.

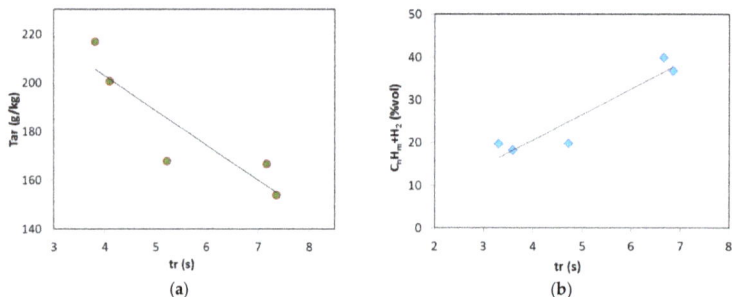

Figure 10. Gasification of hazelnut shells (N): (a) Yield of tar and (b) Incondensable hydrocarbons +H_2 concentration as function of the residence time of the syngas in the bed.

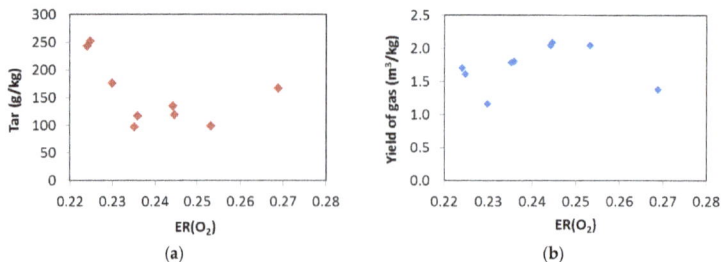

Figure 11. *Cont.*

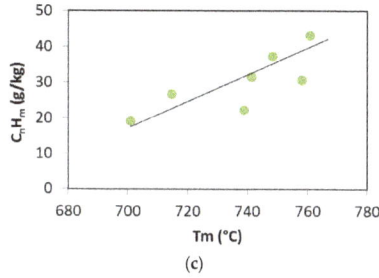

Figure 11. Gasification of almond shells (M): (**a**) Yield of tar versus ER(O$_2$); (**b**) Yield of syngas versus ER(O$_2$); (**c**) Yield of incondensable hydrocarbons versus the average temperature of the bed.

3.3. Process Yields

The thermal conversion of lignocellulosic residues involves the production of incondensable gases, liquid and solid (char or ash), their relative ratio depending on the technology used and the physical-chemical properties of the feedstock. Updraft gasification is characterized by high conversion into gas and liquid. The potential enthalpy of the feedstock is efficiently transferred to gaseous and liquid streams because of the relatively low temperature of the syngas at the exit of the gasifier. Moreover, the recovered ash contains very low residual carbon because of the long residence time of the particles in the gasifier and because the temperature is lower than 1000 °C and avoids the formation of graphitic residues which would be hard to convert. One of the most used parameters in the gasification is cold gas efficiency (CGE) which measures the fraction of energy transferred from the solid state (fed biomass) to the gaseous carrier (clean, dry and cold syngas); it is defined as follows:

$$CGE = \frac{\text{LHV of clean gas } \left[\frac{MJ}{kg}\right] \cdot \text{Flow of clean gas } \left[\frac{kg}{h}\right]}{\text{LHV of feedstock } \left[\frac{MJ}{kg}\right] \cdot \text{Feeding rate } \left[\frac{kg}{h}\right]} \quad (9)$$

Similarly, Cold Liquid Efficiency (CLE) measures the energy converted from the solid to the liquid and is related to the calorific value of condensable organic molecules recoverable from the raw the syngas:

$$CLE = \frac{\text{LHV of condensed organic volatiles } \left[\frac{MJ}{kg}\right] \cdot \text{Flow of condensed organic volatiles } \left[\frac{kg}{h}\right]}{\text{LHV of feedstock } \left[\frac{MJ}{kg}\right] \cdot \text{Feeding rate } \left[\frac{kg}{h}\right]} \quad (10)$$

In an ideal gasification process, the solid is fully converted into gas, therefore CGE would be 1 and CLE 0, while in ideal pyrolysis CLE would be 1 and GCE 0. As LHV of the condensed organics volatiles we used the value of 16 MJ/kg reported for the anhydrous part of the bio-oil obtained from slow pyrolysis [39].

In Figures 12 and 13 and in Table 4 the CGE and CLE are reported together with the net values obtained taking into account the theoretical enthalpy required to produce overheated steam at 160 °C from water at 20 °C. In these cases the net values of CGE and CLE were calculated as follows:

$$\text{net CGE} = \frac{\text{LHV of clean gas } [MJ/kg] \cdot \text{Flow of clean gas } \left[\frac{kg}{h}\right]}{\text{LHV of feedstock } [MJ/kg] \cdot \text{Feeding rate } \left[\frac{kg}{h}\right] + \Delta \text{ entahlpy water to steam } [MJ/h]} \quad (11)$$

$$\text{net CLE} = \frac{\text{LHV of condensed organic volatiles } [MJ/kg] \cdot \text{Flow of condensed organic volatiles } \left[\frac{kg}{h}\right]}{\text{LHV of feedstock } [MJ/kg] \cdot \text{Feeding rate } \left[\frac{kg}{h}\right] + \Delta \text{ entahlpy water to steam } [MJ/h]} \quad (12)$$

For gasification of N, the CGE showed a positive dependence with ER(O$_2$) in the examined range and reached a maximum at ER(O$_2$) of 0.28 (NO1(28/23)) and a minimum value for 0.19, while the dependence on ER(H$_2$O) appeared weaker. The power of the plant, calculated from LHV and the yields of the produced syngas, showed the same trend reaching a maximum value of 0.77 kW for NO1(28/23). The change of oxidant from air to oxygen resulted in an increase of CGE and power, and in both cases the maximum value was obtained with ER(O$_2$) 0.28 and ER(H$_2$O) 0.23. At these values of ER corresponded the lowest tar production and the lowest heat loss in the gas flow at the exit of the gasifier. Moreover, the excess of steam depressed this performance as shown in test NO1(28/23) versus NO2(28/28). For gasification of M, the CGE showed a positive correlation with both ER. Figure 13a shows that the addition of steam improved the efficiency as it appeared from the direct comparison between MA4(24/25) and MAS2(24). Similarly, an excess of steam negatively affected CGE as shown in the tests MA4(24/25) vs. MA4(24/28) and MA3(22/19) vs. MA1(22/24). The use of oxygen and enriched air instead of air as gasifying medium resulted in higher values of efficiency, but the best output in term of efficiency was observed for MA4(24/28). The thermal power of the plant also followed the same trend (Figure 13b). The CLE was proportional to the tar yield and was higher in the tests of air gasification for N and in the tests with lower ER(O$_2$) for M. This parameter is important when all gasification products are considered valuable and the liquid stream is exploited to produce liquid biofuels.

The overall energy conversion efficiency of solid biomass to gaseous and liquid carriers was obtained by summing CGE and CLE. The lowest conversion efficiency was observed for air gasification of almond shells, 0.68 at MAS1(24); while the oxy-steam gasification allowed to reach the highest values up to 0.86 (MO1(23/28)). The net efficiencies followed the same trends but with lower values of 3.7% when using 0.26 kg of steam/kg of biomass (MO1(23/28)) as the production of steam require 3.4% of the LHV available in M (or 3.8% in N). However, supplying steam as co-gasification agent significantly improved the H$_2$ yield and the H$_2$/CO ratio by direct gasification of char and the WGS (NO1(28/23) vs. NO2(28/28) and MA2(24/28) vs. MAS1(24)). Moreover, the addition of steam prevented local overheating mainly near the grate where there is high ash concentration; this avoided ash melting and allowed to tune the thermal profile along the reactor ensuring favorable kinetics and equilibria of the endothermic water gas and Boudouard reactions. Comparing the process yields between M and N at the closest gasification conditions some differences were pointed out. The production of CO$_2$ as well as the CLE were higher for M than for N (test MAO(27/23) vs. NO1(28/23) and MA3(22/18) vs. NA3(22/18)). These findings could be explained by the higher reactivity of M and higher pyrolysis yields that were already highlighted in TGA and explained in terms of different chemical composition. The content of potassium in M was higher than in N (0.35% vs. 0.26%, see Table 1); the catalytic effect of this alkaline metal on combustion and gasification by oxygen-transfer is well known. Biochemical composition revealed the higher content of hemicellulose and cellulose in M that was also envisaged from the ratio C/O calculated by elemental composition analysis.

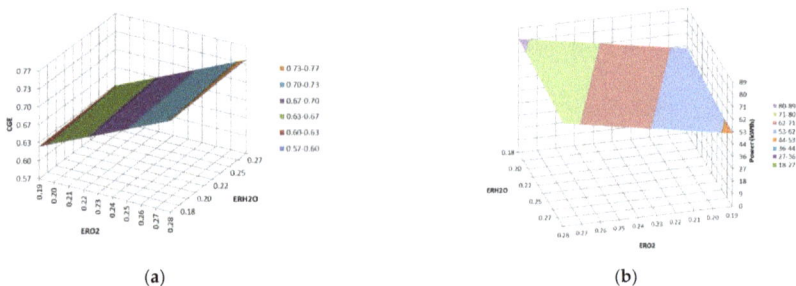

(a)   (b)

Figure 12. *Cont.*

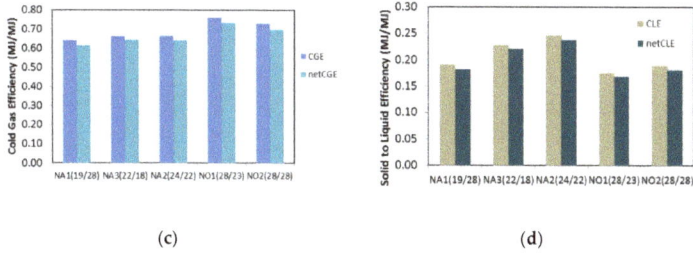

(c)

(d)

Figure 12. Gasification of hazelnut shells (N): (**a**) 3D plot of efficiency of energy conversion in the gasification as function of the equivalence ratio ER(O$_2$) and ER(H$_2$O); (**b**) 3D plot of thermal power of the plant as function of the equivalence ratio ER(O$_2$) and ER(H$_2$O); (**c**) Cold Gas Efficiency (CGE) and the corresponding net values; (**d**) Cold Liquid Efficiency (CLE) and the corresponding net values.

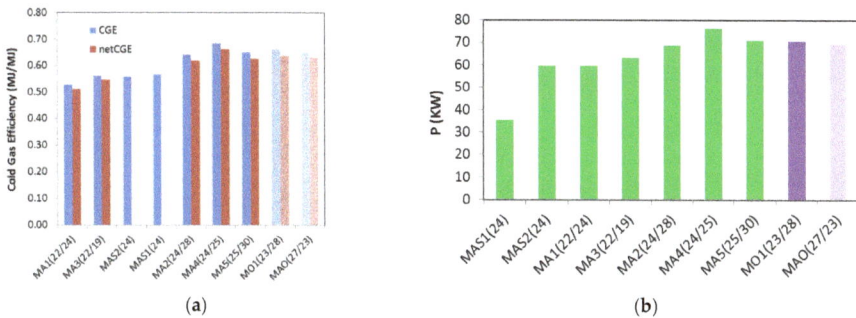

(a)

(b)

Figure 13. Gasification of almond shells (M): (**a**) Efficiency of energy conversion in the gasification; (**b**) Thermal power of the plant.

4. Conclusions

Almond and hazelnuts shells were easily gasified in updraft mode up to the highest conversion degrees because of their low fines content and high permeation of the oxidant gases through the reactive bed. The composition of the obtained syngas could be varied by changing air and steam flows and response surface analysis was able to show functional dependences on ER(O$_2$) and ER(H$_2$O); in particular when steam was added in an optimized ER(H$_2$O), the H$_2$ content increased and the ratio H$_2$/CO reached the significant value of 1.0. Steam also helped for the operational management of the plant by contributing to the temperature control in the lowest zones of the reactor where ash melting is a well-known problem. The cold gas efficiency ranged between 53% and 76% depending on the syngas output (CGE), while the content of condensable organics corresponded to an energy conversion of 11–25% (CLE). The higher content of carbohydrate in M compared to N caused a larger production of organic volatiles through pyrolysis reactions as showed in TGA and confirmed by the correspondent higher yield of tars resulting in the plant gasification tests. The tar yields for M and N were inversely correlated with the residence time of the syngas and explained with a sequence of condensation and cracking inside the reactive bed producing incondensable hydrocarbons and H$_2$.

Author Contributions: The authors equally contributed to the work.

Funding: This research was funded by EU within the project Brisk and Brisk2, Grant 284498 and grant 731101.

Acknowledgments: The authors thank Colomba Di Blasi of the University of Naples for the useful discussion of the experimental data. Maria A. Cerone is gratefully acknowledged for the revision of the manuscript.

Conflicts of Interest: The authors declare no conflict of interest.

References

1. Schulz, H. Short history and present trends of Fischer–Tropsch synthesis. *Appl. Catal. A* **1999**, *186*, 3–12. [CrossRef]
2. Ke, L.; Song, C.; Subramani, V. *Hydrogen and Syngas Production and Purification Technologies*; John Wiley & Sons: Hoboken, NJ, USA, 2009; Volume 104, pp. 113–119. ISBN 978-0-471-71975-5.
3. IEA. Key World Energy Statistics. 2017. Available online: https://www.iea.org/publications/freepublications/publication/KeyWorld2017.pdf (accessed on 4 April 2018).
4. Cerone, N.; Zimbardi, F.; Contuzzi, L.; Prestipino, M.; Carnevale, O.; Valerio, V. Air-steam and oxy-steam gasification of hydrolytic residues from biorefinery. *Fuel Process. Technol.* **2017**, *167*, 451–461. [CrossRef]
5. Elliott, D.C.; Neuenschwander, G.G.; Hart, T.R.; Rotness, L.J.; Zacher, A.H.; Santosa, D.M.; Valkenburt, C.; Jones, S.B.; Tjokro Rahardjo, S.A. *Catalytic Hydrothermal Gasification of Lignin-Rich Biorefinery Residues and Algae Final Report*; Pacific Northwest National Laboratory (PNNL): Richland, WA, USA, 2009.
6. Prins, M.J.; Ptasinski, K.J.; Janssen, F.J. From coal to biomass gasification: Comparison of thermodynamic efficiency. *Energy* **2007**, *32*, 1248–1259. [CrossRef]
7. Brage, C.; Yu, Q.; Chen, G.; Sjöström, K. Tar evolution profiles obtained from gasification of biomass and coal. *Biomass Bioenergy* **2000**, *18*, 87–91. [CrossRef]
8. Galvagno, A.; Prestipino, M.; Zafarana, G.; Chiodo, V. Analysis of an integrated agro-waste gasification and 120 kw SOFC CHP system: Modeling and experimental investigation. *Energy Procedia* **2016**, *101*, 528–535. [CrossRef]
9. Dimitriou, I.; Goldingay, H.; Bridgwater, A.V. Techno-economic and uncertainty analysis of Biomass to Liquid (BTL) systems for transport fuel production. *Renew. Sustain. Energy Rev.* **2018**, *88*, 160–175. [CrossRef]
10. Vera, D.; Jurado, F.; Torreglosa, J.P.; Ortega, M. Biomass Gasification for Power Generation Applications: A Modeling, Economic, and Experimental Study. In *Advances in Renewable Energies and Power Technologie*; Elsevier: New York, NY, USA, 2018; pp. 87–121.
11. Loha, C.; Karmakar, M.K.; De, S.; Chatterjee, P.K. Gasifiers: Types, Operational Principles, and Commercial Forms. In *Coal and Biomass Gasificatio*; Springer: Singapore, 2018; pp. 63–91. [CrossRef]
12. Ding, L.; Yoshikawa, K.; Fukuhara, M.; Kowata, Y.; Nakamura, S.; Xin, D.; Muhan, L. Development of an ultra-small biomass gasification and power generation system: Part 2. Gasification characteristics of carbonized pellets/briquettes in a pilot-scale updraft fixed bed gasifier. *Fuel* **2018**, *220*, 210–219. [CrossRef]
13. Kurkela, E.; Simell, P.; Ståhlberg, P.; Berna, G.; Barbagli, F.; Haavisto, I. *Development of Novel Fixed-Bed Gasification for Biomass Residues and Agrobiofuels*; VTT Research Notes; VTT Technical Research Centre of Finland: Espoo, Finland, 2000; ISBN 951-38-5757-3.
14. James, R.A.M.; Yuan, W.; Boyette, M.D. The Effect of Biomass Physical Properties on Top-Lit Updraft Gasification of Woodchips. *Energies* **2016**, *9*, 283. [CrossRef]
15. Rios, M.L.V.; González, A.M.; Lora, E.E.S.; Del Olmo, O.A.A. Reduction of tar generated during biomass gasification: A review. *Biomass Bioenergy* **2018**, *108*, 345–370. [CrossRef]
16. Narváez, C.R.A.; Blanchard, R.; Dixon, R.; Ramírez, V.; Chulde, D. Low-cost syngas shifting for remote gasifiers: Combination of CO_2 adsorption and catalyst addition in a novel and simplified packed structure. *Energies* **2018**, *11*, 311. [CrossRef]
17. Daniell, J.; Köpke, M.; Simpson, S.D. Commercial biomass syngas fermentation. *Energies* **2012**, *5*, 5372–5417. [CrossRef]
18. Molitor, B.; Richter, H.; Martin, M.E.; Jensen, R.O.; Juminaga, A.; Mihalcea, C.; Angenent, L.T. Carbon recovery by fermentation of CO-rich off gases–turning steel mills into biorefineries. *Bioresour. Technol.* **2016**, *215*, 386–396. [CrossRef] [PubMed]
19. Cerone, N.; Zimbardi, F.; Villone, A.; Strjiugas, N.; Kiyikci, E.G. Gasification of wood and torrefied wood with air, oxygen, and steam in a fixed-bed pilot plant. *Energy Fuels* **2016**, *30*, 4034–4043. [CrossRef]
20. Sánchez-Pérez, R.; Ortega, E.; Duval, H.; Martínez-Gómez, P.; Dicenta, F. Inheritance and relationships of important agronomic traits in almond. *Euphytica* **2007**, *155*, 381–391. [CrossRef]
21. Cristofori, V.; Ferramondo, S.; Bertazza, G.; Bignami, C. Nut and kernel traits and chemical composition of hazelnut (*Corylus avellana* L.) cultivars. *J. Sci. Food Agric.* **2008**, *88*, 1091–1098. [CrossRef]
22. Dogru, M.; Howarth, C.R.; Akay, G.; Keskinler, B.; Malik, A.A. Gasification of hazelnut shells in a downdraft gasifier. *Energy* **2002**, *27*, 415–427. [CrossRef]

23. Rapagna, S.; Latif, A. Steam gasification of almond shells in a fluidised bed reactor: The influence of temperature and particle size on product yield and distribution. *Biomass Bioenergy* **1997**, *12*, 281–288. [CrossRef]

24. Rapagna, S.; Jand, N.; Kiennemann, A.; Foscolo, P.U. Steam-gasification of biomass in a fluidised-bed of olivine particles. *Biomass Bioenergy* **2000**, *19*, 187–197. [CrossRef]

25. Zimbardi, F. Evaluation of Reaction Order and Activation Energy of Char Combustion by Shift Technique. *Combust. Sci. Technol.* **2000**, *156*, 251–269. [CrossRef]

26. Zhang, Y.; Ashizawa, M.; Kajitani, S.; Miura, K. Proposal of a semi-empirical kinetic model to reconcile with gasification reactivity profiles of biomass chars. *Fuel* **2008**, *87*, 475–481. [CrossRef]

27. Feng, D.; Zhao, Y.; Zhang, Y.; Sun, S.; Gao, J. Steam Gasification of Sawdust Biochar Influenced by Chemical Speciation of Alkali and Alkaline Earth Metallic Species. *Energies* **2018**, *11*, 205. [CrossRef]

28. Saddawi, A.; Jones, J.M.; Williams, A.; Wojtowicz, M.A. Kinetics of the thermal decomposition of biomass. *Energy Fuels* **2009**, *24*, 1274–1282. [CrossRef]

29. Cheng, K.; Winter, W.T.; Stipanovic, A.J. A modulated-TGA approach to the kinetics of lignocellulosic biomass pyrolysis/combustion. *Polym. Degrad. Stab.* **2012**, *97*, 1606–1615. [CrossRef]

30. Rabinovich, M.L.; Fedoryak, O.; Dobele, G.; Andersone, A.; Gawdzik, B.; Lindström, M.E.; Sevastyanova, O. Carbon adsorbents from industrial hydrolysis lignin: The USSR/Eastern European experience and its importance for modern biorefineries. *Renew. Sustain. Energy Rev.* **2016**, *57*, 1008–1024. [CrossRef]

31. Demirbas, A. Effects of temperature and particle size on bio-char yield from pyrolysis of agricultural residues. *J. Anal. Appl. Pyrolysis* **2004**, *72*, 243–248. [CrossRef]

32. Taba, L.E.; Irfan, M.F.; Daud, W.A.M.W.; Chakrabarti, M.H. The effect of temperature on various parameters in coal, biomass and CO-gasification: A review. *Renew. Sustain. Energy Rev.* **2012**, *16*, 5584–5596. [CrossRef]

33. Fermoso, J.; Arias, B.; Plaza, M.G.; Pevida, C.; Rubiera, F.; Pis, J.J.; García-Peña, F.; Casero, P. High-pressure CO-gasification of coal with biomass and petroleum coke. *Fuel Process. Technol.* **2009**, *90*, 926–932. [CrossRef]

34. Jayathilake, R.; Rudra, S. Numerical and Experimental Investigation of Equivalence Ratio (ER) and Feedstock Particle Size on Birchwood Gasification. *Energies* **2017**, *10*, 1232. [CrossRef]

35. Saravanakumar, A.; Haridasan, T.M.; Reed, T.B.; Kasturi Bai, R. Experimental investigations of long stick wood gasification in a bottom lit updraft fixed bed gasifier. *Fuel* **2007**, *86*, 2846–2856. [CrossRef]

36. Arena, U. Process and technological aspects of municipal solid waste gasification. A review. *Waste Manag.* **2012**, *32*, 625–639. [CrossRef] [PubMed]

37. Tsai, W.T.; Liu, S.C.; Hsieh, C.H. Preparation and fuel properties of biochars from the pyrolysis of exhausted coffee residue. *J. Anal. Appl. Pyrolysis* **2012**, *93*, 63–67. [CrossRef]

38. Cerone, N.; Zimbardi, F.; Contuzzi, L.; Alvino, E.; Carnevale, O.; Valerio, V. Updraft Gasification at Pilot Scale of Hydrolytic Lignin Residue. *Energy Fuels* **2014**, *28*, 3948–3956. [CrossRef]

39. Czernik, S.; Bridgwater, A.V. Overview of applications of biomass fast pyrolysis oil. *Energy Fuels* **2004**, *18*, 590–598. [CrossRef]

energies

MDPI

Article

Methodical Aspects of Biogas Production in Small-Volume Bioreactors in Laboratory Investigations

Agnieszka Kasprzycka * and Jan Kuna

Institute of Agrophysics, Polish Academy of Sciences, Doświadczalna 4, 20-290 Lublin, Poland; jasiekkuna@gmail.com
* Correspondence: a.kasprzycka@ipan.lublin.pl

Received: 27 April 2018; Accepted: 25 May 2018; Published: 29 May 2018

Abstract: The aim of this study was to develop a methodology to investigate the biofermentation process in small-volume fermenters. Dark serum bottles with a volume of 100–120 mL, tightly sealed with a rubber septum, were used as bioreactors. The optimum measurement conditions in this type of bioreactor comprise: (i) filling two-thirds of the maximum volume with a suspension; (ii) a 2% bioreactor loading (on a dry basis) and; (iii) the daily equalization of pressure by removing the biogas through the septum pierced with a syringe needle and the intensive mixing of the remaining suspension. The methane yield (quantity and dynamics) obtained in this type of bioreactor is analogous to that of industrial bioreactors or large-scale laboratory bioreactors. The use of small-volume bioreactors that can be incubated will facilitate the preliminary selection of analysed systems and provide an indication of those that should be investigated in large-scale bioreactors.

Keywords: biogas; biofermentation method; energy

1. Introduction

Renewable energy is derived from a constantly increasing range of raw materials [1–3] that can be of plant [4–6] or animal [7,8] origin. The technologies developed in this field have already been widely applied in industry [9–11]. The practical application of biogas production installations does not exclude on-going development and improvement, which is evidenced by the latest reports [6,12,13]. Irrespective of the direction of research related to the development of biogas production technologies, three areas of research activity can be distinguished: (i) laboratory-scale experiments [1,8,14], (ii) semi-industrial scale experiments [15] and (iii) pilot experiments in industrial installations [16–18]. Since scientific research is inherently associated with a large number of experiments (multifarious aspects investigated in many repetitions), the technical possibilities of conducting such investigations in large-scale installations pose a problem. The larger the bioreactor, the more similar the research conditions are to industrial-scale bioreactors and the more difficult it is to perform multiple repetitions. Given the necessity to take measurements for many systems in several week-long experiments, the use of large-scale bioreactors is virtually impossible.

It is, therefore, not surprising that researchers use different types and sizes of laboratory bioreactors with a substantially lower volume than that of industrial installations. The data presented in Table 1 confirms that laboratory scale is still important for investigations. Taking into account the volumes of the bioreactors, we divided them into three groups: large volume (>2.5 dm^3), medium volume (≤2.5 dm^3 and >250 mL), and small volume (≤250 mL). However, the volume of the bioreactor is not the only differentiating factor. The construction is usually volume-dependent. The large volume bioreactors (Table 1) are generally miniatures of the full-scale bioreactors. This means that they usually have a similar construction and more or less advanced automation. The medium-scale bioreactors are usually different kinds of bottles or other laboratory vessels. Automation is not often used in such bioreactors and, therefore, all actions undertaken during the biogas production are carried out manually. The advantages of medium scale bioreactors are that they take up far less space and are much cheaper. However, mass measurements are still technically difficult because less space does not mean small space.

Table 1. Review of laboratory investigations and the bioreactors used in these works.

Group of Laboratory Bioreactors	Volume of Used Bioreactor	Publication
big volume	20 dm^3 semi-continuously stirred tank reactors	[19]
	40 dm^3 bioreactor	[20]
	20 dm^3 bioreactor	[21]
	14 dm^3 digesters	[22]
	5,3 dm^3 bioreactor	[23]
	5 dm^3 bioreactor	[24]
	5 dm^3 double glass cylinder	[16]
	3,75 dm^3 clear PVC	[25]
medium volume	2.5 dm^3 batch digester	[26]
	2.3 dm^3 and 1.3 dm^3 glass bottles	[8]
	1 dm^3 serum flasks	[27]
	1.1 dm^3 glass bottles	[28]
	600 cm^3 digesters	[29]
	600 cm^3 PET bottles	[30]
	500 cm^3 digesters	[31]
	500 cm^3 plastic bottle	[32]
small volume	150 cm3 serum bottles	[33]
	119 cm^3 glass bottles	[34]
	100 cm3 serum bottles	[35]
	100 cm^3 glass syringe	[36]
	60 cm^3 vials	[37]
	60 cm^3 dark vials	[38]

The last group of bioreactors presented in Table 1 have a small volume. Serum bottles are used in the vast majority but even smaller vials are also applied. It is obvious that such small-volume vessels must be hermetically sealed and that it is not possible to automate the system. This means that with the assumed frequency, the methane produced has to be released to avoid high pressures in the bottle. The consequence of such a small volume is the necessity to use gas chromatography (GC) for gas quality measurements because the majority of other gas metres require large volumes of the gases. Thus, if small volume bioreactors have such disadvantages the question becomes—why are they used in research in so many laboratories? The answer is quite simple; they allow researchers to carry out mass screening investigations that are quick and cheap. The results obtained in such experiments can demonstrate the most effective method for larger-scale investigations.

As follows from the presented literature review, research on anaerobic digestion in small volumes exists. However, there is no literature that presents a methodology of conducting this type of research.

The purpose of this work is to find an optimal methodology for the production of biogas in small-volume bioreactors.

Hence, the importance of research on the methodology of biogas production in small volumes. An example here may be seen in the latest literature, which addresses the subject of research in the laboratory scale of anaerobic digestion [39–43].

2. Material and Methods

The substrate used for fermentation was chopped maize silage with 32.22% dry matter content. An agricultural biogas plant, using mainly maize silage and beet pulp, provided the bioreactor inoculum. The substrate and the inoculum were mixed at a weight ratio of 1:1 (total solids). The bioreactors prepared in this way were placed in an incubator in the dark at a temperature of 37 °C ± 1 °C. One hundred and twenty millilitre dark serum bottles that were tightly sealed with a rubber septum were used as bioreactors. The bottles were filled to a volume of 75 mL with the substrate and inoculum mixture.

2.1. The First Series of Measurements

The aim of the first series of measurements was to select the optimal loading of the bottle bioreactors. The bioreactor loading was differentiated using a mixture with a concentration corresponding to 2%, 7% and 13% (on a dry basis). This choice of loading was consistent with the information presented in the paper by Hilkiah Igoni et al. [44], who cited data provided by the Oregon State Department of Energy. According to these data, three ranges of loading, i.e., up to 2%, 2–10% and 11–13%, can be applied depending on the type of sludge and bioreactors. The extreme values were chosen for the investigations. A 7% loading value was selected from the 2–10% range.

In order to ensure the greatest similarity between the process occurring in the bottles and process carried out in industrial bioreactors, the produced biogas was removed daily (this procedure was carried out using the needle from a syringe) from the first series of bottles and its volume was measured. The measurements of the concentration of each component of the biogas (methane, carbon dioxide, oxygen and nitrogen) were performed on selected days of the week for three weeks using GC. In the results and discussion sections, this part of the experiment will be referred to as series one. The experiment in this series was carried out in five replications (replications should be understood as parallel measurements in separate bottles).

2.2. The Second Series of Measurements

The aim of the second series of measurements was to compare the methanogenesis yield under the conditions of the daily reduction of pressure (daily gas removal as in series one) and under pressure (no pressure reduction from the incubated bottles). The bottles were filled with silage and inoculum (at the same weight ratio as in series one) at 2% loading, selected on the basis of the results obtained in series one. In the variant with the daily reduction of pressure, measurements were carried out in three bottles throughout the experiment. After incubation, the bottles were opened and the pH was measured.

Only one measurement was taken in the variant without pressure reduction, i.e., on a chosen day of incubation (the same day as the measurements in the reduced-pressure bottles) the gas was removed from each bottle and its volume and composition were determined. Next, the bottle was opened and, after measuring the pH, the biomass was disposed of. This procedure necessitated the preparation of a greater number of bottles.

In series two, the experiment was conducted for three weeks in three replications (three parallel bottles per incubation day).

2.3. The Third Series of Measurements

The results obtained in the other series indicated that the biogas yield in the replications varied largely and the reaction of the suspension after the measurements sometimes exhibited excessive acidification (from the point of view of methane fermentation efficiency). Therefore, the aim of the third series of measurements was to check whether it was possible to adjust the pH at the beginning of the process in order to optimize methanogenesis in the bottles.

From the bioreactor load (2%) indicated by the results obtained in series one, a third series of measurements were performed (with the daily removal of the produced biogas) at different values of the initial pH of the suspension placed in the bottles. In the first variant, the suspension was incubated without any addition (the pH of the input mixture was 7.88). In the second variant, 1 g NaHCO$_3$ (per gram of dry weight) was added to the input suspension. The determination of the biogas composition was carried out on specified days of the week and the incubation lasted for two and a half weeks. Following incubation, the bottles were opened and the pH was measured. The measurements of series three were performed in five replications. In each series, the content of the bottles was intensively stirred after each measurement of the volume of the produced biogas.

The composition of the biogas was determined chromatographically using a Schimadzu-14A gas detector equipped with a thermal conductivity detector (TCD) detector. A detector equipped with a 2 m column and a diameter of 3.2 mm and filled with Porapak Q was used for the determination of the methane content. Helium was used as a carrier gas in the chromatograph. The carrier gas-flow through the column was set at 40 mL·min^{-1}. The temperatures of the column and the detector were 40 °C and 60 °C, respectively.

3. Results and Discussion

3.1. Choice of the Optimal Bioreactor Loading

In the first stage, the optimal loading had to be chosen for this type of bioreactor. The results obtained in the measurements from series one were used for this purpose. The total volume of the biogas obtained for each loading value during the three week incubation is presented in Figure 1. The content of methane in the biogas for all the bioreactor-loading values on the successive incubation days is presented in Figure 2.

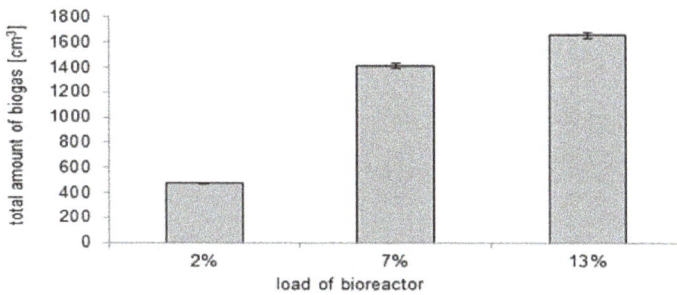

Figure 1. Total volume of biogas obtained during 3 week incubation for different values of bioreactor loading. Results from measurement series 1.

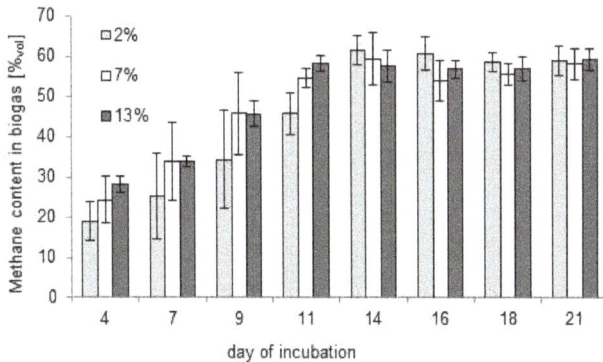

Figure 2. Methane content of biogas on successive incubation days for different bioreactor loading values. Results from measurement series 1.

As might be expected, the highest biogas yield was found in the highest bioreactor loading, i.e., 13% (Figure 1). However, it is worth noting that the difference in biogas yields between 7% and 13% was significantly lower than that between 2% and 7% (244 mL and 937 mL, respectively). This may indicate that the biogas production process at 13% loading was inhibited by pressure formed in the bottle and that the daily reduction thereof was insufficient. Another cause may lie in the fact that microorganisms are not capable of a quicker and efficient reduction of biomass and the production of biogas under excessive loading. This hypothesis is supported by the results presented in the paper by Betts et al. [45]. If the bioreactor is loaded with excessive amounts of biomass, slow growth of methanogenic bacteria can result in a rapid decline in pH throughout the process. This situation may be caused by intermediates produced in previous phases that have not been completely decomposed. Regardless of the actual cause of the decrease in dynamics of gas yields, 13% loading had to be excluded as it was too high. The primary reason for discarding 7% loading was the fact that, similar to 13% loading, the septa sealing the bottles were always bulging and, in some cases, the pressure was so high that the septum lost tightness and gas leaked out.

The total amount of biogas produced at 2% loading was lower than in the case of higher loads (Figure 1). This seems to be obvious: the lower the substrate input; the lower the biogas yield. However, it is worth analysing the methane content in the produced biogas. Analysis of trends on the graph presented in Figure 2 indicates that, at 2% loading, methane concentration in the first biofermentation stage (up to day 11) was lower than at other loading values. However, the analysis did not show statistically significant differences. Moreover, in the second stage (from day 14) of the experiments, concentrations of CH_4 for the 2% loading were highest (there was still no statistical significance between the individual values of loaded biomass). The content of methane at the level of ~60% volume was high and comparable with other investigations in both laboratory [46] and industrial [16,17,47] bioreactors. Therefore, since this provides indirect information about the quality of the biofermentation process, the content of methane in the biogas implied the similar efficiency of the process, irrespective of the loading selected for the experiment. Thus, 2% (per dry weight) was assumed to be the best loading in the case of septum-sealed bottle bioreactors and, therefore, only the results obtained at this loading will be presented and discussed below.

Biogas composition (i.e., primarily methane content) derived in the biofermentation process in bottles with 2% loading did not raise any objections; in contrast, there was a problem of high divergence of results obtained in the parallel replications (Figure 2). This may have been caused by the relatively high variability of the pH value of the suspension.

As indicated in Table 2, in some cases the pH of the suspension dropped below 6.7, i.e., a value regarded as a threshold below which the biofermentation process is inhibited [48,49]. Carbon dioxide, which accumulated in the bottle and dissolved in the solution, forming HCO_3-ions, was found to be a direct cause of suspension acidification [50]. In such cases, methane content in the biogas decreased. Since the same substrate and inoculum mixture was used in all of the replications, the suspension acidification in some bottles can be explained by the heterogeneity of the input mixture, even though it had been vigorously stirred to achieve homogenization [51].

3.2. Inhibition of the Biofermentation Process by Excessive Pressure of Biogas

Gas yields obtained during biofermentation in the bottles in which the gas was removed daily to equalize the pressure to the atmospheric value, and in the bottles without gas removal and increasing pressure, are presented in Figure 3. In both cases, the biomass load was 2%. The content of methane in the biogas on successive days of the series two experiments is shown in Figure 4. The points in the graph correspond to the points in Figure 3. Table 2 presents the pH of the suspension measured in the bottles without gas removal. The measurements were carried out immediately after determination of the biogas yield and composition for the 2% loading.

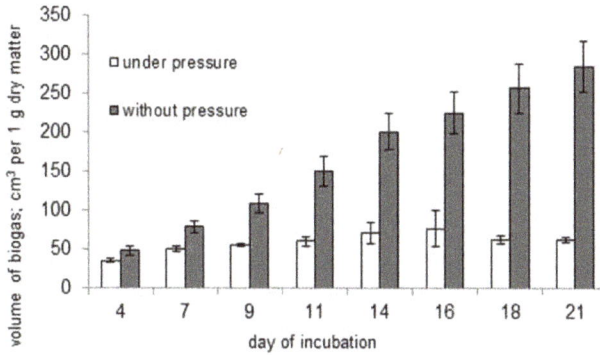

Figure 3. Biogas production on the consecutive days of incubation in the bottles at 2% loading in measurement series 2. Gas volumes are converted into 1 g of dry weight.

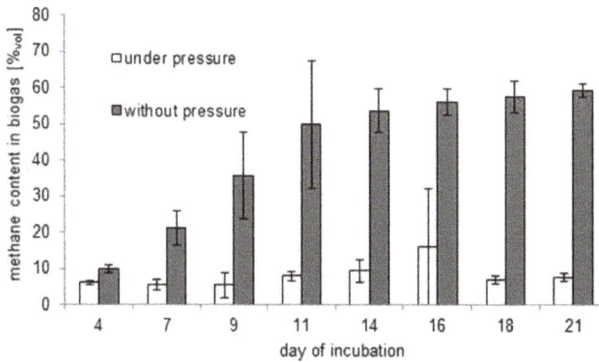

Figure 4. The content of methane in the biogas on successive days of incubation at 2% loading in bottles with daily gas removal and in bottles without gas removal.

Table 2. pH of the suspension in bioreactors (2% loading) without daily gas removal on consecutive incubation days. Results of measurement series 2.

Replication	Day of Incubation								
	3	5	8	10	12	15	17	19	21
Average	6.15	5.77	5.60	5.70	5.86	5.86	5.61	5.49	5.82
Standard deviation	0.02	0.09	0.17	0.08	0.23	0.62	0.03	0.14	0.07

The structure of industrial bioreactors and large-scale laboratory bioreactors ensure the continuous collection of produced biogas. On one hand, this provides safety (elimination of the possibility of an explosion), on the other hand, it shifts the reaction equilibrium towards biogas production. According to Strömberg et al. [52], gas pressure can be one of the most important factors influencing biogas production. The use of small-volume bottles, tightly sealed with septa, gave rise to the problem of excessive pressure. The standard solution adopted in the experiments involved the aforementioned and discussed procedure of the daily removal of gas from the headspace by piercing the septum with a syringe needle (with simultaneous measurements of the volume of the produced biogas). However, we decided to determine the dynamics of the process under excessive pressure. In fact, the probability of eliminating the need for time-consuming pressure equalization was low. Nevertheless, an exploration of the rate of biofermentation inhibition appeared attractive, particularly given the fact that experiments that can provide an answer to this question are not feasible in industrial bioreactors and large-scale laboratory bioreactors for two reasons: the first reason is related to safety (explosion of such a bioreactor would pose a real threat to the environment); the other is economics (the cost of the bioreactor).

Since the pressure in the bottle at the higher bioreactor loadings (7% and 13%) caused bulging of the septum rubber (or unsealed it), even after daily biogas removal, this stage of the investigation was carried out only at 2% loading. Analysis of Figure 3 reveals that biogas production under excessive pressure at this loading was already inhibited from day four. It can be claimed that biogas yield remains virtually unchanged throughout the incubation period and reaches a level of several tens of per cent per gram of dry weight. When pressure is reduced daily, the amount of produced biogas increases throughout the incubation period. A similar relationship can be noted in Figure 4. The amount of methane produced under excessive pressure hardly changed and remained at a level of several per cent. In turn, when pressure was reduced daily, methane production yield had already exceeded 50% by incubation day 11 and had reached almost 60% by the end of the experiment (day 21).

3.3. Stabilisation of the Biogas Production Process with Bicarbonate

The content of methane in variants with and without the addition of a sodium bicarbonate solution to the input mixture is presented in Figure 5.

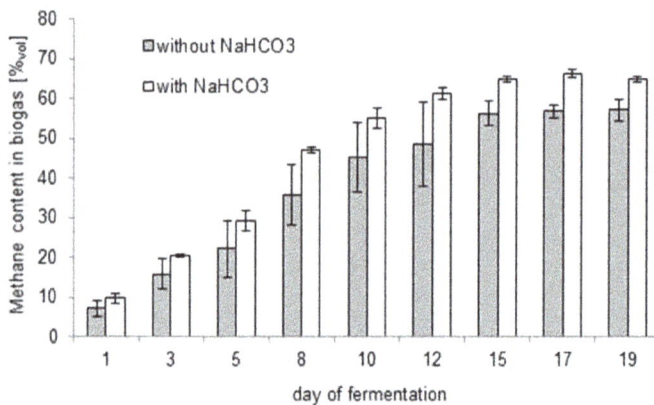

Figure 5. Methane content in the biogas on successive days of incubation in bottles with and without the addition of sodium bicarbonate (series 3).

Analysis of the graph presented in Figure 5 allows the unambiguous conclusion that the addition of bicarbonate not only stabilized the results (with a substantially lower standard deviation for the variant with $NaHCO_3$ supplementation) but also ensured higher methane yields. By day 12 of incubation, methane concentration had already exceeded 60%, i.e., a yield obtained within circa 14 days in the experiment variant presented in Figure 1. Stabilization of pH was confirmed by the results obtained during measurements of the suspension after opening the bottles; pH ranged from 7.04 to 7.48 and did not drop below the threshold of 6.7. The idea of adding a sodium bicarbonate solution to the input suspension was borrowed from the paper by Esposito et al. [14]. The substance is designed to prevent critical pH lowering during the fermentation process, which in extreme cases may lead to the inhibition of methanogenesis [53].

The paper by Mittweg et al. [54] should be mentioned when comparing the results presented above with those reported by other researchers who have used bioreactors with similar volumes.

The authors used 100 mL syringes as fermentation chambers. Their results indicate that the process carried out in such bioreactors can be used in laboratory investigations. The only drawback of this solution is the technical difficulty in reproducing the structure described in their paper. The solution proposed in this study seems to be considerably simpler.

4. Conclusions

Many parameters have to be taken into consideration during research on the methodological aspects of biogas production under laboratory conditions, factors that have undoubtedly influenced the final outcomes of this experiment. Although the highest yield of biogas was reported for the 13% load, in the final results this turned out to be too high for such fixed conditions of the digestion process. The optimal conditions for anaerobic digestion in small bioreactors are to fill two-thirds of the maximum volume with a suspension, using a 2% bioreactor loading (on a dry basis) and to ensure the daily equalization of pressure by the removal of biogas through the septum pierced with a syringe needle followed by intensive mixing of the remaining suspension. The addition of bicarbonate stabilized the results (with a substantially lower standard deviation for the variant with $NaHCO_3$ supplementation) and ensured higher methane yields.

Author Contributions: Concept, research and writing, J.K. and A.K.; Review & Editing, A.K.

Conflicts of Interest: The authors declare no conflict of interest.

References

1. Adebayo, A.; Jekayinfa, S.; Linke, B. Effect of Co-Digestion on Anaerobic Digestion of Cattle Slurry with Maize Cob at Mesophilic Temperature. *J. Energy Technol. Policy* **2013**, *3*, 47–55.
2. Gupta, P.; Gupta, A. Biogas production from coal via anaerobic fermentation. *Fuel* **2014**. [CrossRef]
3. Krzemińska, I.; Pawlik-Skowrońska, B.; Trzcińska, M.; Tys, J. Influence of photoperiods on the growth rate and biomass productivity of green microalgae. *Bioproc. Biosyst. Eng.* **2013**. [CrossRef]
4. Amon, Y.; Amon, B.; Kryvoruchko, V.; Machtmüller, A.; Hopfner-Sixt, K.; Bodiroza, V.; Hrbek, R.; Friedel, J.; Pötsch, E.; Wagentristl, H.; et al. Methane production through anaerobic digestion of various energy crops grown in sustainable crop rotation. *Bioresour. Technol.* **2007**. [CrossRef] [PubMed]
5. Frąc, M.; Jezierska-Tys, S.; Tys, J. Microalgae for biofuels production and environmental applications: A review. *Afr. J. Biotechnol.* **2010**. [CrossRef]
6. Oleszek, M.; Matyka, M.; Lalak, J.; Tys, J.; Paprota, E. Characterization of Sida hermaphrodita as a feedstock for anaerobic digestion process. *J. Food Agric. Environ.* **2013**, *11*, 1839–1841.
7. Chen, Y.; Cheng, J.; Creamer, K. Inhibition of anaerobic digestion precess: A review. *Bioresour. Technol.* **2008**. [CrossRef]
8. Kafle, G.K.; Kim, S.H.; Sung, K.I. Ensiling of fish industry waste for biogas production: A lab scale evaluation of biochemical methane potential (BMP) and kinetics. *Bioresour. Technol.* **2013**. [CrossRef] [PubMed]
9. Pawłowska, M.; Siepak, J. Enhancement of Methanogenesis at a Municipal Landfill Site by Addition of Sewage Sludge. *Environ. Eng. Sci.* **2006**. [CrossRef]
10. Amigun, B.; Sigamoney, R.; Von Blottnitz, H. Commercialisation of biofuel industry in Africa: A review. *Renew. Sustain. Energy Rev.* **2008**. [CrossRef]
11. Lybaek, R.; Christansen, T.; Kjaer, T. Governing innovation for sustainable development in the Danish Biogas Sector—A historical overview and analysis of innovation. *Sustain. Dev.* **2013**. [CrossRef]
12. Gaida, D.; Luis, S.B.A.; Brito, A.; Wolf, C.; Back, T.; Bongards, M.; McLoone, S.F. Optimal control of biogas plant using nonlinear model predictive control. In Proceedings of the ISSC 2011, Dublin, Ireland, 23–24 June 2011.
13. Markard, J.; Stadelmann, M.; Truffer, B. Prospective analysis of technological innovation systems: Identifying technological and organizational development options for biogas in Switzerland. *Res. Policy* **2009**. [CrossRef]
14. Esposito, G.; Frunzo, L.; Liotta, F.; Panico, A.; Pirozzi, F. Bio-Methane potential test to measure the biogas production from the digestion and co-digestion of complex organic substrates. *Open Environ. Eng. J.* **2012**, *5*, 1–8. [CrossRef]

15. Opwis, K.; Mayer-Gall, T.; Gutmann, J.S.; Dammer, C.; Titscher, T.; Nickisch-Hartfiel, A.; Grün, O.; Spurk, C.; Schloderer, C.; Köppe, A.; et al. Semi-industrial production of methane from textile wastewater. *Energy Sustain. Soc.* **2012**. [CrossRef]

16. Kaparaju, P.; Ellegaard, L.; Angelidaki, I. Optimisation of biogas production from manure through serial digestion: Lab-scale and pilot-scale studies. *Bioresour. Technol.* **2009**. [CrossRef] [PubMed]

17. Omil, F.; Garrido, J.; Arrojo, B.; Mendez, R. Anaerobic filter reactor performance for the treatment of complex dairy wastewater at industrial scale. *Water Res.* **2003**. [CrossRef]

18. Singh, K.J.; Sooch, S.S. Comparative study of economics of different models of family size biogas plants for state of Punjab, India. *Energy Convers. Manag.* **2004**. [CrossRef]

19. Wu, S.; Ni, P.; Li, J.; Sun, H.; Wang, Y.; Luo, H.; Dach, J.; Dong, R. Integrated approach to sustain biogas production in anaerobic digestion of chicken manure under recycled utilization of liquid digestate: Dynamics of ammonium accumulation and mitigation control. *Bioresour. Technol.* **2016**. [CrossRef] [PubMed]

20. Montusiewicz, A.; Lebiocka, M. Co-digestion of intermediate landfill leachate and sewage sludge as a method of leachate utilization. *Bioresour. Technol.* **2011**. [CrossRef] [PubMed]

21. Nair, V.V.; Dhar, H.; Kumar, S.; Thalla, A.K.; Mukherjee, S.; Wong, J.W.C. Artificial neural network based modeling to evaluate methane yield from biogas in a laboratory-scale anaerobic bioreactor. *Bioresour. Technol.* **2016**. [CrossRef] [PubMed]

22. Ghanimeh, S.; El Fadel, M.; Saikaly, P. Mixing effect on thermophilic anaerobic digestion of source-sorted organic fraction of municipal solid waste. *Bioresour. Technol.* **2012**. [CrossRef] [PubMed]

23. Asadi, A.; Zinatizadeh, A.A.; Van Loosdrecht, M. A novel continuous feed and intermittent discharge airlift bioreactor (CFIDAB) for enhanced simultaneous removal of carbon and nutrients from soft drink industrial wastewater. *Chem. Eng. J.* **2016**. [CrossRef]

24. Westerholm, M.; Müller, B.; Isaksson, S.; Schnürer, A. Trace element and temperature effects on microbial communities and links to biogas digester performance at high ammonia levels. *Biotechnol. Biofuel* **2015**. [CrossRef] [PubMed]

25. Karim, K.; Hoffmann, R.; Klasson, K.T.; Al-Dahhan, M.H. Anaerobic digestion of animal waste: Effect of mode of mixing. *Water Res.* **2005**. [CrossRef] [PubMed]

26. Getahun, T.; Gebrehiwot, M.; Ambelu, A.; Van Gerven, T.; Van der Bruggen, B. The potential of biogas production from municipal solid waste in a tropical climate. *Environ. Monit. Assess.* **2014**. [CrossRef] [PubMed]

27. Lindmark, J.; Eriksson, P.; Thorin, E. The effects of different mixing intensities during anaerobic digestion of the organic fraction of municipal solid waste. *Waste Manag.* **2014**. [CrossRef] [PubMed]

28. Kim, M.; Ahn, Y.H.; Speece, R.E. Comparative process stability and efficiency of anaerobic digestion; mesophilic vs. thermophilic. *Water Res.* **2002**. [CrossRef]

29. Shen, Y.; Linville, J.L.; Urgun-Demitras, M.; Schoene, R.P.; Snyder, S.W. Producing pipeline-quality biomethane via anaerobic digestion of sludge amended with corn stover biochar with in-situ CO_2 removal. *Appl. Energ.* **2015**. [CrossRef]

30. Durán-García, M.; Ramírez, Y.; Rojas-Solórzano, L. Biogas home-production assessment using a selective sample of organic vegetable waste. A preliminary study. *Interciencia* **2012**, *37*, 128–132.

31. Rojas, C.; Fang, S.; Uhlenhut, F.; Borchert, A.; Stein, I.; Schlaak, M. Stirring and biomass starter influences the anaerobic digestion of different substrates for biogas production. *Eng. Life Sci.* **2010**. [CrossRef]

32. Lo, H.M.; Chiu, H.Y.; Lo, S.W.; Lo, F.C. Effects of micro-nano and non micro-nano MSWI ashes addition on MSW anaerobic digestion. *Bioresour. Technol.* **2012**. [CrossRef] [PubMed]

33. Zhong, W.; Zhang, Z.; Luo, Y.; Qiao, W.; Xiao, M.; Zhang, M. Biogas productivity by co-digesting Taihu blue algae with corn straw as an external carbon source. *Bioresour. Technol.* **2012**. [CrossRef] [PubMed]

34. Kinnunen, H.V.; Koskinen, P.E.P.; Rintala, J. Mesophilic and thermophilic anaerobic laboratory-scale digestion of Nannochloropsis microalga residues. *Bioresour. Technol.* **2014**. [CrossRef] [PubMed]

35. Lee, E.; Cumberbatch, J.; Wang, M.; Zhang, Q. Kinetic parameter estimation model for anaerobic co-digestion of waste activated sludge and microalgae. *Bioresour. Technol.* **2017**. [CrossRef] [PubMed]

36. Kusch, S.; Oechsner, H.; Jungbluth, T. Biogas production with horse dung in solid-phase digestion systems. *Bioresour. Technol.* **2008**. [CrossRef] [PubMed]

37. Kucera, L.; Kurka, O.; Bartak, P.; Bednar, P. Liquid chromatography/high resolution tandem mass spectrometry—Tool for the study of polyphenol profile changes during micro-scale biogas digestion of grape marcs. *Chemosphere* **2017**. [CrossRef]

38. Szafranek-Nakonieczna, A.; Stępniewska, Z. The influence of the aeration status (ODR, Eh) of peat soils on their ability to produce methane. *Wetlands Ecol. Manag.* **2015**. [CrossRef]

39. Kougias, G.P.; Angelidaki, I. Biogas and its opportunities—A review. *Front. Environ. Sci. Eng.* **2018**, *12*, 14. [CrossRef]

40. Schmidt, T.; McCabe, B.; Harris, P. Process Monitoring and Control for an Anaerobic Covered Lagoon Treating Abattoir Wastewater. *Chem. Eng. Technol.* **2018**, *41*, 755–760. [CrossRef]

41. Kozłowski, K.; Dach, J.; Lewicki, A.; Cieślik, M.; Czekała, W.; Janczak, D.; Brzoski, M. Laboratory Simulation of an Agricultural Biogas Plant Start-up. *Chem. Eng. Technol.* **2018**, *41*, 711–716. [CrossRef]

42. Gallegos, D.; Wedwitschka, H.; Moeller, L.; Zehnsdorf, A.; Stinner, W. Effect of particle size reduction and ensiling fermentation on biogas formation and silage quality of wheat straw. *Bioresour. Technol.* **2017**, *245*, 216–224. [CrossRef] [PubMed]

43. Safari, M.; Abdi, R.; Adl, M.; Kafashan, J. Optimization of biogas productivity in lab-scale by response surface methodology. *Renew. Energy* **2018**, *118*, 368–375. [CrossRef]

44. Hilkiah, I.A.; Ayotamuno, M.J.; Eze, C.L.; Ogaji, S.O.T.; Probert, S.D. Designs of anaerobic digesters for producing biogas from municipal solid-waste. *Appl. Energy* **2008**. [CrossRef]

45. Betts, J.I.; Baganz, F. Miniature bioreactors: Current practices and future opportunities. *Microb. Cell Fact.* **2006**. [CrossRef] [PubMed]

46. Oleszek, M.; Tys, J. Lab scale measurement of biogas yield. *Chem. Ind.* **2013**, *92*, 126–130.

47. Lourenço, N.D.; Lopes, J.A.; Almeida, C.F.; Sarraguça, M.C.; Pinheiro, H.M. Bioreactor monitoring with spectroscopy and chemometrics: A review. *Anal. Bioanal. Chem.* **2012**. [CrossRef] [PubMed]

48. Frąc, M.; Ziemiński, K. Methane fermentation process for utilization of organic waste. *Int. Agrophys.* **2012**. [CrossRef]

49. Lalak, J.; Kasprzycka, A.; Martyniak, D.; Tys, J. Effect of biological pretreatment of Agropyron elongatum 'BAMAR' on biogas production by anaerobic digestion. *Bioresour. Technol.* **2016**, *200*, 194–200. [CrossRef] [PubMed]

50. Angelidaki, I.; Sanders, W. Assessment of the anaerobic biodegradability of macropollutants. *Rev. Environ. Sci. Biol. Technol.* **2014**, *3*, 117–129. [CrossRef]

51. Zhang, S.; Zhang, P.; Zhang, G.; Fan, J.; Zhang, Y. Enhancement of anaerobic sludge digestion by high-pressure homogenization. *Bioresour. Technol.* **2012**. [CrossRef] [PubMed]

52. Strömberg, S.; Nistor, M.; Liu, J. Towards eliminating systematic errors caused by the experimental conditions in Biochemical Methane Potential (BMP) tests. *Waste Manag.* **2014**, *34*, 1939–1948. [CrossRef] [PubMed]

53. Piątek, M.; Lisowski, A.; Kasprzycka, A.; Lisowska, B. The dynamics of an anaerobic digestion of crop substrates with an unfavourable carbon to nitrogen ratio. *Bioresour. Technol.* **2016**, *216*, 607–612. [CrossRef] [PubMed]

54. Mittweg, G.; Oechsner, H.; Hahn, V.; Lemmer, A.; Reinhardt-Hanisch, A. Repeatability of a laboratory batch method to determine the specific biogas and methane yields. *Eng Life Sci.* **2016**. [CrossRef]

energies

MDPI

Article

An Assessment of the Sustainability of Lignocellulosic Bioethanol Production from Wastes in Iceland

Sahar Safarian * and Runar Unnthorsson *

Department of Industrial Engineering, Mechanical Engineering and Computer Science, University of Iceland, Hjardarhagi 6, 107 Reykjavik, Iceland
* Correspondence: safarian.sahar@gmail.com (S.S.); runson@hi.is (R.U.)

Received: 30 April 2018; Accepted: 4 June 2018; Published: 7 June 2018

Abstract: This paper describes the development of a model to comprehensively assess the sustainability impacts of producing lignocellulosic bioethanol from various types of municipal organic wastes (MOWs) in Iceland: paper and paperboard, timber and wood and garden waste. The tool integrates significant economic, energy, environmental and technical aspects to analyse and rank twelve systems using the most common pretreatment technologies: dilute acid, dilute alkali, hot water and steam explosion. The results show that among the MOWs, paper and paperboard have higher positive rankings under most assessments. Steam explosion is also ranked at the top from the economic, energy and environmental perspectives, followed by the hot water method for paper and timber wastes. Finally, a potential evaluation of total wastes and bioethanol production in Iceland is carried out. The results show that the average production of lignocellulosic bioethanol in 2015 could be 12.5, 11 and 3 thousand tons from paper, timber and garden wastes, respectively, and that production could reach about 15.9, 13.7 and 3.7 thousand tons, respectively, by 2030.

Keywords: bioethanol; sustainability assessment; lignocellulosic wastes; pretreatment

---

1. Introduction

Negative social, political and environmental impacts of fossil fuels as well as energy security concerns have spurred interest in nonpetroleum energy sources [1]. Among the various alternative energy sources, biomass has garnered substantial interest because it is the only suitable and renewable primary energy resource that can provide alternative transportation fuels [2]. Bioethanol has long been put forward as the most promising biofuel used either as a sole fuel in cars with dedicated engines or as an additive in fuel blends, requiring no engine modifications until the mix reaches 30% [3].

The main feedstocks of bioethanol are sugar- and starch-based materials such as sugarcane and grains. However, there are considerable debates about the sustainability of such feedstocks. Lignocellulosic materials, a third group of feedstocks, represents the most viable option for bioethanol production. Increasing food demand and the need to feed an increasing global population could make conventional agricultural crops less competitive and more costly sources compared to lignocellulosic materials [4–6].

In this context, municipal organic wastes (MOWs) represent one of the most abundant lignocellulosic materials and acquires significant importance for bioethanol production. The replacement of biomass with MOW can provide environmental advantages, particularly with regard to waste management, carbon dioxide, quality and quantity control of water, land use and biodiversity [7]. However, the as yet immature technologies and challenging logistics for sourcing waste pose barriers to utilizing this potential source [4].

Based on these concepts, this paper focusses on bioethanol production from MOWs in Iceland. Iceland possesses no fossil energy resources, but in comparison to its population of 332,529, the country has huge amounts of hydroenergy and geothermal energy. Economic estimations of the country's hydroelectric energy show that about 30 TWh/year may be harnessed, but only 15% is being used. Likewise, geothermal energy has been estimated at about 200 TWh/year, of which only 1% has been harnessed. Approximately 90% of buildings in Iceland are heated with geothermal water, and the aluminum and ferrosilicon industries are powered by hydroelectric energy, consuming 4.25 TWh/year. However, the transportation and fishing sectors are powered totally by imported fossil fuels [8]. Hence, utilization of biofuel can be useful as a sustainable mode for transitioning to green transportation and fishing infrastructure.

An overview of different Icelandic municipalities is shown in Figure 1a, the Capital area (pink), South peninsula (blue), South (yellow), West (purple), East (orange), Northeast (red), Eyjafjörður (turquoise), Northwest (brown) and Westfjords (green), [8]. Although Iceland is large by land mass, more than 60% of the population lives in the capital area and about 80% live in the capital and south regions combined (Figure 1b). However, the capital area occupies only 1% of the total area of Iceland (Figure 1c). This is due to the development of urbanization in this location, poor climate and geographical conditions in other regions and slow and expensive transportation.

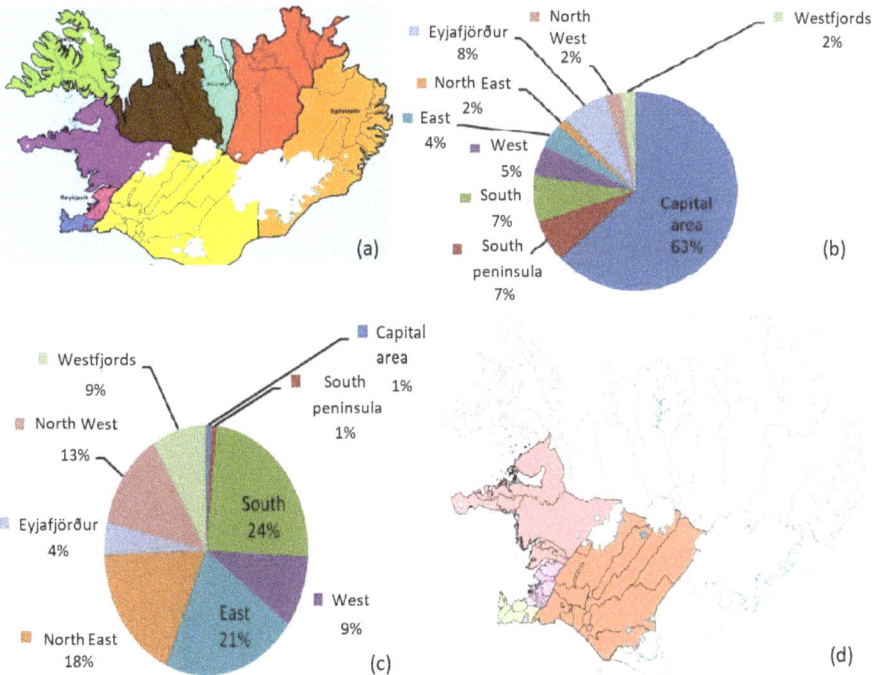

Figure 1. Map and information of Iceland, (**a**) all municipalities, (**b**) population share, (**c**) area share and (**d**) major region.

Climate and weather conditions are more likely to limit farming in Iceland than soil type [9,10]. Even though arable land is available, vast portions of the areas with suitable soil type have climate and weather unsuitable for vegetation growth. In fact, the crop growth window in Iceland is 130 days in the summer (7 May to 15 September). Thus, the only products that can be developed must have a growing time within this period and the temperature required for growth must be compatible with the environment temperature (5–7 °C). This situation means that grasses turn green and tillage can be performed, but little else is grown [9,11].

Based on these conditions, biofuel production efforts in Iceland should focus on MOW as part of the second-generation biomasses. Despite lack of agricultural crops, waste is recognized as a continual source of biomass and is independent of climate conditions and other immutable factors. Furthermore, because the amounts of various wastes produced in different locations are fairly constant and measurable, reasonable estimation of biofuel production capacity should be possible.

To establish sustainable bioethanol production plans, they should be evaluated from different sustainable development perspectives and compared based on different initial feedstocks and applied conversion technologies. Sustainable production scenarios of bioethanol depend on economic, technical, environmental, social and political factors as well as energy balance and are essential elements of sustainable development [12]. The authors are not aware of any published studies assessing bioethanol production systems based on the sustainable development dimensions in Iceland. Therefore, the goals of this study are: (i) to assess 1-ton bioethanol production using three different Icelandic MOWs: paper and paperboard, timber and wood and garden waste; (ii) to develop a model based on energy and material flows to assess the economic (cost and benefit), energy (energy balance, total input/output energy and energy use efficiency) and environmental (GHG emissions, water consumption and supply rate) sustainability criteria; (iii) to analyse and compare the sustainable impacts of different biomass-derived bioethanol, pretreatments and conversion technologies; and (iv) to evaluate the potential of total waste and bioethanol production in Iceland between 2015 and 2030.

2. System Description

The system considered by the model including all the process steps from resources to end products (see Figure 2).

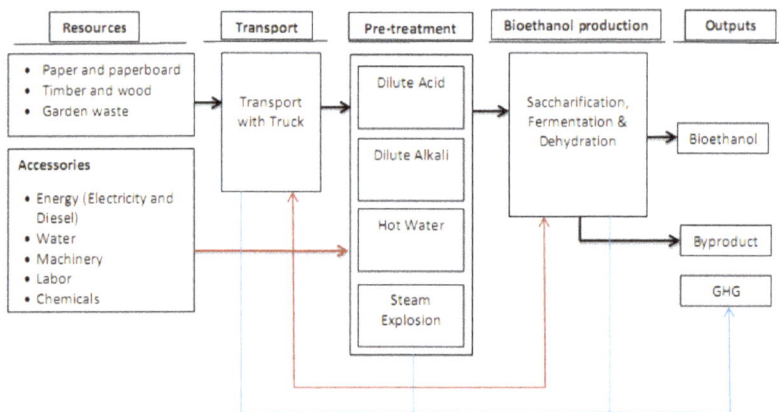

Figure 2. System boundaries considered by in this study including five stages of processes, conversion technologies and associated inputs (energy and material).

2.1. Resource Level

Paper and paperboard, timber and wood and garden waste are considered as the main input resources. According to Icelandic law, all municipalities are obligated to have a waste management plan to organize waste produced in the municipality. Thus, beginning in 2004, common project management groups were established to plan for waste produced in the main municipality parts of Iceland: Western Iceland (pink), Suðurnes (green), the Capital Area (purple) and Southern Iceland (orange) (Figure 1d) [13]. Waste collection and treatment in these places are mostly operated by four Icelandic companies (Table 1).

Table 1. Information regarding waste operation in the main parts of Iceland in 2006.

Participating Areas	Operator Company	No Municipalities	Waste Quantity (1000 tons)
Western Iceland	Sorpurðun Vesturlands	10	29
Suðurnes	Kalka	5	22
Capital area	Sorpa	8	272
Southern Iceland	Sorpstöð Suðurlands	11	33

About 90% of the wastes generated are related to the capital area and are collected by the Íslenska Gámafélagið and Sorpa companies [14,15]. Due to the low number of inhabitants in Iceland, there is limited economic basis for recycling facilities like paper mills. Plastics and paper items are thus shipped abroad, as there is no facility in Iceland to recycle and reuse them. At present, landfilling and incineration are the dominant methods used for the disposal of municipal solid waste in Iceland, and most landfill sites are open dumping areas, which poses serious environmental and social threats. Moreover, the flue gas from waste incineration can contaminate the environment if not handled with appropriate technologies, such as using combined heat and power (CHP) [16] and flue gas cleaning systems. A small incinerator in Skutulsfjörður at Northwest Iceland resulted in the discovery of cancerous and toxic chemicals in Icelandic meat and milk in 2011 [17]. Dioxin levels in the fly-ash from the nearby incinerator were more than 20 times the EU limits [17]. Moreover, soil and incinerator emissions measurements at several sites in Iceland such as Kirkjubaejarklaustur revealed that emissions were 85 times the EU limits [6,18]. These problematic findings resulted in the widespread testing of soil across Iceland [17], shutdown of several incinerators [18], withdrawal of some Icelandic meat and milk from the markets and culling of all the livestock on the farms impacted [17]. In an attempt to reduce these problems, this study considers diverting paper and paperboard, timber and wood and garden waste as portions of MOW from the waste companies (mostly Sorpa) before they are burnt in incinerators. These materials could then be transferred to preprocessing units and conversion plants to produce bioethanol. Lignocellulosic bioethanol is proposed as a viable alternative to MOWs treatment, and may solve the problems of using huge amounts of land for landfill and high emissions from incineration.

In addition, energy, water, labor, chemicals and machinery are considered accessory inputs. Diesel fuel is used by trucks for transportation and electricity is consumed in driving force and heat generation in process units. It has been reported that 25% and 75% of electricity production in Iceland comes from geothermal and hydropower, respectively. Thus, hydro is currently Iceland's main source of clean energy; however, the electricity required in the capital area, as the largest producer of waste, is supplied by geothermal power plants in Hellisheiði and Nesjavellir [19].

2.2. Transport Level

In this study, the transport step includes waste transport from waste operator companies to preprocessing units. As pretreatment plants and bioethanol production centres should be next to each other, transportation of the liquid containing cellulose from preparation facilities to bioethanol production units is not included.

In this study, a 50 km waste transportation distance by truck is used. The data used for fuel calculations was obtained from published studies [20,21]. The energy equivalent for the transportation of waste is about 3 MJ ton^{-1} km^{-1} (0.06 L ton^{-1} km^{-1}).

2.3. Pretreatment Level

Lignocellulosic materials are composed of cellulose, hemicelluloses and lignin in an intricate structure which is resistant to decomposition. One of the best strategies to convert such biomass into sugars is enzymatic saccharification due to its low energy requirement and comparatively smaller environmental impact; however, the main problem is the low accessibility of cellulose because of the rigid association of cellulose with lignin [22,23]. Therefore, removal of lignin, making cellulose more accessible to enzymatic hydrolysis for conversion, is the primary pretreatment process investigated herein.

Pretreatment techniques are mainly classified as physical (e.g., grinding and milling), chemical (e.g., alkali, acid, ammonia percolation), physio-chemical (e.g., steam explosion, ammonia fiber expansion) and biological (e.g., fungi and Actinomycetes) [24,25]. Most of these technologies are not preferred for industrial applications due to high energy demand, inability to remove lignin and substantial chemical requirements. Among the pretreatment techniques, dilute acid (DA), dilute alkali (DAL), hot water (HW) and steam explosion (SE) are modeled in this study and are among the most effective and the most promising for industrial applications [23,25,26].

Dilute acid pretreatment is one of the most frequently employed methods of structurally breaking down lignocellulosic biomass and removing the hemicellulose [27]. During this method, biomass is treated at different combinations of temperatures (100–290 °C), at a pressure of 1.5 bar [28], with residence times ranging from a few seconds to several hours. During hydrothermal pretreatment, most of the hemicellulose is hydrolyzed to sugar monomers and becomes soluble, a fraction of cellulose is depolymerized into glucose and a fraction of lignin is dissolved and distributed.

Dilute alkali pretreatment refers to the application of alkaline solutions such as NaOH, Ca(OH)$_2$ or ammonia to remove lignin and a part of the hemicellulose, increasing the accessibility of enzymes to the cellulose. Pretreatment can be performed at low temperatures but with a relatively long residence time and high base concentration [23,24,29].

In the hot water pretreatment method, water under high pressure penetrates the biomass, hydrating the cellulose and removing the hemicellulose and part of the lignin. The method's major benefits are that is does not require chemicals or corrosion-resistant materials for the hydrolysis reactors. Moreover, feedstock size reduction is a highly energy demanding operation for the huge bulk of materials, while there is no need for size reduction in HW pretreatment [23,30].

Among the physico-chemical processes, steam explosion has received considerable attention as a pretreatment for ethanol production. In steam explosion, the pressure is suddenly reduced, causing the materials to undergo explosive decompression. High pressure and high temperature (160–260 °C) are used in steam explosion for a few seconds to several minutes [23,31]. The process has been used in lab and pilot-scale units by several research groups. Moreover, steam explosion has been found effective on feedstock with large particle sizes, thus reducing the energy required for size reduction, and it can satisfy all the requirements of the pretreatment process [24].

2.4. Bioethanol Production Level

After the pretreatment step, the hydrolyzed stream goes to the saccharification and fermentation plants. First, cellulase enzymes are formed in the fermentation reactor and then the cellulase enzymes are moved to the saccharification reactor, where sugars are formed. Nutrient loadings and the efficiencies of the fermentation sugar conversion (95% for glucose, 85% for xylose and arabinose) are based on the National Renewable Energy Laboratory (NREL) process [32]. At the next step, the enzymes are recycled back to the cellulose reactor and the sugars are sent to a fermentation reactor, where ethanol is produced. The ethanol is then purified by distillation and dehydration in the ethanol recovery unit and then sent to product storage. Bottom effluent from this column contains mostly lignin from biomass and can be combusted using a combined burner, boiler and turbo-generator to produce electricity; while this electricity can be used to supply the ethanol plant, selling this byproduct is considered in our study.

3. Model Description and Data

This paper assessed the lignocellulosic bioethanol supply chain (LBSC) shown schematically in Figure 2. The supply chain consists of five layers: resources, transport, pretreatment, bioethanol production and outputs. The model equations and constraints were structured based on the LBSC flow diagram and they represent the material and energy flows from resources to products. The constraints of the model were described from use sectors back to the resources. This approach reflects the basic concept of the model, which is based on a demand-oriented analytical tool. This approach stresses the point that balance of demand and supply ought to be ascertained [33–35].

3.1. Bioethanol Production Level

Supply and demand balance is a characteristic of the model and it is guaranteed through the demand constraint expressed by Equation (1) [35]:

$$X_{C,be,\tau_C} \geq D_{be} \tag{1}$$

where D_{be} is demand for bioethanol and X_{C,be,τ_C} is exit bioethanol flow from the fermentation unit [36].
The following equation ensures the balance between inputs and outputs of the bioethanol production step:

$$X_{C,be,\tau_C} \leq X_{E,c,\tau_E} \times \eta_{C_{Ferm},c-be,\tau_C} \times \eta_{C_{Dehyd},c-be,\tau_C} \tag{2}$$

Equation (2) is a constraint which shows that the total amount of bioethanol produced from technology τ_C equals the amount of liquid delivered to the refinery from the pretreatment step, multiplied by the associated energy conversion efficiencies of the fermentation and dehydration units, $\eta_{C_{Ferm},c-be,\tau_C}$ and $\eta_{C_{Dehyd},c-be,\tau_C}$ [36].
The quantity of lignin as a byproduct of fermentation is based on inequity expressed by Equation (3):

$$X_{C,bp,\tau_C} \leq X_{T,b,\tau_T} \times x_{b,bp} \times \eta_{E,b-bp,\tau_E} \times \eta_{C,b-bp,\tau_C} \tag{3}$$

However, in addition to the fermentation step, the amount of lignin production is dependent on the type of method used for pretreatment. X_{C,be,τ_C} represents the lignin produced from cellulose to bioethanol conversion technology, X_{T,b,τ_T} is the biomass type b delivered to the preparation unit, $x_{b,bp}$ is the percent lignin for different biomasses, and $\eta_{E,b-bp,\tau_E}$, $\eta_{C,b-bp,\tau_C}$ are lignin removal efficiencies for different pretreatment techniques and conversion steps, respectively.

3.2. Pretreatment Level

The balance between the inputs and outputs of the preparation step is guaranteed by the constraint expressed by Equation (4):

$$X_{E,c,\tau_E} \leq X_{T,b,\tau_T} \times \eta_{E,b-c,\tau_E} \tag{4}$$

This equation indicates that the total amount of liquid containing cellulose extracted under pretreatment processes should equal the amount of feedstock transported to the plant, multiplied by the performance efficiency associated with different pretreatment technologies ($\eta_{E,b-c,\tau_E}$) [36].

3.3. Transport Level

The mass balance of transportation level is given by Equation (5):

$$X_{T,b,\tau_T} \leq X_{A,b} \times (1 - l_{T,b,\tau_T}) \tag{5}$$

where $X_{A,b}$ is the amount of biomass type b shipped from the harvesting facility to the biorefinery plant via transportation mode τ_T; l_{T,b,τ_T} is the fraction of biomass type b lost through transportation; and X_{T,b,τ_T} represents the total amount of delivered biomass type b entering the pretreatment units [37].

3.4. Resource Level

The mass balance for the feedstocks production step is given by Equation (6):

$$X_{A,b} \leq X_{R,b} \times \varepsilon_b \tag{6}$$

where $X_{A,b}$ is the amount of waste feedstock type b that is bought from waste operators; this amount must be equal to or lower than the total waste collected by various waste companies ($X_{R,b}$), taking into account ε_b as the percentage availability of wastes.

Finally, the feasibility of biomass type b is ensured by the constraint expressed by Equation (7):

$$X_{R,b} \leq X_{0,R,b} \tag{7}$$

This equation states that the total amount of collected waste of type b ($X_{R,b}$) must be lower than the total amount of generated resource type b ($X_{0,R,b}$).

3.5. Total Material and Machinery Consumption

The total quantity of different required chemicals (CB_{b-be}) and machinery (M_{b-be}) are calculated based on Equations (8) and (9):

$$CB_{b-be}(kg) = \sum_L \sum_i X_{(L-1),b,\tau_{(L-1)}} \times cb_{L,i,\tau_L} \tag{8}$$

$$M_{b-be}(hr) = \sum_L \sum_i X_{(L-1),b,\tau_{(L-1)}} \times m_{L,b,i,\tau_L} \tag{9}$$

where $X_{(L-1),b,\tau_{(L-1)}}$ is the flow to each level L. cb_{L,i,τ_L} is the specific consumption of chemical type i (sulphuric acid, Ca hydroxide, Diammonium phosphate (DAP), cellulose, yeast) and m_{L,b,i,τ_L} is the specific consumption of machinery type i required for technology τ_L in level L. The specific consumption of different chemicals for various bioethanol production systems is shown in Table 2 [24].

Table 2. Energy and chemicals consumption through the various bioethanol production systems.

Chemicals	Units	Dilute Acid	Dilute Alkali	Hot Water	Steam Explosion
Water	m^3/ton_b	1.422	1.44	1.387	0.934
Sulphuric acid	kg/ton_b	49.04	11.40	0.00	0.00
Ca hydroxide	kg/ton_b	23.88	0.00	0.00	0.00
DAP	kg/ton_b	0.316	0.32	0.32	0.32
Cellulase	kg/ton_b	62.40	72.00	72.00	68.00
Yeast	kg/ton_b	0.79	0.79	0.79	0.79
Sodium hydroxide	kg/ton_b	0.00	0.40	0.00	0.00
Electricity	kwh/ton_b	133.44	123.50	123.28	124.00
Steam	MJ/ton_b	4870.80	4731.60	4936.80	3102.00

3.6. Potential of Bioethanol Production

The total potential of bioethanol production from various types of waste ($X_{C,b-be,\tau_C}$) is calculated by using the total amount of produced waste type b ($X_{A,b}$) and the yield of bioethanol for various feedstocks (y_{b-be}) [8]:

$$X_{C,b-be,\tau_C}(ton) \leq X_{A,b}(ton) \times y_{b-be}\left(\frac{ton_{be}}{ton_b}\right) \tag{10}$$

3.7. Modeling of Sustainability Indicators

In this section, an assessment model is presented that evaluates the sustainability impacts in the bioethanol supply chain. In the initial stages of the study, we reviewed several primary indicators considered in [35,38,39], as well as reviewing other research works [40,41]. Finally, by consulting several studies on the energy sector, we organized five significant sustainability indicators by considering economic, energy, and environmental dimensions that cover most of the aspects mentioned by other literature on this topic.

3.7.1. Economic Indicators

Total production cost, which is the most significant economic impact for economic evaluation of bioethanol systems, is estimated by Equation (11) [35]:

$$C_{total}(\$) = \sum_{L} \left(C_{L,b,fix} + C_{L,b,var} \right) \tag{11}$$

This equation includes fixed and variable setup costs for biomass type b for different supply levels of biomass production, transportation, preparation and conversion costs.

Equation (12) presents the benefit per cost (BPC), which is the ratio between sales income (from produced bioethanol and lignin) and total production costs throughout the entire bioethanol supply chain. The input costs, including capital and operational costs, and output prices for bioethanol production are provided in Table 3 [24,42–44]:

$$BPC = \frac{\text{Total revenue}}{\text{Total production cost}} \quad \frac{X_{C,be,\tau_C} \times Pr_{be} + X_{E,bp,\tau_E} \times Pr_{bp}}{C_{total}} \tag{12}$$

Table 3. Inputs costs and output prices in bioethanol production system.

Items	Unit	Quantity
Waste cost	$/kg	0.014
Chemicals cost		
a. Sulphuric acid	$/kg	0.035
b. Ca hydroxide	$/kg	0.1
c. DAP	$/kg	0.21
d. Cellulase	$/kg	0.52
e. Yeast	$/kg	2.3
f. Sodium hydroxide	$/kg	0.45
Water cost (usage fee)	$/m³	0.32
Energy cost		
a. Electricity	$/kwh	0.03
b. Truck diesel	$/L	1.08
Capital cost of bioethanol production by using:		
a. Dilute acid	$/ton_b	457.52
b. Dilute alkali	$/ton_b	410.93
c. Hot water	$/ton_b	407.78
d. Steam explosion	$/ton_b	363.45
Product price		
Bioethanol	$/kg	0.85
Lignin	$/kg	0.25

3.7.2. Energy Indicators

All stages of the bioethanol production process are based on the consumption of specific sources; therefore, to properly assess energy inputs and outputs, it is necessary to convert all inputs and outputs into their energy equivalents. The energetic efficiency of the bioethanol system is evaluated by the energy ratio between outputs and inputs as shown in Equation (13) [35,45]:

$$\text{EUE} = \frac{\text{Total output energy}}{\text{Total input energy}} = \frac{X_{C,be,\tau_C} \times ec_{be} + X_{E,bp,\tau_E} \times ec_{bp}}{\sum_L \sum_{in} u_{L,b,in,\tau_L} \times ec_{in}} \tag{13}$$

where bioethanol and lignin are considered as output energies and water, fuel, power, chemicals and machinery demand energy input type *in* in supply level *L* for bioethanol production u_{L,b,in,τ_L}). A comprehensive inventory of various inputs for the whole system, based on the studied pretreatment methods, is gathered with the energy coefficients for all inputs (ec_{in}) in Table 4 [24,45].

Table 4. Energy coefficients for different inputs and outputs to the bioethanol production system.

Input	Unit	Energy Coefficient (MJ/unit)
Diesel fuel	L	47.80
Electricity	kWh	11.93
Chemicals and machinery in bioethanol production	ton$_b$	37.5
Bioethanol	kg	29.3
Lignin	kg	21.13

3.7.3. Environmental Indicators

Greenhouse gas (GHG) emissions are one of the most significant environmental indicators, representing the emissions of the major greenhouse gases (CO_2, CH_4 and NO_2) during the unit life cycle. In this study, total GHG emissions over the bioethanol production cycle is calculated by Equation (14) [35]:

$$\text{GHG } (kgCO_2eq) = \sum_L \sum_{in} u_{L,b,in,\tau_L} \times ef_{in} \tag{14}$$

where u_{L,b,in,τ_L} is the required input type *in* for bioethanol production from biomass *b* in level *L* based on technology τ_L, and ef_{in} are the CO_2 emission factors for different inputs; this factor is considered 2.76 kgCO_2eq/L and 0.058 kgCO_2eq/kWh for diesel fuel and electricity generated from geothermal, respectively. Moreover, during fermentation, sugars are converted to ethanol and carbon dioxide. Approximately, 720 kg CO_2 is produced per ton of bioethanol produced [21,46,47]. Firstly, in hydrolysis the cellulose is converted into glucose sugars. Then, sugars are converted to ethanol and carbon dioxide [48].

To facilitate a fair comparison between bioethanol systems, water used over the entire system (W_{b-be}) is measured by the water consumption indicator expressed by Equation (15) [35]:

$$W_{b-be} (m^3) = \sum_L X_{L,b,\tau_L} \times w_{L,b,\tau_L} \tag{15}$$

where X_{L,b,τ_L} is the exit flow from level *L* based on technology type τ_L and w_{L,b,τ_L} is the specific water consumption for each level and technology (which are given in Table 2).

4. Results and Discussion

4.1. Economic Assessment

The model results for the 12 alternatives, ranked according to their contribution to production costs and benefit per cost for 1 ton of bioethanol produced, are shown in Figure 3. This ordering is based on total production cost, which is between 1300 and 2700 $/ton, which are the highest and lowest economically beneficial options, respectively. Moreover, the coloured portions of each bar on the chart indicate the percentage impact of each type of cost for different bioethanol supply levels (BC: biomass cost, TC: transportation cost, CC: conversion cost including pretreatment, fermentation and dehydration costs). Obviously, TC is much lower than the other levels and hence is not shown clearly in Figure 3.

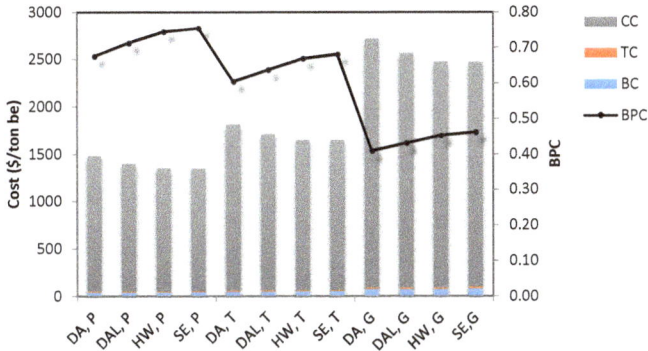

Figure 3. Production costs and BPC for 1 ton bioethanol produced from different wastes; BPC: benefit per cost, BC: biomass cost, TC: transportation cost, CC: conversion cost.

Among the studied wastes, paper and paperboard rank the highest economically; the average total cost of bioethanol production from timber and wood and garden waste is 1.2 and 1.9 times that of paper, respectively. Actually, the higher cellulose percent in lignocellulosic materials means more bioethanol yield can be obtained. In contrast, for a specific amount of bioethanol production, a lower feedstock with a high percentage of cellulose is required. The cellulose percentage of paper, timber and garden wastes are approximately 55%, 45% and 30%, respectively [49,50]. Thus, paper waste is the richest in cellulose and thus has the minimum feedstock requirements for 1 ton of produced bioethanol (Figure 4). Subsequently, this option has the lowest percentage of costs from biomass and transportation due to the lower amount of required feedstock and the higher bioethanol yield. In this way, the costs of waste purchase and transportation for timbers and garden wastes are approximately 1.20 and 1.85 times that for paper.

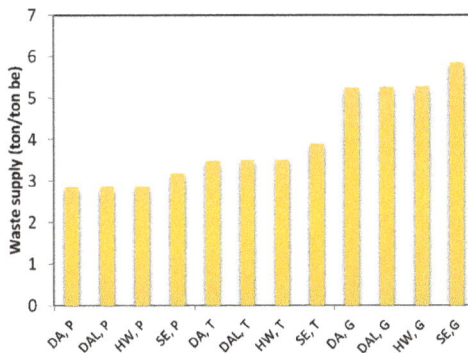

Figure 4. Waste required to produce 1 ton bioethanol; P: paper and paperboard, T: timber and wood, G: garden waste, DA: dilute acid, DAL: dilute alkali, HW: hot water, SE: steam explosion.

As seen in Figure 4, the bioethanol system derived from paper via steam explosion (SE,P) significantly outranks all other systems from the viewpoint of BPC, owing to the favorable results obtained in the economic sector. Although the ethanol yields for steam explosion were relatively low (0.39, 0.32 and 0.21 L_{be}/ton_b for paper, timber and garden wastes, respectively) due to the comparatively low efficiency of cellulose hydrolysis, it has the lowest capital costs because of the high solid loading assumptions during pretreatment and hydrolysis processes. Furthermore, no chemicals are required for SE pretreatment and it requires the lowest amounts of steam and electricity for its various processes.

4.2. Energy Assessment

Total input energy, total output energy and energy ratio between outputs and inputs (EUE) for 1 ton of bioethanol produced are shown in Figure 5. The total output energy for various feedstocks includes the energy content in 1 ton of bioethanol and lignin production. Obviously, due to the higher required amount of timber and garden waste, more lignin is produced. Significantly, the system based on steam explosion and paper ranks the highest among all the systems in terms of energy. This is because no chemicals except water are required in SE and the smallest amount of steam is consumed during this process. On the contrary, the dilute acid method, which requires 1.08 and 1.6 times the amounts of heat and electricity required for SE, also has the highest usage of chemicals and is ranked lowest energetically.

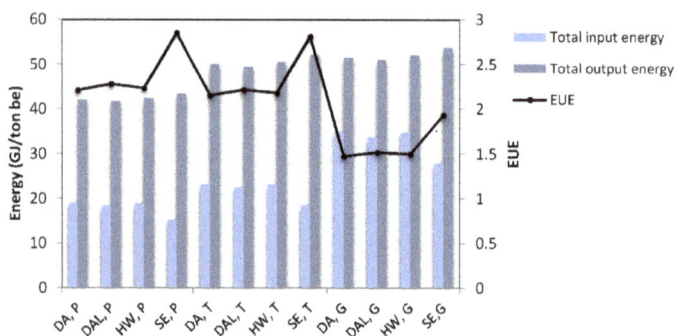

Figure 5. Total input energy, total output energy and EUE for 1 ton bioethanol production from different wastes.

4.3. Environmental Assessment

Figure 6 ranks all the studied bioethanol systems with regard to their contribution to environmental development. This ordering is based on total GHG emission, which is lowest for the steam explosion and hot water processes (about 880 kgCO$_2$eq/ton). The environmental impact of SE and HW are additionally limited because few chemical agents are used in these methods.

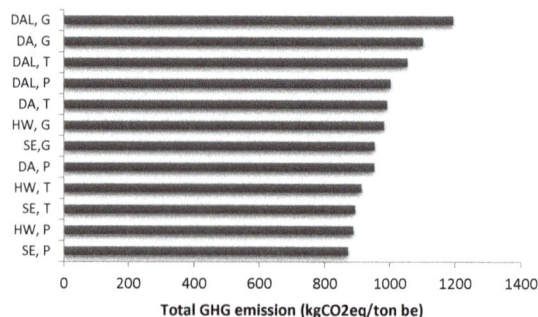

Figure 6. Total GHG emission for 1 ton of bioethanol produced by using different wastes.

Figure 7 shows total water consumption over the entire waste-to-bioethanol process for all the production options. The systems based on steam explosion and paper and timber have the lowest water use, 2.98 and 3.64 m^3/ton, respectively. According to Kumar and Murthy [24], the amount of water required for cooling is lowest for the steam explosion process (401.4 kg/L of ethanol) due to higher solid loading, which decreases the flow rates of streams and energy consumption for cooling. Moreover, bioethanol systems based on dilute alkali and dilute acid for garden wastes require the maximum amounts of water. Obviously these systems would not be practical for regions lacking sufficient available water resources.

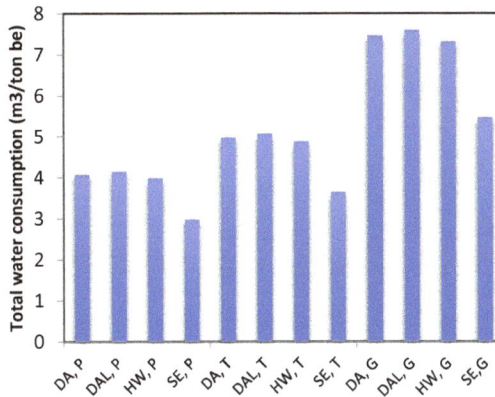

Figure 7. Total water consumption for 1 ton of bioethanol produced by using different wastes.

4.4. Potential of Bioethanol Production

The potential of different wastes and bioethanol production between 2015 and 2030 are shown in Figure 8. The data presented here are drawn from [8]. The main sources of paper and paperboard waste that are suitable for bioethanol production are newspapers, magazines and packaging waste. In addition, the main sources of timber waste are timber from construction, demolition work, packaging waste and pallets. Garden waste can be also defined as grass, branches and other garden waste.

The total amount of paper, timber and garden wastes in Iceland in 2015 can be estimated as approximately 37, 40 and 16 thousand tons, respectively, and they can estimated to increase to about 47, 49 and 20 thousand tons by 2030 assuming 0.8%, 0.6% and 0.6% growth rate per capita, respectively.

Furthermore, Figure 8 shows predictions of ethanol production from various feedstocks. This calculation is based on the average yield of bioethanol production (Equation (10)) calculated as 339, 277 and 185 (kg$_{be}$/ton$_b$) for paper, timber and garden wastes, respectively. The average potential of bioethanol production in Iceland in 2015 was thus approximately 12.5, 11 and 3 thousand tons from paper, timber and garden wastes, respectively, which may reach about 15.9, 13.7 and 3.7 thousand tons by 2030.

Moreover, the potential of bioethanol production employing different pretreatment technologies and various MOWs in 2015 is depicted in Figure 9. It is observed that maximum ethanol production is possible using dilute acid, dilute alkali, hot water and steam explosion. Ethanol yield is relatively low for the steam explosion system due to its comparatively low efficiency of cellulose hydrolysis.

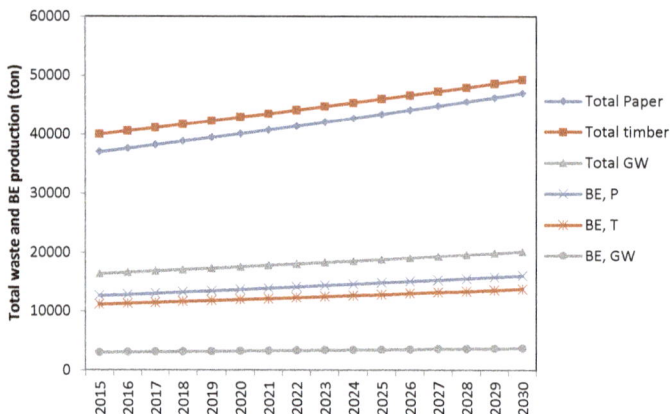

Figure 8. Total amount of various wastes and bioethanol production between 2015 and 2030.

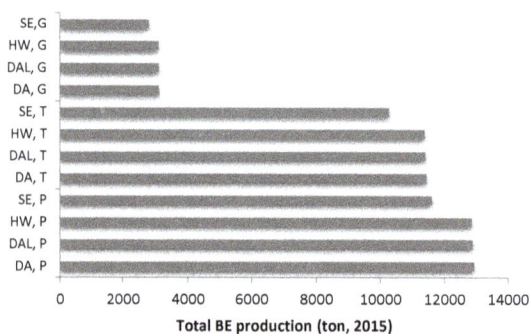

Figure 9. Total bioethanol production in 2015.

4.5. Comparison of Alternatives

The rank orders for the 12 alternative bioethanol systems considered herein across the studied sustainability indicators are shown in Table 5.

Table 5. Rank order of bioethanol systems based on sustainability indicators.

Systems	BPC	EUE	GHG Emission	Water Consumption	Production Yield
SE,P	1	1	1	1	4
HW,P	2	4	2	3	3
DAL,P	3	2	9	5	2
DA,P	5	5	5	4	1
SE,T	4	3	3	2	8
HW,T	6	7	4	6	7
DAL,T	7	6	10	8	6
DA,T	8	8	8	7	5
SE,G	9	9	6	9	12
HW,G	10	11	7	10	11
DAL,G	11	10	12	12	10
DA,G	12	12	11	11	9

SE,P is the top ranked system for nearly all impact indicators, except for production yield, for which it is in fourth place. DA, DAL and HW for garden wastes have the lowest ranks for almost all of the indicators. In addition to SE,P, more economic and environmentally oriented policies should favor HW,P and SE,T. Moreover, DAL,P ranks high for most indicators except for environmental, due to its high GHG emissions and moderate water consumption. DA,P has ranks low for most of the indicators, except for technically occupying the highest position for production yield. Finally, among the MOWs, paper and paperboard rank highest in most assessments.

5. Conclusions

This study presented a comprehensive sustainability assessment model covering the entire lifecycle of bioethanol production (resources, transport, pretreatment and production) to produce lignocellulosic bioethanol from various types of MOWs in Iceland: paper and paperboard, timber and wood and garden waste. Moreover, we carried out an analysis and comparison of the 12 systems based on sustainability impacts (economic, energy, environmental and technical) using the most common pretreatment technologies: dilute acid, dilute alkali, hot water and steam explosion. Based on the analysis conducted, the following conclusions can be reached:

(1) Among the MOWs, paper and paperboard rank highest in most assessments. The steam explosion option is the most beneficial production technology from an economic, energy and environmental perspective; thus, it would be the most promising technique for producing bioethenal from the studied wastes.

(2) The dilute acid method also ranks highly, technically; however, this it is not a desirable option due to its low energy performance, high production costs and high greenhouse gasses emission.

(3) In addition to steam explosion, more economic and environmentally oriented policies should favor hot water technology, as no chemicals are required for treatment and the hydrolysis yield is similar to that of other pretreatment methods.

(4) The dilute alkali technique for deriving bioethanol from paper also ranks highly for most of the indicators, except for environmental, due to its high GHG emissions and moderate water consumption.

(5) Finally, estimates of the potential of different wastes and bioethanol production types in 2015 and 2030 were calculated. The total amount of paper, timber and garden wastes in Iceland in 2015 can be estimated as approximately 37, 40 and 16 thousand tons, respectively, which could increase to about 47, 49 and 20 thousand tons by 2030. Thus, the average potential of bioethanol production in 2015 was approximately 12.5, 11 and 3 thousand tons from paper, timber and garden wastes, which may likewise increase to about 15.9, 13.7 and 3.7 thousand tons in 2030.

Author Contributions: Data curation, S.S. And R.U.; Formal analysis, S.S.; Investigation, S.S.; Methodology, S.S.; Supervision, R.U.; Validation, S.S.; Writing—original draft, S.S.; Writing—review & editing, R.U.

Conflicts of Interest: The authors declare no conflicts of interest

References

1. Rajaeifar, M.A.; Akram, A.; Ghobadian, B.; Rafiee, S.; Heijungs, R.; Tabatabaei, M. Environmental impact assessment of olive pomace oil biodiesel production and consumption: A comparative lifecycle assessment. *Energy* **2016**, *106*, 87–102. [CrossRef]

2. Jahirul, M.I.; Rasul, M.G.; Chowdhury, A.A.; Ashwath, N. Biofuels production through biomass pyrolysis—A technological review. *Energies* **2012**, *5*, 4952–5001. [CrossRef]

3. Talebnia, F.; Karakashev, D.; Angelidaki, I. Production of bioethanol from wheat straw: An overview on pretreatment, hydrolysis and fermentation. *Bioresour.Technol.* **2010**, *101*, 4744–4753. [CrossRef] [PubMed]

4. Hirschnitz-Garbers, M.; Gosens, J. Producing bio-ethanol from residues and wastes. A technology with enormous potential in need of further research and development. *Policy Brief* **2015**. Available online: http:

//ec.europa.eu/environment/integration/green_semester/pdf/Recreate_PB_2015_SEI.PDF (accessed on 2 December 2017).

5. Michailos, S.; Parker, D.; Webb, C. Design, sustainability analysis and multiobjective optimisation of ethanol production via syngas fermentation. *Waste Biomass Valoriz.* **2017**, 1–12. [CrossRef]

6. Liguori, R.; Soccol, C.R.; Porto de Souza Vandenberghe, L.; Woiciechowski, A.L.; Faraco, V. Second generation ethanol production from brewers' spent grain. *Energies* **2015**, *8*, 2575–2586. [CrossRef]

7. Miezah, K.; Obiri-Danso, K.; Kádár, Z.; Heiske, S.; Fei-Baffoe, B.; Mensah, M.; Meyer, A.S. Municipal solid waste management in a low income economy through biogas and bioethanol production. *Waste Biomass Valoriz.* **2017**, *8*, 115–127. [CrossRef]

8. Sundberg, M.; Guðmundsson, J.; Guðmundsson, M. *Biofuel Production in Iceland: Survey of Potential Raw Materials and Yields to 2030*; MANNVIT Engineering: Kópavogur, Iceland, 15 October 2010. Available online: http://www.lifeldsneyti.is/resources/Files/Lifeldsneyti/Lifeldsneyti-Arsskyrsla-2010/Fylgiskjal_1_Mannvit.pdf (accessed on 2 December 2017).

9. Hálfdánarson, H.E. Ethanol Production from Timothy (*Phleum pratense* L.). Master' Thesis, Agricultural University of Iceland, Hvanneyri, Island, June 2015.

10. Snæbjörnsson, A.; Hjartardóttir, D.; Blöndal, E.; Pétursson, J.; Eggertsson, Ó.; Halldórsson, O. *Skýrsla Nefndar um Landnotkun-Athugun á Notkun og Varðveislu Ræktanlegs Lands*; Ministry of Fisheries and Agriculture: Reykjavík, Iceland, 2010.

11. Dell, R.; Unnthorsson, R.; Wei, C.; Foley, W. Waste geothermal hot water for enhanced outdoor agricultural production. In Proceedings of the ASME 2013 Power Conference, Boston, MA, USA, 29 July–1 August 2013; American Society of Mechanical Engineers: New York, NY, USA, 2013; p. V002T009A013. [CrossRef]

12. Lin, J.; Babbitt, C.W.; Trabold, T.A. Life cycle assessment integrated with thermodynamic analysis of bio-fuel options for solid oxide fuel cells. *Bioresour. Technol.* **2013**, *128*, 495–504. [CrossRef] [PubMed]

13. *Strategic Planning for the Management of Wastes 2009–2020*; Mannvit Engineering: Reykjavik, Iceland, 2015.

14. Íslenska Gámafélagið. Available online: http://www.gamur.is/ (accessed on 2 December 2017).

15. Ng, S.Y.; Ong, S.Y.; Ng, Y.Y.; Liew, A.H.; Ng, D.K.; Chemmangattuvalappil, N.G. Optimal design and synthesis of sustainable integrated biorefinery for pharmaceutical products from palm-based biomass. *Process Integr. Optim. Sustain.* **2017**, *1*, 135–151. [CrossRef]

16. Holmgren, K. Waste incineration in swedish municipal energy systems. *WIT Trans. Ecol. Environ.* **2006**, *92*. [CrossRef]

17. Halldorsson, T.I.; Auðunsson, G.A.; Guicharnaud, R.; Dýrmundsson, Ó.R.; Hansson, S.Ö.; Hreinsson, K. Contamination of livestock due to the operation of a small waste incinerator: A case incident in skutulsfjörður, iceland, in 2010. *Acta Vet. Scand.* **2012**, *54*, S4. [CrossRef]

18. *Niðurstöður úr Mælingum á Díoxínum í jarðvegi*; Umhverfisstofnun: Reykjavík, Iceland, 2011. Available online: https://www.ust.is/einstaklingar/frettir/frett/2011/07/06/Nidurstodur-ur-maelingum-a-dioxinum-i-jardvegi/ (accessed on 2 December 2017).

19. Gunnarsson, A. Maintenance of the Steam Turbines at Hellisheiði Power Plant. Master' Thesis, University of Iceland, Reykjavik, Iceland, May 2013.

20. Rajaeifar, A.M.; Ghobadian, B.; Davoud Heidari, M.; Fayyazi, E. Energy consumption and greenhouse gas emissions of biodiesel production from rapeseed in iran. *J. Renew. Sustain. Energy* **2013**, *5*, 063134. [CrossRef]

21. Mousavi-Avval, S.H.; Rafiee, S.; Jafari, A.; Mohammadi, A. Energy flow modeling and sensitivity analysis of inputs for canola production in iran. *J. Clean. Prod.* **2011**, *19*, 1464–1470. [CrossRef]

22. Mood, S.H.; Golfeshan, A.H.; Tabatabaei, M.; Jouzani, G.S.; Najafi, G.H.; Gholami, M.; Ardjmand, M. Lignocellulosic biomass to bioethanol, a comprehensive review with a focus on pretreatment. *Renew. Sustain. Energy Rev.* **2013**, *27*, 77–93. [CrossRef]

23. Taherzadeh, M.J.; Karimi, K. Pretreatment of lignocellulosic wastes to improve ethanol and biogas production: A review. *Int. J. Mol. Sci.* **2008**, *9*, 1621–1651. [CrossRef] [PubMed]

24. Kumar, D.; Murthy, G.S. Impact of pretreatment and downstream processing technologies on economics and energy in cellulosic ethanol production. *Biotechnol. Biofuels* **2011**, *4*, 27. [CrossRef] [PubMed]

25. Talebnia, F.; Bafrani, M.P.; Lundin, M.; Taherzadeh, M. Optimization study of citrus wastes saccharification by dilute acid hydrolysis. *BioResources* **2007**, *3*, 108–122.

26. Behera, S.; Arora, R.; Nandhagopal, N.; Kumar, S. Importance of chemical pretreatment for bioconversion of lignocellulosic biomass. *Renew. Sustain. Energy Rev.* **2014**, *36*, 91–106. [CrossRef]

27. Agbor, V.B.; Cicek, N.; Sparling, R.; Berlin, A.; Levin, D.B. Biomass pretreatment: Fundamentals toward application. *Biotechnol. Adv.* **2011**, *29*, 675–685. [CrossRef] [PubMed]
28. Sritrakul, N.; Nitisinprasert, S.; Keawsompong, S. Evaluation of dilute acid pretreatment for bioethanol fermentation from sugarcane bagasse pith. *Agric. Natl. Resour.* **2018**. [CrossRef]
29. McIntosh, S.; Vancov, T. Enhanced enzyme saccharification of sorghum bicolor straw using dilute alkali pretreatment. *Bioresour. Technol.* **2010**, *101*, 6718–6727. [CrossRef] [PubMed]
30. Hu, G.; Heitmann, J.A.; Rojas, O.J. Feedstock pretreatment strategies for producing ethanol from wood, bark, and forest residues. *BioResources* **2008**, *3*, 270–294.
31. Viola, E.; Zimbardi, F.; Cardinale, M.; Cardinale, G.; Braccio, G.; Gambacorta, E. Processing cereal straws by steam explosion in a pilot plant to enhance digestibility in ruminants. *Bioresour. Technol.* **2008**, *99*, 681–689. [CrossRef] [PubMed]
32. Wang, L.; Littlewood, J.; Murphy, R.J. Environmental sustainability of bioethanol production from wheat straw in the uk. *Renew. Sustain. Energy Rev.* **2013**, *28*, 715–725. [CrossRef]
33. Safarian, S.; Khodaparast, P.; Kateb, M. Modeling and technical-economic optimization of electricity supply network by three photovoltaic systems. *J. Sol. Energy Eng.* **2014**, *136*, 024501. [CrossRef]
34. Safarian, S.; Saboohi, Y.; Kateb, M. Evaluation of energy recovery and potential of hydrogen production in iranian natural gas transmission network. *Energy Policy* **2013**, *61*, 65–77. [CrossRef]
35. Safarian, S.; Sattari, S.; Hamidzadeh, Z. Sustainability assessment of biodiesel supply chain from various biomasses and conversion technologies. *Biophys. Econ. Resour. Qual. (BERQ)* **2018**, *3*, 6. [CrossRef]
36. Azadeh, A.; Arani, H.V.; Dashti, H. A stochastic programming approach towards optimization of biofuel supply chain. *Energy* **2014**, *76*, 513–525. [CrossRef]
37. You, F.; Tao, L.; Graziano, D.J.; Snyder, S.W. Optimal design of sustainable cellulosic biofuel supply chains: Multiobjective optimization coupled with life cycle assessment and input–output analysis. *AIChE J.* **2012**, *58*, 1157–1180. [CrossRef]
38. Sheinbaum-Pardo, C.; Ruiz-Mendoza, B.J.; Rodríguez-Padilla, V. Mexican energy policy and sustainability indicators. *Energy Policy* **2012**, *46*, 278–283. [CrossRef]
39. Tsai, W.-T. Energy sustainability from analysis of sustainable development indicators: A case study in taiwan. *Renew. Sustain. Energy Rev.* **2010**, *14*, 2131–2138. [CrossRef]
40. Angelis-Dimakis, A.; Arampatzis, G.; Assimacopoulos, D. Monitoring the sustainability of the greek energy system. *Energy Sustain. Dev.* **2012**, *16*, 51–56. [CrossRef]
41. How, B.S.; Lam, H.L. Pca method for debottlenecking of sustainability performance in integrated biomass supply chain. *Process Integr. Optim. Sustain.* **2018**, 1–22. [CrossRef]
42. Atlantsolía ehf. Available online: https://www.atlantsolia.is/stodvaverd/ (accessed on 2 December 2017).
43. Veitur ohf. Available online: https://www.veitur.is/verdskrar/vatnsveita (accessed on 2 December 2017).
44. Foreign Big Company in Crash with Electricity Prices. Available online: https://kjarninn.is/skodun/erlent-storfyrirtaeki-i-feluleik-med-raforkuverd/ (accessed on 2 December 2017).
45. Mousavi-Avval, S.H.; Rafiee, S.; Jafari, A.; Mohammadi, A. Optimization of energy consumption for soybean production using data envelopment analysis (dea) approach. *Appl. Energy* **2011**, *88*, 3765–3772. [CrossRef]
46. Klein, S.J.; Whalley, S. Comparing the sustainability of us electricity options through multi-criteria decision analysis. *Energy Policy* **2015**, *79*, 127–149. [CrossRef]
47. Rajaeifar, M.A.; Ghobadian, B.; Safa, M.; Heidari, M.D. Energy life-cycle assessment and co2 emissions analysis of soybean-based biodiesel: A case study. *J. Clean. Prod.* **2014**, *66*, 233–241. [CrossRef]
48. Hamelinck, C.N.; Van Hooijdonk, G.; Faaij, A.P. Ethanol from lignocellulosic biomass: Techno-economic performance in short-, middle-and long-term. *Biomass Bioenergy* **2005**, *28*, 384–410. [CrossRef]
49. Shahzadi, T.; Mehmood, S.; Irshad, M.; Anwar, Z.; Afroz, A.; Zeeshan, N.; Rashid, U.; Sughra, K. Advances in lignocellulosic biotechnology: A brief review on lignocellulosic biomass and cellulases. *Adv. Biosci. Biotechnol.* **2014**, *5*, 246–251. [CrossRef]
50. Kumar, S.; Shukla, S. A review on recent gasification methods for biomethane gas production. *Int. J. Energy Eng.* **2016**, *6*, 32–43.

![energies logo] *energies*

MDPI

Article

Increasing Profits in Food Waste Biorefinery— A Techno-Economic Analysis

Juan-Rodrigo Bastidas-Oyanedel * and Jens Ejbye Schmidt

Chemistry Department, Khalifa University of Science and Technology, Masdar Campus,
P.O. Box 54224 Abu Dhabi, United Arab Emirates; jschmidt@masdar.ac.ae
* Correspondence: yanauta@gmail.com; Tel.: +97-150-502-8842

Received: 7 May 2018; Accepted: 21 May 2018; Published: 13 June 2018

Abstract: The present manuscript highlights the economic profit increase when combining organic waste anaerobic digestion with other mixed culture anaerobic fermentation technologies, e.g., lactic acid fermentation and dark fermentation. Here we consider the conversion of 50 tonnes/day of food waste into methane, power generation (from CHP of biomethane), lactic acid, polylactic acid, hydrogen, acetic acid and butyric acid. The economic assessment shows that the basic alternative, i.e., anaerobic digestion with methane selling to the grid, generates 19 USD/t_VS (3 USD/t_foodwaste) of profit. The highest profit is obtained by dark fermentation with separation and purification of acetic and butyric acids, i.e., 296 USD/t_VS (47 USD/t_foodwaste). The only alternative that presented losses is the power generation alternative, needing tipping fees and/or subsidy of 176 USD/t_VS (29 USD/t_foodwaste). The rest of the alternatives generate profit. From the return on investment (ROI) and payback time, the best scenario is the production of polylactic acid, with 98% ROI, and 7.8 years payback time. Production of butyric acid ROI and payback time was 74% and 9.1 years.

Keywords: food waste; anaerobic digestion; lactic acid fermentation; dark fermentation; poly-lactic acid; butyric acid

1. Introduction

Unlocking value from organic waste is a feasible idea, in contrast to the disposal of these organic wastes into landfills, that has an associated cost ranging from 40–400 USD/t [1–3]. Instead, the organic wastes (residues) can be converted into bio-products and/or bioenergy, creating economic value rather than costs, generating value and benefits for the society.

Based on the characteristics of organic wastes/residues, they can be characterized according to their saccharides, lignin, lipids and protein content [2]. The source of the waste is also classified into: agricultural waste, food waste, and municipal waste. Here we focus in the creations of economic value from complex organic wastes, e.g., food waste, by anaerobic digestion processes. The creation of value from non-complex residues, e.g., citrus peels, coffee spends, for the extraction of pigments has been reviewed elsewhere [2–4].

For the treatment of complex organic waste, anaerobic digestion has been historically the chosen technology. Anaerobic digestion converts the complex wastes into biogas, containing methane (bioenergy) and a digestate that can be valorize as soil improver. However, as noted by Pfaltzgraff et al. [3], the conversion of biomass to bulk chemicals is 3.5 to 7.5 times more profitable than its conversion to fuels/energy. This is the main motivation for this techno-economic analysis.

In recent years, several mixed culture anaerobic technologies, different from the conventional anaerobic digestion for biogas production, has emerged. Among these technologies are dark fermentation and mixed culture lactic acid fermentation. The interest in these "new" technologies is their value products, with market prices more attractive than methane and digestate. Their average prices are 400 USD/t and 15 USD/t, respectively [5–12].

Figure 1 presents the market price range for the products yielded by anaerobic digestion, dark fermentation, and mixed culture lactic acid fermentation. Dark fermentation has been extensively reviewed for the production of value products [13–25]. Here it is presented the prices of hydrogen, acetic acid, ethanol, propionic acid, butyric acid, and caproic acid as dark fermentation products.

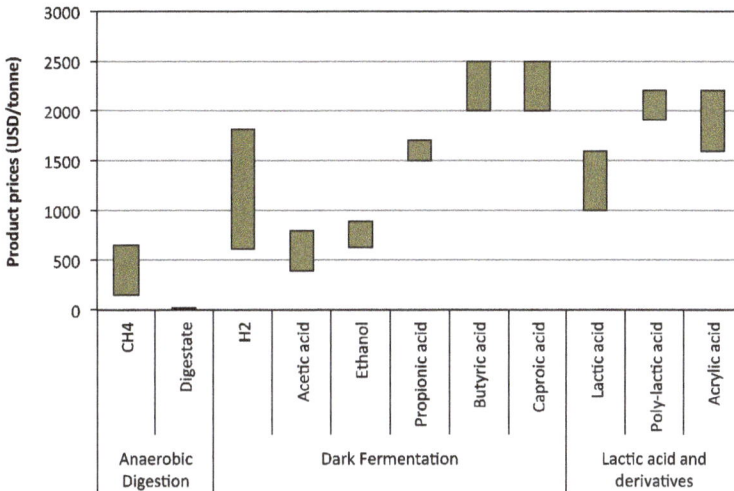

Figure 1. Market price ranges of main anaerobic-fermentation-based products.

Hydrogen price (600–1800 USD/t_H$_2$) is based on the production costs of hydrogen from natural gas. Hence, hydrogen price depends on natural gas market price. Here the hydrogen price was given when the natural price is in the range of 3–6 USD/GJ [26]. As described by Padro and Putsche [26], hydrogen production costs from natural gas are the cheapest when compare to other sources, i.e., coal, biomass, electrodialysis. To illustrate this, Clarke and Alibardi [7] reported a biomass-based bio-hydrogen price of 4700 USD/t_H$_2$.

Market price ranges for acetic acid [7,13] and ethanol [7,13,27] are closer to the methane price range, i.e., 400–900 USD/t, which makes it less attractive to spend effort trying to improve the production of either acetic acid and ethanol from organic wastes by anaerobic technologies. In contrast, price ranges for propionic acid, butyric acid and caproic acid [13], from 1500–2500 USD/t, are an incentive to optimize and improve the production of these organic acids by dark fermentation.

Lactic acid and polylactic acid relevance in this assessment is expressed in that that technology has been commercialized and has demonstrated its sustainability and profitability [2]. The conversion of food-waste into lactic acid by uncontrolled pH mixed culture fermentation, has been shown to be feasible [28–31], achieving lactic acid concentrations of 30 g/L with lactic acid selectivity of 93% (in COD base) over other organic acids [29,32]. Lactic acid market price range, 1000–1600 USD/t [13,27,33,34], and its derivatives, i.e., poly-lactic acid and acrylic acid, 1600–2200 USD/t [27,33,35] makes them attractive alternatives to biogas. Below is presented a techno-economic assessment of these three routes, i.e., anaerobic digestion, mixed culture lactic acid fermentation, and dark fermentation, in order to compare their potential revenues, costs and profit.

2. Methodology

The present techno-economic assessment is based on relevant literature data, and using conservative assumptions. The assessment compares 3 main routes: anaerobic digestion, mixed culture lactic acid fermentation, and dark fermentation. Each route has been divided into two sub-routes: (A1) Anaerobic digestion where the produced and upgraded methane is sold to the grid; (A2) Methane is used to in-site power generation; (B1) Mixed culture lactic acid fermentation, where lactic acid is produced, separated and purified, and the residues from this process are converted into methane by anaerobic digestion; (B2) As in B1, adding a conversion of lactic acid to poly-lactic acid step; (C1) Dark fermentation producing an upgraded hydrogen stream sold to the grid, where the residues of this process are converted into methane by anaerobic digestion; (C2) Dark fermentation producing an upgraded hydrogen stream, and producing, separating and purifying acetic and butyric acid streams, where the residues are converted into methane by anaerobic digestion. Table 1 presents the capital and operational costs for each route/sub-route, considered in this techno-economic analysis. Capital and operational costs were obtained from literature [5,8,13,35–37]. For all the scenarios, it was considered a project time of 20 years, with an annual interest of 5%.

Figure 2 presents a scheme of the three main routes and sub-routes. The different assessed scenarios are as follow: Anaerobic digestion: (**A1**) methane (to the gas grid) and digestate; (**A2**) Electricity and digestate. Lactic acid fermentation combined with anaerobic digestion: (**B1**) lactic acid, methane, and digestate; (**B2**) Poly-lactic acid, methane, and digestate. Dark fermentation combined with anaerobic digestion: (**C1**) hydrogen, methane and digestate; (**C2**) hydrogen, acetic acid, butyric acid, methane and digestate. For the economic assessments, it was assumed a well segregated food waste, from the hospitality and catering sectors, and a biorefinery "valorization" plant of 50 t/day of food waste capacity, which is in the range of what has been reported in the anaerobic digestion economic assessment literature [5,6,12,36]. Segregated food waste is a realistic assumption where post-harvest activities, and processed food industries, e.g., breweries, fruit pulp/juice production, are additional sources of segregated food waste, avoiding the use of municipal organic solid waste (contaminated with plastics, glass, and metals). It was assumed a food waste composition (w/w) of 13% carbohydrates, 1% fats, 2% proteins, 3% ashes, with a total solids (TS) composition of 19% and a volatile solid (VS) of 16% [29,38]. Conversion yields, revenues and costs are based on tonnes of volatile solid content of food waste (t_VS_fw). The costs include investment and operational costs. The assessments are detailed below.

2.1. Anaerobic Digestion

For the assumed food waste composition, it was estimated a methane yield over the volatile solid content of food waste (VS_fw) after anaerobic digestion and biogas upgrading of 0.33 t/t_VS_fw, and a digestate (solid) yield of 0.25 t/t_VS_fw [39,40]. It was considered that the final treated digestate (sold as soil improver) has a moisture content of 55%. The price of methane sold to the grid was estimated at 207 USD/t [9], and 5 USD/t of digestate [5]. For the power generation by Combined Heat and Power (CHP) it was assumed a methane energy content of 10.35 kWh/m^3, 35% electricity generation efficiency and 65% heat production [40,41]. It was considered that 60% from the total electricity produced is sold as electricity surplus [40,41]. The selected price of electricity was 0.1 USD/kWh [10]. Heat was assumed to be used in the plant, with no market price as discussed by Gebrezgabher [6]. The investment and operational cost for the sub-route A1, production of methane (to the grid) and digestate as soil improver, were set as 53 USD/t_VS_fw [8]. For A2, power generation and digestate, was 280 USD/t_VS_fw [36].

Table 1. Capital and operational costs, obtained from literature, for the different assessed scenarios. All scenarios considered a project time of 20 years, with an annual interest of 5%, and a designed capacity of 50 t/day of food waste, with 16% volatile solids composition.

Route	Annualized Capital Cost (USD/t_VS_fw/year) *	Annual Operational Cost (USD/t_VS_fw/year)	Annualized Total Investment (USD/t_VS_fw/Year)	Capital Cost as Present Value (USD)	Total Investment as Present Value (USD)
(A1) Anaerobic digestion—methane sold to the grid	42 [a,b]	11 [a,b]	53	1,528,365	1,928,652
(A2) Anaerobic digestion—power generation	167 [c]	113 [c]	280	6,091,628	10,189,103
(B1) Mixed culture lactic acid fermentation	108 [a,b,d]	55 [a,b,d]	163	3,930,083	5,931,514
(B2) Polylactic acid production	114 [a,b,d]	59 [a,b,d]	173	4,148,421	6,295,410
(C1) Dark fermentation—hydrogen and methane sold to the grid	47 [a,b,d,e,f]	13 [a,b,d,e,f]	60	1,710,314	2,183,379
(C2) Dark Fermentation—Acetic and butyric acids purified	252 [a,b,d,e,f]	148 [a,b,d,e,f]	400	9,170,193	14,555,862

* t_VS_fw: tonnes of volatile solids of food waste, in this review it was used volatile solid composition of 16%. (A1) The costs considered an anaerobic digestion reactor, a digestate solid composting facility, and a methane upgrading facility; (A2) as in A1, plus a combined heat and power generator; (B1) Considers a mixed culture lactic acid reactor, a lactic acid separator and purification system, an anaerobic digestion reactor, a digestate solid composting facility, and a methane upgrading facility; (B2) as in B1, plus a lactic acid polymerization facility; (C1) Considers a dark fermentation reactor, a hydrogen upgrading facility, an anaerobic digestion reactor, a digestate solid composting facility, and a methane upgrading facility; (C2) as in C1, plus a acetic acid separation and purification facility, and a Butyric acid separation and purification facility. [a] Kim et al., 2016; [b] Whyte and Perry, 2001; [c] Moriarty, 2013; [d] Nampoothiri et al., 2010; [e] Bastidas-Oyanedel et al., 2015; [f] Bonk et al., 2015.

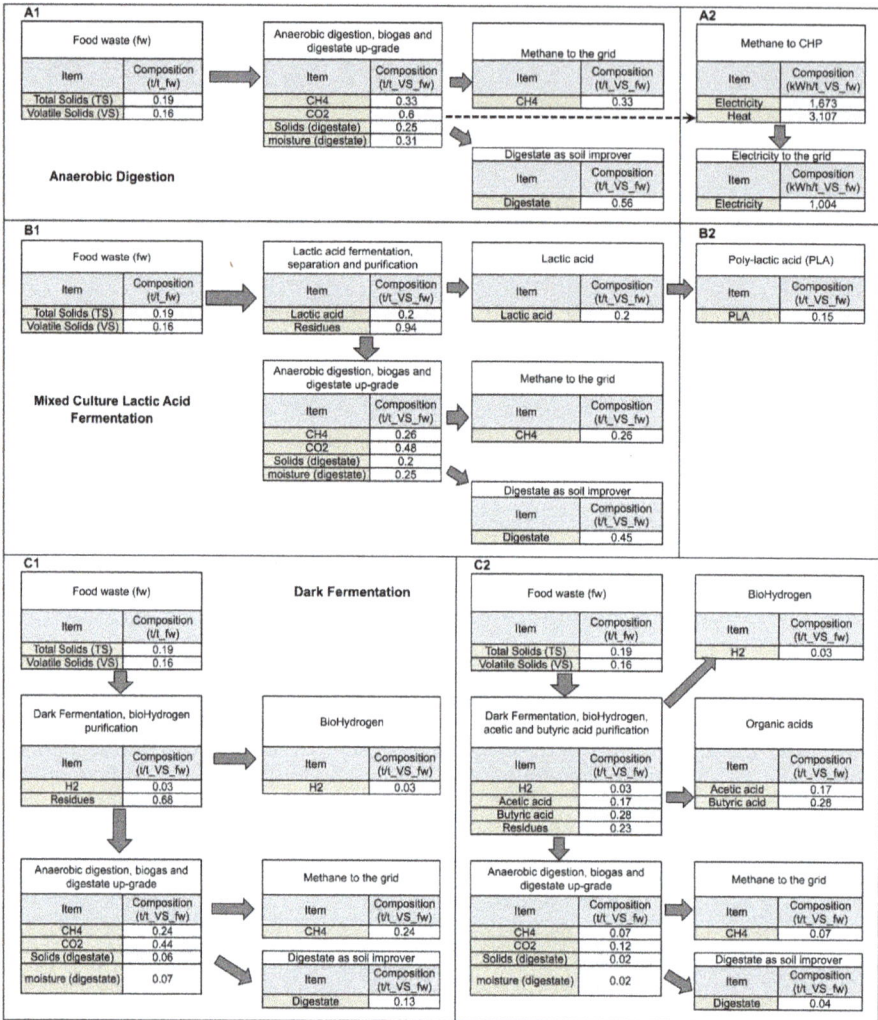

Figure 2. Mass flow diagram (t/t_VolatileSolid_foodwaste) for the different assessed scenarios. Anaerobic digestion: (**A1**) methane (to the gas grid) and digestate; (**A2**) Electricity and digestate. Lactic acid fermentation combined with anaerobic digestion: (**B1**) lactic acid, methane, and digestate; (**B2**) Poly-lactic acid, methane, and digestate. Dark fermentation combined with anaerobic digestion: (**C1**) hydrogen, methane and digestate; (**C2**) hydrogen, acetic acid, butyric acid, methane and digestate.

2.2. Mixed Culture Lactic Acid Fermentation

A yield of 0.2 t_lacticacid/t_VS_fw was used for the conversion, separation and purification of lactic acid [29,32,42–44]. This overall yield takes into account the fermentative production of lactic acid from food-waste [29,32], and the conventional downstream process-train, i.e., coarse separation of suspended solids from the broth, acidification of the broth with strong acid (H_2SO_4), removal of gypsum from the lactic acid solution, and distillation [42–44]. It was assumed that the residues of the lactic acid process were used to produce methane and digestate. Methane and digestate production and upgrading yield, from the lactic acid residue, were estimated as 0.2 and 0.45 t/t_VS_fw, respectively.

Poly-lactic acid (PLA) yield from the purified lactic acid was 0.75 t_PLA/t_LA [45,46]. In this process, purified lactic acid is first converted to lactide by a chemical process, and this molecule is further polymerized into PLA [45,46] Market prices for lactic acid and polylactic acid were assumed as 1000 and 1900 USD/t, respectively [33,35]. Prices for methane and digestate are as in Section 2.1. Cost for the production of lactic acid and anaerobic digestion of residues with methane and digestate as products was estimated as 163 USD/t_VS_fw and 173 USD/t_VS_fw for the scenario when lactic acid is converted to poly-lactic acid, converting the residues to methane and digestate [8,35]. These values are based on the targeted cost of lactic acid and polylactic acid reported by Nampoothiri et al. [35] of 0.55 USD/kg_LA and 0.8 USD/kg_PLA, respectively.

2.3. Dark Fermentation

Biohydrogen production and purification yield were assumed as 0.03 t/t_VS_fw [13,37], this is close to the maximum hydrogen yield. This was chosen, in order to discuss the huge effort in recent years regarding the biohydrogen production optimization from organic wastes by dark fermentation [16,47–52]. Purified hydrogen was assumed to be sold to the grid at a price of 1800 USD/t [26], to make it competitive versus natural gas-based hydrogen.

During the sub-route "hydrogen, methane and digestate" the residues from the biohydrogen production are converted to methane and digestate with 0.24 and 0.13 t/t_VS_fw yields, respectively. The cost of the dark fermentation and hydrogen upgrade was assumed to be 7.5 USD/t_VS_fw. This cost was based on the hydraulic retention time ratio of dark fermentation (2 days) [13] over anaerobic digestion (14 days) [17], i.e., the cost of dark fermentation was estimated as 1/7 the cost of the anaerobic digestion is as in Section 2.1, i.e., 53 USD/t_VS_fw [8]. The total cost for this sub-route was estimated as 60 USD/t_VS_fw.

The second sub-route "hydrogen, acetic acid, butyric acid, methane and digestate" assumed that acetic acid and butyric acid are separated and purified from the dark fermentation process. The combined production and purification yields of acetic and butyric acids used were 0.17 and 0.28 t/t_VS_fw, respectively [13,37,42–44]. These overall yields take into account the fermentative production of acetic and butyric acid from food waste [13,37], and the conventional downstream processing for other organic acid, i.e., lactic acid [42–44]. The residues are converted to methane and digestate by anaerobic digestion. Methane and digestate yields of 0.07 and 0.04 t/t_VS_fw were used. Prices for acetic and butyric acid considered were 400 and 2000 USD/t, respectively [13,37]. Cost for dark fermentation and anaerobic digestion are as in the previous sub-route. Separation and purification costs of 170 USD/t_VS_fw for acetic acid were assumed. The same cost was assumed for butyric acid. These values are based on the target production cost of polylactic acid [35], these costs can be decreased considerably for organic acids as discussed by Bonk et al. [37]. The total cost for this sub-route was estimated as 400 USD/t_VS_fw.

3. Results and Discussion

3.1. Economic Assessment Results

Figure 3 shows the revenues, cost and profit of the different assessed scenarios. For the first route, anaerobic digestion only, the food waste valorization plant, producing methane and digestate has profit of 19 USD/t_VS_fw. When considering the conversion into heat and power, the venues from electricity and digestate are close to 100 USD/t_VS_fw, but generating electricity at the assessed plant scale, i.e., 50 tonnes of food waste per day, considerably increases the total costs, 275 USD/t_VS_fw versus 50 USD/t_VS_fw when selling the methante to the grid. The conversion of methane into heat and power at the plant is only economically feasible when considering a minimum tipping fees and/or subsidies of 176 USD/t_VS_fw, i.e., 28 USD/t_foodwaste. This tipping-fees/subsidies are in the range of what has been reported in the literature for tipping fees, 40–60 USD/t_foodwaste [36], and considerable lower than the food waste landfilling costs, 40–400 USD/t_foodwaste [1–3].

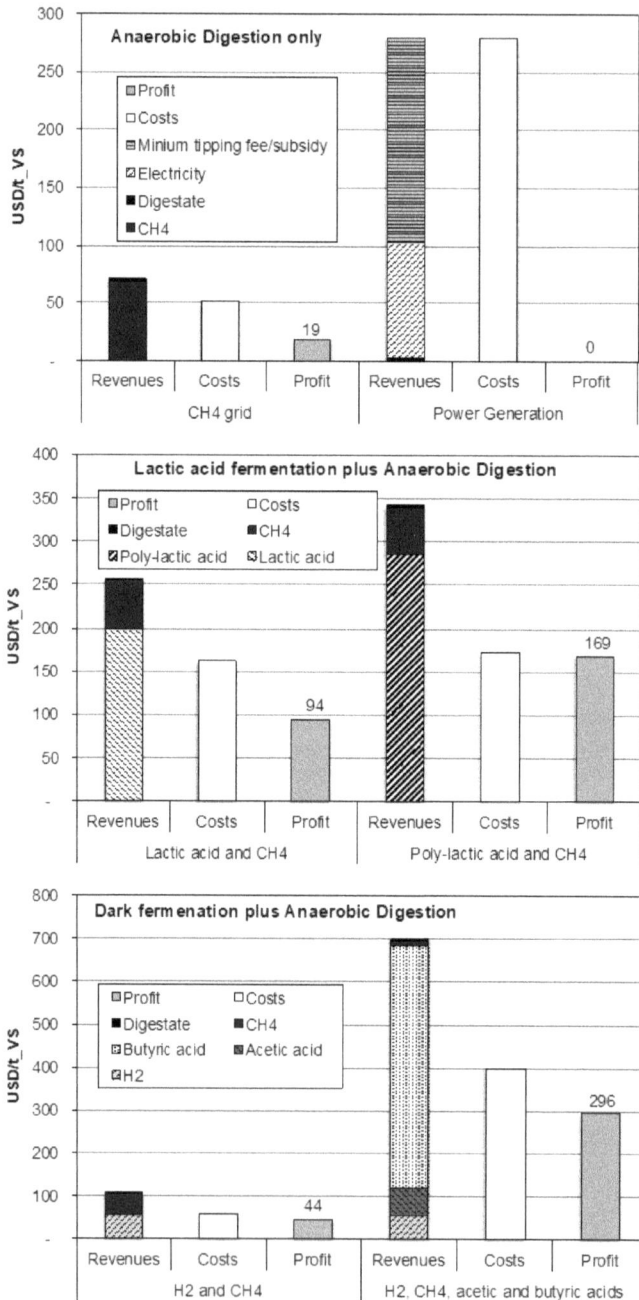

Figure 3. Economic assessment results for the different assessed scenarios.

The mixed culture lactic acid fermentation and anaerobic digestion scenario assessment resulted in higher profit, compared to the anaerobic digestion only scenario. Revenues for the production of lactic acid, methane (to the grid) and digestate were estimated to be 94 USD/t_VS_fw, with total cost of 162 USD/t_VS_fw. The revenues for the production of poly-lactic acid, methane and digestate were 169 USD/t_VS_fw, with total cost of 172 USD/t_VS_fw. This implies an 80% revenues increase, with a 6% increase in the costs from lactic acid to poly-lactic acid.

The dark fermentation with production of hydrogen, methane and digestate resulted in total revenues of 95 USD/t_VS_fw, where hydrogen represented 57% of the revenues, followed by methane (42%) and digestate (0.3%). This sub-route generated a profit of 44 USD/t_VS_fw, with a cost of 60 USD/t_VS_fw. This profit is considerably increased by 570% to 296 USD/t_VS_fw when producing hydrogen, acetic acid, butyric acid, methane and digestate. The main source of revenues was butyric acid, 80% of the revenues, followed by acetic acid (10%), hydrogen (7%), methane (2%) and digestate (0.01%). This assessment suggests that the huge scientific effort toward biohydrogen production optimization from organic waste [16,47,48,53–55] may be re-focused into the production, separation, and purification of the organic acids produced during dark fermentation [13,19,21,56–58]. It should be noted that the cost of the organic acids sub-route has also increased by 570% to 400 USD/t_VS_fw, when compared to the 60 USD/t_VS_fw of the previous sub-route. This increase is due to the high acetic and butyric separation-purification costs assumed in this assessment, 170 USD/t_VS_fw for each of the organic acids.

As discussed before, the revenues/cost from the organic acids can be improved as the scientific effort could be (re)-directed into finding more environmentally friendly processes and lowering the cost of organic acid separation-purification [38,59]. In this regard, several techniques has been investigated for the recovery of organic acids from fermentation broths, including adsorption [60], solvent extraction [61], membrane-based solvent extraction [62], electro dialysis [63,64], and membrane separation [65].

In general, the conversion of food waste into valuable chemicals was more profitable than its conversion to fuels (methane and hydrogen), as noted by Pfaltzgraff et al. [3]. In our assessment the lactic and poly-lactic scenarios were 5 and 9 times more profitable than the methane (to the grid) scenario, respectively, while the acetic-butyric acid scenario was 16 times more profitable. Also, as has been discussed by Belasri et al. [66] fuel and energy generation from biomass, will not match the total requirements of the society.

Figure 4 presents a sensitivity analysis on profit variation for the assessed scenarios. The sensitivity analysis was performed using a 15% decrease/increase in selected parameters. In the case of scenarios (A1) biogas to grid, the parameters that strongly affect the profit variation, i.e., more than 20% variation, are capital cost, methane price, methane yield, methane upgrading yield, and volatile solid content. In the case of (A2) biogas to power, only the volatile solid content produces a variation over 20% in the minimum tipping fees/subsidy. For scenario (B1) lactic acid, lactic acid price, and combined lactic acid yield and separation efficiency generates profit variations above 20%. (B2) polylactic acid, the parameters that produce profit variations higher than 20% are polylactic acid price, polylactic acid yield, and combined lactic acid yield and separation efficienty. For scenario (C1) hydrogen and methane, none of the parameters produced variations higher than 20% on the profit. The parameters that produced at least 10% profit variation were capital cost, methane price, hydrogen price, combined methane yield and upgrade efficiency, combined hydrogen yield and upgrade efficiency. Case (C2) acetic and butyric acids, butyric acid price, and combined butyric acid yield separation efficiency produced more than 20% profit variation.

Figure 5 shows the return on investment (ROI) and the payback time for all the assessed scenarios. From the ROI perspective scenario (B2) polylactic acid, generates the highest ROI, 98%. From all the assessed scenarios, only (A2) biogas to power, does not generates ROI, assuming that tipping fees/subsidies are minimal. This is due to the high cost of the combined heat and power generator and the low prices for electricity and digestate sold as soil improver [36]. All the other scenarios present ROI higher than 30%.

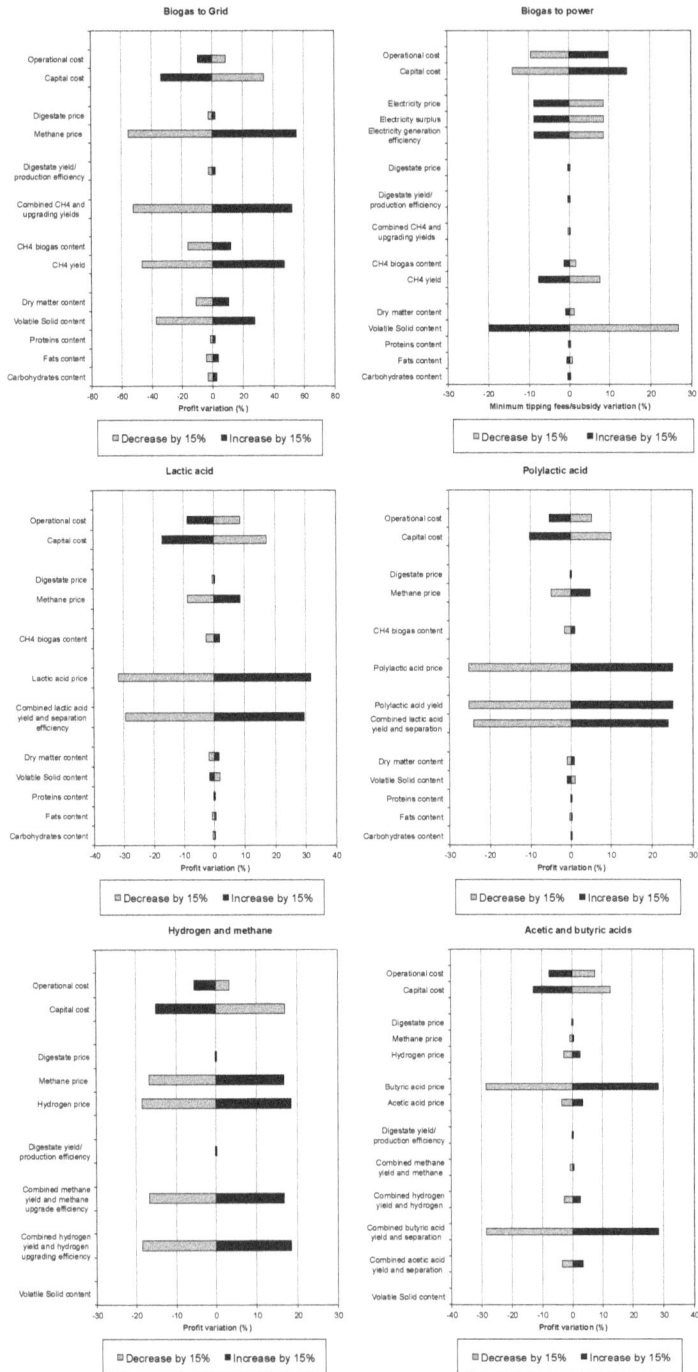

Figure 4. Economic sensitivity analysis results for the different assessed scenarios. All the scenarios where evaluated in a time frame of 20 years, with an annual interest of 5%. Costs include the capital cost, converted to annual, plus the annual operational cost.

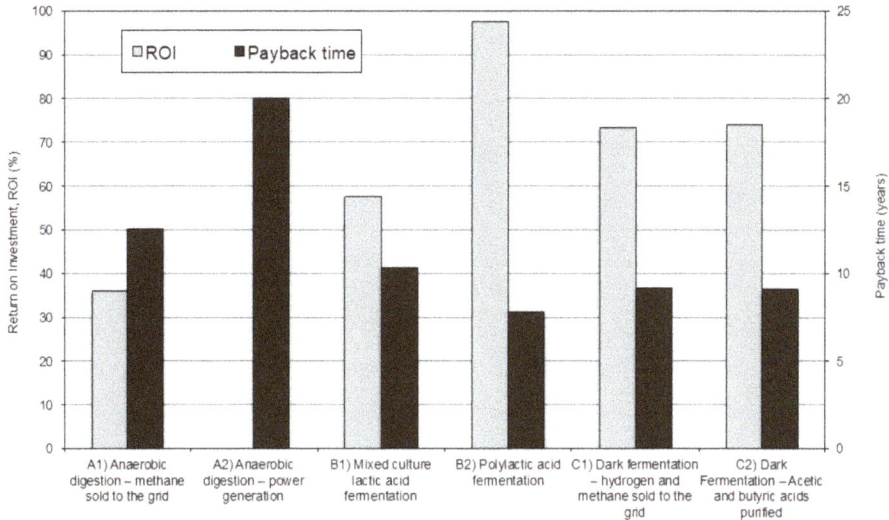

Figure 5. Return on investment (ROI) and payback time for the different assessed scenarios. All the scenarios where evaluated in a time frame of 20 years, with an annual interest of 5%.

The best pay back time was obtained for scenario (B2) polylactic acid, 7.8 years. Both scenarios (C1) hydrogen-methane and (C2) acetic-butyric acids, have payback times of 9.1 years. Scenarios (A1) biogas to grid and (B1) lactic acid, have payback times of 12.6 and 10.3 years, respectively.

3.2. Emerging Mixed Culture Technologies

The added benefits of mixed culture fermentation technologies, versus pure culture fermentations (where the feedstock has to be sterilized to prevent microbial contamination of the pure culture) has been partially discussed previously in the literature [13]. Here these benefits are expanded to: (1) use of a complex feedstock as food waste, (2) no feedstock sterilization, e.g., autoclaving, which reduces the process investment costs (no autoclave facilities are required) and operational costs (no energy is required for autoclaving). In the other hand, pure cultures are known for their high product selectivity, yielding higher product efficiency, and well controlled by environmental parameters [13]. However, new technologies in mixed culture dark fermentation are being developed to revert this trend. High product selectivity of one organic acid from the mixed culture dark fermentation has been reported for propionate using glycerol [56], propionate by controlling ammonium levels in the culture broth [57], or lactic acid using food waste in an uncontrolled pH mode [29,32]. High selectivity is pursued in order to make separation and purification a less complicated task. In this regard, elongation of carboxylic acids has been explored as an alternative to this issue. As an example, promoting the chain elongation of acetic and butyric acid can enhance caproic acid production [67–69]. Caproic acid solubility, 10.8 g/L in water [68] is low when compared to the miscibility of acetic and butyric acid [70]. Carboxilic acids esterification has been also reported to be improve separation, and reducing costs [71]. Electro-dialysis has been explored in order to increase carboxylic acid yield in dark fermentation [72–74], and for the separation of lactic acid from the culture broth [75]. In a complete different perspective, the production of carboxylic acids from syngas using anaerobic biofilms has been reported [76,77]. This is relevant, since anaerobic technologies can be used in combination to thermochemical processes for the valorization of plastics and lignocellulosic wastes, for the production of value chemicals.

4. Conclusions

The present techno-economic analysis has shown that profitability of food waste conversion to bulk chemicals, e.g., lactic acid or butyric acid, can be increased 5 to 16 times when compared to the base scenario, i.e., production of methane (sold to the grid). From the discussed scenarios, the highest profit is obtained by dark fermentation with separation and purification of butyric acid, 296 USD/t_VS (47 USD/t_foodwaste). From the return on investment (ROI) and payback time, the best scenario is the production of polylactic acid, with 98% ROI, and 7.8 years payback time. Production of butyric acid ROI and payback time was 74% and 9.1 years. From these profit, ROI, and payback time perspectives, the present techno-economic analysis suggests a change in focus from biogas/biohydrogen into butyric acid and polylactic acid production from food waste. These results suggest that industry may refocus effort on bulk chemicals, e.g., butyric acid and/or polylactic acid, rather than only focusing on biofuels, as H_2 and CH_4.

Author Contributions: Conceptualization, J.-R.B.-O.; Methodology, J.-R.B.-O. and Schmidt; Formal Analysis, J.-R.B.-O.; Writing-Original Draft Preparation, J.-R.B.-O.; Writing-Review & Editing, J.E.S.; Supervision, J.E.S.

Conflicts of Interest: The authors declare no conflict of interest.

References

1. Diggelman, C.; Ham, R.K. Household food waste to wastewater or to solid waste? That is the question. *Waste Manag. Res.* **2003**, *21*, 501–514. [CrossRef] [PubMed]
2. Tuck, C.O.; Perez, E.; Horvath, I.T.; Sheldon, R.A.; Poliakoff, M. Valorization of Biomass: Deriving More Value from Waste. *Science* **2012**, *337*, 695–699. [CrossRef] [PubMed]
3. Pfaltzgraff, L.A.; De bruyn, M.; Cooper, E.C.; Budarin, V.; Clark, J.H. Food waste biomass: A resource for high-value chemicals. *Green Chem.* **2013**, *15*, 307–314. [CrossRef]
4. Arancon, R.A.D.; Lin, C.S.K.; Chan, K.M.; Kwan, T.H.; Luque, R. Advances on waste valorization: New horizons for a more sustainable society. *Energy Sci. Eng.* **2013**, *1*, 53–71. [CrossRef]
5. Whyte, R.; Perry, G. A rough guide to anaerobic digestion costs and MSW diversion. *Biocycle* **2001**, *42*, 30–33.
6. Gebrezgabher, S.A.; Meuwissen, M.P.M.; Prins, B.A.M.; Lansink, A.G.J.M.O. Economic analysis of anaerobic digestion—A case of Green power biogas plant in the Netherlands. *NJAS-Wagen. J. Life Sci.* **2010**, *57*, 109–115. [CrossRef]
7. Clarke, W.P.; Alibardi, L. Anaerobic digestion for the treatment of solid organic waste: What's hot and what's not. *Waste Manag.* **2010**, *30*, 1761–1762. [CrossRef] [PubMed]
8. Kim, M.S.; Na, J.G.; Lee, M.K.; Ryu, H.; Chang, Y.K.; Triolo, J.M.; Yun, Y.M.; Kim, D.H. More value from food waste: Lactic acid and biogas recovery. *Water Res.* **2016**, *96*, 208–216. [CrossRef] [PubMed]
9. EIA United States Natural Gas Industrial Price. Available online: https://www.eia.gov/dnav/ng/hist/n3035us3m.htm (accessed on 3 July 2017).
10. Eurostat Energy Price Statistics. Available online: http://ec.europa.eu/eurostat/statistics-explained/index.php/Energy_price_statistics#Natural_gas_prices_for_industrial_consumers (accessed on 3 July 2017).
11. Indexmundi Natural Gas Price. Available online: http://www.indexmundi.com/commodities/?commodity=natural-gas (accessed on 3 July 2017).
12. Koupaie, E.H.; Barrantes Leiva, M.; Eskicioglu, C.; Dutil, C. Mesophilic batch anaerobic co-digestion of fruit-juice industrial waste and municipal waste sludge: Process and cost-benefit analysis. *Bioresour. Technol.* **2014**, *152*, 66–73. [CrossRef] [PubMed]
13. Bastidas-Oyanedel, J.-R.; Bonk, F.; Thomsen, M.H.; Schmidt, J.E. Dark fermentation biorefinery in the present and future (bio)chemical industry. *Rev. Environ. Sci. Bio/Technol.* **2015**, *14*, 473–498. [CrossRef]
14. Coma, M.; Martinez Hernandez, E.; Abeln, F.; Raikova, S.; Donnelly, J.; Arnot, T.C.; Allen, M.; Hong, D.D.; Chuck, C.J. Organic waste as a sustainable feedstock for platform chemicals. *Faraday Discuss.* **2017**. [CrossRef] [PubMed]
15. Kleerebezem, R.; Joosse, B.; Rozendal, R.; Van Loosdrecht, M.C.M. Anaerobic digestion without biogas? *Rev. Environ. Sci. Bio/Technol.* **2015**. [CrossRef]

16. Ghimire, A.; Frunzo, L.; Pirozzi, F.; Trably, E.; Escudie, R.; Lens, P.N.L.; Esposito, G. A review on dark fermentative biohydrogen production from organic biomass: Process parameters and use of by-products. *Appl. Energy* **2015**, *144*, 73–95. [CrossRef]

17. Aceves-Lara, C.A.; Trably, E.; Bastidas-Oyanedel, J.R.; Ramirez, I.; Latrille, E.; Steyer, J.P. Bioenergy production from waste: Examples of biomethane and biohydrogen. *J. Soc. Biol.* **2008**, *202*, 177–189. [CrossRef] [PubMed]

18. Birgitte, K. Ahring Biochemical Production and Separation of Carboxylic Acids for Biorefinery Applications Biochemical Production and Separation of Carboxylic Acids for Biorefinery Applications. *Fermentation* **2017**, *3*, 22. [CrossRef]

19. Agler, M.T.; Wrenn, B.A.; Zinder, S.H.; Angenent, L.T. Waste to bioproduct conversion with undefined mixed cultures: The carboxylate platform. *Trends Biotechnol.* **2011**, *29*, 70–78. [CrossRef] [PubMed]

20. Zhou, M.; Yan, B.; Wong, J.W.C.; Zhang, Y. Enhanced volatile fatty acids production from anaerobic fermentation of food waste: A mini-review focusing on acidogenic metabolic pathways. *Bioresour. Technol.* **2017**. [CrossRef] [PubMed]

21. Jankowska, E.; Chwiałkowska, J.; Stodolny, M.; Oleskowicz-Popiel, P. Volatile fatty acids production during mixed culture fermentation—The impact of substrate complexity and pH. *Chem. Eng. J.* **2017**. [CrossRef]

22. Chang, H.N.; Kim, N.J.; Kang, J.; Jeong, C.M. Biomass-derived volatile fatty acid platform for fuels and chemicals. *Biotechnol. Bioprocess Eng.* **2010**, *15*, 1–10. [CrossRef]

23. Kiran, E.U.; Trzcinski, A.P.; Liu, Y. Platform chemical production from food wastes using a biorefinery concept. *J. Chem. Technol. Biotechnol.* **2015**, *90*, 1364–1379. [CrossRef]

24. Cope, J.L.; Hammett, A.J.M.; Kolomiets, E.A.; Forrest, A.K.; Golub, K.W.; Hollister, E.B.; DeWitt, T.J.; Gentry, T.J.; Holtzapple, M.T.; Wilkinson, H.H. Evaluating the performance of carboxylate platform fermentations across diverse inocula originating as sediments from extreme environments. *Bioresour. Technol.* **2014**, *155*, 388–394. [CrossRef] [PubMed]

25. Queir, D.; Sousa, R.; Pereira, S.; Serafim, S. Valorization of a Pulp Industry By-Product through the Production of Short-Chain Organic Acids. *Fermentation* **2017**, *3*, 20. [CrossRef]

26. Padro, C.E.G.; Putsche, V. *Survey of the Economics of Hydrogen Technologies*; National Renewable Energy Lab.: Golden, CO, USA, 1999.

27. Beerthuis, R.; Rothenberg, G.; Shiju, N.R. Catalytic routes towards acrylic acid, adipic acid and ε-caprolactam starting from biorenewables. *Green Chem.* **2015**, *17*, 1341–1361. [CrossRef]

28. Gao, M.T.; Shimamura, T.; Ishida, N.; Takahashi, H. PH-uncontrolled lactic acid fermentation with activated carbon as an adsorbent. *Enzyme Microb. Technol.* **2011**, *48*, 526–530. [CrossRef] [PubMed]

29. Bonk, F.; Bastidas-Oyanedel, J.-R.; Yousef, A.F.; Schmidt, J.E. Exploring the selective lactic acid production from food waste in uncontrolled pH mixed culture fermentations using different reactor configurations. *Bioresour. Technol.* **2017**, *238*, 416–424. [CrossRef] [PubMed]

30. Tang, J.; Wang, X.; Hu, Y.; Zhang, Y.; Li, Y. Lactic acid fermentation from food waste with indigenous microbiota: Effects of pH, temperature and high OLR. *Waste Manag.* **2016**, *52*, 278–285. [CrossRef] [PubMed]

31. Li, X.; Chen, Y.; Zhao, S.; Chen, H.; Zheng, X.; Luo, J.; Liu, Y. Efficient production of optically pure L-lactic acid from food waste at ambient temperature by regulating key enzyme activity. *Water Res.* **2015**, *70*, 148–157. [CrossRef] [PubMed]

32. Yousuf, A.; Bastidas-Oyanedel, J.-R.; Schmidt, J.E. Effect of total solid content and pretreatment on the production of lactic acid from mixed culture dark fermentation of food waste. *Waste Manag.* **2018**. [CrossRef] [PubMed]

33. Datta, R.; Henry, M. Lacic acid: Recent advances in products processes and technologies—A review. *J. Chem. Technol. Biotechnol.* **2006**, *81*, 1119–1129. [CrossRef]

34. González, M.I.; Álvarez, S.; Riera, F.; Álvarez, R. Economic evaluation of an integrated process for lactic acid production from ultrafiltered whey. *J. Food Eng.* **2007**, *80*, 553–561. [CrossRef]

35. Nampoothiri, K.M.; Nair, N.R.; John, R.P. An overview of the recent developments in polylactide (PLA) research. *Bioresour. Technol.* **2010**, *101*, 8493–8501. [CrossRef] [PubMed]

36. Moriarty, K. *Feasibility Study of Anaerobic Digestion of Food Waste in St. Bernard, Louisiana*; National Renewable Energy Lab.: Golden, CO, USA, 2013.

37. Bonk, F.; Bastidas-Oyanedel, J.-R.; Schmidt, J.E. Converting the organic fraction of solid waste from the city of Abu Dhabi to valuable products via dark fermentation—Economic and energy assessment. *Waste Manag.* **2015**, *40*, 82–91. [CrossRef] [PubMed]

38. Yousuf, A.; Bonk, F.; Bastidas-Oyanedel, J.R.; Schmidt, J.E. Recovery of carboxylic acids produced during dark fermentation of food waste by adsorption on Amberlite IRA-67 and activated carbon. *Bioresour. Technol.* **2016**. [CrossRef] [PubMed]

39. Nielfa, A.; Cano, R.; Fdz-Polanco, M. Theoretical methane production generated by the co-digestion of organic fraction municipal solid waste and biological sludge. *Biotechnol. Rep.* **2015**, *5*, 14–21. [CrossRef] [PubMed]

40. Weiland, P. Biogas production: Current state and perspectives. *Appl. Microbiol. Biotechnol.* **2010**, *85*, 849–860. [CrossRef] [PubMed]

41. Walla, C.; Schneeberger, W. The optimal size for biogas plants. *Biomass Bioenergy* **2008**, *32*, 551–557. [CrossRef]

42. Joglekar, H.G.; Rahman, I.; Babu, S.; Kulkarni, B.D.; Joshi, A. Comparative assessment of downstream processing options for lactic acid. *Sep. Purif. Technol.* **2006**, *52*, 1–17. [CrossRef]

43. Åkerberg, C.; Zacchi, G. An economic evaluation of the fermentative production of lactic acid from wheat flour. *Bioresour. Technol.* **2000**, *75*, 119–126. [CrossRef]

44. Vaidya, A.N.; Pandey, R.A.; Mudliar, S.; Kumar, M.S.; Chakrabarti, T.; Devotta, S. Production and Recovery of Lactic Acid for Polylactide—An Overview. *Crit. Rev. Environ. Sci. Technol.* **2005**, *35*, 429–467. [CrossRef]

45. Gruber, P.R. Commodity polymers from renewable resources: Polylactic acid. In *Carbon Management: Implications for R&D in the Chemical Sciences and Technology: A Workshop Report to the Chemical Sciences Roundtable*; National Academies Press (US): Washington, DC, USA, 2001; pp. 111–127, ISBN 0309503051.

46. Subramanian, M.R.; Talluri, S.; Christopher, L.P. Production of lactic acid using a new homofermentative Enterococcus faecalis isolate. *Microb. Biotechnol.* **2015**, *8*, 221–229. [CrossRef] [PubMed]

47. Ntaikou, I.; Antonopoulou, G.; Lyberatos, G. Biohydrogen production from biomass and wastes via dark fermentation: A review. *Waste Biomass Valorization* **2010**, *1*, 21–39. [CrossRef]

48. Palomo-Briones, R.; Razo-Flores, E.; Bernet, N.; Trably, E. Dark-fermentative biohydrogen pathways and microbial networks in continuous stirred tank reactors: Novel insights on their control. *Appl. Energy* **2017**, *198*, 77–87. [CrossRef]

49. Elbeshbishy, E.; Dhar, B.R.; Nakhla, G.; Lee, H.-S. A critical review on inhibition of dark biohydrogen fermentation. *Renew. Sustain. Energy Rev.* **2017**, *79*, 656–668. [CrossRef]

50. Poggi-Varaldo, H.M.; Munoz-Paez, K.M.; Escamilla-Alvarado, C.; Robledo-Narvaez, P.N.; Ponce-Noyola, M.T.; Calva-Calva, G.; Rios-Leal, E.; Galindez-Mayer, J.; Estrada-Vazquez, C.; Ortega-Clemente, A.; et al. Biohydrogen, biomethane and bioelectricity as crucial components of biorefinery of organic wastes: A review. *Waste Manag. Res.* **2014**, *32*, 353–365. [CrossRef] [PubMed]

51. Bundhoo, M.A.Z.; Mohee, R.; Hassan, M.A. Effects of pre-treatment technologies on dark fermentative biohydrogen production: A review. *J. Environ. Manag.* **2015**, *157*, 20–48. [CrossRef] [PubMed]

52. Abreu, A.A.; Tavares, F.; Alves, M.M.; Pereira, M.A. Boosting dark fermentation with co-cultures of extreme thermophiles for biohythane production from garden waste. *Bioresour. Technol.* **2016**, *219*, 132–138. [CrossRef] [PubMed]

53. Bastidas-Oyanedel, J.-R.; Mohd-Zaki, Z.; Zeng, R.J.; Bernet, N.; Pratt, S.; Steyer, J.-P.; Batstone, D.J. Gas controlled hydrogen fermentation. *Bioresour. Technol.* **2012**, *110*, 503–509. [CrossRef] [PubMed]

54. Zhang, F.; Zhang, Y.; Chen, M.; Zeng, R.J. Hydrogen supersaturation in thermophilic mixed culture fermentation. *Int. J. Hydrogen Energy* **2012**, *37*, 17809–17816. [CrossRef]

55. Yeshanew, M.M.; Frunzo, L.; Pirozzi, F.; Lens, P.N.L.; Esposito, G. Production of biohythane from food waste via an integrated system of continuously stirred tank and anaerobic fixed bed reactors. *Bioresour. Technol.* **2016**, *220*, 312–322. [CrossRef] [PubMed]

56. Chen, Y.; Wang, T.; Shen, N.; Zhang, F.; Zeng, R.J. High-purity propionate production from glycerol in mixed culture fermentation. *Bioresour. Technol.* **2016**, *219*, 659–667. [CrossRef] [PubMed]

57. Chen, Y.; Shen, N.; Wang, T.; Zhang, F.; Zeng, R.J. Ammonium level induces high purity propionate production in mixed culture glucose fermentation. *RSC Adv.* **2017**, *7*, 518–525. [CrossRef]

58. Den Boer, E.; Lukaszewska, A.; Kluczkiewicz, W.; Lewandowska, D.; King, K.; Reijonen, T.; Kuhmonen, T.; Suhonen, A.; Jaaskelainen, A.; Heitto, A.; et al. Volatile fatty acids as an added value from biowaste. *Waste Manag.* **2016**, *58*, 62–69. [CrossRef] [PubMed]

59. Bastidas-Oyanedel, J.R.; Fang, C.; Almardeai, S.; Javid, U.; Yousuf, A.; Schmidt, J.E. Waste biorefinery in arid/semi-arid regions. *Bioresour. Technol.* **2016**, *215*, 21–28. [CrossRef] [PubMed]

60. Zhou, J.; Wu, J.; Liu, Y.; Zou, F.; Wu, J.; Li, K.; Chen, Y.; Xie, J.; Ying, H. Modeling of breakthrough curves of single and quaternary mixtures of ethanol, glucose, glycerol and acetic acid adsorption onto a microporous hyper-cross-linked resin. *Bioresour. Technol.* **2013**, *143*, 360–368. [CrossRef] [PubMed]

61. Alkaya, E.; Kaptan, S.; Ozkan, L.; Uludag-Demirer, S.; Demirer, G.N. Recovery of acids from anaerobic acidification broth by liquid-liquid extraction. *Chemosphere* **2009**, *77*, 1137–1142. [CrossRef] [PubMed]

62. Choudhari, S.K.; Cerrone, F.; Woods, T.; Joyce, K.; Flaherty, V.O.; Connor, K.O.; Babu, R. Pervaporation separation of butyric acid from aqueous and anaerobic digestion (AD) solutions using PEBA-based composite membranes. *J. Ind. Eng. Chem.* **2014**, *23*, 163–170. [CrossRef]

63. Lopez, A.M.; Hestekin, J.A. Separation of organic acids from water using ionic liquid assisted electrodialysis. *Sep. Purif. Technol.* **2013**, *116*, 162–169. [CrossRef]

64. Prochaska, K.; Woźniak-Budych, M.J. Recovery of fumaric acid from fermentation broth using bipolar electrodialysis. *J. Membr. Sci.* **2014**, *469*, 428–435. [CrossRef]

65. Xiong, B.; Richard, T.L.; Kumar, M. Integrated acidogenic digestion and carboxylic acid separation by nanofiltration membranes for the lignocellulosic carboxylate platform. *J. Membr. Sci.* **2015**, *489*, 275–283. [CrossRef]

66. Belasri, D.; Sowunmi, A.; Bastidas-Oyanedel, J.R.; Amaya, C.; Schmidt, J.E. Prospecting of renewable energy technologies for the Emirate of Abu Dhabi: A techno-economic analysis. *Prog. Ind. Ecol.* **2016**, *10*, 301–318. [CrossRef]

67. Leng, L.; Yang, P.; Mao, Y.; Wu, Z.; Zhang, T.; Lee, P.H. *Thermodynamic and Physiological Study of Caproate and 1,3-Propanediol Co-Production through Glycerol Fermentation and Fatty Acids Chain Elongation*; Elsevier Ltd.: New York, NY, USA, 2017; Volume 114, ISBN 8522766606.

68. Liu, Y.; He, P.; Shao, L.; Zhang, H.; Lü, F. Significant enhancement by biochar of caproate production via chain elongation. *Water Res.* **2017**, *119*, 150–159. [CrossRef] [PubMed]

69. Chen, W.S.; Strik, D.P.B.T.B.; Buisman, C.J.N.; Kroeze, C. Production of Caproic Acid from Mixed Organic Waste- An Environmental Life Cycle Perspective. *Environ. Sci. Technol.* **2017**. [CrossRef] [PubMed]

70. López-Garzón, C.S.; Straathof, A.J.J. Recovery of carboxylic acids produced by fermentation. *Biotechnol. Adv.* **2014**, *32*, 873–904. [CrossRef] [PubMed]

71. Cabrera-Rodriguez, C.I.; Moreno-Gonzalez, M.; de Weerd, F.A.; Viswanathan, V.; van der Wielen, L.A.M.; Straathof, A.J.J. Esters production via carboxylates from anaerobic paper mill wastewater treatment. *Bioresour. Technol.* **2016**. [CrossRef] [PubMed]

72. Jones, R.J.; Massanet-Nicolau, J.; Mulder, M.J.J.; Premier, G.; Dinsdale, R.; Guwy, A. Increased biohydrogen yields, volatile fatty acid production and substrate utilisation rates via the electrodialysis of a continually fed sucrose fermenter. *Bioresour. Technol.* **2017**, *229*, 46–52. [CrossRef] [PubMed]

73. Huang, C.; Xu, T.; Zhang, Y.; Xue, Y.; Chen, G. Application of electrodialysis to the production of organic acids: State-of-the-art and recent developments. *J. Membr. Sci.* **2007**, *288*, 1–12. [CrossRef]

74. Villano, M.; Paiano, P.; Palma, E.; Micheli, A.; Majone, M. Electrochemically-driven fermentation of organic substrates with undefined mixed microbial cultures. *ChemSusChem* **2017**. [CrossRef] [PubMed]

75. Yi, S.S.; Lu, Y.C.; Luo, G.S. Separation and concentration of lactic acid by electro-electrodialysis. *Sep. Purif. Technol.* **2008**, *60*, 308–314. [CrossRef]

76. Chen, X.; Ni, B.J. Anaerobic conversion of hydrogen and carbon dioxide to fatty acids production in a membrane biofilm reactor: A modeling approach. *Chem. Eng. J.* **2016**, *306*, 1092–1098. [CrossRef]

77. Gildemyn, S.; Molitor, B.; Usack, J.G.; Nguyen, M.; Rabaey, K.; Angenent, L.T. Upgrading syngas fermentation effluent using Clostridium kluyveri in a continuous fermentation. *Biotechnol. Biofuels* **2017**, *10*, 83. [CrossRef] [PubMed]

![energies logo]

Article

Fly Ash Formation and Characteristics from (co-)Combustion of an Herbaceous Biomass and a Greek Lignite (Low-Rank Coal) in a Pulverized Fuel Pilot-Scale Test Facility

Aaron Fuller [1,*], Jörg Maier [1], Emmanouil Karampinis [2,3], Jana Kalivodova [4,5], Panagiotis Grammelis [2,3], Emmanuel Kakaras [2,3] and Günter Scheffknecht [1]

[1] Institute of Combustion and Power Plant Technology, University of Stuttgart, Pfaffenwaldring 23, 70569 Stuttgart, Germany; joerg.maier@ifk.uni-stuttgart.de (J.M.); guenter.scheffknecht@ifk.uni-stuttgart.de (G.S.)
[2] Centre for Research & Technology Hellas/Chemical Process and Energy Resources Institute (CERTH/CPERI), Egialias 52, 15125 Marousi, Athens, Greece; karampinis@certh.gr (E.K.); grammelis@certh.gr (P.G.); kakaras@certh.gr (E.K.)
[3] National Technical University of Athens/Department of Mechanical Engineering/Laboratory of Steam Boilers and Thermal Plants (NTUA/LSBTP), Heroon Polytechniou 9, 15780 Zografou, Athens, Greece
[4] The Energy Research Center of The Netherlands (ECN), Biomass, Coal and Environmental Research, Heat and Power Generation, P.O. Box 1, 1755 ZG Petten, The Netherlands; Jana.Kalivodova@cvrez.cz
[5] Research Centre Rez, Husinec-_Re_z _c.p. 130, 250 68 Husinec e_Re_z, Czech Republic
* Correspondence: aaron.fuller@ifk.uni-stuttgart.de; Tel.: +49-711-685-63763; Fax: +49-711-685-63491

Received: 1 May 2018; Accepted: 5 June 2018; Published: 15 June 2018

Abstract: The lignite boilers are designed for lower quality fuels, and often the ash is not utilized. This work assessed the impact of combustion of an herbaceous biomass with a low-quality Greek lignite on the quality of the resulting fly ash. Test results were compared with those of fly ash samples from an industrial facility using the same fuel qualities. Inductively coupled plasma-optical (ICP) emission spectroscopy, X-ray powder diffraction (XRD), and scanning electron microscope (SEM) analyses were performed on the collected samples. Despite the significantly higher contents of K, Na and S in the biomass, at a 50% co-firing thermal share, the major and minor oxides in the fly ash were comparable to the lignite fly ash quality. This is attributed to the high ash content of the lignite, the low ash content of the biomass, and the much higher heating value of the biomass. There were improvements in fly ash performance characteristics with the herbaceous biomass in the fuel blend. The initial setting time and volume stability evaluations were improved with the biomass in the fuel blend. The work supports efforts of good practices in ash management, social responsibility, a circular economy, power plant renewable energy operations, and co-firing herbaceous biomass fuels in lignite power plants.

Keywords: bioenergy; co-firing; renewable energy; biomass fly ash; herbaceous biomass; by-products; recycled wastes; sustainable energy

1. Introduction

Over the last decades there has been increased interest in developing bioenergy technologies worldwide due to declining energy supplies and severe environmental constraints associated with fossil fuel use [1]. The goal of the European Union is to increase the percentage of biomass fuels used in the primary energy consumption, offering a reasonable, acceptable option to reduce greenhouse emissions [1], mainly CO_2. However, there are still concerns regarding the acceptability of coal and biomass fuel blends' ash materials, i.e., by-product quality, current utilization, and disposal routes.

It is not possible to correctly predict properties of ash material generated from combustion of a fuel mixture from known characteristics of ash produced from each fuel [1]. The interactions between fuels are poorly understood, and unanticipated ash properties may arise when combusting fuel mixtures [1], so, it is important to fully characterize the ashes produced from burning and co-firing any relatively new biomass fuel.

The reaction mechanisms of ash formation involve melting of most of the mineral components in the fuel, the formation of small fused drops, and a subsequent sudden cooling that transforms the merged drops partly or entirely into spherical glass particles [2]. Smaller fly ash particles may cool down more rapidly during the cooling process resulting in a more glassy state and giving rise to a higher reactivity [3]. The fly ash yield is the inorganic residue resulting from the complete combustion (or oxidation) of fuel (e.g., coal or biomass), and it consists of original and new inorganic mineral phases formed from the inorganic, organic, and fluid components in the fuel [4]. An overview of the physical transformation involved in the formation of ash during coal combustion is depicted in Figure 1.

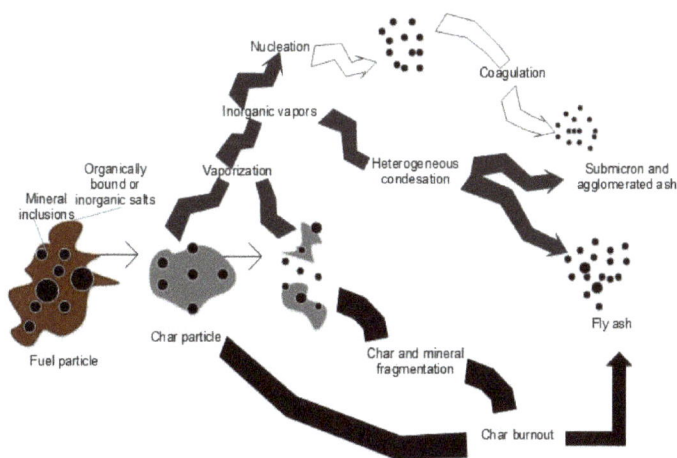

Figure 1. Physical transformation in ash formation during combustion of coal/biomass, Reproduced with permission from Cieplik, M.K.; Fryda, L.E.; van de Kamp, W.L.; Kiel, J.H.A. *Solid Biofuels for Energy: A Lower Greenhouse Gas Alternative*; Springer: London, UK, 2011 (Ref. [5], p. 198).

Lignite fly ash is subject to comparatively larger fluctuations compared to coal ashes [6]. Moreover, lignite boilers are designed for lower quality fuels, and often the ash is not utilized, making co-firing in lignite boilers an attractive solution to improving the ash quality [7]. Co-firing ash generated with ratios within the interest of commercial operation at power plants has proven to be applicable in concrete [8]. Those co-firing experiences mostly employed bituminous coals. Co-firing experiments with herbaceous or agricultural biomass in pulverized fuel power plants or test facilities focusing on fly ash quality are limited [9].

Lignite fly ashes have been used for refilling open cast mines, underground mining, surface restoration, soil amendment (mechanical stabilization and stabilization or strengthening of a soil), immobilization purposes, hydraulic road binders, plaster and mortar, road construction, asphalt filler, blended cement (contributing to CO_2 reduction during the production process), concrete addition (only after receiving technical approval), and various environmental uses (retention of pollutants and soil rehabilitation) [6]. Lignite fly ashes have been successfully tested or used in self-consolidating concrete [10], an adsorbent for SO$_2$ removal (after separation of enriched carbon fractions) [11], and in combination with a biochar or other organic amendments for improving soil properties [12]. Biochar and fly ash has the potential to improve soil quality, crop yield, and expand the carbon pool for soils [12]. Lignite coal fly ash blended with waste glass was successfully tested in the manufacture of lightweight aggregates [13]. Lignite fly ash with a high calcium content and used in a fly ash reinforced aluminum based metal matrix composite developed a good hardness due to the increased complex Si- and Ca-Si-crystalline phases produced [14]. A high Ca fly ash (CaO > 30%) has been used as a starting material when a hydrothermal calcination process is used as the pathway of the belite synthesis [15]. Thus, there are various options to use lignite fly ashes beyond the most stringent material provision requirements, i.e., beyond concrete addition. Special topics needed to be addressed by producers and users of combustion residues include, among others, co-combustion of coal with biomass that lead to changes in fly ash quality and its value as a marketable product [16].

There is a need for the development of new applications for lignite coal fly ash in order to reduce reliance on landfills [13]. More limited is the literature on the use of herbaceous biomass for co-firing with lignite. Thus, more work is needed that report the potential impact on ash quality that affects its sale or disposal routes due to a compositional changes from herbaceous or agricultural biomass co-fired with lignite.

Many of the problems associated with mono-combustion of biomass fuels may be ameliorated by co-combustion with coal, but the effect on the ash quality may be considered a downside of co-firing [16]. A major challenge for researchers, produces, users, and regulators is to improve the utilization of all grades of fly ash [16]. In this work the ash quality from co-firing an herbaceous biomass with a low-quality Greek lignite was evaluated. The species *Cynara cardunculus* shows significant potential as an energy crop under the semiarid conditions of the southern EU countries [17]. There is a need to provide solutions to the fly ash management problem from lignite coal power plants in many countries, e.g., Greece [13]. Fly ash should undergo a broad and critical assessment before being disposed, with an aim to provide one option for beneficial use in construction materials, which reduces disposal costs that account for short-term and long-term challenges [18]. The work reported in this paper elucidates the chemical, mineralogical, mobility, and some mechanical properties of the fly ash samples, as motivation for their potential utilization. The co-firing results are compared to the combustion of the unblended fuels to identify interactions between the lignite and biomass that affect the fly ash characteristics from the fuel blend. The data obtained provide valuable information on the ash quality in the area of co-combustion of agricultural or herbaceous biomass for heat and power production in a pulverized fuel facility and the by-product quality. The quality of an ash has meaning when viewed in the context of a certain application, as varying qualities are suitable for different applications [16]. Therefore, co-firing of a low-quality lignite with a herbaceous biomass may led to results warranting regulations requiring a certain co-firing ratio in older lignite power plants, if handling, transporting, storage, milling, feeding, boiler operations, and flue gas cleaning devices are not negatively impacted.

2. Materials and Methods

2.1. Facility

Figure 2 shows a diagram of the pilot-scale test facility. The facility consists of a cylindrical combustion chamber with a top-fired swirl burner, a selective catalytic reduction (SCR) catalyst, an electrostatic precipitator (ESP), and a fabric filter. In the inner annulus of a swirl burner, the coal is injected with primary air, and a center pipe allows injecting of secondary fuels. The computer aided monitoring system for the mass flow of the secondary fuel dosing system allows for a precise tuning of the desired biomass mass fraction in the fuel feed. The maximum thermal load of the facility is 0.5 MW$_{th}$. A more detailed description is provided in [19].

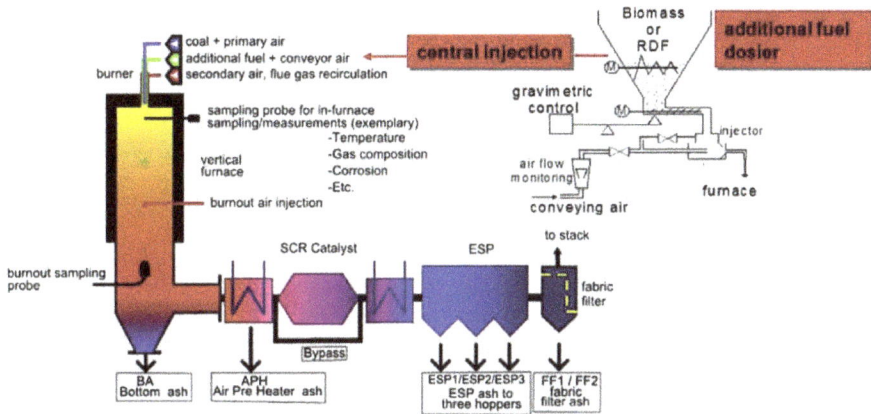

Figure 2. Institute of combustion and power plant technology (IFK) 0.5 MWth pulverized fuel pilot-scale test facility (KSVA).

2.2. Materials

The biomass type *Cynara cardunculus* L. (*Cynara*), which is characteristic of the Mediterranean countries, such as an agricultural or fast growing energy crop, and a low-quality Greek lignite were selected for this study. Also, blends of the lignite/biomass fuels were used in biomass percentages of 10% (typical of co-firing applications worldwide) and 50% on a thermal basis. A 1 mm sieve was used for grinding the *Cynara* pellets, as this is the size where larger particles are problematic for industrial pulverized fuel boilers. Preparation of the lignite occurred by using the fan beater mill that is part of the pulverized fuel pilot-scale test facility at the Institute of Combustion and Power Plant Technology (IFK). Because of the low heating value and high moisture content of the lignite, natural gas was used to support the flame in the near-burner region. The D_{10}, D_{50}, and D_{90} values of the *Cynara* were 33.72 μm, 244.01 μm and 838.34 μm, respectively. The lignite had a D_{50} of about 105 μm and D_{90} of about 475 μm. The larger particle size for *Cynara* is indicative of the need for a longer residence time to achieve a good burnout, which could influence fly ash formation, particle morphology, and mineral phases in the fly ash. The larger size will also impact the amount of ash at the various collection points, potentially affecting ash handling needs.

The lignite (low-rank coal) came from the open-cast mine near the Kardi Power Plant located in the Region of West Macedonia, near the cities of Ptolemaida and Kozani in Greece. The low quality of the lignite necessitated the use of natural gas, as shown in Table 1 of the following section.

The characterization of biomass fuels were performed according to the European standards drawn up by the Technical Committee CEN/TC 335 for solid biomass fuels. The determination of the proximate analysis was according to EN 51718, EN 51719 and EN 51720 for water, ash, and volatiles, respectively. Biomass ashes were produced using the standard at 550 °C. Determination of the ultimate analysis was according to EN 51732.

Ordinary Portland Cement of the class CEM I 42.5 R was used to prepare the concrete samples for strength testing. The sand (fine aggregate) used in making the test specimens was CEN-Norms and according to EN 196-1 conforming to ISO 679. The sand was stored at room temperature before use and in dry conditions. The fly ash was added as a partial replacement of cement at three levels: 10%, 20% and 30% by mass fraction.

Table 1. *Cynara* pilot scale tests' parameters.

Th_{share} (%) Target	*Cynara*	0	10.0	50.0	100
	lignite	100	90.0	50.0	0
Thermal Input P [kW$_{el}$]	*Cynara*	0	11.4	54	169
	lignite	106	94	58	0
	natural gas	200	180	149	119
$\dot{M}$ [kg/h]	*Cynara*	0	2.70	12.8	44.2
	lignite	73	71.4	44.4	0
Mass Percent of Fuel in Blend [%]	*Cynara*	0	3.6	22.4	100
	lignite	100	96.4	77.6	0
$\dot{V}$ [m^3/h] (Gas)	natural gas	19.3	17.4	14.4	10.5
ϑ [°C] (Primary Air)		35	35	35	35
ϑ [°C] (Secondary Air)		200	200	200	200
$\dot{V}$ [Nm^3/h] (Primary Air)		32	32	32	32
$\dot{V}$ [Nm^3/h] (Secondary Air 1)		81	83	83	82
$\dot{V}$ [Nm^3/h] (Secondary Air 2)		187	150	153	141
η			1.3		

2.3. Tests Operations

The direct co-combustion method was used during experimental testing [20,21]. The bio-fuel was handled and milled in a separate biomass milling facility, and it was injected into the pulverized fuel pipework before the burner inlet entrance. The feeding option utilized allows high co-firing ratios. Gas was used during the experiments due to the high water and ash contents (shown later) of the raw lignite, as it is widely known that those two fuel parameters adversely impact ignition and combustion stability. The use of gas ensured approximately a 300 kW$_{el}$ power output. Table 1 below lists the different test settings.

2.4. Ash Analyses

During the experiments, fly ash samples were collected from each ESP hopper. A representative fly ash sample of each test scenario was obtained from ash balance calculations of the hoppers. The fly ash was stored in a dry condition before testing.

2.4.1. Chemical Analyses

For fly ash analyses, major and minor elemental composition (expressed as oxides) were analyzed by inductively coupled plasma-optical emission spectroscopy (ICP-OES) to determine concentrations of the major and minor elements by EN 15290. Loss on ignition was identified for fly ash at 950 °C according to EN 196-2. Determination of trace elements for fly ash was according to EN 15297 or EN 22022-1-6. Determination of chlorine for fly ash was according to DIN 38414 and EN ISO 10304. The determination of leachable heavy metals and leachable S, K, Na, P and Cl was according to EN 15290, EN 15296 and DIN 38414 part 4. Biomass fuel ash oxide analysis was taken at 550 °C in accordance CEN/TC 335.

2.4.2. Scanning Electron Microscope (SEM)

Samples for SEM are generally prepared by dispersing the sample onto a conductive carbon tape (sample plate). Because of the insulating effect of the embedding material, samples are coated with gold to prevent charging. Afterwards, samples are placed in the SEM chamber. Images are taken with a phosphorus scintillation detector (secondary electron images) and a quadrupole backscatter detector. The addition of an X-ray spectrometer (normally of energy-dispersive type) makes it capable of chemical analyses, and spot analyses can be obtained [22]. Each fly ash sample was characterized by randomly selecting 3–4 fields of view and examining the morphology of each field. The elemental composition and morphology were worked out further for all fields for the different ESP ash hoppers for each test scenario.

2.4.3. X-ray Powder Diffraction Analyses (XRD)

Fly ash samples for XRD analyses were first finely ground. Fine grinding of the powder ensures there are sufficient amount of crystals to generate detectable signals at all angles and that the back-ground noise is kept to a minimum as possible [23]. A D8 Advance powder diffractometer (Bruker Advanced X-ray Solutions GmbH, Karlsruhe, Germany, with headquarters in Billerica, MA, USA) equipped with a Bragg-Brentano geometry and scintillation detector and CuKα radiation wavelength 1.5418 Å was used to determine mineral phases. The tube voltage was 40 kV, and the tube current was 40 mA. X-ray powder diffraction scans were performed at a diffraction angle range from 11° 2θ to 65° 2θ and with a step size of 0.02° 2θ and 1 s counting time per step. The evaluation of the X-ray results was done using the software DIFFRAC^{Plus} EVA Version 5.0 Revision 1, produced by Bruker Advanced X-ray Solutions GmbH (Karlsruhe, Germany).

2.4.4. Particle Size Distribution Determination

The particle sizes for the fly ash samples were measured in Malvern laser particle analyzers, model number 2000 and 2600 (Malvern, Malvern Hills, UK). A powdery sample is transported pneumatically by the light beam of a laser. The measuring range of the Malvern is from 25 to 500 μm.

2.4.5. Ash Enrichment Factors

An enrichment factor determines the relative enrichment of an element in the fly ash when compared to the amount of the element in the fuel ash. The enrichment factor is defined below (Equation (1)) as reported in [24]:

$$EF_i = \frac{x_{d,i}}{x_{a,i}} \tag{1}$$

where, EF_i is the enrichment factor, $x_{d,i}$ is the mass fraction of element i in the fly ash on a carbon-free basis and $x_{a,i}$ is the mass fraction of element in the fuel ash or fuel ash blend.

2.4.6. Volume Stability Testing (Soundness)

Some constituents in fly ash, e.g., free lime (if no aeration converts it to carbonate or calcium hydroxide), sulfate, and periclase, can participate in reactions that lead to undesirable volume changes (unsoundness) in hardened concrete, causing deterioration and failure [25]. Volume stability testing was performed at the Material Testing Institute (MPA Stuttgart, Otto-Graf-Institute (FMPA), at the University of Stuttgart, Stuttgart, Germany). The soundness testing was conducted according to EN450-1 requirements that stipulate to follow procedures in EN 196-3. The test was done using the LeChatelier test method principal.

2.4.7. Initial Setting Time Testing

Setting is the process where a fresh cement paste of freely flowing or plastic consistency is converted into a set material, which has lost its unlimited deformability and crumbles under a sufficiently great external force [26]. Admixtures, to include fly ash, influence the setting times by either accelerating or retarding it [26]. The initial time of set was determined at the Material Testing Institute (MPA Stuttgart, Otto-Graf-Institute (FMPA)). The setting time is one of the most important properties of concrete [27]. The initial time of set test was conducted according to EN 450-1 requirements that stipulate to follow procedures in EN 196-3 for preparing the cement paste and adhering to requirements in EN 197-1. The results are from the penetration of a standard needle into a cement paste or cement and other admixture mortar, such as fly ash. The test involves measuring a penetration resistance of the needle after a period of time.

2.4.8. Compressive Strength Development

Fly ash was used to partially replace ordinary Portland cement in amounts of 10%, 20% and 30% by mass fraction. The compressive strength property was evaluated and determined according to EN450-1 following procedures in EN 196-1. The cured specimens tested were 40 × 40 × 160 (W × H × L) mm prismatic samples. The cured specimen was loaded until bending failure to obtain two halves. The two halves obtained were tested in compression until failure, and the compressive strength was determined according to (Equation (2)) below:

$$f_c = \frac{F}{A} \tag{2}$$

where, f_c is the compressive strength, in megapascals [MPa]; F is the maximum load at fracture, in Newtons [N]; and A is the cross sectional area, in square millimeters [mm^2]. Compressive strength was measured at various days from 1 day up to 28 days after curing of test specimens. The 28-day strength corresponds to about 80% of concrete's strength under short-term loading [28].

3. Results and Discussion

3.1. Chemical Characterization of Fuels

Table 2 shows the fuels proximate and ultimate analyses. As expected, there were noticeable differences in the moisture and ash contents of the lignite and the volatile content of the *Cynara*.

Table 2. Characteristics and energy contents of the fuels.

Parameter	Cynara	Lignite
γ_{H_2O} [mass fraction percent-as received]	11.40	53.30
Proximate analysis [mass fraction percent]		
γ_{VM} (daf)	83.67	68.67
γ_A (d)	7.73	31.40
$\gamma_{fixed-C}$ (daf)	16.36	31.34
Ultimate analysis [mass fraction percent (daf)]		
γ_C	50.72	58.46
γ_H	5.26	4.91
γ_N	0.77	1.44
γ_S	0.22	0.34
γ_O (diff.)	43.04	34.85
Chloride [mass fraction percent (daf)]		
γ_{Cl}	0.19	<0.01
Heating values		
H_L [kJ/kg, as received]	15,198	4776
H_L [kJ/kg, dry]	17,632	13,014

The most noticeable difference of the fuels in Table 2 is that the herbaceous biomass has a significantly lower water content and ash yield, and a correlating higher heating value. The *Cynara* has a higher volatile content, about a 17% increase and a higher oxygen content. The *Cynara* in the fuel blend indicates a much-improved regulation of combustion.

The ultimate analysis shows that the percentage of hydrogen and elemental oxygen are higher and the carbon lower in the *Cynara* biomass when compared to the lignite. The lower value of nitrogen in the biomass shows that co-firing of *Cynara* with the lignite will have the advantage of reducing the stack emissions of the conventional pollutant NOx that is regulated. Also, the lower nitrogen content of *Cynara* means that co-firing may reduce the need for reagents to be injected into an selective non-catalytic reduction catalyst (SNCR) or high dust SCR equipment to reduce NOx emissions, if already occurring, which would decrease the propensity to absorb ammonia salt on fly ash (ammonia slip), as ammonia contamination of fly ash creates a host of potential odor issues when used in Portland cement-based projects [29].

The calorific values illustrate the influence of the change in moisture content, and to a lesser extent, the volatile matter amount had on the heating value of the fuels. The net calorific value of the *Cynara* is comparable to the values measured for other biomass types [21]. Co-firing of the fuels would cause parameters to vary in proportion to their contribution to the fuel blend, e.g., energy and ash contents. Thus, the co-firing of *Cynara* with the illustrated lignite quality may not only lead to improvements in the combustion performance, but also potentially in the ash quality. Figure 3 shows the major and minor oxides of the fuels on a carbon free basis.

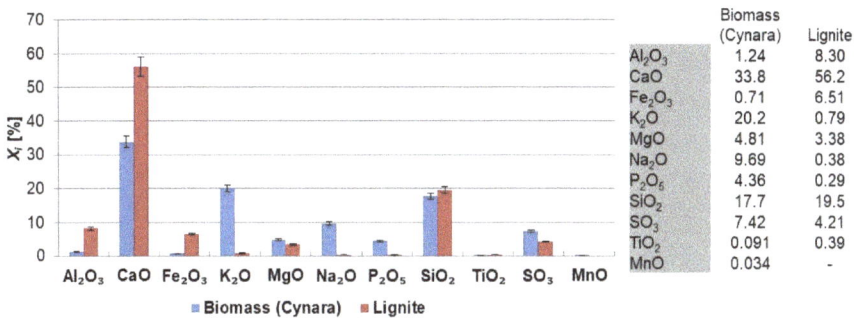

	Biomass (Cynara)	Lignite
Al_2O_3	1.24	8.30
CaO	33.8	56.2
Fe_2O_3	0.71	6.51
K_2O	20.2	0.79
MgO	4.81	3.38
Na_2O	9.69	0.38
P_2O_5	4.36	0.29
SiO_2	17.7	19.5
SO_3	7.42	4.21
TiO_2	0.091	0.39
MnO	0.034	-

Figure 3. Major and minor oxides of the herbaceous biomass (*Cynara*) and Greek lignite.

The herbaceous biomass has less CaO than the lignite, but more K_2O, Na_2O, SO_3 and P_2O_5 contents. Generally, this trend is reported in literature for biomass fuels when compared to coal or lignite [30], except for the SO_3 trend. The oxides of alkali, sulfur, and phosphorous are limited in concrete standards for applications in construction. However, when combusting the fuel with a lignite, having a significantly higher ash content, and at moderate thermal shares, the impact on the ash quality may be negligible too little. The SiO_2 contents in the fuels are comparable. The silica value for the *Cynara* is aligned with values for herbaceous and agricultural biomass [30]. The SiO_2 value in the lignite may reflect the quality from Greece, as the CaO oxide content is very high. Higher co-firing shares could change the ash quality for use in concrete, but masonry or mortar construction applications may be valuable options, as many of the concerns for concrete use with strict material provisions for conservative design are not warranted, and requirements are usually more performance based. Table 3 shows values for the major and minor elements in the fuels.

Table 3. Values for major and minor elements in the fuels.

Fuel	[%]									
	Al	Ca	Fe	K	Mg	Na	P	Si	Ti	S
Lignite	4.39	40.2	4.55	0.66	2.03	0.28	0.13	9.11	0.23	1.68
Cynara	0.66	24.2	0.50	16.8	2.89	7.18	1.90	8.27	0.05	2.97

Most of the ash-forming elements in annual plant biomass are noted as nutrients required for plant growth with inorganic impurities, occurring as a result of biomass being contaminated with soil during collection or handling [31]. The diverse ash forming matter in biomass fuels depends on the biomass type, soil type, and harvesting conditions, with Ca, K, Na, Si and P being major ash forming inorganic chemical elements due to some of these needed as nutrients in biomass [32]. Potassium and phosphorous are primary plant nutrients while sulfur, calcium, and magnesium are secondary plant nutrients [33], so an increasing share of co-fired *Cynara* would likely produce an ash quality very suitable for uses other than concrete, such as in low technical demand masonry applications, e.g., trafficked areas, parking lots, and in soil amelioration applications too.

3.2. Ash Loadings (Removal Rate) Trend

Ash was collected along the flue gas stream at different locations. The collected ash was completely weighted and finally related to the collection period for different scenarios. Figure 4 shows the different removal rates at the different collection points.

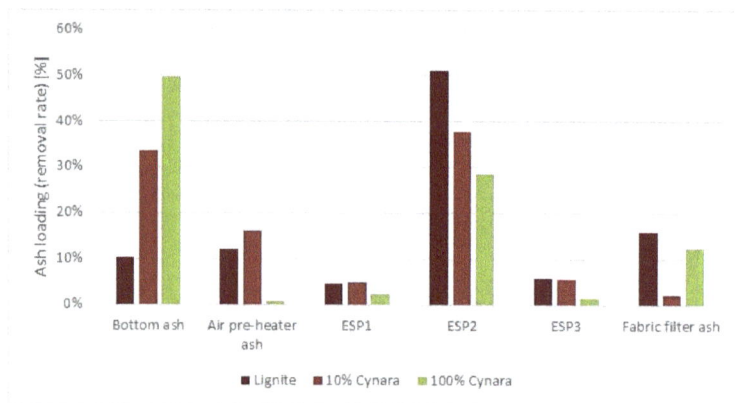

Figure 4. Ash loadings in the furnace and collection points downstream.

Based upon the operational design of the ESP, different size fractions are collected in the three hoppers. In the first ESP hopper, the largest ash fractions are collected. In the second ESP hopper, the middle size ash fractions are collected, which is about 80% of the total ESP ash. In the last ESP hopper, the finer ash fractions are collected, which is the least amount of the size fractions.

With the addition of *Cynara* in the fuel blend, there is an increase in fuel particle size distribution. Thus, the ash loading at the bottom of the combustion chamber increases due to the larger particles not being entrained in the flue gas stream exiting the furnace. Therefore, less ash is collected at points downstream the furnace. Consequently, there may be a significant increase in the bottom ash amount that needs to be addressed in the ash management procedures. There may be a need to find additional utilization markets for the bottom ash or apply post-treatment processes for utilization options.

3.3. Particle Size Distribution of Ashes

Fineness may be considered the most important property of fly ash in relation to its use in concrete, with the general rule the finer the better [16]. The fineness of fly ash used as part of cement mixtures influences the rate of hydration since hydration occurs at the interface with water [34,35]. The fineness mainly impacts the short-term strength and only slightly affect the long-term strength [36], which is due to the proposed occurrence of a more uniform microstructure expected for an increase in fineness [35]. The pozzolanic quality of fly ash correlates to fineness and its glassy content (amorphous matter) [37]. The strength increases as the granular distribution become narrow, which is predominantly an effect of a faster hydration under the conditions [35].

Compared in Figure 5 is the particle size distribution (PSD) of the representative bulk fly ash samples from each *Cynara* experimental trial. The pure *Cynara* combustion fly ash is noticeably finer than the other ashes, while the other co-firing cases are similar to the lignite case.

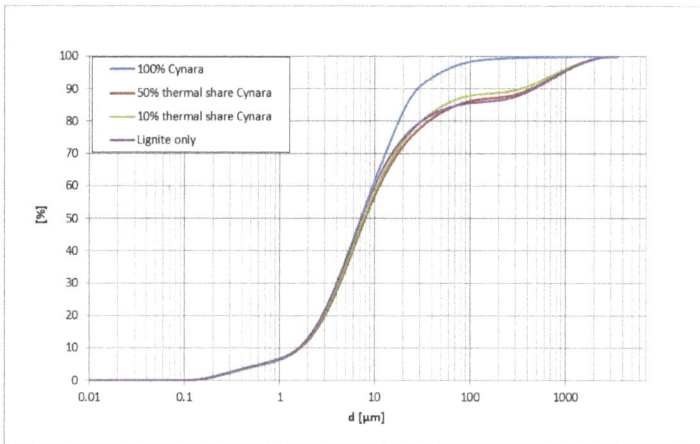

Figure 5. Particle size distribution of representative bulk fly ash samples from *Cynara* co-firing tests.

The comparability of the co-fired fly ash particle sizes to the lignite case may be attributed to the percentage of *Cynara* ash being much lower in the fuel mixture. In Table 4 is shown the size distributions for D_{10}, D_{50} and D_{90} values.

Table 4. Particle size distributions of fly ashes from test experiments.

	Lignite	10% *Cynara*	50% *Cynara*	100% *Cynara*
D_{10}	1.59	1.62	1.66	1.61
D_{50}	7.22	7.70	7.96	7.17
D_{90}	460	329	431	28.3

The particle size distribution of the co-fired ashes are similar to the lignite case, and the D_{90} value is improved. Thus, the ash reactivity is not negatively affected by the increased *Cynara* thermal share in the fuel blend, regarding particle size distribution. The particle size is the most important physical characteristic determining the reactivity of fly ash [3]. Fly ash particles in the range of 10 µm to 50 µm mainly are void-space fillers in concrete and particle sizes smaller than 10 µm are classified as pozzolanic reactive (hardening when reacted with $Ca(OH)_2$) [3]. Thus, the larger biomass fuel particle size should not lead to reductions in the ash reactivity from the particle size distribution.

With the higher heating value of the *Cynara* fuel, it is likely a rise in temperature may have contributed to improvements in char and mineral fragmentation and ultimately smaller particle sizes for the 100% *Cynara* test scenario compared to the other cases. Furthermore, the alkali species are known to lower the eutectic temperature of compounds [38] creating melts, which would be enhanced with the increase in the furnace temperature, and this likely creates an improved thermal contact due to melting of alkali compounds that enhances the heat transfer rate to other compounds [23], e.g., via heterogeneous condensation. The *Cynara* ash had much more K_2O content than the lignite. Thus, the herbaceous biomass fuel likely had a better particle fragmentation throughout the burning process. This would likely contribute to a finer particle size distribution of the fly ash produced from that test scenario. This needs further studying.

The very close gradation curves of the fly ash from the co-fired *Cynara* tests compared to the lignite fly ash indicates no negative concerns for water requirement or cohesiveness of the fly ash from *Cynara* in the fuel blend. The particle size distribution actually got somewhat finer with the 10% co-fired share of *Cynara*. This slight change in fineness indicates a possible improvement in the workability when used in mixtures, as finer round shape particles can reduce the friction between ordinary Portland cement particles and among fly ash particles [10].

It is believed that the D_{90} value for the 10% *Cynara* is caused by imperfect mixing of the ash before determination of the D_{90}. The ash formation for the co-firing of *Cynara* is believed to be driven by the ash formation from the lignite, i.e., mineral fragmentation, coagulation, agglomeration, heterogeneous condensation, etc. Also, the burning of pure *Cynara*, which has much lower ash and fixec-carbon contents along with a higher volatile matter amount, due to evaporation of alkali compounds and other organic matter the ash is finer. The PSD from the pilot-scale tests were compared to results from industrial experimental tests. This supports evaluations for the scale-up of pilot-scale test results. The industrial tests were conducted at the Kardia Power Plant located in Ptolemedia, Greece, and is described in the literature in [39]. Table 5 shows those comparisons.

Table 5. Comparisons of sieve results from industrial and pilot-scale test results.

Sample	% Passing by Mass				
	3 [μm]	32 [μm]	45 [μm]	90 [μm]	200 [μm]
* Lignite fly ash	7.59	72.3	85.2	99.5	100
* Lignite ground fly ash	22.3	88.7	97.0	100	100
* 10% *Cynara* thermal share fly ash	8.08	73.6	86.5	99.8	100
* 10% *Cynara* thermal share ground fly ash	23.9	89.6	97.3	100	100
** KSVA fly ash	19.89	100	100	100	100

* Industrial fly ash samples; ** pilot-scale fly ash sample.

Table 5 shows a good correlation among the particle sizes from the pilot-scale fly ash and the industrial ground fly ashes. The pilot-scale results are very close to the ground industrial fly ashes, which is a practice in the industry. Thus, the pilot-scale observations can be considered to be transferable to industrial scale, regarding the PSD of the fly ashes, which has a very significant impact on the reactivity of ashes, and is one characteristic relied upon to assess fly ash pozzolanic behavior.

3.4. Chemical Analysis of Ashes

3.4.1. Chemical Composition of Fly Ashes

Figures 6 and 7 show the chemical analysis of major and minor inorganic contents in the ashes. The trends are as expected.

Fly ash generated from combustion of lignite is a calcareous fly ash (sulfur-calcitic fly ash) mostly comprising chemical components like silicates, aluminates, calcium oxide, and sulfates [40]. Generally, compared to coal ash, biomass ash from natural biomass usually are enriched in descending order in K, P, Cl, Ca, Na and Mg and depleted in descending order in Al, Ti, Fe, Si and S [30]; however, this may not be the case for certain biomass fuel qualities, so it is only a guideline. The Figures 6 and 7 above show that the co-fired ashes had increased oxide concentrations of K_2O, P_2O_5, Na_2O and CaO while having depletions in oxides of SiO_2 and Al_2O_3. The majority of the trends in the ash oxides in this study are supported by results noted in the literature.

The most noticeable change in the oxides with an increasing thermal share of *Cynara* reflect the contribution of each fuel in proportion to their percentage in the fuel feed blend. Those are expected based on the fuel properties. Otherwise, the oxide contents are comparable for each test case. So, for average lignite fly ash uses, a high co-fired thermal share of *Cynara*, up to 50%, in the fuel blend does not potentially alter any market applications practices for ash utilization, despite the increased amount of problematic species in the *Cynara*. The seemingly acceptable fly ash quality up to 50% thermal share of *Cynara* may be due to the much lower ash content in the fuel along with the higher calorific value of the fuel and the higher ash content of the lignite.

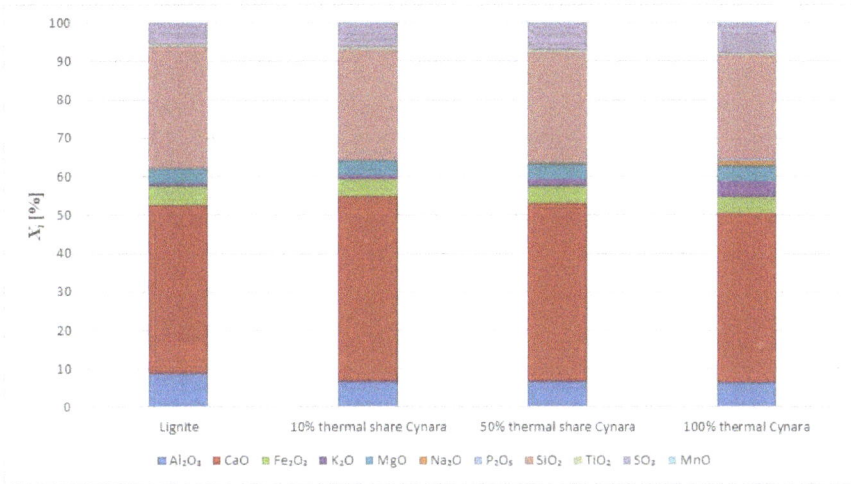

Figure 6. Major and minor oxides of fly ashes, I.

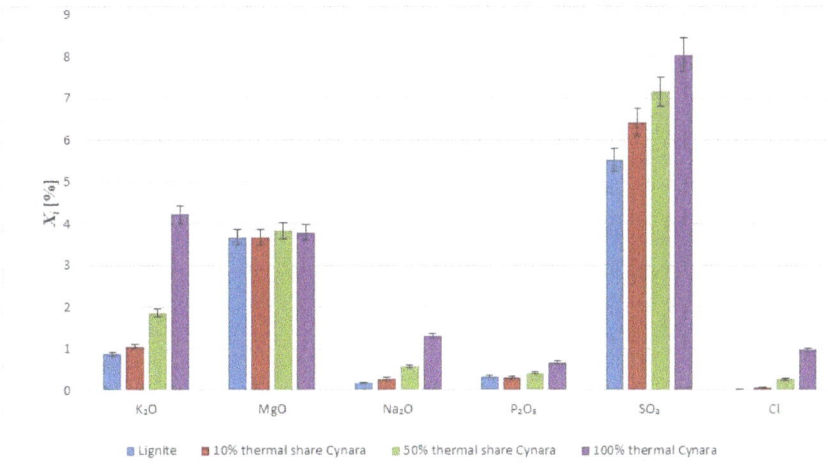

Figure 7. Major and minor oxides of fly ashes, II.

Fly ash quality concerns could be the increasing contents of alkalis and sulfur and the reduction in alumina and silica. Silica and alumina along with iron oxide comprise the total oxides in EN 450-1 to be greater than 70% on a mass basis [41] to ensure a good pozzolanic reactive fly ash. For lignite ashes, the summation is greater than 50% on a mass basis according to ASTM C618 to ensure a good reactive lignite fly ash.

The most common form of chemical attack in concrete is sulfate attack, leading to erosion of the concrete that can be caused by sodium sulfate, potassium sulfate, calcium sulfate, magnesium sulfate, etc. [42]. Sulfate SO_3, can cause long-term expansion [43] from sulfates reacting with calcium hydroxide released during hydration and with calcium aluminates in the cement [44]. The sulfate reactions create a range of sulfate compounds (e.g., sulfur-aluminates) that occupy a greater volume than the original concrete components, causing expansion and eventual failure of the concrete [44]. The SO_3 content for coal fly ash is limited to 3% by mass in EN450-1 for use in concrete. So, there is a potential need to consider the impact of the sulfur content on possible uses involving construction applications with concrete. Meanwhile, ageing (both hydrated and naturally aged fly ash) comprises mineralogical changes in the fly ash that may include reactions as oxidation, carbonation, neutralization of pH, dissolution, precipitation, and absorption, towards the formation of stable secondary minerals [45]. Soluble salts concentrations and trace elements decreased in fly ash due to weathering (aging) during natural leaching that mitigated their detrimental effects over time [46]. Thus, allowing fly ash to undergo aging (maturation) would lead to lower sulfur contents, improving its quality. On the other hand, if there is an increased sulfur amount combined with calcium, i.e., anhydrite ($CaSO_4$), in the fly ash, the mineral is very active [30], which is desirable in masonry and other low technical utilization market applications.

Alkalis (i.e., alkalis (K_2O and Na_2O), as sodium oxide equivalent) limits are stated to minimize the effect of alkalis reacting with aggregates, containing a reactive form of silica (alkali-silica reaction, or ASR), in the presence of moisture [47]. The reaction under those conditions produces a gel that can absorb water and swell, generating a pressure that causes expansion and cracking [47]. The alkali content is below the 5% by mass limit for coal fly ash as stated in EN 450-1 for concrete use. So, despite the high potassium content of the biomass, the fly ash quality did not have an adverse impact regarding unwanted alkali levels. Theoretical fly ash oxide calculations, assuming 70% of fly ash collection in the ESP as 5% each goes to the economizer ash and air pre-heater ash [48], were near to measured values, excluding the pure *Cynara* combustion.

The chloride contents in the lignite, 10% *Cynara*, 50% *Cynara* and 100% *Cynara* thermal share ashes were 0.0025%, 0.0061%, 0.026% and 0.097%, respectively, on a mass basis. Chloride can lead to corrosion of steel reinforcement in concrete [47], which is used to provide concrete with tensile strength [47]. That occurs due to Cl ions undermining the protective effect of passivation and encouraging the process of corrosion in concrete [49]. The values in the tests are below the 0.1% by mass in EN 450-1 for coal fly ash. With pure *Cynara* combustion unlikely to be practiced in large boilers, e.g., 300 MW, due to reasons of logistics, e.g., fuel supply, milling, feeding, storage, etc., the possible amount co-fired share of *Cynara*, 10–50% on a thermal share basis, does not lead to an adverse fly ash quality regarding chloride content.

The loss on ignition (LOI) used to assess the unburnt carbon in ash for the lignite, 10% *Cynara*, and 50% *Cynara* thermal share ashes were 9.48%, 14.64% and 13.7%, respectively. In biomass combustion, most of the mass is lost during devolatilization with char formation being only about 10% and unburned carbon in ash is not such an issue [50]. The occurrence is the opposite during the co-combustion trails observed. So, the increase in the LOI with co-firing is likely impacted in a similar manner in this work. In EN 450-1 the maximum LOI limit on a mass basis is 9%. The high values indicated in these tests would likely warrant concern for various reasons. High LOI values hinder the strength gain in concrete [47] and require the use of more water in concrete mixtures [32]. Higher amounts of carbon have a propensity to adsorb significant quantities of water and specialty surfactants or chemical admixtures in concrete (e.g., air-entraining admixtures (AEA), water-reducing admixtures, and retarders) [32]. Also, high carbon content can result in discoloration of the hardened concrete and to mixture segregation [51,52]. However, industrial technologies have been developed to reduce the carbon content in fly ash. Commercial practices include sieving, thermal post-treatment of fly ash [52], triboelectric separation [52,53], froth flotation [53], and re-burning [54]. So, increased carbon concerns can be mitigated, improving the marketability of the ash.

The chemical composition of the fly ashes are comparable and very near to lignite fly ashes successfully tested in a blend of fly ash and waste glass, 60% and 40%, respectively, to produce lightweight aggregates [13]. Thus, the production of successfully performing lightweight aggregates from the fly ash qualities produced is a very relevant utilization option that requires further studies. Also, the high Ca content along with the Si content in the fly ashes make them suitable for studies of high calcium fly ashes in aluminum based metal matrix composites. Work done has revealed that the strengthening of aluminum alloys with fine ceramic particles (such as fly ash) strongly increases their potential in wear resistance and structural applications [14].

3.4.2. Mobility of Some Species

Fly ash disposal is associated with ecological risks due to acidification and the infiltration of heavy metals and radioactive components that can be released into the soil [55]. The environmental impact of fly ash, either used in an application or sent to a landfill depends to a large extent on the mobility of the polluting components in the ash [16]. The subject of leaching behavior and mobility of elements in co-combustion ashes is less common, resulting in not many works dealing with the topic [56]. Biomass ash does not contain the toxic metals like in coal ash [32]. One problem with biomass ashes is attributed to the fact that trace elements are in more mobile compounds, where the fraction can be up to 61% more in biomass ash compared to 0.2% to 7.2% in coal ashes [57]. Table 6 shows the influence of an increasing co-firing share on the mobility of some nutrient element species and some heavy metals in the fly ash. Values for Cu, As, Pb, Ni and Cd were below the detection limit of the equipment used.

Table 6. Influence of an increasing Cynara thermal share on mobility of some species in the fly ashes.

Parameter	Thermal Share			
	Lignite	10% *Cynara*	50% *Cynara*	100% *Cynara*
K [mg/L]	24 ± 2.4	61 ± 6.1	543 ± 54.3	2660 ± 266
P [mg/L]	0.016 ± 0.0016	0.021 ± 0.0021	0.020 ± 0.002	0.013 ± 0.0013
Na [mg/L]	7.4 ± 0.74	20 ± 2	121 ± 12.1	593 ± 59.3
$SO_4{}^{2-}$ [mg/L]	1300 ± 130	1370 ± 137	1700 ± 170	3600 ± 360
Cl^- [mg/L]	22.7 ± 2.27	61.4 ± 6.14	260 ± 26	970 ± 97
Zn [mg/L]	0.08 ± 0.008	0.07 ± 0.007	0.08 ± 0.008	0.03 ± 0.003
Cr_{total} [mg/L]	2.18 ± 0.218	2.59 ± 0.259	2.67 ± 0.267	6.98 ± 0.698

Table 6 shows that the increasing co-fired share of *Cynara* did not lead to a significant increase in the mobility of trace elements observed. Heavy metals are correlated to the ash content of a fuel, a higher ash content mean higher levels. The lignite had considerable higher amounts of heavy metal elements than the *Cynara*. Cr increased in the fly ash with higher co-firing shares of *Cynara*, which is an widely known occurrence in wood combustion. It is likely that the association of Cr is more organic in biomass fuels, causing higher mobility, or catalytic effects occur from other biomass species, or changes in ash compounds impacting mineral phases that effect the mobility of Cr. There is the possibility of Cr being introduced from the pelletizing process that contributes to the Cr level observations in biomass combustion. Further work on the issue would clarify the cause. On the other hand, from the results observed, environmental concerns of the fly ash quality from co-firing of *Cynara* probably would not be warranted. The leachability of sulfate, chloride, sodium, and potassium increased correspondingly to the *Cynara* thermal share in the fuel blend. Those increases suggest more soluble salts concentrations in the fly ash. Regarding potential soil applications, high soluble salt levels in a green fly ash could limit its use, as an increase in the ground salinity is harmful to a successful re-vegetation goal [58]. Excessive salts in the soil reduce the water potential for plants causing water stress in the plant due to the ground becoming more saline and ceasing to support agricultural production [59]. On the other hand, the bioavailability of nutrient species is a concern. With increases in the co-firing share of *Cynara*, the results show that there is qualitatively a high potential of K being available for uptake as a nutrient in soils, and along with no negative impact from the majority of heavy metal leaching results, this is added value as a soil ameliorant on a qualitative basis. The observances indicate that at the 10% and 50% co-firing shares of *Cynara* produce fly ashes of sound environmental and ecological qualities. Figure 8 highlights the potential impact of increased thermal shares on the potential availability of some elements.

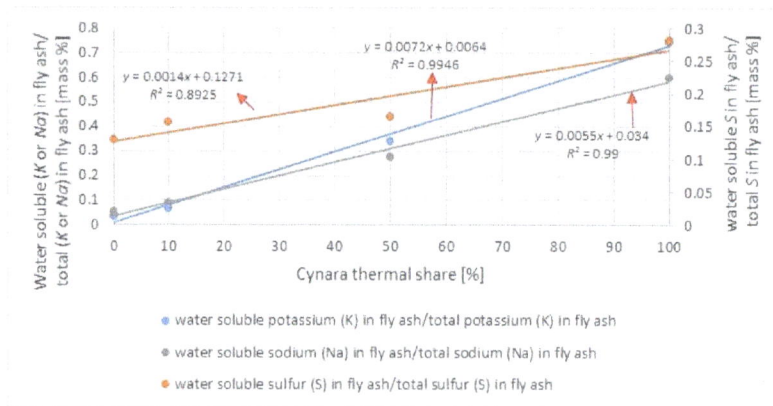

Figure 8. The ratio of water soluble S in the fly ash to total S in the fly ash and the ratio of water soluble alkali (K and Na) in the fly ash to total alkali (K and Na) in the fly ash as a function of the thermal share of *Cynara* in the fuel blend.

The regression observations reveal that the soluble fractions of potential soil nutrients were significantly associated with an increasing share of *Cynara* in the fuel blend. Therefore, there is a high potential of the co-fired fly ashes, when used in combination with organic materials, to substantially improve soil properties. Organic composts with high organic carbon and micronutrients in combination with nitrogen fertilizer and fly ash, lead to improved soil properties and slowly released nutrients in accordance with the demands of growing crops; the blended materials lead to reduced bioavailability of metals supplied by fly ash too [12]. The increased soluble fractions of the basic elements, K and Na, with increasing *Cynara* co-firing shares would raise the pH of the fly ash. Biochar combined with a lignite fly ash and applied as an amendment to an acidic soil lead to improved soil parameters, e.g., pH, microbial biomass carbon, etc. [12].

3.4.3. Enrichment of Major and Minor Elements

When coal is combusted, a concentration process occurs, which can result in the ash element concentration being higher than those in the coal [16]. During the cooling of the flue gas along the path to the dust removal facility (from approximately 1600 °C to 140 °C) the dew point will be passed an condensation will begin on the surface of fly ash particles [16]. Elements in the ash after combustion can be enriched, depending on the type of ash and the particular element [16]. Figures 9–12 show the enrichment of major and minor elements in parity diagrams for the lignite, 10% thermal share of *Cynara*, 50% thermal share of *Cynara*, and 100% *Cynara* combustion, respectively. The effect of the higher heating value of the biomass and the high ash content of the lignite is observable.

All fly ashes are shown to be enriched in Si. All fly ashes had a slight, comparable enrichment of sulfur. All fly ashes, excluding the 100% thermal share of *Cynara*, are shown to be depleted in Ca. The 100% *Cynara* combustion scenario produced fly ashes enriched in the refractory elements, but significantly depleted in the elements K, Na and P, which is an unexpected observation.

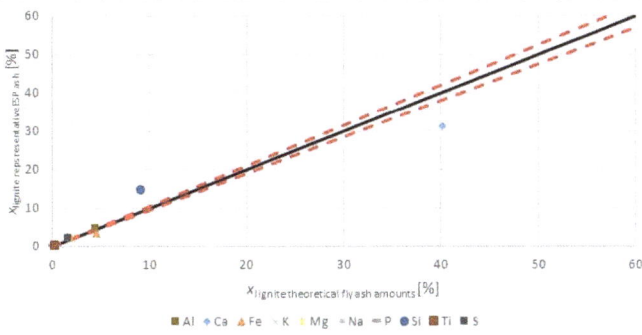

Figure 9. Enrichment of elements in the fly ash for lignite combustion; fuel ash plotted on the abscissa and representative electrostatic precipitator (ESP) ash plotted on the ordinate; dashed lines estimate the uncertainty due to analyses variability (±5%).

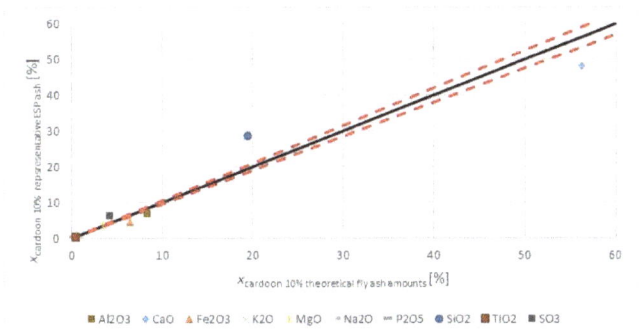

Figure 10. Enrichment of elements in the fly ash for 10% thermal share of *Cynara*; fuel ash blend plotted on the abscissa and representative ESP ash plotted on the ordinate; dashed lines estimate the uncertainty due to analyses variability (±5%).

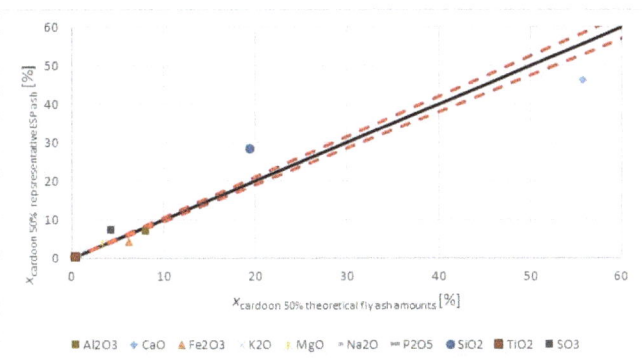

Figure 11. Enrichment of elements in the fly ash for 50% thermal share of *Cynara*; fuel ash blend plotted on the abscissa and representative ESP ash plotted on the ordinate; dashed lines estimate the uncertainty due to analyses variability (±5%).

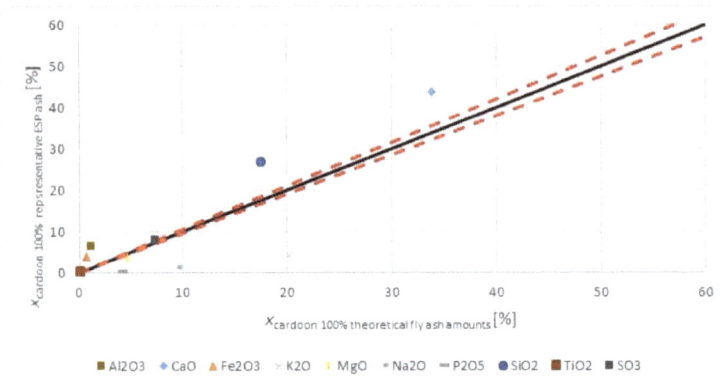

Figure 12. Enrichment of elements in the fly ash for 100% thermal share of Cynara; fuel ash plotted on the abscissa and representative ESP ash plotted on the ordinate; dashed lines estimate the uncertainty due to analyses variability (±5%).

Ca is among the group of elements considered not volatile [16] that can condense on bottom ash particles. The group of compounds containing calcium are noted to not vaporize during combustion and their concentration, i.e., enrichment, generally, is considered the same in all ash types and is independent of size [16]. The enrichment factors of Ca for the bottom ashes were 0.78, 0.63 and 1.2 for lignite, 10% *Cynara* thermal share, and 100% *Cynara*, respectively. Those observations correlate to the fly ashes. With the introduction of biomass in the fuel blend, the combustion behavior was improved, which may have led to more Ca being volatile or the association of calcium in biomass fuel makes it more volatile than in coal, e.g., organic associated, etc., rather than clay associated. However, the enrichment of Ca is apparently the same in all ash types, as is the case reported as noted above. The depletion of K, Na and P in the fly ash could be a result of their more volatile nature in biomass fuels and along with a reduced Si content, which may contribute to more calcium sulfate compounds being formed, can lead to a lower heterogeneous condensation of those species. Thus, they remained in the vapor phase after the primary dust collection point (electrostatic precipitator), leading to fly ash enrichment factors being low (<<1), and are collected with the finer dust particles (fabric filter) and emitted to the stack. The enrichment factors for K, Na and P in the fabric filter were 0.86, 1 and 0.23, respectively. This indicates that most of K and Na was captured in the fabric filter, but P was emitted via the stack.

The interactions of the co-fired fuel ashes resulted in the much higher contents of problematic species (K, Na, P, S), for ash behavior in a pulverized fuel facility, not being a significant concern in the *Cynara* for the fly ash quality. Calcium and magnesium, which do not condense of bottom ash particles, may react with alumina and silica not condensed on bottom ash particles. The reaction of calcium with alumina silicates (Equation (3)) and silica (Equation (4)), as in the addition of an alumina-silicate additive in the fuel blend in the gas phase is shown below [7]:

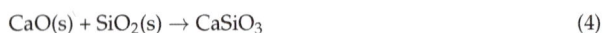

$$CaO(s) + SiAl_2O_5 \rightarrow CaAl_2SiO_6 \tag{3}$$

$$CaO(s) + SiO_2(s) \rightarrow CaSiO_3 \tag{4}$$

Thus, the above mechanism may have contributed to enrichment of silica in the fly ashes. This may have contributed to more volatile species leaving with stack emissions or collected in ash deposition within the boiler in other compounds besides silicates.

The increase in SO_3 in the co-fired ashes is a trend that is not usually found in increasing co-firing shares of biomass. Co-firing of natural gas with coal can lead to increases of sulfur retained in the ash. It was reported that from a field test of co-firing natural gas with coal, SO_2 levels showed a 18% reduction for 11% gas co-firing [60]. It was believed that a mechanism exists wherein there is a shift of fuel sulfur conversion from SO_2 to SO_3 with gas co-firing [60]. This would lead to more alkaline species reacting with the SO_3 to form sulfates. Also, co-firing of natural gas may increase the reactivity of sorbent metals in the fly ash due to a combination of activation and increasing ash surface area [61]. Because of the formation of oxygen radicals a natural gas flame may lead to higher concentrations of the more reactive and easily captured SO_3 species above equilibrium levels [61]. In addition, the flame from gas co-firing may led to an increase in the heating rate of coal particles, which increases the internal surface area, causing a higher reactivity [61]. Mechanisms of formation of super-equilibrium concentrations of SO_3 when SO_2 is exposed to a hydrocarbon flame is due to excess O_2 forming oxygen radicals (Equation (5)), and the oxygen radicals combining with SO_2 to form SO_3 (Equation (6)) [61]:

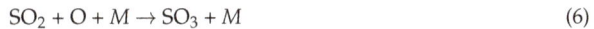

$$O_2 + M \rightarrow O + O + M \tag{5}$$

$$SO_2 + O + M \rightarrow SO_3 + M \tag{6}$$

where M equals metals. Because of the relatively slow reaction rate of (Equation (7)), SO_3 concentrations remain at super-equilibrium levels for several hundred milliseconds after the sulfur gas passes through the flame, despite once sulfur gas has passed through the flame, the SO_3 concentrations begin to decrease via reaction with the oxygen radicals (Equation (7)) [61]:

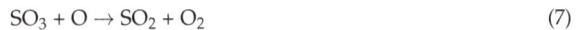

$$SO_3 + O \rightarrow SO_2 + O_2 \tag{7}$$

Previous tests showed that without co-firing of methane, the measured sulfur in the ash was 6.4% of the original amount in the coal, but with 4% methane addition, the measured sulfur in the ash was 58.4% of the original sulfur in the coal [61]. Therefore, it is believed that an effect of the natural gas co-firing was more calcium reacting with sulfur in the furnace, causing less sulfur available for reactions with volatile species in the cooler areas, i.e., convection pass. This may have contributed to no enrichment of the alkali species and P in the fly ashes, albeit the undesirable concentrations in the *Cynara*. There may have been positive synergistic effects between calcium reacting with alumina-silicates or silicates and afterwards with sulfur that led to no enrichments of the volatile species in the fly ash Thus, an optimized co-firing scenario of a low-quality lignite and an herbaceous biomass may be with the addition of natural gas. Further work is needed to elucidate the observations.

Thus, it is apparent that the use of natural gas enhanced the desulfurization from calcium in the reactor, as sulfur is among the group of elements that are the most volatile and expected to not be enriched in fly ashes [16]. The critical temperature range of sulfur reaction with calcium is 900 °C to 1250 °C with the lower limit being 870 °C to 900 °C and unstable above 1250 °C [62]. With the use of natural gas, there may be a longer temperature time history at higher temperatures in the boiler and higher kinetic reaction rates. It was reported that tests done with gas co-firing led to slagging, which heat transfer models reveled that an increase in the combustion zone temperature of approximately 38 °C and the flue gas exit temperature of about 7 °C could have contributed to the problem [60]. In this work, for example with a power output of 300 kW_{el}, for a 10% thermal share of *Cynara* and with the lignite considered to be pre-dried to a 15% moisture content, air pre-heated to 200 °C, and an excess air ratio of 1.2, the adiabatic flame temperature was calculated as approximately 1720 °C. With the use of natural gas at a 20% substitution of the heat rate, the adiabatic flame temperature was calculated as approximately 1840 °C. Thus, the use of natural gas in the co-firing fuel blend would increase temperatures, potentially allowing for a longer time at higher temperatures and reactions of calcium with more sulfur oxidized to the more reactive SO_3 species.

In the temperature range of 1700 °C to 1300 °C potassium species can react with alumina and silica to form potassium alumina-silicates, and the reaction dominates over sulfating reactions of potassium [7]. However, available Ca and Mg takes precedence with silica and alumina for reactions, limiting available species for reactions with potassium [7]. Potassium sulfate is expected to nucleate at about 800 °C [63] and sulfating potassium reactions are favored at low temperatures in the range 800 °C to 900 °C, and they are very slow below 800 °C [64]. However, all potassium is predicted to appear as K_2SO_4 below 700 °C [7], so the lack of enrichment of potassium in the fly ashes, when compared to the fuel content, especially for the pure *Cynara* tests, supports the assertion that the natural gas in the fuel blend created gas atmospheres favoring reactions of sulfur with calcium. Therefore, it is believed that the natural gas co-firing led to higher temperatures and more sulfur being captured by calcium in the higher temperature range. This would reduce the sulfur available for forming lower temperature compounds with K, Na, etc. Figure 13 compares the enrichment of elements in the fly ash in the 10% thermal share of *Cynara* scenario with the 50% thermal share of *Cynara* case.

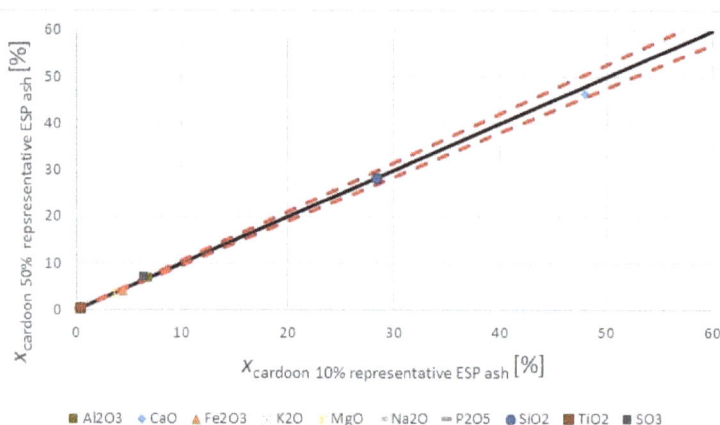

Figure 13. Enrichment of elements in the fly ash compared for 10% thermal share of *Cynara* and 50% thermal share of *Cynara*; 10% thermal share of *Cynara* plotted on the abscissa and 50% thermal share of *Cynara* plotted on the ordinate; dashed lines estimate the uncertainty of analyses variability (±5%).

The above Figure 13 shows that despite the unusually high contents of K_2O, Na_2O, P_2O_5 and SO_3 in the biomass fuel, especially compared to the lignite, there was no change in the fly ash quality between a 10% thermal share and a 50% thermal share of *Cynara*. Along with the impact of natural gas, this may be attributed to the higher heating value of the biomass fuel, requiring much less mass for a thermal share substitution, which allows the ash in the lignite to have a relevant influence on the ash formation process. Thus, there is an apparent benefit of co-firing natural gas with a low quality lignite and an herbaceous biomass fuel. However, this has to be thoroughly assessed for the impact on boiler efficiency from the heat of vaporization of the hydrogen in the natural gas fuel and the higher cost of the premium fuel that can make it expensive during winter months [60].

3.5. Scanning Electron Microscope (SEM) Analyses and EDX Results

In the figures in this section, ESP hopper samples are numbered from 1 to 3, and a sample is identified as the bulk ash. The first three fly ash samples correspond to the three batches of ash precipitated at the three hoppers in the ESP (ranked from the coarser fraction, number 1, to fine the fraction, number 3). The bulk sample is a mixed representative fly ash of the ESP for each firing scenario. Figure 14 shows the general overview of collected fly ash samples during pure lignite combustion tests. The chemical composition of the fly ash from the three hoppers collected during pure lignite combustion from energy dispersive X-ray (EDX) analyses is given in the figure too. The EDX analyses are normalized on carbon and oxygen free basis, expressed in oxides.

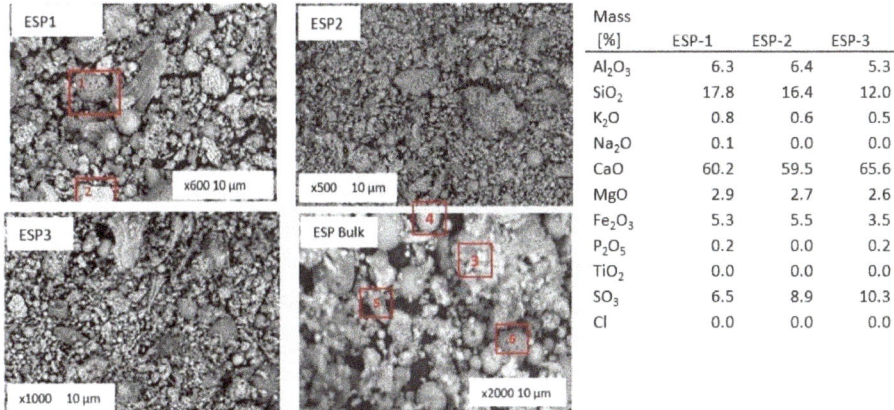

Mass [%]	ESP-1	ESP-2	ESP-3
Al_2O_3	6.3	6.4	5.3
SiO_2	17.8	16.4	12.0
K_2O	0.8	0.6	0.5
Na_2O	0.1	0.0	0.0
CaO	60.2	59.5	65.6
MgO	2.9	2.7	2.6
Fe_2O_3	5.3	5.5	3.5
P_2O_5	0.2	0.0	0.2
TiO_2	0.0	0.0	0.0
SO_3	6.5	8.9	10.3
Cl	0.0	0.0	0.0

Figure 14. Backscattered electron images and energy dispersive X-ray (EDX) analyses of fly ash from pure lignite combustion.

An amorphous particle, number 1, represents an alumina-silicate compound enriched in iron and some potassium along with smaller amounts of calcium and magnesium. Particle number 2 is an example of a particle when calcium was primarily associated with oxygen and some sulfur.

The median particle size varied between 6 and 80 μm, with the larger particle sizes generally corresponding to fused molten agglomerates, see particle numbered 3. There is no significant difference in morphology of the particles collected at various hoppers, apart from the particle size. Obviously ash particles in the sample number 3 (from the third hopper) are smaller (see magnification). From the bulk sample, both mineral (particle number 4) and amorphous components (particle number 5) can be identified. Spherically and colorless particles indicate glassy textures or predominance of crystallinity and spherical, rounded light colored particles indicate a glassy surface [65]. Rounded particles indicate a glassy particle, irregular shapes indicate partly crystalline particles, and angular shapes indicate a crystalline nature [65,66]. As determined by EDX, the predominant elements in the fly ash samples were calcium, silicon, aluminum, iron, magnesium and sulfur in various compounds. Calcium was observed to primarily be associated with oxygen, sulfur, and silica, or possibly with carbon (calcite). Particle number 6 shows a particle with holes, which represent such particles in the fly ash. Volatiles trapped in a particle cause it to have swelled surfaces, varying the shape of the particles and bubbles, which alters its surface [67]. The bubbles formed may be attributed to the release of volatile matter from the internal zones of the particle as the solid particle's surface is softening and melting [67]. The bubbles can generate large cavities under the surface, which are smooth zones and considered to be less reactive for the subsequent steps of oxidation [67]. This indicates a continuing combustion, which may have undesirable effects on the fly ash quality due to the presence of such apparent uncomplete combusted particles. Such particles were qualitatively observed to be more prevalent in the fly ash from the lignite scenario. This suggests the lignite case had the least optimal combustion performance.

Figure 15 shows the general overview of collected fly ash samples during 10% thermal share co-firing of *Cynara*. The chemical composition of the fly ash from the three hoppers collected during 10% thermal share co-firing of *Cynara* from EDX analyses is given in the figure too. There is no significant difference in morphology of the particles collected at various hoppers.

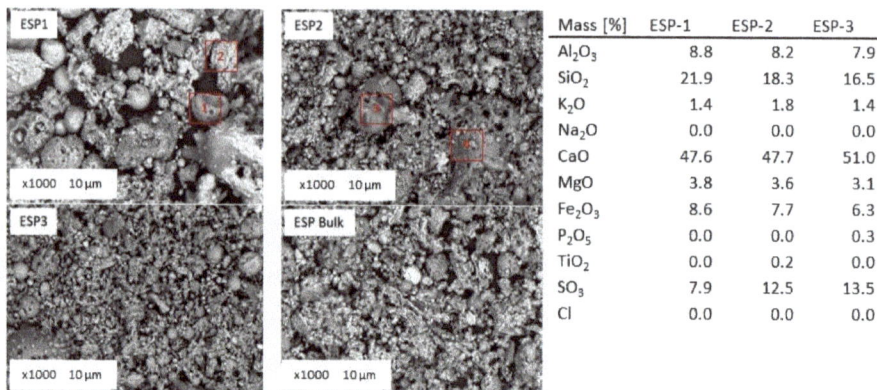

Mass [%]	ESP-1	ESP-2	ESP-3
Al_2O_3	8.8	8.2	7.9
SiO_2	21.9	18.3	16.5
K_2O	1.4	1.8	1.4
Na_2O	0.0	0.0	0.0
CaO	47.6	47.7	51.0
MgO	3.8	3.6	3.1
Fe_2O_3	8.6	7.7	6.3
P_2O_5	0.0	0.0	0.3
TiO_2	0.0	0.2	0.0
SO_3	7.9	12.5	13.5
Cl	0.0	0.0	0.0

Figure 15. Backscattered electron images of fly ash from 10% thermal share *Cynara* co-firing.

The coarser fraction (20–90 μm), in ESP1, is mainly composed of alumina silicates particles with potassium, calcium rich crystalline phase, and iron and magnesium oxides The SEM-EDX data showed calcium is associated with sulfur or oxygen in distinct particles, (see Figure 15, particle number 2). Sulfur captured by lime, which is originating from fuel, is widely known as a self-desulfurization process. This process leads to a spontaneous reduction of SO_2 emissions in industrial boilers. The observation support the assertion noted in Section 3.4.3, regarding the enrichment of elements in the fly ashes.

Particle 1 is an example of an amorphous alumina-silicate sphere. The majority of the iron-rich spheres observed consisted of an iron oxide mixed with amorphous alumina-silicate. Compared to the previous fly ash sample (100% lignite combustion), the fly ash samples from 10% thermal share of *Cynara* co-firing contain less calcium, but more aluminum, silica, iron, and sulfur. Char particles with holes were detected in this fly ash too, see particle number 3 and particle number 4. Those particles indicate a swelling phenomenon and structural changes after devolatilization that led to melting and fusion into a mass with smooth surfaces. A particle with irregular and indented surfaces expose more active sites for gaseous reactant than a smooth surface, which influences the reactivity of the char [67]. Thus, those char particles were not as reactive, which apparently led to their presence in the fly ash.

Necked and interlocked particles appear to be more prevalent in the 10% *Cynara* bulk ash compared to the lignite scenario, which generally increased in the bulk ash with the increase in the co-firing share of *Cynara*. The agglomeration phenomena may be the result of low melting compounds, mainly, alkali in biomass ashes, than can stick contacting surfaces of separate particles, causing them to cluster [67]. Moreover, for this fly ash, the increase in irregular, elongated, flat, etc. particles may be associated with more particles considered as fissures. Fissures, holes, and superficial porosity represent the way volatile gases escape and for gaseous reactants to reach the active sites inside the char particle for oxidation [67]. Fissures, which are associated with a scabrous particle surface with a wide distribution, lead to a more reactive particle [67]. Thus, the apparent qualitative increase of irregular particles in this ash compared to the lignite scenario may suggest a better burning behavior of the fuel blend.

The general overview of collected fly ash samples during 50% thermal share co-firing of *Cynara* is shown in Figure 16. Mineral (particle numbers 1 and 2) and amorphous (particle numbers 3 and 4) components can be identified. The chemical composition of the fly ash from the three hoppers collected during 50% thermal share co-firing of *Cynara* from EDX analyses is given in the figure too. There is no significant difference in the morphology or in the particle sizes between various hopper ashes.

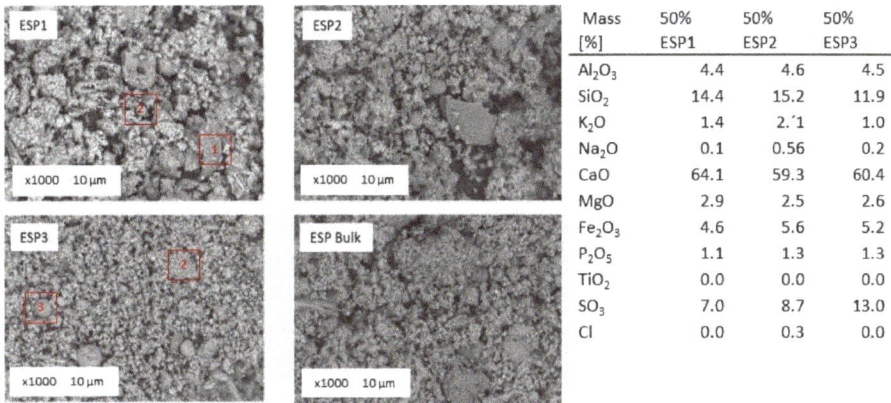

Mass [%]	50% ESP1	50% ESP2	50% ESP3
Al_2O_3	4.4	4.6	4.5
SiO_2	14.4	15.2	11.9
K_2O	1.4	2.´1	1.0
Na_2O	0.1	0.56	0.2
CaO	64.1	59.3	60.4
MgO	2.9	2.5	2.6
Fe_2O_3	4.6	5.6	5.2
P_2O_5	1.1	1.3	1.3
TiO_2	0.0	0.0	0.0
SO_3	7.0	8.7	13.0
Cl	0.0	0.3	0.0

Figure 16. Backscattered electron images of FA from 50% thermal share *Cynara* co-combustion.

The median particle size varied between 1 and 30 μm that means somewhat smaller sizes than in the lignite and 10% thermal share fly ashes, which is different than the particle size distribution of the bulk ash shown in Section 3.3. The predominant elements in the fly ash samples were calcium, silicon, aluminum, iron, magnesium, sulfur and potassium in various compounds. Aluminum was observed with silicon and potassium. Apart from alumina-silicates, silica has been identified as a quartz too. Calcium was primarily associated with oxygen, sulfur, and phosphate. Again, this observation supports the assertion of the ash formation process in Section 3.4.3 from the addition of natural gas.

For this fly ash quality, there was an apparent increase in irregular, fused, etc. particles. This may be attributed to more biomass particles in the fuel blend, giving rise to a higher temperature. At higher heating rates, from a higher temperature, particles can partially or completely melt, losing their distinction, fusing together, and forming completely or partially hollow particles and smoothed surface particles too [68]. Also, a higher temperature implies not only a higher heating rate but also a rapid evolution of gases that build up pressure during volatile release, which may change the shape of some parts of the particle structure and cause some particle breaking (fragmentation) [69]. The increase in pyrolysis pressure has the potential to lead to the formation of larger particles due to swelling as well as formation of particle clusters as a result of melting and subsequent fusion of particles [68]. This is likely the mechanism of the clusters of particles formed with an increasing share of *Cynara* in the fuel blend. This may support a better overall burning of the chars for co-firing cases, as the 50% thermal share of *Cynara* had a lower LOI than the 10% thermal share of Cynara.

Figure 17 shows fly ash samples from the ESP hoppers and the mixed bulk fly ash collected during the 100% thermal share *Cynara* tests. The chemical composition of the fly ash from the three hoppers collected during the 100% thermal share *Cynara* tests is given in the Figure too from EDX analyses.

The fly ash contains amorphous (point numbers 1, 2 and 3) and crystalline ash particles of various shapes. The median particle size varied between 4 μm and 80 μm, with the larger particle sizes (particles numbers 4, 5 and 6,) generally corresponding to fused molten agglomerates. The sphere-shaped ash particles are composed mainly of aluminum and silicon, with potassium and calcium. The detection of alumina-silicate with potassium, as it was not detected in the other fly ash samples, could be attributed to the lower calcium content in the biomass fuel. Calcium is also the main ash forming element in the 100% *Cynara* fly ash samples. Next to alumina-silicate cenospheres, similar particles, but then enriched in iron (ferrospheres) (particle number 7) were identified. Next to particle number 7, it can be seen a hollow ferrosphere as a result of the possible expansion of trapped volatile matter. Ferrospheres may be associated with heavy metals, such as Cu, Cr, Pb, and Zn due to inorganic,

organic, and intermediate affinity of trace elements, which may make environmental pollution risks a paramount importance [70]. The identification of ferrospheres give rise to a multi-utilization of the fly ash as they can be economically extracted by magnetic separation techniques that have industrial possibilities in metallurgy, ore and coal dressing processes, and dense concrete production [70]. The iron oxide exhibits various textures even the dendritic pattern was found. The fine fraction of the ash deposit from the 100% *Cynara* scenario is mainly associated with calcium, but also with potassium and sodium oxides, sulfates, phosphates and chlorides. Several crystalline particles containing alkali metals and sulfur with chlorine were identified. Since alkali chlorides crystal structure is octahedral as calcium oxide, it is difficult to clearly distinguish between various compounds. Nevertheless, the fly ash samples contain significantly higher amount of sulfur, potassium, and chloride compared to the other fly ashes. Compared to the 50% *Cynara* co-firing scenario, there is more potassium, sodium, and chloride in the fly ash from 100% *Cynara* combustion. The concentration of the sulfur remains comparable. Without the interaction of the lignite ash, i.e., more alumina-silicates and calcium available, the volatile species were more readily able to react with compounds of sulfur, silica, etc. that contributed to their capture in the fly ash. This is supported by the enrichment of elements for the 100% *Cynara* combustion test scenario shown in Section 3.4.3.

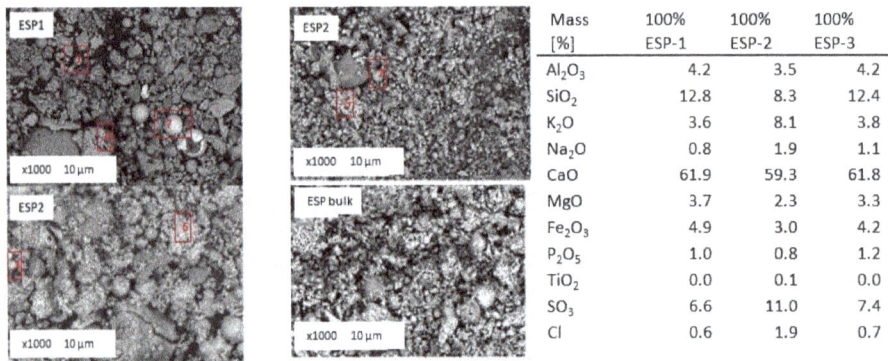

Mass [%]	100% ESP-1	100% ESP-2	100% ESP-3
Al_2O_3	4.2	3.5	4.2
SiO_2	12.8	8.3	12.4
K_2O	3.6	8.1	3.8
Na_2O	0.8	1.9	1.1
CaO	61.9	59.3	61.8
MgO	3.7	2.3	3.3
Fe_2O_3	4.9	3.0	4.2
P_2O_5	1.0	0.8	1.2
TiO_2	0.0	0.1	0.0
SO_3	6.6	11.0	7.4
Cl	0.6	1.9	0.7

Figure 17. Backscattered electron images of fly ash from 100% thermal share *Cynara* combustion.

It is noted that for the influence of aggregates on concrete (an engineering construction material) shrinkage, flaky, elongated, angular, and rough aggregate particles have high voids, requiring more material to fill them, which increases water demand and shrinkage [71]. Spherical and cubical aggregates with their less specific surface area than flat and elongated particles require less water for use, and they lead to lower shrinkage than flaky and elongated aggregates [71]. The irregular particles can be overcome by applying a grinding fly ash post-treatment process to reduce the particles that are fused and create more spherical particles [72]. Thus, the inclusion of *Cynara* in the fuel blend may necessitate a small grinding post-treatment process to obtain an optimal fly ash for use in construction engineering materials, i.e., concrete and mortar for various masonry applications.

From the SEM analyses, it is seen that the fly ash samples consist of particles of various shapes and sizes. At temperatures in a pulverized fuel boiler, generally in excess of 1400 °C, the mineral matter within the fuel oxidizes, decompose, fuse, undergo fragmentation, agglomerate, etc. During the burning of coal, rapid cooling in the post-combustion zone results in the formation of spherical, amorphous (non-crystalline) particles, ($\leq$90%), or glass and a small amount of crystalline material [3]. Shown in Section 3.4, the average chemical composition was relatively consistent for the pure lignite, 10% *Cynara* thermal share, and the 50% *Cynara* thermal share. The major element in all fly ash samples is calcium. The relatively bigger molten agglomerates are composed mainly of alumina silicates with large proportions of potassium. The coarse fly ash fraction may also contain significant amounts of calcium sulfates. Next to the coarse fraction, agglomerates of very fine particles (<<1 μm) clustered together were observed too. The fine fly ash fraction (ESP3) is formed from volatilized ash material that condenses in the cooling flue gas. The 100% *Cynara* combustion test resulted in alkali salts formation in submicron particulates. Moreover, with the increasing share of *Cynara* in the fuel blend, there was a reduction of amorphous particles (mainly associated with spherical particles) and an increase in irregular particles. Thus, the effects of irregular particles with more inner voids may need to be considered when deciding the utilization route of fly ash produced from *Cynara* in the fuel blend, which will be dependent upon the application.

3.6. X-ray Powder Diffraction Analyses of the Ashes

Most elements in fly ash occur in both organic and inorganic matter, and each has a dominant association and affinity with some mineral phases [73]. Elements in fly ash can occur in distinct phases that occur in different content, size, morphology, association, and generation in fly ashes, namely active, semi-active, pozzolanic, or inert mineral phases [73]. Each crystal phase group has individual behaviors regarding hydration-dehydration and hydroxylation-dehydroxylation processes in fly ashes, which have a leading role in the production of construction materials that is missing information from the bulk chemical composition of fly ashes [73]. Characterizing, identifying, and quantifying the phase mineral composition, along with the chemical content, of biomass ashes, are initial and necessary steps for assessing their utilization [57]. Minerals are also essential for the building industry, where fly ash is likely to have a significant use, with lime and siliceous acid being among the most important for this industry [74]. The elements will form different minerals in co-fired biomass ash compared to coal ash or lignite ash. X-ray powder diffraction scans of the fly ashes from the different scenarios were compared among each other, as shown in Figure 18. Table 7 shows the minerals detected in the fly ashes for all scenarios.

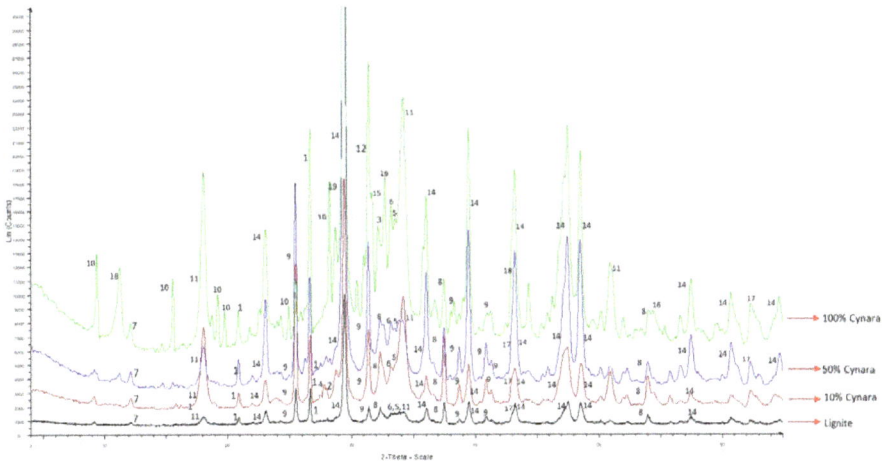

Figure 18. Comparison of X-ray powder diffraction scans of the fly ashes; numbers correspond to mineral phases in Table 7.

Table 7. X-ray powder diffraction analyses of the fly ashes from experimental tests.

No.	Mineral Phase	Lignite	10% *Cynara*	50% *Cynara*	100% *Cynara*
1	Quartz (SiO_2)	+++	++	++	++
2	Anorthite ($CaAl_2Si_2O_8$)	-	+	-	-
3	Larnite (Ca_2SiO_4)	-	-	-	+
4	Protoenstatite ($MgSiO_3$)	-	+	-	-
5	Merwinite ($Ca_3Mg(SiO_4)_2$)	++	++	+	-
6	Hematite (Fe_2O_3)	++	++	+	+
7	Iron oxide calcium oxide ($Fe_2O_3(CaO)_2$)	+	+	+	
8	Lime (CaO)	++	++	+	+++
9	Anhydrite ($CaSO_4$)	+++	+++	+++	+++
10	Syngenite ($K_2Ca(SO_4)_2 \cdot H_2O$)	-	-	-	++
11	Calcium hydroxide ($Ca(OH)_2$)	++	+++	+	+++
12	Buetschliite ($K_2Ca(CO_3)_2$)	-	-	-	++
14	Calcite ($CaCO_3$)	++++	++++	++++	+++
15	Dolomite ($CaMg(CO_3)_2$)	-	-	-	++
16	Magnesite ($Mg(CO_3)$)	-	-	-	++
17	Periclase (MgO)	++	+	+	-
18	Iron oxide chloride (FeOCl)	-	-	-	+
19	Sylvite (KCl)	-	-	-	+

++++ significant; +++ high; ++ moderate; + minuscule; - not present.

Table 7 shows that fly ash qualities are dominated by calcium minerals, as expected. Dominated by calcium minerals are lignite fly ashes [75]. The mineral phases in fly ashes are categorized as noted below, with calcium based minerals having active qualities and the most desirable in fly ashes for masonry uses. The following list highlight those categories [57]:

- Active (lime, periclase, anhydrite, bassanite, Ca and Ca-Mg silicates and alumina-silicates, ca-enriched glass, among others);
- Semi-active (portlandite, brucite, gypsum, carbonates, clay and mica minerals, feldspars, Fe oxides, Ca-containing glass, among others),
- Pozzolanic (glass); and
- Inert or inactive (quartz, mullite, some char) behavior during hydration-dehydration and hydroxylation-dehydroxylation processes of biomass ashes.

The XRD analyses showed that there were no alkali-sulfates detected in the fly ashes from the co-firing trials. The K in a sulfur phase detected in the 100% *Cynara* combustion is in agreement with SEM-EDX observations. Thus, it is maintained that the inclusion of natural gas in the co-firing fuel blend led to more sulfur being transformed to the more reactive SO_3 species, as explained in Section 3.4.3. The effect reduced the available sulfur to form alkali sulfates. This may be a major reason why the lignite ash and the co-fired ashes are more similar than anticipated. Therefore, an optimal co-firing of a low-quality lignite and an herbaceous biomass fuel may be with natural gas in order to mitigate ash related problems. More work is needed in this area.

It's necessary to classify fly ashes according to their potential pozzolanic activity and cementitious properties, as well as others if needed, such as alkali-aggregate reactivity and sulfate resistance [53]. Fly ash produced from combustion of low-rank lignites and sub-bituminous coals have both cementitious (self-hardening when reacted with water) and pozzolanic properties [73]. The pozzolanic mineral phases undergo pozzolanic reactions, which are mainly diffusion controlled and begin at a much later period in the hardening system [57]. Fly ashes with medium to little pozzolanic tendencies and generated from fuels with a variable rank, but mostly lignite, to a lesser extent bituminous coals and rarely anthracites, are active types [73]. The primary key to the pozzolanic reaction is the structure of the silica, which must be in a glassy or amorphous phase with a disordered structure, being formed, like in the rapid cooling of a volcanic magma [28]. If the silica is in a uniform crystalline structure, like found in silica sand, it is noted to not be chemically active [28]. The fly ash qualities produced in this work are considered to have both cementitious and pozzolanic qualities as it is expected to be some amount of amorphous matter in fly ash produced in pulverized fuel boilers, as quenching retains high temperature phases and melts will appear as glass [76], but below about 5%, it may not be detected by XRD analyses.

Primary minerals present do not undergo transformations during combustion, such as stable silicates [73]. Quartz, which is a primary mineral and can be essentially non-reactive in combustion processes due to its high fusion temperature [77]. Also, the quartz mineral, which may form by partial crystallization of the glassy phases in fly ash, decrease expansion from of alkali-aggregate (i.e., silica in the aggregates) reactions in concrete, and they are non-reactive [49]. Quartz was found in all ash samples but had a qualitative decrease with increasing co-firing shares, which was an observance from peak intensities.

Anhydrite ($CaSO_4$), forming from the reaction of CaO, SO_2 and O_2 in the furnace or flue, plays a role in concrete's hydration behavior [75]. Shown below are the reactions of forming anhydrite (Equations (8) and (9)) [78]:

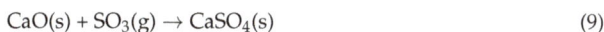

$$CaO(s) + SO_2(g) + 1/2O_2(g) \rightarrow CaSO_4(s) \tag{8}$$

$$CaO(s) + SO_3(g) \rightarrow CaSO_4(s) \tag{9}$$

Anhydrite may also be formed from the dehydration of gypsum [1,77], dehydration of bassanite ($CaSO_4 \cdot 1/2H_2O$), or between a sulfuric acid (generated during sulfide oxidation) and calcium carbonates in the coal/lignite. Shown below is the reaction (Equation (10)) of gypsum decomposition [1].

$$CaSO_4 \cdot 2H_2O \rightarrow CaSO_4(s) + 2H_2O(g) \tag{10}$$

Anhydrite along with other soluble aluminates lead to the formation of ettringite (trisulfate of calcium aluminate) immediately upon addition of water to fly ash; the initial hydration reaction contributes much to the self-hardening characteristics of fly ash, which can participate and control the solubility of potentially hazardous trace elements [75]. The decrease of quartz and increase of anhydrite, found in all samples, with the increasing co-firing share, implies an increased reactive fly ash quality. The lime decreased in fly ash with co-firing of *Cynara*, which is a result of the lower level in the biomass. It is likely formed via the reaction of calcium in the fuel with oxygen, see below (Equation (11)) [79]:

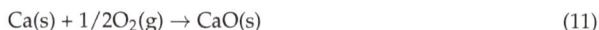

$$Ca(s) + 1/2O_2(g) \rightarrow CaO(s) \tag{11}$$

The lime may also be formed by the mechanism of the breakdown of calcium carbonate in the fuel. The strongly endothermic process leads to the breaking up of the chemical bond in the carbonate ion and the rearrangement of ions to form CaO crystals and the diffusion of CO_2 out of the cluster of lime crystals, as shown below (Equation (12)) [25,80]:

$$CaCO_3(s) + O_2(g) + heat\ energy \rightarrow CaO(s) + CO_2(g) \tag{12}$$

The lime may also be formed from calcium incorporated into the organic matter [77]. The decreae in calcium oxide implies a decrease in any free lime (CaO) potential. Free CaO in excess amounts must be limited or unsoundness (undesirable volume change [25]), i.e., expansion, could occur during hydration processes in concrete [81]. Furthermore, free CaO in ash may react with water forming $Ca(OH)_2$, causing an increased volume and structural destruction of concrete [82]. Probably formed via the reaction shown below (Equation (13)) was the portlandite mineral [83]. The mineral's presence in the fly ash increased with the 10% *Cynara* co-firing, but decreased with the 50% share:

$$CaO(s) + H_2O(g) \rightarrow Ca(OH)_2(s) \tag{13}$$

Formed via the reaction of CaO with CO_2 is calcite as shown below (Equation (14)) [25,84]. The reaction is widely known:

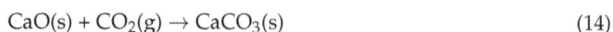

$$CaO(s) + CO_2(g) \rightarrow CaCO_3(s) \tag{14}$$

The periclase mineral may be formed via the general reaction shown below (Equation (15)) for the decomposition of carbonates in the fuel [38]:

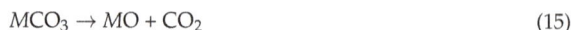

$$MCO_3 \rightarrow MO + CO_2 \tag{15}$$

where M equals Fe, Mg or Mn. The periclase may also be formed form the reaction indicated below (Equation (16)) [85]:

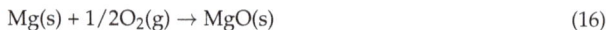

$$Mg(s) + 1/2O_2(g) \rightarrow MgO(s) \tag{16}$$

Also, the periclase may be formed from incorporation into the organic matrix [77]. The increasing co-firing shares led to a reduction of periclase, and it was not detected in fly ash from 100% *Cynara* combustion. Deduced is the association of the mineral with the lignite fuel, and its absence in higher co-firing *Cynara* shares suggests an improved fly ash quality. Free MgO present above a certain size [34] slowly hydrates yielding brucite ($Mg(OH)_2$), causing vast and localized volume increases, leading to deleterious expansion [34,75].

The primary agents for hardened concrete's expansion and cracking from sulfate attack are alumina-bearing hydrates (like calcium monosulfo-aluminate and calcium aluminate hydrate) [32]. Those hydrates are attacked by the sulfate ion forming ettringite or calcium tri-sulfoaluminate [32], a high sulfate form [86], or by forming gypsum or wollastonite [42]. The gypsum formed reacts with calcium aluminate hydrates ($4CaO \cdot Al_2O_3 \cdot 19H_2O$) and calcium alumino-sulfate hydrates ($3CaO \cdot CaSO_4 \cdot 18H_2O$), producing the insoluble ettringite that can lead to expansion (from reactions causing expandable pressure) causing deterioration of already hardened concrete due to high tensile stresses, according to the reactions that follow (Equations (17) and (18)) [42]:

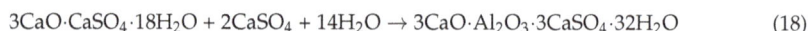

$$4CaO \cdot Al_2O_3 \cdot 19H_2O + 3CaSO_4 + 14H_2O \rightarrow 3CaO \cdot Al_2O_3 \cdot 3CaSO_4 \cdot 32H_2O + Ca(OH)_2 \tag{17}$$

$$3CaO \cdot CaSO_4 \cdot 18H_2O + 2CaSO_4 + 14H_2O \rightarrow 3CaO \cdot Al_2O_3 \cdot 3CaSO_4 \cdot 32H_2O \tag{18}$$

Thus, the co-firing of *Cynara* with natural gas is believed to reduce the risks associated with calcium aluminates in the fly ash, as there were no detection of alkali sulfates, despite the extremely high amount of alkalis in the biomass fuels. Anhydrite also participates in the problematic reaction for concrete, but in masonry applications, anhydrite is an important mineral, e.g., for render, mortars, or binders on building sites [74], as adhesion and binding properties have a precedence.

For the 50% thermal share of *Cynara* co-firing, it was observed that mineral phases were similar to the 10% thermal share of fly ash. This is considered to be a significant finding due to the much higher content of problematic ash species in the biomass fuel. However, the merwinite and andradite phases had a relative decrease from peak intensities in the 50% *Cynara* co-firing fly ash. Both of those phases are calcium bearing and would have contributions to the reactivity of the fly ash. A relative decrease with a relative increase in calcite may not change the fly ash reactive nature to any significant extent. Merwinite is formed by the reaction below, (Equation (19)) [1]:

$$3CaO(s) + MgO(s) + 2SiO_2(s) \rightarrow Ca_3Mg(SiO_4)_2(s) \tag{19}$$

Lime had a somewhat lower peak intensity in the 50% *Cynara* co-firing case compared to the other scenarios. The observation may be due to interactions of the fuels leading to reduced reaction rates for this thermal share impacting conversion of species or attributed to the increase and formation of other minerals involving Ca and Si. Further work is needed to clarify the observation.

Dolomite and sylvite were detected in the pure biomass combustion fly ash. The presence of dolomite in the fly ash from the pure combustion of *Cynara* is a testament to the fuel's larger particle sizes that require a longer residence time to burnout than design of the boiler for coal particles. Dolomite is formed by the reaction below (Equation (20)) [87]:

$$CaCO_3(s) + MgO(s) + CO_2(g) \rightarrow CaMg(CO_3)_2(s) \qquad (20)$$

The presence of sylvite in the pure *Cynara* combustion is due to the high amount of the species in the biomass and their subsequent reaction in the furnace and condensation in cooler regions. Also, the presence of KCl is attributed to the reduction of ash amount in the fuel blend containing silicates to form compounds with K. Sylvite is formed by the reaction sequences below (Equations (21) and (22)) [21,88]:

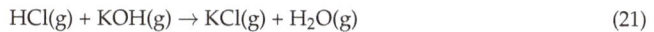

$$HCl(g) + KOH(g) \rightarrow KCl(g) + H_2O(g) \qquad (21)$$

$$KCl(g) \rightarrow KCl(s) \text{ (condensation of aerosols)} \qquad (22)$$

Ash composition changes during storage and under varying environmental conditions as carbon dioxide and moisture react with ash to form carbonates, bicarbonates, and hydroxides [89]. Absorption of water on a molecular level is noted to be possibly due to the hydrophilic nature of ash and to small charge imbalances [89]. Thus, the hydroxide mineral phases for the 100% *Cynara* are attributed to the ash being hydrophilic.

The co-fired fly ash qualities generated may be useful were setting times are of concern, as the self-hardening properties can improve the setting time. Suitable applications may be pavements, parking lots, road embankments, and other trafficked areas due to self-hardening characteristics that improve the setting time [54], bonding, and possibly adhesion qualities, characteristics associated with calcium minerals [74]. The fly ashes appear to have a high value in masonry applications, such as making bricks, etc. Overall, co-firing a herbaceous biomass up to 50% thermal share with a low-quality lignite that has a high ash content and with natural gas in the fuel blend positively influence the fly ash quality.

3.7. Behavior Characteristics from Performance Test Applications of Some Fly Ashes

The technically quality of how a fly ash performs in a given application is generally good if the performance is equal to or better than the commercial available material for the same application [16]. A key aspect in accepting the use of an industrial mineral like fly ash is its technical suitability and the ability of a power plant to supply a constant-quality product, which is guaranteed by specifications and certifications [16]. Thus, the co-fired ash performance in certain applications was compared to the lignite ash.

Soundness testing gives an evaluation of possible impacts on durability in concrete from using the fly ash quality. This assessment of volume expansion (instability) is necessary when assessing long-term performance considerations of using the fly ash, for example, in concrete. Table 8 shows results of soundness testing of ashes from co-firing 10% and 50% thermal shares of *Cynara* compared to the lignite only case.

Table 8. Soundness testing results for some fly ashes.

Scenario	L, Expansion [mm]	L, Limit [mm]
Greek lignite fly ash	1	≤10
10% thermal share *Cynara*	1	≤10
50% thermal share *Cynara*	0	≤10

All fly ash samples tested met the limit for expansion. Shown is the addition of *Cynara* does not have a negative impact on the long-term stability characteristic compared to the lignite fly ash, and the 50% thermal share even improve the performance compared to the lignite. The apparent improvement shows that a 50% thermal share of co-firing *Cynara* with a low-quality lignite can improve the fly ash quality. Further long-term studies are needed.

Another technical parameter of importance is the initial setting time. The initial setting time is necessary for assessing when a concrete mixture can be expected to become rigid. The time of set of concrete mixes impacts a construction project schedule. Many factors influence the initial time of setting. Some factors are listed below [49]:

- The fly ash may contain high calcium amounts (self-hardening);
- Fly ash may contain sulfates that react with cement;
- The mixture may contain less water due to the presence of fly ash, and this will influence the process of hardening;
- The fly ash may impede surface-active agents added to modify the rheology (water reducers) of concrete, and again this impacts the stiffness of mortar; and
- Fly ash particles may act as nuclei for crystallization of cement hydration products.

In Table 9 are shown initial setting times results of some tested fly ashes. A control sample is shown for enhancing comparisons. The impact of *Cynara* in the fuel blend is apparent.

Table 9. Initial setting time results for some fly ashes.

Scenario	t [min]	t, Limit [min]
Control sample	180	≥ 45
Greek lignite fly ash	240	≤ 360
10% thermal share *Cynara*	195	≤ 360

Table 9 shows that co-firing *Cynara* improved the setting time performance of the fly ashes. The improvement is also expected with the 50% thermal share, although not tested due to the unavailable material. The improvement in setting time with *Cynara* addition could be a result of reduced compounds in the fly ash that do not have good self-hardening properties an enhanced burning of the fuel blend with the biomass that contribute to an increase in mineral phases with hydraulic properties, i.e., anorthite ($CaAl_2Si_2O_8$) was shown in Table 7, or an improved reactivity of the fly ash from a finer particle size distribution, as was shown in Section 3.3.

Strength is the most important parameter used to characterize cement-based products; thus, the determination of compressive strength is the most important property of construction materials [90], mainly concrete as a structural material [27]. Compressive strength is widely used as an index of all other types of strength categories [91]. The experience with using fly ash in structural concrete has seen about 30% substitution of the binder, where the possibility is shown for a similar or improved performance in comparison to OPC concrete of equivalent 28-day strength [32]. The development of compressive strength on the fly ash generated from the lignite baseline case and the 10% thermal share of *Cynara* are evaluated in Figure 19 for a 10% cement substitution. Also, a comparison of the results to the same type of tests for fly ash from the industrial Kardia Power Plant was performed. Characterization of strength development occurred for 1, 2, 7 and 28 days.

Figure 19 shows that the 10% cement substitution of *Cynara* fly ash from a 10% thermal share produces a fly ash with nearly the same early strength development behavior and a similar 28-day strength development compared with the lignite fly ash. After grinding the fly ashes, the 28-day strength for the 10% thermal share of *Cynara* fly ash at a 10% cement substitution has nearly the same strength value as the pure lignite combustion fly ash. Also, after grinding, the 28-day strength for the 10% thermal share of *Cynara* at a 10% cement substitution has a 28-day strength development very close to the pilot-scale 10% thermal share *Cynara* fly ash at a 10% cement substitution.

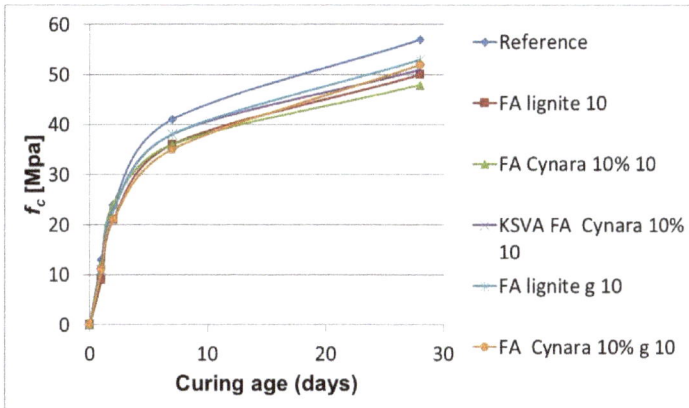

Figure 19. Comparison of strength development for 10% thermal share of *Cynara* fly ash from KSVA with industrial fly ashes to include 10% thermal share of *Cynara*, with a 10% cement substitution; g = ground fly ash, FA = fly ash.

Grinding of fly ashes is a way to reveal their pozzolanic and hydraulic properties [92]. Results indicate that a 10% thermal share of *Cynara* in the fuel blend for a low-quality lignite would have the same behavior as a cement substitution where a lignite only fly ash applies. Also, from the results, ash obtained from the pilot-scale tests are considered to be representative of industrial fly ash behaviors. Figure 20 shows strength development with substituting 20% of the cement with fly ash.

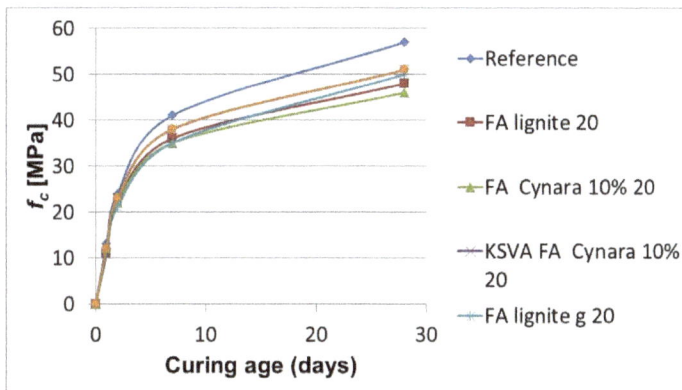

Figure 20. Comparison of strength development for 10% thermal share of *Cynara* fly ash from KSVA with industrial fly ashes to include 10% thermal share of *Cynara*, with 20% cement substitution; g = ground fly ash, FA = fly ash.

The 20% cement substitution in Figure 20 shows the same trend for strength development as the 10% cement replacement. Also, the values are very close, but with a noticeable slight increased variation. After grinding, the performance of the 10% thermal share fly ash at 20% cement substitution is the same as the lignite fly ash. Again, this shows the added value in a small co-firing proportion of a herbaceous biomass with a low-quality lignite on the fly ash quality. A higher cement substitution rate contributes more to mitigating CO_2 emissions from the manufacture of cement. Figure 21 shows the comparison of strength development for 10% thermal share of *Cynara* fly ash from KSVA with industrial fly ashes with a 30% cement substitution.

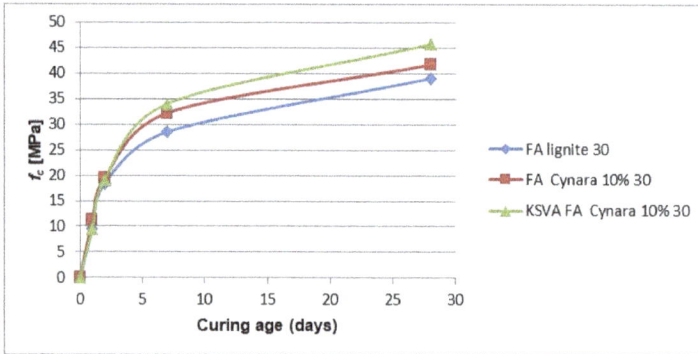

Figure 21. Comparison of strength development for 10% thermal share of *Cynara* fly ash from KSVA with industrial fly ashes to include 10% thermal share of *Cynara*, with 30% cement substitution; FA = fly ash.

The values for the 30% cement substitution are lower than the other values, as expected due to the presence of less cement in the mix for hydration to supply $Ca(OH)_2$ needed to react with the pozzolanic fractions of the ash. So, it will take a longer time to start the reactions with fly ash, as the pH required is reached a later time compared to the other lower substitution rates. The industrial fly ash was not ground and compared, but the trend of exact values for the pilot-scale (KSVA) results should be the same, like the results for the other substitutions. An optimal replacement amount for utilization in concrete applications (with a technical approval) is likely at about 25% cement replacement, which is the practice according to EN 450-1 for bituminous coal ashes or ashes from co-firing particular green biomass fuel with bituminous coals. On the other hand, many masonry applications do not have the strict material provisions and conservative design like EN 450-1. Many masonry applications allow one to choose the components of a binder to meet specific performance requirements, e.g., EN 998-1, focusing on binding and adhesion properties. So, performance studies of the fly ash qualities in masonry applications could result in a significant higher amount of cement being replaced by the fly ash qualities. That would be a huge benefit as the total energy required based on kJ/kg of material is 372 and 0 for cement and fly ash, respectively [75].

From the results, a high thermal share of *Cynara* does not appear to change the fly ash quality for economic utilization applications, despite the significantly higher contents of problematic ash components. At 50% thermal share of *Cynara*, the amount is likely able to meet a good fuel sustainable supply, easy handling and storage capabilities, efficient milling and feeding capabilities, and a normal boiler operation and performance. Thus, it is believed that a co-firing 50% thermal share of *Cynara* can be employed based upon the overall presented results without having negative impacts on the fly ash quality. This would also stimulate local and regional economies and may be a catalyst to promote trade between countries for such a biomass type to co-fire with lignite fuels. Quality fly ash standards requiring a minimum of co-firing of an herbaceous biomass with a low-quality lignite fuel on a thermal share basis would have far reaching positive benefits across environmental sustainability practices in lignite fuel combustion and biomass ash management. Such legislation can be employed regionally, nationally, EU-wide, or even internationally through European and ISO standards. However, such a practice needs assessment on a case basis for herbaceous biomass fuel types, since logistics and fuel supply would take priority, along with boiler performance concerns.

The largest use of ashes in Greece is in the cement industry to replace cement clinker and production of special cements, which include successful tests in road construction, mortars, waste treatment, embankments, and cement grouting [92]. From results, it is believed a co-firing share of up to 50% of *Cynara* biomass does not change those utilization options when potentially substituted up to 30% by mass of the cement content. Further long-term studies are needed.

From the technical performance test results, it is believed that co-firing of the *Cynara* biomass with a low-quality lignite would be added value to a lignite power plant not operating at boiler load swings, which would be enhanced at a boiler de-rate with the inclusion of natural gas into the co-firing fuel mix. However, at lower loads there may be an impact on fuel/gas mixing due to changes in velocity and momentum [60,93]. Changes in running conditions of boilers result in variations of fly ash quality, as adjustments to the loading of boilers regularly have to take place to meet the changing demand for electricity, and the coexistence of alternative electricity generating options give rise to the policy of rapidly changing between supply sources for an optimal economy [16]. Co-firing of *Cynara* with a low-quality lignite in a boiler subjected to fluctuating operational conditions can possibly lead to keeping the variation of inhomogeneous fly ash to an acceptable condition, as the co-fired fly ashes had an equally comparable or better technical performance than the compared commercial material or baseline fly ash quality. This may be enhanced with the addition of natural gas into the fuel blend due to higher temperatures reached and a potential reduction of LOI values in the fly ash.

4. Conclusions

Pilot-scale test experiments of co-firing a low-quality Greek lignite and an herbaceous biomass along with the mono-combustion of each fuel were conducted to evaluate the effect of co-firing on the fly ash quality (ash formation, changes, and characteristics). Also, natural gas was used in the tests due to the low-quality of the lignite. Co-firing thermal shares were 10%, 50%, and pure biomass combustion. Results were compared to the same fuel qualities for industrial tests of lignite and 10% thermal share of *Cynara*.

Co-firing of a low-quality lignite with *Cynara* and natural gas in the fuel blend is believed to lead to a better conversion of SO_2 to the more reactive SO_3 in the furnace, resulting in more fuel self-desulfurization in the furnace with earth alkaline metals. This causes less alkali sulfates to be formed in the convection pass, which releases alkalis via the stack in the gas phase. Thus, more of the problematic ash components, i.e., alkalis and phosphates, are not significantly captured in the fly ashes due to a lower availability of sulfur. This potentially improves the quality of the fly ash for utilization in low technical applications and as a binder, as $CaSO_4$ is a widely known self-hardening compound in building materials, especially in masonry applications. Thus, it appears that using natural gas in the co-fired fuel blend enhances the fly ash quality for uses not subjected to stringent material provisions, as the two fuels used during tests, generally, are considered problematic within their respective fuel categories.

With *Cynara* in the fuel blend, the fly ash qualities showed performance improvements over the lignite only fly ash when tested for initial setting times and volume stability. Those improvements were obtained without the need to grind the fly ash to break up the irregular, fused, angular, etc. particles. Thus, the co-firing of *Cynara* led to positive changes in the fly ash quality; however, further process analyses and techno-economic studies are necessary before *Cynara* can be considered as a sustainable co-firing component with lignite. Also, the reduction of periclase with an increasing *Cynara* share indicate an improved ash quality for application in constructions, as the mineral participates in unwanted reactions causing a destruction of hardened concrete. This is further added value in co-firing *Cynara* with a low-quality lignite. Furthermore, the particle size distribution of the fly ash with *Cynara* in the fuel blend was very similar to or better than the lignite fly ash, which is attributed to the higher heating value of the biomass and lower ash and water contents that can lead to a better ignition and combustion of the fuel. There would also be an enhanced particle fragmentation, which could be supported by the higher contents of alkalis in the *Cynara* that can improve heat transfer during combustion via more melt phases.

Scalability evaluations showed that the pilot-scale fly ashes are reliable to qualities produced in industrial facilities, from both particle size distribution and concrete specimens' mechanical strength testing. Indirectly correlated to many other parameters of hardened concrete is the compressive strength and mineral phases have a significant influence on the occurrence due to their different reactivity and hydration properties. The similarities of compressive strength development in the fly ashes from the pilot-scale and industrial-scale testing indicate the reliability of pilot-scale results for observing interactions between the fuels, leading to various fly ash qualities.

In regards to reproducibility, the scalability observations show the potential of all the pilot-scale results to be reproduced with fly ash obtained at an industrial scale pulverized fuel facility. This suggests that the results can be reproduced with similar fuel qualities and the same thermal shares in a large-scale pulverized fuel facility with fly ash collected by an ESP. Moreover, all the methods and protocols followed are noted to facilitate the reproducibility of the results.

The co-firing results reported indicate that there is potentially no negative impact for uses that primarily access environmental concerns, regarding the possible use of the ashes for soil amelioration. This is based upon a likely sustainable co-firing share of up to 50% from fuel supply sustainability and operational capabilities. There may be an improvement in the ash quality for soil stabilization due to the increased calcium minerals, increased carbon in the ash, and an increase of mobility of nutrient species.

The co-firing of *Cynara* with a low-quality lignite would not only lead to a real benefits in the fly ash qualities but also contribute to reduced emissions of unwanted gasses species, such as CO_2 and NOx, and sustainable energy production. The paper demonstrates that there is potentially no negative effect on ash quality up to 50% thermal share of *Cynara* co-fired with a low-quality lignite, and the observed improvement of the fly ash quality supports the assertion from literature that co-firing is a potentially attractive option to obtain a better fly ash quality in lignite boilers. This work validates the importance of co-firing *Cynara* or another herbaceous biomass with a low quality lignite, which may also have a positive contribution to local socio-economic developments. The results support efforts of biomass co-combustion in not only conventional air-firing, but also in lignite power plants that may be subjected to dispatch load requirements to support other renewable energy intermittent sources, e.g., wind and solar. as well as future oxy-fuel power plants, as long as sulfur concentrations are acceptable in the flue gas (risk of corrosion), contributing to measures to enhance operational flexibility and economic competitiveness. Moreover, the results could be very useful in future energy engineering projects that target not only the operational flexibility but also widening the fuel flexibility in flexibly operated lignite power plants.

Author Contributions: All authors contributed to the work presented in this paper. The writing of the original draft preparation was done by A.F. All authors were involved in preparing the final manuscript. All authors have read and approved the manuscript.

Funding: This research was funded by the European Seventh Framework Program (FP7/2007-2013) under grant agreement number 218968. The authors acknowledge the Demonstration of Large Scale Biomass Co-firing and Supply Chain Integration (DEBCO) project partners.

Acknowledgments: The authors acknowledge personnel at the Institute of Combustion and Power Plant Technology, the Materials Testing Institute (MPA Stuttgart, Otto-Graf-Institute (FMPA)), the Institute of Crystal Chemistry Faculty of Chemistry, all located at the University of Stuttgart, Germany, and the Laboratory of Inorganic and Analytical Chemistry, School of Chemical Engineers, National Technical University located in Athens, Greece for their collaboration and support.

Conflicts of Interest: The authors declare no conflict of interest.

Energies **2018**, *11*, 1581

Abbreviations

Symbol Unit	Value
d µm	particle size
D_{10} µm	the diameter having 10% of the distribution as smaller particle sizes and 90% as larger particle sizes
D_{50} µm	the diameter having 50% of the distribution as smaller particle sizes and 50% as larger particle sizes
D_{90} µm	the diameter having 90% of the distribution as smaller particle sizes and 10% as larger particle sizes
f_c MPa kJ/kg	characteristic compressive strength
H_L	lower calorific value
L mm	length
$\dot{M}$ [kg/h]	mass flow
η	air ratio
P [kW$_{el}$]	power
t minutes	time
$\dot{V}$ [m^3/h]	volumetric flow
X %	mass percent
γ_i	fuel mass fraction of the substance i (C, N, O, S, H, H_2O, A (=ash))
ϑ [°C]	temperature
ASTM	American Society of Testing and Materials
CEN	European Committee for Standardization
d	dry
daf	dry-ash-free
DIN	Deutsches Institut für Normung e.V. (German Institute for Standardization)
EN	European Norms
ESP	electrostatic precipitator
FA	fly ash
ISO	International Standard Organization
KSVA	Kohlenstaubmahl- und Verbrennungsanlage (pilot-scale test facility)
RDF	refuse derived fuel
SCR	selective catalytic reduction
SNCR	selective non-catalytic reduction
TC	technical committee

References

1. Vamvuka, D.; Kakaras, E. Ash properties and environmental impact of various biomass and coal fuels and their blends. *Fuel Process. Technol.* **2011**, *92*, 570–581. [CrossRef]
2. Massazza, F. Pozzolana and Pozzolanic Cements. In *Lea's Chemistry of Cement and Concrete*, 4th ed.; Hewlett, P.C., Ed.; Arnold: London, UK, 1998; pp. 471–631.
3. Xu, A. Fly Ash in Concrete. In *Waste Materials Used in Concrete Manufacturing*; Chandra, S., Ed.; Noyes Publications: Westwood, NJ, USA, 1997; pp. 142–183.
4. Vassilev, S.V.; Baxter, D.; Andersen, L.K.; Vassileva, C.G. An overview of the chemical composition of biomass. *Fuel* **2010**, *89*, 913–933. [CrossRef]
5. Cieplik, M.K.; Fryda, L.E.; van de Kamp, W.L.; Kiel, J.H.A. Ash Formation, Slagging and Fouling in Biomass Co-firing in Pulverised-fuel Boilers. In *Solid Biofuels for Energy: A Lower Greenhouse Gas Alternative*; Grammelis, P., Ed.; Springer: London, UK, 2010; pp. 197–217.
6. Feuerborn, H.J.; Müller, B.; Walter, E. *Use of Calcareous Fly Ash in Germany. Ash and Slaga Handling, 3.7 Analytics*; EUROCOALASH 2012; Industrial By-Products Research and Development Association (EVIPAT): Thessaloniki, Greece, 2012.
7. Zheng, Y.; Jensen, P.A.; Jensen, A.D.; Sander, B.; Junker, H. Ash transformation during co-firing coal and straw. *Fuel* **2007**, *86*, 1008–1020. [CrossRef]
8. Wang, G.; Shen, L.; Sheng, C. Characterization of Biomass Ashes from Power Plants Firing Agricultural Residues. *Energy Fuels* **2012**, *26*, 102–111. [CrossRef]
9. Sarabèr, A. Co-combustion and its impact on fly ash quality; pilot-scale experiments. *Fuel Process. Technol.* **2012**, *104*, 105–114. [CrossRef]

10. Sua-iam, G.; Makul, N. Utilization of high volumes of unprocessed lignite-coal fly ash and rice husk ash in self-consolidating concrete. *J. Clean. Prod.* **2014**, *78*, 184–194. [CrossRef]

11. Kisiela, A.M.; Czajka, K.M.; Moroń, W.; Rybak, W.; Andryjowicz, C. Unburned carbon from lignite fly ash as an adsorbent for SO_2 removal. *Energy* **2016**, *116*, 1454–1463. [CrossRef]

12. Masto, R.E.; Ansari, M.A.; George, J.; Selvi, V.A.; Ram, L.C. Co-application of biochar and lignite fly ash on soil nutrients and biological parameters at different crop growth stages of *Zea mays*. *Ecol. Eng.* **2013**, *58*, 314–322. [CrossRef]

13. Kourti, I.; Cheeseman, C.R. Properties and microstructure of lightweight aggregate produced from lignite coal fly ash and recycled glass. *Resour. Conserv. Recycl.* **2010**, *54*, 769–775. [CrossRef]

14. Moutsatsou, A.; Itskos, G.; Vounatsos, P.; Koukouzas, N.; Vasilatos, C. Microstructural characterization of PM-Al and PM-Al/Si composites reinforced with lignite fly ash. *Mater. Sci. Eng. A* **2010**, *527*, 4788–4795. [CrossRef]

15. Pimraksa, K.; Hanjitsuwan, S.; Chindaprasirt, P. Synthesis of belite cement from lignite fly ash. *Ceram. Int.* **2009**, *35*, 2415–2425. [CrossRef]

16. Cox, M.; Nugteren, H.; Janssen-Jurkovicova, M. (Eds.) *Combustion Residues: Current, Novel and Renewable Applications*; John Wiley: Chichester, UK, 2008.

17. Llorente, M.F.; García, J.C. Suitability of thermo-chemical corrections for determining gross calorific value in biomass. *Thermochim. Acta* **2008**, *468*, 101–107. [CrossRef]

18. Fava, G.; Naik, T.R. Compressive Strength and Leaching Behavior of Mortars with Biomass Ash. In *By-Products Utilization Sustainable Construction Materials and Technologies: Conference Proceedings Downloads and Books*; Claisse, P., Ganjian, E., Naik, T., Eds.; The University of Wisconsin Milwaukee: Milwaukee, WI, USA, 2016.

19. Heinzel, T.; Lopez, C.; Maier, J.; Spliethoff, H.; Hein, K.R.G. Ash deposit and corrosion characteristics of coal-biomass blends in a 0.5 MW pulverized fuel test facility. In *International Conference: "Effects of Coal Quality on Power Plant Management: Ash Problems, Management and Solutions"*; Mehta, A.K., Ed.; United Engineering Foundation: New York, NY, USA, 2001; pp. 2–27.

20. Spliethoff, H. *Power Generation from Solid Fuels*; Springer: Heidelberg, Germany, 2010.

21. Van Loo, S.; Koppejan, J. *The Handbook of Biomass Combustion and Co-Firing*; Earthscan: London, UK, 2010.

22. Mineralogical Society (Great Britain). *Microprobe Techniques in the Earth Sciences*; The Mineralogical Society of Great Britain and Ireland: London, UK, 1995.

23. Misra, M.K.; Ragland, K.W.; Baker, A.J. Wood ash composition as a function of furnace temperature. *Biomass Bioenergy* **1993**, *4*, 103–116. [CrossRef]

24. Robinson, A.L.; Junker, H.; Baxter, L.L. Pilot-Scale Investigation of the Influence of Coal−Biomass Cofiring on Ash Deposition. *Energy Fuels* **2002**, *16*, 343–355. [CrossRef]

25. Bye, G.C. *Portland Cement: Composition Production and Properties*, 1st ed.; Pergamon Press: Oxford, UK, 1983.

26. Ramachandran, V.S.; Feldman, R.E. Concrete Science. In *Concrete Admixtures Handbook: Properties, Science, and Technology*, 2nd ed.; Ramachandran, V.S., Ed.; Noyes Publications: Park Ridge, NJ, USA, 1995; pp. 1–66.

27. Abolpour, B.; Mehdi Afsahi, M.; Hosseini, S.G. Statistical analysis of the effective factors on the 28 days compressive strength and setting time of the concrete. *J. Adv. Res.* **2015**, *6*, 699–709. [CrossRef] [PubMed]

28. Gehlen, C.; Eligehausen, R. *Engineering Materials: Concrete: COMMAS-C7*; University of Stuttgart: Stuttgart, Germany, 2006.

29. University of Kentucky. *Techniques for Measuring Ammonia in Fly Ash, Mortar, and Concrete*; University of Kentucky Center for Applied Eneregy Research: Lexington, KY, USA, 2003.

30. Vassilev, S.V.; Baxter, D.; Andersen, L.K.; Vassileva, C.G. An overview of the composition and application of biomass ash. Part 1. Phase–mineral and chemical composition and classification. *Fuel* **2013**, *105*, 40–76. [CrossRef]

31. Knudsen, J.N.; Jensen, P.A.; Dam-Johansen, K. Transformation and Release to the Gas Phase of Cl, K, and S during Combustion of Annual Biomass. *Energy Fuels* **2004**, *18*, 1385–1399. [CrossRef]

32. Ahmaruzzaman, M. A review on the utilization of fly ash. *Prog. Energy Combust. Sci.* **2010**, *36*, 327–363. [CrossRef]

33. Yara. Yara Fertilizer Industry Handbook February 2012. Available online: http://www.yara.com/doc/ 37694_2012%20Fertilizer%20Industry%20Handbook%20wFP.pdf (accessed on 11 January 2013).

34. Lawrence, C.D. The Constitution and specification of Portland Cements. In *Lea's Chemistry of Cement and Concrete*, 4th ed.; Hewlett, P.C., Ed.; Arnold: London, UK, 1998; pp. 131–193.

35. Lawrence, C.D. Physicochemical and Mechanical Properties of Portland Cements. In *Lea's Chemistry of Cement and Concrete*, 4th ed.; Hewlett, P.C., Ed.; Arnold: London, UK, 1998; pp. 343–419.

36. Odler, I. Hydration, Setting and Hardening of Portland Cement. In *Lea's Chemistry of Cement and Concrete*, 4th ed.; Hewlett, P.C., Ed.; Arnold: London, UK, 1998; pp. 242–297.

37. Ramachandran, V.S. (Ed.) Alkali-Aggregate Expansion and Corrosion Inhibiting Admixtures. In *Concrete Admixtures Handbook: Properties, Science, and Technology*, 2nd ed.; R Noyes Publications: Park Ridge, NJ, USA, 1995; pp. 878–938.

38. Gupta, R.; Wall, T.; Baxter, L. *Impact of Mineral Impurities in Solid Fuel Combustion*; Kluwer Academic: New York, NY, USA; London, UK, 1999.

39. Karampinis, E.; Nikolopoulos, N.; Nikolopoulos, A.; Grammelis, P.; Kakaras, E. Numerical investigation Greek lignite/cardoon co-firing in a tangentially fired furnace. *Appl. Energy* **2012**, *97*, 514–524. [CrossRef]

40. European Committee for Standardization. *Specifications—Part 4: Fly Ash for Hydraulically Bound Mixtures*; 93.080.20 (EN 14227-4:2013:E); CEN: Brussels, Belgium, 2013.

41. European Committee for Standardization. *Fly Ash for Concrete—Part 1: Definition, Specifications and Conformity Criteria*, 5th ed.; 91.100.30(EN 450-1); CEN/TC 104; Technical Committee: Brussels, Belgium, 2012.

42. Gu, X.; Jin, X.; Zhou, Y. *Basic Principles of Concrete Structures*; Springer: New York, NY, USA, 2015.

43. Gu, X.; Jin, X.; Zhou, Y. *Basic Principles of Concrete Structures*, 1st ed.; Springer: Berlin/Heidelberg, Germany, 2016.

44. Butler, W.B. Sulphate Attack on Concrete: What It Is and How to Stop It. Available online: http://www.adaa.asn.au/documents/Technical_Notes_2.pdf (accessed on 1 April 2013).

45. Zacco, A.; Borgese, L.; Gianoncelli, A.; Struis, R.P.; Depero, L.E.; Bontempi, E. Review of fly ash inertisation treatments and recycling. *Environ. Chem. Lett.* **2014**, *12*, 153–175. [CrossRef]

46. Pandey, V.C.; Singh, N. Impact of fly ash incorporation in soil systems. *Agric. Ecosyst. Environ.* **2010**, *136*, 16–27. [CrossRef]

47. Jackson, P.J. Portland Cement: Classification and Manufacture. In *Lea's Chemistry of Cement and Concrete*, 4th ed.; Hewlett, P.C., Ed.; Arnold: London, UK, 1998; pp. 25–94.

48. Singer, J.G. *Combustion: Fossil Power Systems a Reference Book on Fuel Burning and Steam Generation*, 3rd ed.; Combustion Engineering: Windsor, CT, USA, 1981.

49. Ramezanianpour, A.A. *Cement Replacement Materials: Properties, Durability, Sustainability*; Springer: Berlin, Germany, 2014.

50. Backreedy, R.I.; Jones, J.M.; Pourkashanian, M.; Williams, A. Burn-out of pulverised coal and biomass chars. *Fuel* **2003**, *82*, 2097–2105. [CrossRef]

51. Blissett, R.S.; Rowson, N.A. A review of the multi-component utilisation of coal fly ash. *Fuel* **2012**, *97*, 1–23. [CrossRef]

52. Pedersen, K.; Jensen, A.; Skjothrasmussen, M.; Damjohansen, K. A review of the interference of carbon containing fly ash with air entrainment in concrete. *Prog. Energy Combust. Sci.* **2008**, *34*, 135–154. [CrossRef]

53. Manz, O.E. Coal fly ash: A retrospective and future look. *Fuel* **1999**, *78*, 133–136. [CrossRef]

54. American Coal Ash Association. *Fly Ash Facts for Highway Engineers*; National Technical Information Service: Springfield, VA, USA, 2003.

55. Boycheva, S.; Zgureva, D.; Vassilev, V. Kinetic and thermodynamic studies on the thermal behaviour of fly ash from lignite coals. *Fuel* **2013**, *108*, 639–646. [CrossRef]

56. Kalembkiewicz, J.; Chmielarz, U. Ashes from co-combustion of coal and biomass: New industrial wastes. *Resour. Conserv. Recycl.* **2012**, *69*, 109–121. [CrossRef]

57. Vassilev, S.V.; Baxter, D.; Andersen, L.K.; Vassileva, C.G. An overview of the composition and application of biomass ash. *Fuel* **2013**, *105*, 19–39. [CrossRef]

58. Ram, L.C.; Masto, R.E. An appraisal of the potential use of fly ash for reclaiming coal mine spoil. *J. Environ. Manag.* **2010**, *91*, 603–617. [CrossRef] [PubMed]

59. Decagon Devices. Electrical Conductivity of Soil as a Predictor of Plant Response. Available online: http://old.decagon.com/assets/Uploads/Electrical-Conductivity-of-Soil-as-a-predictor-of-Plant-Response.pdf (accessed on 15 December 2014).

60. Dusatko, G.; Brown, R. *Gas Cofiring Assessment for Coal-Fired Utility Boilers: Technical Report*; Final Report; Electric Power Research Institute (EPRI): Palo Alto, CA, USA, 2000.

61. Bayless, D.J.; Schroeder, A.R.; Olsen, M.G.; Johnson, D.C.; Peters, J.E.; Krier, H.; Buckius, R.O. The effects of cofiring natural gas and coal on sulfur retention in ash. *Combust. Flame* **1996**, *106*, 231–240. [CrossRef]

62. Muzio, L.J.; Often, G.R. Assessment of Dry Sorbent Emission Control Technologies Part I. Fundamental Processes. *JAPCA* **1987**, *37*, 642–654. [CrossRef]

63. Jiménez, S.; Ballester, J. Influence of operating conditions and the role of sulfur in the formation of aerosols from biomass combustion. *Combust. Flame* **2005**, *140*, 346–358. [CrossRef]

64. Zeuthen, J.H.; Jensen, P.A.; Jensen, J.P.; Livbjerg, H. Aerosol Formation during the Combustion of Straw with Addition of Sorbents. *Energy Fuels* **2007**, *21*, 699–709. [CrossRef]

65. Watt, J.D.; Thorne, D.J. The composition and pozzolanic properties of pulverised fuel ashes: III. Pozzolanic Properties of Fly Ashes as Determined by Chemical Methods. *J. Appl. Chem.* **1996**, *16*, 33–39. [CrossRef]

66. Watt, J.D.; Thorne, D.J. The composition and pozzolanic properties of pulverised fuel ashes: II. Pozzolanic Properties of Fly Ashes as Determined by Chemical Methods. *J. Appl. Chem.* **1996**, *15*, 585–604. [CrossRef]

67. Biagini, E.; Narducci, P.; Tognotti, L. Size and structural characterization of lignin-cellulosic fuels after the rapid devolatilization. *Fuel* **2008**, *87*, 177–186. [CrossRef]

68. Cetin, E.; Moghtaderi, B.; Gupta, R.; Wall, T.F. Influence of pyrolysis conditions on the structure and gasification reactivity of biomass chars. *Fuel* **2004**, *83*, 2139–2150. [CrossRef]

69. Davidsson, K.O.; Pettersson, J.B.C. Birch wood particle shrinkage during rapid pyrolysis. *Fuel* **2002**, *81*, 263–270. [CrossRef]

70. Xue, Q.-F.; Lu, S.-G. Microstructure of ferrospheres in fly ashes: SEM, EDX and ESEM analysis. *J. Zhejiang Univ. Sci. A* **2008**, *9*, 1595–1600. [CrossRef]

71. Tia, M.; Subramanian, R.; Brown, D.; Broward, C. Evaluation of Shrinkage Cracking Potential of Concrete Used in Bridge Decks in Florida. Available online: http://www.fdot.gov/research/completed_proj/summary_smo/fdot_bc354_26_rpt.pdf (accessed on 10 September 2016).

72. Härdtl, R. Effectiveness of Fly Ash Processing Methods in Improving Concrete Quality. In *Waste Materials in Construction*, 1st ed.; Aalbers, T.G., Goumans, J.J.J.M., van der Sloot, H.A., Eds.; Elsevier Science Ltd.: Amsterdam, The Netherlands, 1991; pp. 399–406.

73. Vassilev, S.V.; Vassileva, C.G. A new approach for the classification of coal fly ashes based on their origin, composition, properties, and behaviour. *Fuel* **2007**, *86*, 1490–1512. [CrossRef]

74. Berge, B. *The Ecology of Building Materials*; Architectural Press: Oxford, UK, 2000.

75. Lohtia, R.P.; Joshi, R.C. Mineral Admixtures. In *Concrete Admixtures Handbook: Properties, Science, and Technology*, 2nd ed.; Ramachandran, V.S., Ed.; Noyes Publications: Park Ridge, NJ, USA, 1995; pp. 657–739.

76. Putnis, A. *Introduction to Mineral Sciences*; BaAS Printers Ltd.: Hampshire, UK, 1992.

77. Koukouzas, N.; Ward, C.R.; Papanikolaou, D.; Li, Z.; Ketikidis, C. Quantitative evaluation of minerals in fly ashes of biomass, coal and biomass-coal mixture derived from circulating fluidised bed combustion technology. *J. Hazard. Mater.* **2009**, *169*, 100–107. [CrossRef] [PubMed]

78. Miller, B.G. *Clean Coal Engineering Technology*; Elsevier Butterworth-Heinemann: Amsterdam, The Netherlands; Heidelberg, Germany, 2011.

79. Valued Gateway Client. Ionic Molecules and the Born-Haber Cycle: Chapter 9 Looks at Forces That Keep Atoms Together. Available online: http://www2.onu.edu/~k-broekemeier/BornHaber.pdf (accessed on 7 November 2017).

80. Narayanan, S. Principles of Sustainable Building Design. In *Green Building with Concrete: Sustainable Design and Construction*; Sabnis, G.M., Ed.; CRC Press: Boca Raton, FL, USA, 2015; pp. 35–87.

81. Macphee, D.E.; Lachowski, E.E. Cement Components and Their Phase Relations. In *Lea's Chemistry of Cement and Concrete*, 4th ed.; Hewlett, P.C., Ed.; Arnold: London, UK, 1998; pp. 95–129.

82. Grammelis, P.; Skodras, G.; Kakaras, E. Effects of biomass co-firing with coal on ash properties. Part I: Characterisation and PSD. *Fuel* **2006**, *85*, 2310–2315. [CrossRef]

83. El-Mahallawy, F.M.; Habik, S.E.-D. *Fundamentals and Technology of Combustion*; Elsevier: Amsterdam, The Netherlands; London, UK, 2002.

84. Ridha, F.N.; Lu, D.; Macchi, A.; Hughes, R.W. Combined calcium looping and chemical looping combustion cycles with CaO–CuO pellets in a fixed bed reactor. *Fuel* **2015**, *153*, 202–209. [CrossRef]

85. Edwards, K. Enthalpy of Formation of MgO: Electronic Laboratory Notebook (ELN) Instructions. Available online: http://faculty.sites.uci.edu/chem11/files/2013/11/C01aMANDHMgO.pdf (accessed on 7 November 2017).

86. Lamond, J.F.; Pielert, J.H. *Significance of Tests and Properties of Concrete & Concrete-Making Materials*; ASTM: Philadelphia, PA, USA, 2006.

87. Graf, D.L.; Goldsmith, J.R. Dolomite-magnesian calcite relations at elevated temperatures and CO_2 pressure. *Geochim. Cosmochim. Acta* **1955**, *7*, 109–128. [CrossRef]

88. Boström, D.; Skoglund, N.; Grimm, A.; Boman, C.; Ohman, M.; Brostrom, M.; Backman, R. Ash Transformation Chemistry during Combustion of Biomass. *Energy Fuels* **2012**, *26*, 85–93. [CrossRef]

89. Etiégni, L.; Campbell, A.G. Physical and chemical characteristics of wood ash. *Bioresour. Technol.* **1991**, *37*, 173–178. [CrossRef]

90. Nazari, A.; Sanjayan, J.G. Johnson–Mehl–Avrami–Kolmogorov equation for prediction of compressive strength evolution of geopolymer. *Ceram. Int.* **2015**, *41*, 3301–3304. [CrossRef]

91. Jankovic, K.; Nikolic, D.; Bojovic, D.; Loncar, L.; Romakov, Z. The estimation of compressive strength of normal and recycled aggregate concrete. *Facta Univ.-Ser. Archit. Civ. Eng.* **2011**, *9*, 419–431. [CrossRef]

92. Skodras, G.; Grammelis, P.; Kakaras, E.; Karangelos, D.; Anagnostakis, M.; Hinis, E. Quality characteristics of Greek fly ashes and potential uses. *Fuel Process. Technol.* **2007**, *88*, 77–85. [CrossRef]

93. Kitto, J.B.; Stultz, S.C. *Steam, Its Generation and Use*, 41st ed.; Babcock & Wilcox: Barberton, OH, USA, 2005.

![energies logo] *energies*

MDPI

Review

Biological Pretreatment Strategies for Second-Generation Lignocellulosic Resources to Enhance Biogas Production

Andreas Otto Wagner *, **Nina Lackner, Mira Mutschlechner, Eva Maria Prem, Rudolf Markt and Paul Illmer**

Department of Microbiology, Universität Innsbruck, Technikerstraße 25d, A-6020 Innsbruck, Austria; nina.lackner@uibk.ac.at (N.L.); mira.mutschlechner@uibk.ac.at (M.M.); eva.prem@uibk.ac.at (E.M.P.); rudolf.markt@uibk.ac.at (R.M.); paul.illmer@uibk.ac.at (P.I.)
* Correspondence: Andreas.Wagner@uibk.ac.at; Tel.: +43-512-507-51342

Received: 8 June 2018; Accepted: 3 July 2018; Published: 9 July 2018

Abstract: With regard to social and environmental sustainability, second-generation biofuel and biogas production from lignocellulosic material provides considerable potential, since lignocellulose represents an inexhaustible, ubiquitous natural resource, and is therefore one important step towards independence from fossil fuel combustion. However, the highly heterogeneous structure and recalcitrant nature of lignocellulose restricts its commercial utilization in biogas plants. Improvements therefore rely on effective pretreatment methods to overcome structural impediments, thus facilitating the accessibility and digestibility of (ligno)cellulosic substrates during anaerobic digestion. While chemical and physical pretreatment strategies exhibit inherent drawbacks including the formation of inhibitory products, biological pretreatment is increasingly being advocated as an environmentally friendly process with low energy input, low disposal costs, and milder operating conditions. Nevertheless, the promising potential of biological pretreatment techniques is not yet fully exploited. Hence, we intended to provide a detailed insight into currently applied pretreatment techniques, with a special focus on biological ones for downstream processing of lignocellulosic biomass in anaerobic digestion.

Keywords: pretreatment; biological pretreatment; anaerobic digestion; biogas; methane

---

1. Introduction

Although it is known that CO_2 production from fossil fuel combustion is a major contributor to global warming, these energy carriers are still the most important resources for global energy generation [1]. Great efforts have been devoted to increasing energy production from nonfossil fuels and to replacing climate-change-relevant energy sources by renewable ones. Hydropower, wind, and solar energy are probably the most promising alternative energy resources but can exhibit limitations concerning flexible energy production, storage and/or backup, transportation, and land requirements [2].

Biogas production from anaerobic digestion (AD) processes is considered as an attractive source for green energy [3,4] and, therefore, endeavors have been made to increase the share of biogas in global energy production. During anaerobic digestion, organic feedstocks are converted into biogas containing methane (CH_4) as a valuable end-product. The energy input for biogas production is calculated to be lower than in current ethanol production, leading to a higher energy output-to-input ratio [5]. However, the expanded production of biogas was often achieved by the utilization of energy crops directly competing with food crop farming (first-generation biofuels). Therefore, the exploration of lignocellulosic materials (second-generation biofuels) for bio-methane production was substantially accelerated during the past years, thus offering ecological as well as economic advantages [6]. However, lignin resists (complete) degradation under anaerobic conditions, posing a challenge regarding the overall degradability of lignocellulose in AD. In this context, enhancing the substrate conversion to overcome the degradation resistance of lignocellulosic resources is of utmost importance to achieving environmentally friendly and economically feasible processes [7,8]. Hence, effective pretreatment methods are needed, particularly because lignocellulosic biomass has been evaluated as an attractive renewable energy source due to its inexhaustible, ubiquitous character [2,9].

The main objectives of this work are, therefore, (i) to present a short update on the currently available pretreatment strategies for enhanced disintegration of lignocellulosic resources and their application, and (ii) to review biological pretreatments currently applied for enhanced biogas production.

2. Lignocelluloses

Lignocellulose is the most abundant renewable biomass [10], with a worldwide annual production of an estimated 1000 Gt [11], including wheat straw, sugarcane bagasse, corn stalks, rye straw, rice straw, and barley straw as well as various types of organic waste (fractions). For data on the composition of different feedstocks, please refer to, e.g., Dahadha et al. [12], or Paudel et al. [13].

Lignocellulose contains up to 45% cellulose as the main component, 30% hemicellulose, and 25% lignin, although the composition varies considerably among different plants [14,15]. With about from 50% to 80% of organic material deriving from photosynthetic processes, lignocellulose represents one of the main components of global biomass [16,17]. Therefore, lignocellulose plays a major role as a constituent of biological resources and represents the most abundantly available raw material for the generation of renewable primary products and energy [18].

In the following section the chemical structure and characteristics of the most important fibrous components of lignocellulosic resources are described briefly.

2.1. Cellulose

Cellulose is the major component of plant matter and, therefore, a valuable source of biomass storing an enormous quantity of energy conserved by photosynthesis. It is a fibrous, hard, and water-insoluble substance that can be found in the wooden part of plant tissue. As a linear polymer, it comprises from 3000 to 14,000 glucose monomers, which are linked via β-1,4 glycosidic bonds. Approximately 60–70 of those cellulose polymers are interconnected by hydrogen bonds, forming so-called elementary fibrils, which themselves build up microfibrils. Multiple of those collocated chains can form a network of stable supramolecular fibers, with high tensile strength and a partially crystalline structure [19]. In plants, cellulose molecules are synthesized individually, which then undergo immediate self-assembly [20] probably regulated by hemicelluloses [21]. An important feature of cellulose is its crystalline structure, with the degree of crystallinity being highly variable depending on the type of plant tissue [19]. While the crystalline structure of cellulose fibers hinders degradation, various types of irregularities—pits, pores, and capillaries—increase the surface area of cellulose molecules [22]. This results in at least partially hydrated areas when being immersed in water, thus permitting access for enzymatic attack, including cellulases.

2.2. Hemicellulose

Hemicellulose is the term for branched heteropolysaccharides—mostly matrix polysaccharides—including monomers like glucose, mannose, galactose, xylose, and arabinose. Although similar enzymes are involved in cellulose and hemicellulose decomposition, complete hemicellulose degradation requires more enzymes due to its greater chemical and structural heterogeneity [23]. Hemicellulose is degraded to monomeric sugars and acetic acid [18], with the latter being of special interest for anaerobic digestion, representing the dominant methane precursor [24].

2.3. Lignin

Lignin is an aromatic polymer synthesized of phenylpropanoid precursors. Lignin is predominantly found in combination with cellulose (and hemicellulose), the so-called lignocellulose. Therein, lignin is encrusting both cellulose and hemicellulose, forming a physical seal, and is an impenetrable barrier in the plant cell wall. This polymer is synthesized by the generation of free radicals, which are released in the peroxide-mediated dehydrogenation of three phenylpropionic alcohols [18]. Lignin breakdown is necessary to facilitate the access to cellulose and hemicellulose but can, however, only occur via co-metabolism [18].

3. Biodegradation of Lignocellulose

Bioconversion of lignocellulosic residues is predominantly carried out by fungi. Because of the insolubility of cellulose, hemicellulose, and lignin, it occurs exocellularly, in association with the outer cell envelope, or extracellularly [18]. Two types of fungal enzymes are known to break down lignocellulose: (i) the hydrolytic system that produces hydrolases responsible for polysaccharide degradation and (ii) a unique oxidative and extracellular ligninolytic system degrading lignin by opening phenyl rings [18,25].

The ability to digest cellulose is widely distributed among many genera in the domain of Bacteria and in fungal groups within the domain of Eukarya [26], whereas cellulolytic organisms in the domain of Archaea have not (yet) been identified. Specialized groups of fungi are further able to attack lignin-encrusted cellulose. Generally, a 10- to 100-fold higher productivity of fungal compared with bacterial enzymes was assessed for cellulases [25].

Concerning the eubacteria, the ability to decompose cellulose is widespread in bacteria within the predominantly aerobic order Actinomycetales (Actinobacteria) and the anaerobic order Clostridiales (Firmicutes) [19,27]. Mechanisms of bacterial decomposition differ significantly from those of their fungal counterparts. Within cellulolytic clostridia, the breakdown of cellulose is organized in the so-called cellulosome [28,29], which is attached to the cell surface, contains all necessary enzymes, and forms a bridge between the cell and the insoluble cellulose components [16]. In anaerobic digestion systems, cellulose-degrading bacteria play an important role regarding the interaction between several groups of organisms, resulting in a complete conversion into carbon dioxide, methane, and water [30]. However, due to the small amount of energy that can be preserved in anaerobic processes and the lower productivity of bacterial cellulases compared with fungal ones [25], the degradation of cellulose is significantly slower under anoxic than under oxic conditions.

A specialized group within the Neocallimastigomycota called "anaerobic fungi", commonly found in ruminants, is able to degrade cellulose and hemicellulose under strictly anaerobic conditions [31,32]. The use of anaerobic fungi for an improved anaerobic digestion was taken into account, e.g., by Dollhofer et al. [33] or Leis et al. [34]. Also, Nakashimada et al. [35] investigated methane production from cellulose as a substrate with defined mixed cultures using the cellulolytic *Neocallimastix frontalis* and methanogens.

In contrast to anaerobic fungi, the direct application of aerobically growing fungi in anaerobic systems is completely hampered by their oxygen demand. Among fungi, there are a number of representatives, e.g., of the genera Fusarium and Chaetomonium that also target lignin-encrusted cellulose. In particular, so-called white rot fungi can effectively degrade lignin using an oxidative process with phenol oxidases as key enzymes [36], including *Phanerochaete chrysosporium* and *Trametes versicolor*, representing the most extensively studied members [37]. As the degradation of lignin is hardly possible under anoxic conditions, aerobic pretreatment prior to anaerobic digestion is of special interest [38–40].

4. Concepts of Pretreatment

Pretreatment strategies commonly comprise physical, chemical, and biological methods [40], and are applied in various fields of bioenergy and biofuel generation including biogas, bioethanol, biohydrogen, and hythane ($H_2 + CH_4$) production. Since lignocellulose materials represent the largest fraction of waste generated by modern society, increasing scientific interest is orientated towards combined cellulose waste management and energy resources [41]. Factors for ecological and economical feasible application of pretreatment strategies include low capital and energy investments, applicability over a wide variety of substrates, and high product yields to enhance revenues along with low waste treatment costs [7].

Among all bioconversion technologies for energy production, anaerobic digestion seems to be the most cost-effective that has been implemented worldwide for commercial production of electricity, heat, and compressed natural gas [40]. Anaerobic digestion has been adopted for bioenergy production from different organic feedstocks, such as forestry and agricultural residues, animal manures, organic fractions of municipal solid wastes, food wastes, and energy crops [42], answering the increasing demand for renewable energy sources. For recalcitrant substrates such as lignocellulosic resources, conventional anaerobic digestion cannot maximize the substrate conversion into biogas [43]. Thus, the application of biological pretreatments has gained significant importance in the past few years because of (i) the complex composition of lignocellulosic resources persistent in anaerobic environments; (ii) the desire to reduce hydraulic retention times; and (iii) the wish to increase the net carbon conversion rates. The latter is characterized by an enhanced total biogas and methane yield, representing the ultimate goal for any pretreatment strategy.

4.1. Physical and Chemical Pretreatment

Physical and chemical pretreatments are the most widespread strategies to improve the substrate quality designated for anaerobic digestion. They are often designed to improve the general digestibility and do not specifically target a certain compound of the substrate matrix. Physical strategies comprise comminution, heat and/or pressure treatment, steam explosion, liquid hot water, extrusion, and irradiation as well as ultrasonic and microwave technologies. Chemical ones include the use of acids or bases, catalyzed steam explosion, ozonization, oxidation, organosolve methods, and ionic liquid extraction [40]. A combination of physical and/or chemical methods is often applied. The main disadvantages of physical/chemical pretreatments are high energy and/or chemical demands with possible quality reductions of the digestion residues, thus hampering the subsequent use as biological fertilizer, accompanied by increasing costs for their disposal [44,45]. For a detailed description of physical and chemical pretreatment strategies, please refer to the respective review papers.

4.2. Biological Pretreatment

An effective biological pretreatment requires no preceding mechanical size reduction, preserves the pentose (hemicellulose) fractions, avoids the formation of degradation products that inhibit growth of fermentative microorganisms, minimizes energy demands, and limits costs. Therefore, a major objective of biological pretreatment is to break and remove the lignin seal and to disrupt the crystalline structure of cellulose to make it (more) susceptible to an enzymatic or microbial attack, while minimizing the loss of carbohydrates for anaerobic digestion [40,46,47]. The delignification and the decomposition of hemicellulose enhance the availability of cellulose and resultant monomers, which can boost the overall anaerobic digestion process. The choice of application is mainly dependent on the chemical composition of the substrate; however, in practice, structural and economic factors like available facilities or excess energy can often play an equally important role. Biological pretreatment techniques for enhanced biogas production have mainly focused on the use of fungal and bacterial strains or microbial consortia under both aerobic and anaerobic conditions, as well as on enzymes, with the latter being less important [40]. Therefore, this review is focusing on pretreatment strategies using active microorganisms.

The advantages of biological pretreatments compared with nonbiological procedures are the potential production of useful by-products, reduced formation of inhibitory substances due to milder operation conditions, the minimization of applied chemicals and energy input, and lower costs for waste deposit [44,45]. Beside the hydrolysis of (ligno)cellulose during pretreatment, microorganisms can further be used to upgrade the quality of certain substrates by removing undesired, potentially inhibitory substances. However, the efficiency of biological pretreatment is limited by the rate of microbial growth and the utilization of readily available sugars by the engaged organisms [48].

4.2.1. Micro-Aerobic Pretreatment

Micro-aeration during anaerobic processes is known to increase microbial activity during the initial hydrolysis phase [49]. Pretreatments using different doses of oxygen during anaerobic digestion can also be ascribed to biological pretreatments since the oxygen input alters the microbial community; however, these methods are mainly applied in waste water treatment plants [49,50]. The goal of micro-aeration is to stimulate microbial growth and activity during hydrolysis, the rate-limiting step in anaerobic digestion [51]. Up to now, this method has been successfully applied for brown water and/or food waste [52–54] as well as for energy crops [43] and agricultural residues [55,56].

4.2.2. Ensiling, Composting

Another microbiological pretreatment strategy originated from the necessity to store and stabilize lignocellulosic resources to guarantee a whole year's substrate supply for anaerobic digestion facilities. Cui et al. [57] investigated a wet storage technique via ensiling with simultaneous chemical and fungal pretreatment which could increase glucose and xylose yields 2.9- and 3.9-fold, respectively. Sugar beet pulp silage in different maturity stages was evaluated concerning its methane potential by Heidarzadeh et al. [58], indicating a positive trend but also the risk of energy loss if the ensiling was not conducted properly. Papinagsorn et al. [59] successfully tested ensiled napier grass in combination with chemical pretreatment for co-digestion with cow dung yielding up to 8.34 kJ·g^{-1} VS. Vervaeren et al. [60] investigated maize silage additives for enhanced biogas production and could verify up to a 14.7% increase in biogas production for certain additives. Wagner et al. [61] used composting as a treatment strategy to enhance methane production from digestate and showed a positive impact of composting with increased biogas and methane yields in a subsequent anaerobic digestion process.

4.2.3. Physical Separation of Digestion Phases or Microbial Consortia

Efforts have been made earlier to separate the different phases of anaerobic digestion to increase total biogas and methane yields. These methods are often referred to special digestion systems but not to pretreatment technologies. Since almost all methods are at least some kind of upstream treatment prior to anaerobic digestion per se, they can also be seen as pretreatment methods. As this type of technology does not aim to increase degradation rates of one or more specific components of substrates, the mode of action is rather unspecific. The physical separation of the hydrolytic and methanogenic phase can further be used to apply suitable conditions (e.g., temperature, pH) for each step. Quin et al. [62] investigated the effect of a thermophilic and hyperthermophilic anaerobic treatment prior to mesophilic anaerobic digestion. In this context, a preceding hyperthermophilic step increased the hydrolytic activity of the engaged microorganisms and resulted in a higher organic solids reduction rate [63,64]. Thermophilic or hyperthermophilic conditions are further beneficial to pathogen removal [65–67]. During aerobic hyperthermophilic pretreatment, *Geobacillus stearothermophilus*, for instance, turned out to be important for downstream processing [68].

4.2.4. Aerobic Pretreatment with Defined Fungal Cultures

Due to their powerful enzymatic capabilities, fungi offer great potential for biotechnological application. Lignocellulose-degrading fungi can be classified into white rot, brown rot, and soft rot fungi. White and soft rot fungi are known to attack lignin and, to a certain extent, also cellulose, while brown rots mainly target cellulose [69]. Nevertheless, white rot fungi are the preferred pretreatment organisms as they mandate highly efficient delignification enzyme equipment [70,71], with basidiomycetes being supremely effective [18,72]. The degree of delignification using white rots varies among applied fungi and depends on various factors such as pH and available N sources [70]. In contrast to anaerobic fungi, higher decomposition rates can be achieved due to aerobic conditions, which lead to higher energy yields for the engaged microorganisms and can therefore result in faster turnover rates. However, the application of fungi to pretreat substrates for anaerobic digestion is rather new [70]. Nevertheless, aerobic pretreatment using fungi was described for various species with diverse outcomes and is summarized in Table 1. A comparison concerning the effectivity and efficiency is rather difficult due to different experimental setups, substrates, inocula, inoculation rates, incubation times, and conditions, etc.; however, the table is intended to give a helpful overview. However, in more than 60 percent of the reviewed publications dealing with fungal pretreatment, white rot fungi were applied with pretreatment periods extending over approximately 3–4 weeks, which is notably longer than for other fungal pretreatments. In most cases, different organic waste fractions were used as substrate; to a minor extent, energy crops were also applied.

Liu et al. [73] found a positive impact of pretreatment on methane production potential using forest residues as co-substrate inoculated with ligninolytic *Ceriporiopsis subvermispora*, but the effect was dependent on the basic substrate used. Ge et al. [74] incubated Albizia (silk tree) biomass, a forestry waste, with the same organism and were also able to increase the cumulative methane yield. From sisal (agave) leaf decortication residues pretreated with two fungal strains including *Trichoderma reesei*, an increased methane production was observed [75]. In comparison with other strategies, Take et al. [76] found a positive effect of biological pretreatment with *Cyathus stercoreus* and *Trametes hirsute* on subsequent biogas production using cedar wood chips as substrate, whereas the pretreatment with *Ceriporiopsis subvermispora* positively influenced the methane yield in another study using the same substrate but with a nutritional supplement [77]. Phutela et al. [78] pretreated paddy straw with *Fusarium* sp. and observed decreased lignin and cellulose contents in the substrate and an improved digestibility. However, the application of a facultative pathogenic microorganism seems to be problematic. Wheat straw was incubated with *Polyporus brumalis* BRFM 985, a white rot fungi, and in combination with metal amendment with the result that the treatment positively influenced the methane potential [79]. In another study by Vasmara et al. [80], 4- and 10-week incubation of wheat straw with 7 different fungal isolates was investigated regarding increased methane yields in a subsequent anaerobic digestion step. A positive effect was found enhancing the methane yield by an optimized treatment up to 16% for the 4-week and up to 37% for the 10-week pretreatment. *Pleurotus ostreatus* and *Trichoderma reesei* pretreatment of rice straw resulted in a 120% increase in methane yields in a study by Mustafa et al. [81]. Moreover, Mustafa et al. [82] found increased delignification and methane production from rice straw by pretreatment with the fungus *Pleurotus ostreatus*. Yard trimmings were subjected to a pretreatment using the white rot *Ceriporiopsis subvermispora* [83]; the enhanced methane production was attributed to an increased delignification by the fungus. Mutschlechner et al. [84] inoculated a similar substrate containing high portions of grass and tree cut with *Trichoderma viride* and could secure increased methane production. This organism was also used to pretreat raw bio-waste with a positive effect on the methane production potential of the substrate [85]. *Phanerochaete chrysosporium* was used during solid-stage fermentation of corn stover to successfully enhance methane production in a subsequent anaerobic digestion step [86]. Tisma et al. [87] observed a positive effect of *Trametes versicolor* pretreatment on the biogas productivity of corn silage. Pretreating orange processing waste with strains of *Sporotrichum*, *Aspergillus*, *Fusarium*, and *Penicillium*, Srilatha et al. [88] observed a positive effect on biogas production and biogas potential. Mackul'ak et al. [89] inoculated sweet chestnut leaves and hay with the fungus *Auricularia auricula-judae* and observed an increase in methane productivity. Parthiba Karthikeyan et al. [48] found that biological pretreatments are not yet available for food wastes and demand urgent need for further research.

Table 1. Comparison of different fungal pretreatment strategies for enhanced biogas production.

Pretreatment Organism (Type of Fungus¹)	Substrate	Pretreatment Incubation Conditions²	Additional Information on Fungal Pretreatment Process³	Anaerobic Digestion Conditions⁴	Impact of Pretreatment on Substrate	Impact of Pretreatment on Biogas Production	Reference
Ceriporiopsis subvermispora (wrf)	Japanese cedar wood	8 weeks a, 28 °C b, 70% c	orig, hyphal biomass grown on agar added, substrate supplemented with 10% wheat bran.	batch, mp, t 60	28% lignin removal in initial substrate, ~75% cleavage of ß-O-4 aryl ether	35% and 5% conversion of holocellulose to methane with and without pretreatment, respectively	[77]
Ceriporiopsis subvermispora (wrf)	Albizia biomass (forestry waste)	48 days a, 28 °C b, 60% c	e, autoc	batch, mp, ssAD, t 58	24% lignin removal of initial substrate, 4-fold increase in xylose and glucose production after 72 h of enzymatic hydrolysis	3.7-fold increase in methane production	[74]
Ceriporiopsis subvermispora (wrf)	Hazel and acacia branches, barley straw, and sugarcane bagasse	28 days a, 28 °C b	e, autoc, grinded substrate	batch, mp	2- to 4-fold increase in enzymatic cellulose degradability for hazel and bagasse, decrease for straw and acacia	Increase of biomethane potential (BMP) for hazel (60%), loss of BMP for acacia (34%), straw and sugarcane bagasse	[73]
Phanerochaete chrysosporium (wrf)	Corn stover silage	30 days a, 28 °C b, Stable ambient d	f, autoc, washed substrate	batch, mp, t 30	39% lignin removal of initial substrate, improved degradation of substrate cell wall components	19.6–32.6% increase in methane production compared with controls	[86]
Fusarium sp. (wrf)	Paddy straw	10 days a, 30 °C b, 70% c	g, orig	batch, mp, t 35	17.1% decrease in lignin content, 10.8% decrease in silica content compared with controls	53.8% increase in biogas production	[78]
Trametes versicolor (wrf)	Corn silage	7 days a, 27 °C b, 70–80% c	g, orig	cont, mp, co-digestion with cow manure	70% increase in lignin degradation compared with control approach	Increased pH stability and biogas productivity, enhanced anaerobic degradation	[87]
Ceriporiopsis subvermispora (wrf)	Yard trimmings	30 days a, 28 °C b, 60% c	e, autoc	batch, mp, ssAD, t 40	20.9% degradation of initial lignin content	54% increase in methane production compared with controls, increased cellulose degradation	[83]
Polyporus brumalis (wrf)	Wheat straw	12.5 to 20 days a, 20–30 °C b, wet weight to initial solid ratio of 2.1 to 4.5	e, autoc, addition of metal supplement solution	batch, mp, t 57		Decrease in methane production compared with the control. Within fungal pretreatment, best methane production after 12.5 days incubation at 30 °C at 3.7 ww/ts ratio	[79]

Table 1. *Cont.*

Pretreatment Organism (Type of Fungus [1])	Substrate	Pretreatment Incubation Conditions [2]	Additional Information on Fungal Pretreatment Process [3]	Anaerobic Digestion Conditions [4]	Impact of Pretreatment on Substrate	Impact of Pretreatment on Biogas Production	Reference
Pleurotus ostreatus (wrf) *Trichoderma reesei* (srf [5])	Rice straw	20 days a 28 °C b 75% c	g, autoc	batch, mp, ssAD, t 45	33% lignin removal of initial substrate with wrf and 23.6% with brf Lignin-to-cellulose ratio after treatment: wrf 4.2, brf 2.88	20% increase in methane production with wrf and 21.7% decrease for brf treatment	[81]
CCHT-1 (wrf) *Trichoderma reesei* (srf [5])	Sisal leaf decortication residues	4 + 8 days a 28 °C b	g, orig, two fungal stages: wrf followed by brf	batch, mp, t 84	22.5%, decrease in neutral detergent fiber content, 21% increase in cellulose content	30–101% increase in biogas production compared to control	[75]
Sporotrichum sp. *Aspergillus* sp. *Fusarium* sp. *Penicillium* sp.	Orange processing waste	3 days a 30 °C b 65% c	g, orig, mixed culture pretreatment.	cont, mp, t 25	Reduction in inhibitory limonene content in the substrate.	Pretreatment leads to higher possible organic loading rates that improve overall productivity	[88]
Trichoderma viride (srf [5])	Organic waste	4 days a 25 °C b	e, orig	batch, tp, t 18	Increased cellulase activity during pretreatment compared with controls	Up to 400% increase in methane production compared with controls	[85]
Trichoderma viride (srf [5])	Organic waste	10 days a 22 °C b 70% c	f, orig	batch, tp, t 14	Increased cellulase and dehydrogenase activity compared to control	More than 2-fold increase in methane production	[84]

[1] **wrf**: white rot fungi; **brf**: brown rot fungi; **srf**: soft rot fungi; [2] **a**: incubation period, **b**: incubation temperature, **c**: moisture content in %, **d**: humidity in %; [3] inoculation with **e**: submerged fungal culture, **f**: fungal spores, **g**: autoclaved substrate overgrown by fungal mycelia; **autoc**: autoclaved substrate, **orig**: original, unmodified substrate; [4] **batch**: batch system, **cont**: continuous system, **ssAD**: solid-state anaerobic digestion, **mp**: mesophilic conditions, **tp**: thermophilic conditions, **t**: anaerobic incubation period or hydraulic retention time in days; [5] according to Klein and Eveleigh [90].

5. By-Product Formation

Pretreatment of recalcitrant material enhances the availability of substrates but can also result in the formation of various inhibitory or even toxic substances. While biological pretreatments are less rigid, physico-chemical ones can be problematic. For example, although high glucose yields are achieved by acid treatment of lignocellulosic biomass, this procedure also leads to hydroxymethylfurfural (HMF) formation, one of the most unwanted pretreatment by-products [10,91]. Moreover, toxic and highly corrosive heavy metal ions like copper, nickel, chromium, and iron are released due to acid application [10]. In contrast, biological pretreatments apply milder conditions, tend to be less corrosive, and release fewer inhibitory substances [92]. Inhibitory substances introduced with the initial substrate can even be degraded during biological pretreatment, leading to an increase in substrate quality. In this context, lignocellulose pretreatment for biofuel production using the fungus *Coniochaeta ligniaria* NRRL30616 led to a degradation of various undesired by-products including phenolic compounds, furfural, and HMF along with an assimilation of these inhibitory substances in the cells and/or a release of less toxic intermediates into the liquid phase [93]. In another study, pretreatment of oil palm mill effluent using thermophilic bacteria resulted in the removal of unwanted phenols coevally improving the anaerobic digestion performance [94,95].

Inhibitory substances can have an adverse effect on the engaged microorganisms involved in biological pretreatment and downstream anaerobic degradation [96] or on pretreatment facilities by corrosion [91,97]. Synergistic toxic effects are known for lignocellulosic hydrolysates, meaning that the toxicity of two or more toxic substances combined (on yeasts) can be higher than their sum [98].

However, a more profound knowledge of inhibitory substances is urgently needed for anaerobic degradation processes including ethanol/biofuel production, waste water treatment, and, especially, biogas production.

6. Closing Remarks—Conclusions

Various pretreatment strategies—physical, chemical, and biological—have been developed to overcome the inherent resistance of lignocellulose to anaerobic degradation. Biological pretreatment strategies, however, outcompete other pretreatments due to the application of milder conditions, and lower by-product formation and corrosiveness. The variety of applied techniques comprises micro-aerobic treatments, ensiling or composting, the separation of digestion stages, and pretreatments using various fungi. Fungal pretreatments have achieved particular success using various white, brown, and soft rot fungi, or a combination of these. Pretreatment processes applying white rot fungi from the genera Ceripoioposis, Phanerochaete, Fusarium, Trametes, Polyporus, and Pleurotus target cellulose as well as lignin, allowing the use of recalcitrant, second-generation substrates for biogas production. Therefore, biological pretreatment strategies offer great potential to improve the digestibility of different biogas substrates; however, detailed investigations of the mode of action, the application of different substrates, full-scale implementation, and possible by-product formation are still needed.

Author Contributions: N.L., M.M., E.M.P. and R.M. contributed equally.

Funding: This review was financially supported by the Austrian Science Fund (FWF), projects P29143 and P29360, and Ph.D. grants of the University of Innsbruck.

Conflicts of Interest: The authors declare no conflict of interest.

References

1. Sawatdeenarunat, C.; Surendra, K.C.; Takara, D.; Oechsner, H.; Khanal, S.K. Anaerobic digestion of lignocellulosic biomass: Challenges and opportunities. *Bioresour. Technol.* **2015**, *178*, 178–186. [CrossRef] [PubMed]

2. Ellabban, O.; Abu-Rub, H.; Blaabjerg, F. Renewable energy resources: Current status, future prospects and their enabling technology. *Renew. Sustain. Energy Rev.* **2014**, *39*, 748–764. [CrossRef]

3. Martínez-Gutiérrez, E. Biogas production from different lignocellulosic biomass sources: Advances and perspectives. *3 Biotech* **2018**, *8*, 233–286. [CrossRef] [PubMed]

4. Amin, F.R.; Khalid, H.; Zhang, H.; Rahman, S.U.; Zhang, R.; Liu, G.; Chen, C. Pretreatment methods of lignocellulosic biomass for anaerobic digestion. *AMB Express* **2017**, *7*, 72. [CrossRef] [PubMed]

5. Börjesson, P.; Mattiasson, B. Biogas as a resource-efficient vehicle fuel. *Trends Biotechnol.* **2008**, *26*, 7–13. [CrossRef] [PubMed]

6. Zhen, G.; Lu, X.; Kato, H.; Zhao, Y.; Li, Y.-Y. Overview of pretreatment strategies for enhancing sewage sludge disintegration and subsequent anaerobic digestion: Current advances, full-scale application and future perspectives. *Renew. Sustain. Energy Rev.* **2017**, *69*, 559–577. [CrossRef]

7. Wyman, C.E.; Dale, B.E.; Elander, R.T.; Holtzapple, M.; Ladisch, M.R.; Lee, Y.Y. Coordinated development of leading biomass pretreatment technologies. *Bioresour. Technol.* **2005**, *96*, 1959–1966. [CrossRef] [PubMed]

8. Capolupo, L.; Faraco, V. Green methods of lignocellulose pretreatment for biorefinery development. *Appl. Microbiol. Biotechnol.* **2016**, *100*, 9451–9467. [CrossRef] [PubMed]

9. Zieminski, K.; Romanowska, I.; Kowalska, M. Enzymatic pretreatment of lignocellulosic wastes to improve biogas production. *Waste Manag.* **2012**, *32*, 1131–1137. [CrossRef] [PubMed]

10. Alvira, P.; Tomás-Pejó, E.; Ballesteros, M.; Negro, M.J. Pretreatment technologies for an efficient bioethanol production process based on enzymatic hydrolysis: A review. *Bioresour. Technol.* **2010**, *101*, 4851–4861. [CrossRef] [PubMed]

11. Hermosilla, E.; Schalchli, H.; Mutis, A.; Diez, M.C. Combined effect of enzyme inducers and nitrate on selective lignin degradation in wheat straw by *Ganoderma lobatum*. *Environ. Sci. Pollut. Res. Int.* **2017**, *24*, 21984–21996. [CrossRef] [PubMed]

12. Dahadha, S.; Amin, Z.; Bazyar Lakeh, A.A.; Elbeshbishy, E. Evaluation of Different Pretreatment Processes of Lignocellulosic Biomass for Enhanced Biomethane Production. *Energy Fuels* **2017**, *31*, 10335–10347. [CrossRef]

13. Paudel, S.R.; Banjara, S.P.; Choi, O.K.; Park, K.Y.; Kim, Y.M.; Lee, J.W. Pretreatment of agricultural biomass for anaerobic digestion: Current state and challenges. *Bioresour. Technol.* **2017**, *245*, 1194–1205. [CrossRef] [PubMed]

14. Howard, R.I.; Abotsi, E.; van Jansen Regensburg, E.L.; Howard, S. Lignocellulose biotechnology: Issues of bioconversion and enzyme production. *Afr. J. Biotechnol.* **2003**, *2*. [CrossRef]

15. Prassad, S.; Singh, A.; Joshi, H.C. Ethanol as an alternative fuel from agricultural, industrial and urban residues. *Resour. Conserv. Recycl.* **2007**, *50*, 1–15. [CrossRef]

16. Pérez, J.; Muñoz-Dorado, J.; de La Rubia, T.; Martínez, J. Biodegradation and biological treatments of cellulose, hemicellulose and lignin: An overview. *Int. Microbiol. Off. J. Span. Soc. Microbiol.* **2002**, *5*, 53–63. [CrossRef] [PubMed]

17. Tuomela, M.; Vikman, M.; Hatakka, A.; Itvaara, M. Biodegradation of lignin in a compost environment: A review. *Bioresour. Technol.* **1999**, *72*, 169–178. [CrossRef]

18. Sánchez, C. Lignocellulosic residues: Biodegradation and bioconversion by fungi. *Biotechnol. Adv.* **2009**, *27*, 185–194. [CrossRef] [PubMed]

19. Lynd, L.R.; Weimer, P.J.; van Zyl, W.H.; Pretorius, I.S. Microbial cellulose utilization: Fundamentals and biotechnology. *Microbiol. Mol. Biol. Rev.* **2002**, *66*, 506–577. [CrossRef] [PubMed]

20. Brown, R.M., Jr.; Saxena, I.M. Cellulose biosynthesis: A model for understanding the assembly of biopolymers. *Plant Physiol. Biochem.* **2000**, *38*, 57–67. [CrossRef]

21. Atalla, R.H.; Hackney, J.M.; Uhlin, I.; Thompson, N.S. Hemicelluloses as structure regulators in the aggregation of native cellulose. *Int. J. Biol. Macromol.* **1993**, *15*, 109–112. [CrossRef]

22. Fan, L.T.; Lee, Y.-H.; Beardmore, D.H. Mechanism of the enzymatic hydrolysis of cellulose: Effects of major structural features of cellulose on enzymatic hydrolysis. *Biotechnol. Bioeng.* **1980**, *22*, 177–182. [CrossRef]

23. Malherbe, S.; Cloete, T.E. Lignocellulose biodegradation: Fundamentals and applications. *Rev. Environ. Sci. Biotechnol.* **2002**, *1*, 105–114. [CrossRef]

24. Gerardi, M.H. *The Microbiology of Anaerobic Digesters*; Wiley: New York, NY, USA, 2003.

25. Adney, W.S.; Rivard, C.J.; Ming, S.A.; Himmel, M.E. Anaerobic digestion of lignocellulosic biomass and wastes: Cellulases and related enzymes. *Appl. Biochem. Biotechnol.* **1991**, *30*, 165–183. [CrossRef] [PubMed]

26. Dashtban, M.; Schraft, H.; Qin, W. Fungal Bioconversion of Lignocellulosic Residues; Opportunities & Perspectives. *Int. J. Biol. Sci.* **2009**, *5*, 578–595. [PubMed]

27. Shrestha, S.; Fonoll, X.; Khanal, S.K.; Raskin, L. Biological strategies for enhanced hydrolysis of lignocellulosic biomass during anaerobic digestion: Current status and future perspectives. *Bioresour. Technol.* **2017**, *245*, 1245–1257. [CrossRef] [PubMed]

28. Desvaux, M. The cellulosome of Clostridium cellulolyticum. *Enzym. Microbiol. Technol.* **2005**, *37*, 373–385. [CrossRef]

29. Vodovnik, M.; Marinsek-Logar, R. Cellulosomes—Promising supramolecular machines of anaerobic cellulolytic microorganisms. *Acta Chem. Slov.* **2010**, *57*, 767–774.

30. Leschine, S.B. Cellulose degradiation in anaerobic environment. *Annu. Rev. Microbiol.* **1995**, *49*, 399–426. [CrossRef] [PubMed]

31. Orpin, C.G. Studies on rumen flagellate *Neocallimastix frontalis*. *J. Gen. Microbiol.* **1975**, *91*, 249–262. [CrossRef] [PubMed]

32. Teunissen, M.J.; Dencamp, H.J.M.O.; Orpin, C.G.; Veld, J.H.; Vogels, G.D. Comparsion of growth-characteristics of anaerobic fungi isolated from ruminant and non-ruminant herbivores during cultivation in a defined medium. *J. Gen. Microbiol.* **1991**, *137*, 1401–1408. [CrossRef] [PubMed]

33. Dollhofer, V.; Podmirseg, S.M.; Callaghan, T.M.; Griffith, G.W.; Fliegerova, K. Anaerobic Fungi and Their Potential for Biogas Production. *Adv. Biochem. Eng. Biotechnol.* **2015**, *151*, 41–61. [CrossRef]

34. Leis, S.; Dresch, P.; Peintner, U.; Fliegerová, K.; Sandbichler, A.M.; Insam, H.; Podmirseg, S.M. Finding a robust strain for biomethanation: Anaerobic fungi (Neocallimastigomycota) from the Alpine ibex (*Capra ibex*) and their associated methanogens. *Anaerobe* **2014**, *29*, 34–43. [CrossRef] [PubMed]

35. Nakashimada, Y.; Srinivasan, K.; Murakami, M.; Nishio, N. Direct conversion of cellulose to methane by anaerobic fungus *Neocallimastix frontalis* and defined methanogens. *Biotechnol. Lett.* **2000**, *22*, 223–227. [CrossRef]

36. Rabinovich, M.L.; Bolobova, A.V.; Vasilchenko, N. Fungal decomposition of natural aromatic structures and xenobiotics: A review. *Appl. Biochem. Microbiol.* **2004**, *40*, 1–17. [CrossRef]

37. Manzanares, P.; Fajardo, S.; Martin, C. Production of ligniolytic activities when treating paper pulp effluents by Trymes versicolor. *J. Biotechnol.* **1995**, *43*, 125–132. [CrossRef]

38. Brown, M.E.; Chang, M.C.Y. Exploring bacterial lignin degradation. *Curr. Opin. Chem. Biol.* **2014**, *19*, 1–7. [CrossRef] [PubMed]

39. Ko, J.-J.; Shimizu, Y.; Ikeda, K.; Kim, S.-K.; Park, C.-H.; Matsui, S. Biodegradation of high molecular weight lignin under sulfate reducing conditions: Lignin degradability and degradation by-products. *Bioresour. Technol.* **2009**, *100*, 1622–1627. [CrossRef] [PubMed]

40. Zheng, Y.; Zhao, J.; Xu, F.; Li, Y. Pretreatment of lignocellulosic biomass for enhanced biogas production. *Prog. Energy Combust. Sci.* **2014**, *42*, 35–53. [CrossRef]

41. Bayer, E.A.; Lamed, R.; Himmel, M.E. The potential of cellulases and cellulosomes for cellulosic waste management. *Curr. Opin. Biotechnol.* **2007**, *18*, 237–245. [CrossRef] [PubMed]

42. Sawatdeenarunat, C.; Sung, S.; Khanal, S.K. Enhanced volatile fatty acids production during anaerobic digestion of lignocellulosic biomass via micro-oxygenation. *Bioresour. Technol.* **2017**, *237*, 139–145. [CrossRef] [PubMed]

43. Kaparaju, P.; Serrano, M.; Thomsen, A.B.; Kongjan, P.; Angelidaki, I. Bioethanol, biohydrogen and biogas production from wheat straw in a biorefinery concept. *Bioresour. Technol.* **2009**, *100*, 2562–2568. [CrossRef] [PubMed]

44. Zhong, W.; Zhang, Z.; Luo, Y.; Sun, S.; Qiao, W.; Xiao, M. Effect of biological pretreatments in enhancing corn straw biogas production. *Bioresour. Technol.* **2011**, *102*, 11177–11182. [CrossRef] [PubMed]

45. Zimbardi, F.; Viggiano, D.; Nanna, F.; Demichele, M.; Cuna, D.; Cardinale, G. Steam explosions of straw in batch and continous systems. *Appl. Biochem. Biotechnol.* **1999**, *77*, 117–125. [CrossRef]

46. Mosier, N.; Wyman, C.; Dale, B.; Elander, R.; Lee, Y.Y.; Holtzapple, M.; Ladisch, M. Features of promising technologies for pretreatment of lignocellulosic biomass. *Coord. Dev. Lead. Biomass Pretreat. Technol.* **2005**, *96*, 673–686. [CrossRef] [PubMed]

47. Karimi, K.; Taherzadeh, M.J. A critical review of analytical methods in pretreatment of lignocelluloses: Composition, imaging, and crystallinity. *Bioresour. Technol.* **2016**, *200*, 1008–1018. [CrossRef] [PubMed]

48. Parthiba Karthikeyan, O.; Trably, E.; Mehariya, S.; Bernet, N.; Wong, J.W.C.; Carrere, H. Pretreatment of food waste for methane and hydrogen recovery: A review. *Bioresour. Technol.* **2018**, *249*, 1025–1039. [CrossRef] [PubMed]

49. Jensen, T.R.; Lastra Milone, T.; Petersen, G.; Andersen, H.R. Accelerated anaerobic hydrolysis rates under a combination of intermittent aeration and anaerobic conditions. *Water Sci. Technol. J. Int. Assoc. Water Pollut. Res.* **2017**, *75*, 1944–1951. [CrossRef] [PubMed]

50. Montalvo, S.; Vielma, S.; Borja, R.; Huiliñir, C.; Guerrero, L. Increase in biogas production in anaerobic sludge digestion by combining aerobic hydrolysis and addition of metallic wastes. *Renew. Energy* **2018**, *123*, 541–548. [CrossRef]

51. Henze, M. *Wastewater Treatment. Biological and Chemical Processes*, 3rd ed.; Springer: Berlin, Germany; London, UK, 2011.

52. Lim, J.W.; Wang, J.-Y. Enhanced hydrolysis and methane yield by applying microaeration pretreatment to the anaerobic co-digestion of brown water and food waste. *Waste Manag.* **2013**, *33*, 813–819. [CrossRef] [PubMed]

53. Ni, Z.; Liu, J.; Zhang, M. Short-term pre-aeration applied to the dry anaerobic digestion of MSW, with a focus on the spectroscopic characteristics of dissolved organic matter. *Chem. Eng. J.* **2017**, *313*, 1222–1232. [CrossRef]

54. Sarkar, O.; Venkata Mohan, S. Pre-aeration of food waste to augment acidogenic process at higher organic load: Valorizing biohydrogen, volatile fatty acids and biohythane. *Bioresour. Technol.* **2017**, *242*, 68–76. [CrossRef] [PubMed]

55. Tsapekos, P.; Kougias, P.G.; Vasileiou, S.A.; Lyberatos, G.; Angelidaki, I. Effect of micro-aeration and inoculum type on the biodegradation of lignocellulosic substrate. *Bioresour. Technol.* **2017**, *225*, 246–253. [CrossRef] [PubMed]

56. Xu, W.; Fu, S.; Yang, Z.; Lu, J.; Guo, R. Improved methane production from corn straw by microaerobic pretreatment with a pure bacteria system. *Bioresour. Technol.* **2018**, *259*, 18–23. [CrossRef] [PubMed]

57. Cui, Z.; Shi, J.; Wan, C.; Li, Y. Comparison of alkaline- and fungi-assisted wet-storage of corn stover. *Bioresour. Technol.* **2012**, *109*, 98–104. [CrossRef] [PubMed]

58. Heidarzadeh Vazifehkhoran, A.; Triolo, J.; Larsen, S.; Stefanek, K.; Sommer, S. Assessment of the Variability of Biogas Production from Sugar Beet Silage as Affected by Movement and Loss of the Produced Alcohols and Organic Acids. *Energies* **2016**, *9*, 368. [CrossRef]

59. Prapinagsorn, W.; Sittijunda, S.; Reungsang, A. Co-Digestion of Napier Grass and Its Silage with Cow Dung for Methane Production. *Energies* **2017**, *10*, 1654. [CrossRef]

60. Vervaeren, H.; Hostyn, K.; Ghekiere, G.; Willems, B. Biological ensilage additives as pretreatment for maize to increase the biogas production. *Renew. Energy* **2010**, *35*, 2089–2093. [CrossRef]

61. Wagner, A.O.; Janetschek, J.; Illmer, P. Using Digestate Compost as a Substrate for Anaerobic Digestion. *Chem. Eng. Technol.* **2018**, *41*, 747–754. [CrossRef]

62. Qin, Y.; Higashimori, A.; Wu, L.-J.; Hojo, T.; Kubota, K.; Li, Y.-Y. Phase separation and microbial distribution in the hyperthermophilic-mesophilic-type temperature-phased anaerobic digestion (TPAD) of waste activated sludge (WAS). *Bioresour. Technol.* **2017**, *245*, 401–410. [CrossRef] [PubMed]

63. Carrère, H.; Dumas, C.; Battimelli, A.; Batstone, D.J.; Delgenès, J.P.; Steyer, J.P.; Ferrer, I. Pretreatment methods to improve sludge anaerobic degradability: A review. *J. Hazard. Mater.* **2010**, *183*, 1–15. [CrossRef] [PubMed]

64. Carrere, H.; Antonopoulou, G.; Affes, R.; Passos, F.; Battimelli, A.; Lyberatos, G.; Ferrer, I. Review of feedstock pretreatment strategies for improved anaerobic digestion: From lab-scale research to full-scale application. *Pretreat. Biomass* **2016**, *199*, 386–397. [CrossRef] [PubMed]

65. Lu, J.; Gavala, H.N.; Skiadas, I.V.; Mladenovska, Z.; Ahring, B.K. Improving anaerobic sewage sludge digestion by implementation of a hyper-thermophilic prehydrolysis step. *J. Environ. Manag.* **2008**, *88*, 881–889. [CrossRef] [PubMed]

66. Wagner, A.O.; Malin, C.; Gstraunthaler, G.; Illmer, P. Survival of selected pathogens in diluted sludge of a thermophilic waste treatment plant and in NaCl-solution under aerobic and anaerobic conditions. *Waste Manag.* **2009**, *29*, 425–429. [CrossRef] [PubMed]

67. Wagner, A.O.; Gstraunthaler, G.; Illmer, P. Survival of bacterial pathogens during the thermophilic anaerobic digestion of biowaste: Laboratory experiments and in situ validation. *Anaerobe* **2008**, *14*, 181–183. [CrossRef] [PubMed]

68. Hasegawa, S.; Shiota, N.; Katsura, K.; Akashi, A. Solubilization of organic sludge by thermophilic aerobic bacteria as a pretreatment for anaerobic digestion. *Water Sci. Technol.* **2000**, *41*. [CrossRef]

69. Rodriguez, C.; Alaswad, A.; Benyounis, K.Y.; Olabi, A.G. Pretreatment techniques used in biogas production from grass. *Renew. Sustain. Energy Rev.* **2017**, *68*, 1193–1204. [CrossRef]

70. Rouches, E.; Herpoël-Gimbert, I.; Steyer, J.P.; Carrere, H. Improvement of anaerobic degradation by white-rot fungi pretreatment of lignocellulosic biomass: A review. *Renew. Sustain. Energy Rev.* **2016**, *59*, 179–198. [CrossRef]

71. Hattaka, A.I. Pretreatment of wheat straw by white-rot fungi for enzymatic saccharification of cellulose. *Appl. Microbiol. Biotechnol.* **1983**, *82*, 131–137.

72. Isroi; Millati, R.; Syamsiah, S.; Niklasson, C.; Cahyanto, M.N.; Lundquist, K.; Taherzadeh, M.J. Biological pretreatment of lignocelluloses with white-rot fungi and its applications: A review. *BioResources* **2011**, *6*, 5224–5259.

73. Liu, X.; Hiligsmann, S.; Gourdon, R.; Bayard, R. Anaerobic digestion of lignocellulosic biomasses pretreated with *Ceriporiopsis subvermispora*. *J. Environ. Manag.* **2017**, *193*, 154–162. [CrossRef] [PubMed]

74. Ge, X.; Matsumoto, T.; Keith, L.; Li, Y. Fungal Pretreatment of Albizia Chips for Enhanced Biogas Production by Solid-State Anaerobic Digestion. *Energy Fuels* **2015**, *29*, 200–204. [CrossRef]

75. Muthangya, M.; Mshandete, A.M.; Kivaisi, A.K. Two-stage fungal pre-treatment for improved biogas production from sisal leaf decortication residues. *Int. J. Mol. Sci.* **2009**, *10*, 4805–4815. [CrossRef] [PubMed]

76. Take, H.; Andou, Y.; Nakamura, Y.; Kobayashi, F.; Kurimoto, Y.; Kuwahara, M. Production of methane gas from Japanese cedar chips pretreated by various delignification methods. *Biochem. Eng. J.* **2006**, *28*, 30–35. [CrossRef]

77. Amirta, R.; Tanabe, T.; Watanabe, T.; Honda, Y.; Kuwahara, M.; Watanabe, T. Methane fermentation of Japanese cedar wood pretreated with a white rot fungus, *Ceriporiopsis subvermispora*. *J. Biotechnol.* **2006**, *123*, 71–77. [CrossRef] [PubMed]

78. Phutela, U.G.; Sahni, N. Effect of *Fusarium* sp. on Paddy Straw Digestibility and Biogas Production. *J. Adv. Lab. Res. Biol.* **2012**, *3*, 9–12.

79. Rouches, E.; Zhou, S.; Sergent, M.; Raouche, S.; Carrere, H. Influence of white-rot fungus *Polyporus brumalis* BRFM 985 culture conditions on the pretreatment efficiency for anaerobic digestion of wheat straw. *Biomass Bioenergy* **2018**, *110*, 75–79. [CrossRef]

80. Vasmara, C.; Cianchetta, S.; Marchetti, R.; Galletti, S. Biogas produciton from wheat straw pre-treated with ligninlytic fungi and co-digestion with pig slurry. *Environ. Eng. Manag. J.* **2015**, *14*, 1751–1760.

81. Mustafa, A.M.; Poulsen, T.G.; Sheng, K. Fungal pretreatment of rice straw with *Pleurotus ostreatus* and *Trichoderma reesei* to enhance methane production under solid-state anaerobic digestion. *Appl. Energy* **2016**, *180*, 661–671. [CrossRef]

82. Mustafa, A.M.; Poulsen, T.G.; Xia, Y.; Sheng, K. Combinations of fungal and milling pretreatments for enhancing rice straw biogas production during solid-state anaerobic digestion. *Bioresour. Technol.* **2017**, *224*, 174–182. [CrossRef] [PubMed]

83. Zhao, J.; Zheng, Y.; Li, Y. Fungal pretreatment of yard trimmings for enhancement of methane yield from solid-state anaerobic digestion. *Bioresour. Technol.* **2014**, *156*, 176–181. [CrossRef] [PubMed]

84. Mutschlechner, M.; Illmer, P.; Wagner, A.O. Biological pre-treatment: Enhancing biogas production using the highly cellulolytic fungus *Trichoderma viride*. *Waste Manag.* **2015**, *43*, 98–107. [CrossRef] [PubMed]

85. Wagner, A.O.; Schwarzenauer, T.; Illmer, P. Improvement of methane generation capacity by aerobic pre-treatment of organic waste with a cellulolytic *Trichoderma viride* culture. *J. Environ. Manag.* **2013**, *129*, 357–360. [CrossRef] [PubMed]

86. Liu, S.; Li, X.; Wu, S.; He, J.; Pang, C.; Deng, Y.; Dong, R. Fungal pretreatment by *Phanerochaete chrysosporium* for enhancement of biogas production from corn stover silage. *Appl. Biochem. Biotechnol.* **2014**, *174*, 1907–1918. [CrossRef] [PubMed]

87. Tišma, M.; Planinić, M.; Bucić-Kojić, A.; Panjičko, M.; Zupančič, G.D.; Zelić, B. Corn silage fungal-based solid-state pretreatment for enhanced biogas production in anaerobic co-digestion with cow manure. *Bioresour. Technol.* **2018**, *253*, 220–226. [CrossRef] [PubMed]

88. Srilatha, H.R.; Nand, K.; Sudhakar, K.; Madhukara, K. Fungal pretreatment of orange processing waste by solid-state fermentation for improved production of methane. *Process Biochem.* **1995**, *30*, 327–331. [CrossRef]

89. Mackuľak, T.; Prousek, J.; Švorc, Ľ.; Drtil, M. Increase of biogas production from pretreated hay and leaves using wood-rotting fungi. *Chem. Pap.* **2012**, *66*, 3498. [CrossRef]

90. Klein, D.; Eveleigh, D.E. Ecology of Trichoderma. In *Trichoderma & Gliocladium, Basic Biology, Taxonomy and Genetics*; Kubicek, C.P., Harmna, G.E., Eds.; Taylor and Francis: London, UK, 1998; Volume 1, pp. 57–69. ISBN 0-7484-0572-0.

91. Jönsson, L.J.; Martín, C. Pretreatment of lignocellulose: Formation of inhibitory by-products and strategies for minimizing their effects. *Bioresour. Technol.* **2016**, *199*, 103–112. [CrossRef] [PubMed]

92. Sindhu, R.; Binod, P.; Pandey, A. Biological pretreatment of lignocellulosic biomass—An overview. *Bioresour. Technol.* **2016**, *199*, 76–82. [CrossRef] [PubMed]

93. Cao, G.; Ximenes, E.; Nichols, N.N.; Zhang, L.; Ladisch, M. Biological abatement of cellulase inhibitors. *Bioresour. Technol.* **2013**, *146*, 604–610. [CrossRef] [PubMed]

94. Chantho, P.; Musikavong, C.; Suttinun, O. Removal of phenolic compounds from palm oil mill effluent by thermophilic *Bacillus thermoleovorans* strain A2 and their effect on anaerobic digestion. *Int. Biodeterior. Biodegrad.* **2016**, *115*, 293–301. [CrossRef]

95. Borja, R.; Martin, A.; Alonso, V.; Garcia, I.; Banks, J. Influence of different aerobic pretreatments on the kinetics of anaerobic digestion of olive mill wastewater. *Water Res.* **1995**, *29*, 489–495. [CrossRef]

96. Panagiotou, G.; Olsson, L. Effect of compounds released during pretreatment of wheat straw on microbial growth and enzymatic hydrolysis rates. *Biotechnol. Bioeng.* **2006**, *96*, 250–258. [CrossRef] [PubMed]

97. Haghighi Mood, S.; Hossein Golfeshan, A.; Tabatabaei, M.; Salehi Jouzani, G.; Najafi, G.H.; Gholami, M.; Ardjmand, M. Lignocellulosic biomass to bioethanol, a comprehensive review with a focus on pretreatment. *Renew. Sustain. Energy Rev.* **2013**, *27*, 77–93. [CrossRef]

98. Pienkos, P.T.; Zhang, M. Role of pretreatment and conditioning processes on toxicity of lignocellulosic biomass hydrolysates. *Cellulose* **2009**, *16*, 743–762. [CrossRef]

![energies logo] *energies*

MDPI

Article

Performance Evaluation of Mesophilic Anaerobic Digestion of Chicken Manure with Algal Digestate

Na Duan [1,*], Xia Ran [1], Ruirui Li [1], Panagiotis G. Kougias [2], Yuanhui Zhang [1,3], Cong Lin [1] and Hongbin Liu [4,*]

[1] College of Water Resources and Civil Engineering, China Agricultural University, Beijing 100083, China; roxanna1991@163.com (X.R.); liruirui2007.love@163.com (R.L.); yzhang1@illinois.edu (Y.Z.); lincong@cau.edu.cn (C.L.)
[2] Department of Environmental Engineering, Technical University of Denmark, Kgs. DK-2800 Lyngby, Denmark; panak@env.dtu.dk
[3] Department of Agricultural and Biological Engineering, University of Illinois at Urbana-Champaign, Urbana, IL 61801, USA
[4] Key Laboratory of Nonpoint Source Pollution Control, Ministry of Agriculture, Beijing 100081, China
* Correspondence: duanna@cau.edu.cn (N.D.); liuhongbin@caas.cn (H.L.); Tel.: +86-10-6273-7329 (N.D.); +86-10-8210-8763 (H.L.)

Received: 4 June 2018; Accepted: 27 June 2018; Published: 12 July 2018

Abstract: Dilution is considered to be a fast and easily applicable pretreatment for anaerobic digestion (AD) of chicken manure (CM), however, dilution with fresh water is uneconomical because of the water consumption. The present investigation was targeted at evaluating the feasibility and process performance of AD of CM diluted with algal digestate water (AW) for methane production to replace tap water (TW). Moreover, the kinetics parameters and mass flow of the AD process were also comparatively analyzed. The highest methane production of diluted CM (104.39 mL/g volatile solid (VS)) was achieved with AW under a substrate concentration of 8% total solid (TS). The result was markedly higher in comparison with the group with TW (79.54–93.82 mL/gVS). Apart from the methane production, considering its energy and resource saving, nearly 20% of TW replaced by AW, it was promising substitution to use AW for TW to dilute CM. However, the process was susceptible to substrate concentration, inoculum, as well as total ammonia and free ammonia concentration.

Keywords: anaerobic digestion; chicken manure; biogas; algal digestate; mass flow

1. Introduction

Livestock manure without appropriate management can cause serious problems to the environment, such as odor, attraction of insects, rodents, and other pests, release of animal pathogens, as well as surface and groundwater pollution [1]. Anaerobic digestion (AD) is considered to be an attractive and efficient technology for livestock manure treatment, apart from the main target of organic matter removal and environment pollution control, simultaneously producing biogas for local energy needs [2]. Chicken manure (CM) with an original dry matter of 20–25% or more, has a high fraction of biodegradable organic matter [3,4]. Thus, conversion of the organic matter of CM to renewable energy through the AD process will not only reduce the adverse impact on the environment, but will also make great contributions to the energy supply [5].

Although AD technology in livestock manure treatment for biogas production is very mature and considerable research has been intensively conducted [6,7], limited studies can be found on the AD of CM, especially mono-digestion [4,8]. The AD of original CM with a low carbon to nitrogen (C/N) ratio of 5–10 usually ends up with reactor instability, and even failure, due to its inactive enzymes, affecting material transportation and inhibiting methanogenic microflora as a result of free ammonia (FA) accumulation [9–12]. Total ammonia nitrogen (TAN), produced through biological degradation of nitrogenous matter, was made up of ammonium ions (NH_4^+) and FA. Both forms can directly and indirectly lead to inhibition in the AD process [12]. For one thing, the partitioning between these forms was related to temperature and pH [10]. Ammonia inhibition was proven to occur in the range of 1500–3000 mg/L TAN with pH above 7.4, whereas if the TAN concentrations were in excess of 3 g/L, ammonia was claimed to be toxic irrespective of pH [5,13]. In addition, acclimated or unacclimated inoculum used was also an important factor for ammonia tolerance [14]. TAN concentration of 1700–1800 mg/L was completely inhibitory with unacclimated inoculum under mesophilic conditions [12]. However, with acclimation, inhibitory TAN levels could increase up to 7000 mg/L or more [15], and 100% ammonia inhibition occurred in the range of 8000–13,000 mg/L depending on the acclimatization condition and the pH of the reactor [16]. Thus, it is important to state the temperature, pH conditions, and inoculum quality while reporting the ammonia inhibition thresholds.

Several attempts were proposed to avoid the ammonia accumulation during the AD process of CM. Co-digestion with different carbon-rich biomasses to achieve a favorable C/N ratio was tested, such as co-digestion with hog wastes [6], municipal solid waste [17], and chicken processing waste [18]. Dilution with water to reduce high TS concentration of the original CM (20–25%) to a lower and appropriate one (0.5–3%) was investigated to eliminate ammonia inhibition [19], which has been widely used in the actual applications. However, dilution with fresh water is considered to be uneconomical due to the water consumption and the subsequent treatment of large amounts of effluent [20,21]. Meanwhile, fresh water cannot enhance the feedstock biodegradability or the operational efficiency of the AD process [10]. Therefore, wastewater containing certain amounts of organic matter has been used to dilute organic waste to produce biogas because the goals of enhancing methane production, recycling wastewater, and water saving can be achieved simultaneously, such as seawater for camel excrement [22] and sugar mill wastewater for rice straw and cow dung [19]. However, there is a lack of knowledge about the effect of different diluents on the methane production and stability in mono-digestion of CM at different substrate concentration levels.

Liquid digestate, as an important byproduct of AD process, is rich in recalcitrant organic compounds (total nitrogen of 139–3456 mg/L, total phosphorus of 7–381 mg/L) and its management is considered as the bottleneck for the biogas industry [23]. Microalgae can utilize carbon, nitrogen, phosphorus, and other nutrients from digestate, enhancing microalgae growth, as well as reducing cultivation costs and environmental impacts [24]. *Chlorella* 1067 can recover most of the nutrients in CM-based digestate with its high protein content of 47.34%TS and a relatively low C/N of 5.97 [25]. However, the separation of *Chlorella* 1067 and digestate involves additional energy consumption and the substantial amounts of effluent should be further treated.

In this study, two kinds of diluents (tap water and algal digestate water) were investigated in terms of the methane production potential and process performance. Results were modelled with different kinetic models to validate the effect of different diluents on the ultimate methane yield. Mass flow analysis was also performed to assess and compare the feasibility for further application. Moreover, the multi-inhibited factors in the AD process of diluted CM were investigated, such as ammonia, substrate concentration, and inoculum.

2. Results and Discussion

2.1. Methane Production

The maximal methane production (104.39 mL/gVS) of diluted CM was obtained with AW at TS of 8% (Figure 1), which was 11.27% higher than that of the best performed TW group at TS of 10% (93.82 mL/gVS). Apart from the treatment of AW at TS of 8%, the methane production of the AW group were all less than that of the TW group under the same TS levels. The AD process was found to be very sensitive to the feeding substrate composition [26]. For diluted CM, it was proved that the characteristics of different diluents, especially the nutrient composition, could affect the AD process. Compared with TW, AW, as a mixture of digestate and *Chlorella* 1067, contains a certain amount of organic matter (TS: 0.21–0.47%, VS: 44.66–52.62%TS), as well as a number of metal elements and trace elements. It is worth mentioning that algae biomass had been widely used for co-substrate in the AD process, and it demonstrated that synergism had occurred while *Chlorella* 1067 was used as a co-feedstock for CM [25]. In this study, the synergism might be achieved at the low substrate concentration of 8% TS. In addition, high nitrogen substrates, like CM, can pose an inhibitory effect in the AD process through NH_3 accumulation [18]. While the ammonia inhibition occurs, the pathway of acetate shifts from acetoclastic to hydrogenotrophic methanogenesis [10,27]. However, for AW, it can provide the lacking nutritional elements to synthesize the enzymes needed in syntrophic hydrogenotrophic methanogenesis, especially at the essential stage for propionic acid degradation [10].

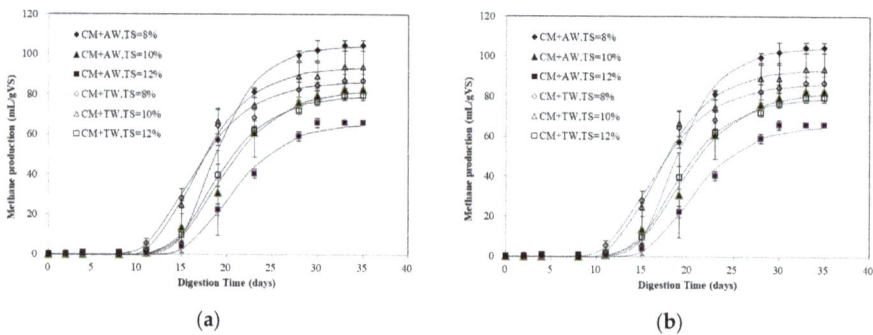

Figure 1. Methane accumulation fitted by the modified Gompertz model (**a**) and logistic model (**b**).

The methane yield of diluted CM obtained in the present study was lower than those reported in the literature (270–400 mL/gVS) [20,21,28]. It should be pointed out that anaerobic digestion is a dynamic process influenced by several parameters, like inoculum source, inoculum substrate ratio, nutrient, and others. Moset et al. [29] reported that the inoculum source had a significant impact on methane potential and the inoculum to substrate ratio (ISR) was dependent on the substrate. Due to the long time for the microbial population to adapt and degrade the substrate, the longer lag phase was observed at lower ISR (0.25). This was in line with the result of the present study (Figure 1). Li et al. [28] pointed out that the suitable substrate-to-inoculum ratio for mono-digestion of CM was 1.5. In this study, a low C/N ratio of CM, high ammonia concentration, poorly-adapted inoculum, and low ISR (<0.25) may be the reason for the low methane production. The amount of microorganisms in the digester was increased with higher ISR and it was beneficial for a well-balanced startup of the anaerobic reaction and inhibition resilience [18,30]. Thus, it is very important to provide the necessary microorganisms with a high ISR and a low substrate concentration to reduce feeding shock and inhibition risk, as well as shorten the adaptation time of the microbial population.

2.2. Kinetic Analysis

The experimental data were modelled with the modified Gompertz model and logistic model (Figure 1) and the results are summarized in Table 1. The results of the two models and the experimental data were in high agreement ($p > 0.05$). The R^2 values of the modified Gompertz model were all higher than 0.992, with the difference between the measured and predicted methane production ranging from <0.40% to <1.85%. For the logistic model, the R^2 values and the difference percentage were 0.9908–0.9970 and 0.08–0.76%, respectively.

Table 1. Accumulated methane production, methane production rate (R_m), and duration of the lag phase (λ) predicted by the modified Gompertz model and the logistic model.

Kinetic Parameters	Unit	CM + AW, TS			CM + TW, TS		
		8%	10%	12%	8%	10%	12%
				Modified Gompertz Model			
R_m	mL/(gVS.d)	12.07 ± 0.35	8.54 ± 1.46	6.22 ± 0.02	7.81 ± 0.35	8.96 ± 0.54	7.71 ± 0.51
λ	d	14.69 ± 0.10	14.96 ± 2.71	15.73 ± 0.19	11.37 ± 0.40	12.17 ± 0.23	14.03 ± 0.85
R^2		>0.9973	>0.9963	>0.9957	>0.9924	>0.9947	>0.9984
Measured methane yield	mL/(gVS)	104.39 ± 2.54	82.74 ± 0.44	65.89 ± 0.02	86.77 ± 3.63	93.82 ± 7.97	79.54 ± 2.23
Predicted methane yield	mL/(gVS)	104.03 ± 2.02	82.15 ± 0.34	64.67 ± 0.03	86.33 ± 2.86	93.39 ± 6.30	78.54 ± 1.68
Difference	%	<0.40	<0.74	<1.85	<0.64	<0.69	<1.44
				Logistic Model			
R_m	mL/(gVS.d)	11.56 ± 0.30	8.54 ± 1.53	5.90 ± 0.08	7.80 ± 0.55	8.87 ± 0.60	7.48 ± 0.61
λ	d	14.72 ± 0.09	15.11 ± 2.67	15.80 ± 0.16	11.48 ± 0.35	12.25 ± 0.18	14.08 ± 0.84
R^2		>0.9920	>0.9962	>0.9970	>0.9908	>0.9912	>0.9966
Measured methane yield	mL/(gVS)	104.39 ± 2.54	82.74 ± 0.44	65.89 ± 0.02	86.77 ± 3.63	93.82 ± 7.97	79.54 ± 2.23
Predicted methane yield	mL/(gVS)	104.32 ± 2.06	82.62 ± 0.35	65.39 ± 0.02	86.71 ± 2.95	93.75 ± 6.46	79.21 ± 1.76
Difference	%	<0.08	<0.15	<0.76	<0.09	<0.13	<0.49

An obvious lag phase was detected since less than 10% of the final cumulative methane production of diluted CM was obtained within the first ten days (Figure 1). Obviously, the inoculum was not adapted to afford digesting the diluted CM, so the AD process was in an inhibited steady-state or inhibited stage within the first several days, and it was expected to observe an initial lag phase in each trial. Additionally, uric acid breakdown causes the ammonia accumulation. CM contains a high content of the undigested proteins, uric acid, and nitrogen contents [5]. While anaerobically decomposed, the uric acid could significantly contribute to the formation of ammonia: 0.75 parts methane, 4.25 parts carbon dioxide, and four parts of ammonia would be produced by one part uric acid [4]. It should be noted that the AW group showed a lag phase of 14–16 days for both the modified Gompertz model and the logistic model, longer than that of the TW group (Table 1). *Chlorella* 1067, cultivated in digestate, was also a nitrogen-rich biomass like CM [25]. Thus, with dilution with AW (TAN 268.24–485.50 mg/L), some additional nitrogen might have been added and increased the inhibition risk. The methane production gradually increased after an obvious lag phase. The maximal methane production of CM was achieved with AW at TS of 8%, indicating that AW and CM had a positive effect on the methane production at a low substrate concentration. In addition, for nitrogen-rich CM with a high substrate concentration, the AD process was hampered by FA inhibition resulting in a poor methane yield and, consequently, a long retention time. These results were consistent with the study of Bujoczek et al. on the high solid AD of CM, and it was reported that highest gas production was obtained at a TS level of 5% [4].

2.3. pH, TAN, FA Concentration, and C/N of Liquid Digestate

The TAN concentration of diluted CM with AW and TW at different substrate concentration was presented in Figure 2a. Under mesophilic conditions, TAN concentrations continued to rise within 17 days and had a range of 1970 to 2460 mg/L at the 17th day. It was confirmed that the process was not adapted to high TAN concentrations as evidenced by the long lag phase and low methane production in the early stage (Figure 1 and Table 1). Due to the characteristics of AW, the initial TAN concentration of the AW group was similar, or slightly higher, than that of the TW group at the same TS levels. It was interesting that the TAN concentration was found to be similar for both the AW group and the TW group as the AD process went on. In addition, it was found that the higher substrate concentration, the higher the TAN concentration, so dilution was an alternative to reduce the risk of TAN inhibition during CM mono-digestion. Additionally, the low substrate concentration was a benefit for startup, process stability, diluting ammonia toxicity, and even methane production.

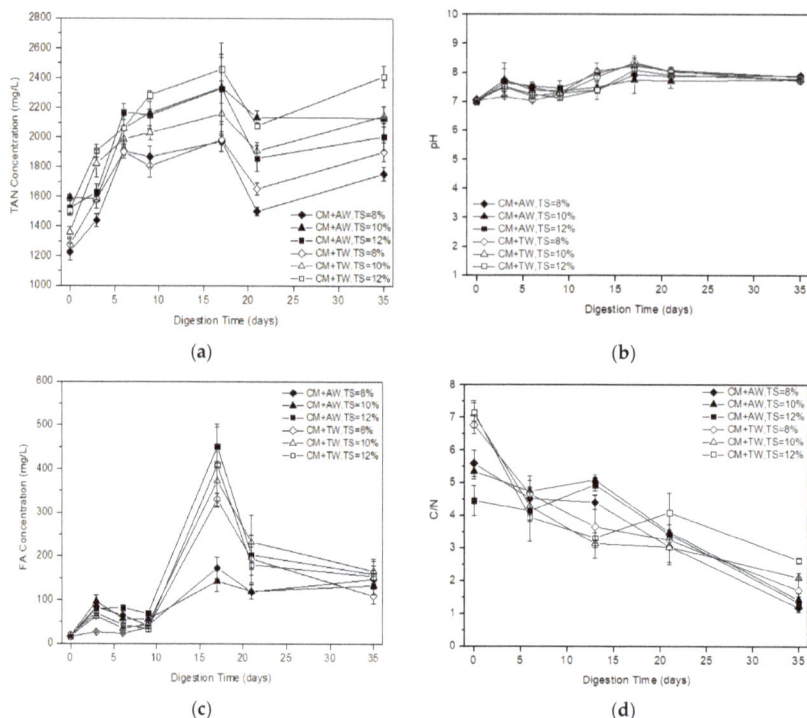

Figure 2. TAN (**a**), pH (**b**), FA (**c**), and C/N (**d**) profile with digestion time for diluted CM.

In the first nine days, FA concentration gradually increased within the range of 100 mg/L, and then increased significantly with the increasing pH value and TAN concentration (Figure 2a–c). As the TAN and FA increased, the pH value also increased, which was also reported by Hassan et al. [31]. The maximal FA values for all the treatments were found in the 17th day. Apart from the FA concentration (110–200 mg/L) of the AW group with substrate concentrations of 8% and 10% TS, that of the other treatments were all more than 300 mg/L. Different studies reported that below 99–150 mg/L was the acceptable limit for the methanogens [32,33]. Therefore, the FA concentrations of all the treatments were over the inhibition threshold, which might result in an unstable AD process due to the loss of methanogenic activity [34]. After 17 days, the microorganisms gradually adapt to the environment, the FA concentration began to decrease and the methane yield rapidly increased. At the end of the experiment, the FA concentration was decreased at about 150 mg/L for each treatment.

The methane production of diluted CM was low in this study. The effect would be carried out from two negative aspects. One was the higher FA concentration caused by the raw material characteristics, high initial substrate concentration, and low ISR. The other was the low C/N ratio of the liquid fractions of the digested feedstock (Figure 2d). Specifically, the initial C/N ratio of the AW group (4.46–5.60) was lower than that of the TW group (6.77–7.16), while there was only the diluent variable at this time due to the high TN concentration of AW. However, they were both outside the ideal range of the C/N ratio of 20–30 [35]. This was mainly caused by intrinsic characteristics of CM. In addition, the best performed methane production and degradation efficiency of diluted CM with AW was mainly attributable to the synergetic effect and the nutrient composition of CM and AW, so the AD process was not only influenced by the C/N ratio. The results were in line with the study of Wang et al. [33]. During the AD process, the TOC concentration decreased and the TAN concentration increased, resulting in the decreased C/N ratio.

2.4. TOC Concentration and SCOD Removal

It was obvious that TOC concentration of the AW group and the TW group had a similar trend (Figure 3a). The TOC concentration of each treatment was nearly over 10,000 mg/L within the first 13 days. Subsequently, the TOC concentration began to rapidly decrease and the methane yield increased (Figure 1). At the end of the experiment, the lowest TOC concentration of diluted CM was obtained with AW at a TS of 8% and kept at about 3850 mg/L. TOC removal of each treatment are more than 60%.

Results showed that the SCOD removal decreased with the increasing substrate concentration (Figure 3b). The maximum SCOD removal of 79.79% was achieved by the diluted CM with AW of 8%TS, which was in line with the results of methane production (Figure 1). Compared with other treatments, it had relatively high AD efficiency and less residual organic matter for further treatment. In addition, the importance of the substrate concentration for methane production was proved, as well as for organic matter degradation.

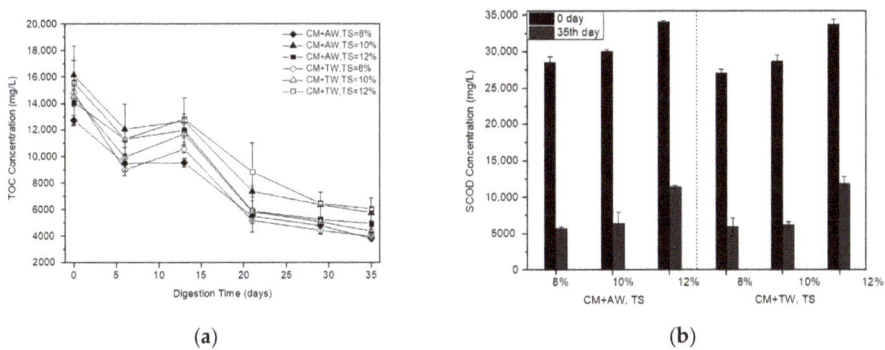

(a)   (b)

Figure 3. TOC concentration (**a**) and SCOD removal (**b**) profile with the digestion time for diluted CM.

2.5. Mass Flow and Water Substitution

Based on the experimental data, the feasibility of using AW to dilute CM for anaerobic digestion was determined. A closed loop process could be formed, including digestate reuse and diluting CM with algae digestate water for biogas production (Figure 4). CM was firstly fed for AD process to produce biogas. Then, the liquid digestate was recycled as media to cultivate *Chlorella* 1067 and the unseparated algae digestate water was then used as the diluent for regulating CM to the targeted feeding concentration.

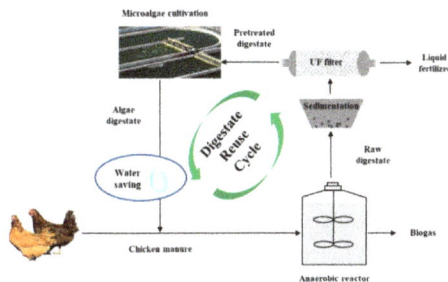

Figure 4. A closed loop system of CM diluted by algae digestate water.

A hypothesized process with 1.0 t/day of CM and a substrate concentration of 8%TS was used for calculation. The values of TS, VS, and methane production were all based on the experimental data. From a mass flow perspective, using AW in place of TW to dilute CM was advantageous (Figure 5). The AD process generated 0.04 t biogas, 0.32 t solid digestate and 2.34 t liquid digestate. For using TW, 1.70 t tap water was needed to dilute CM achieving the substrate concentration of 8%TS per day. However, using AW, 0.34 t liquid digestate and 1.36 t tap water would be used, equivalently using 14.53% of liquid digestate and saving 20% of tap water per day. Hence, the current observations indicated that using AW for diluting CM was feasible for methane production coupled with digestate reuse and saving clean water. For the whole year, 124.1 t of liquid digestate would be utilized, simultaneously saving the same amount of clean water.

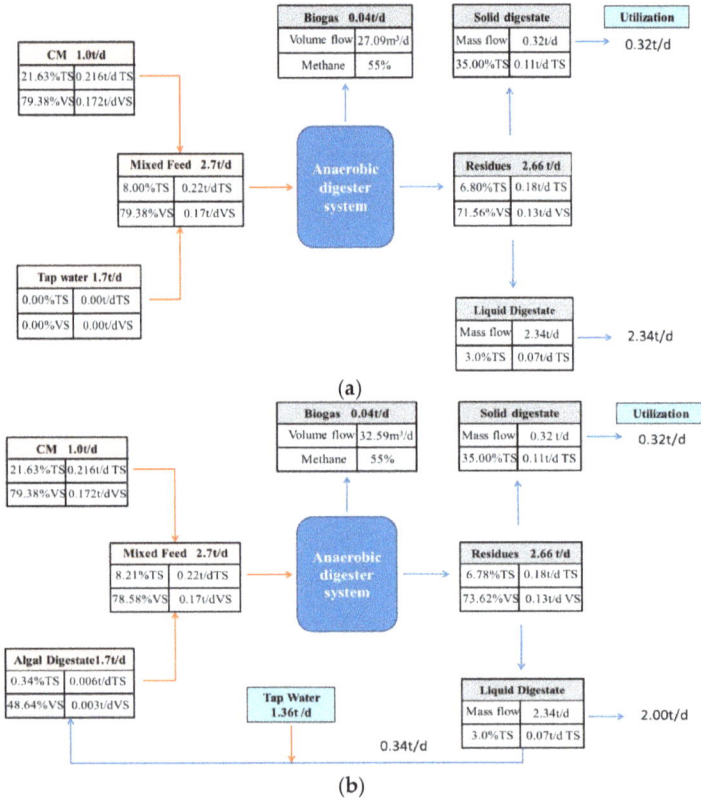

Figure 5. Mass flow of CM digestion with TW (**a**) and AW (**b**).

3. Materials and Methods

3.1. Substrate, Inoculum, and Diluent Characteristics

The original CM (TS: 21.63%; VS: 79.38%TS) was obtained from Minhe Co., Ltd. (Penglai, Shandong, China), and stored at 4 °C before use. The C/N ratio of the CM was 9.28. Two different diluents were used: one was tap water (TW) and the other was algal digestate water (AW). *Chlorella* 1067 was cultivated in the CM-based liquid digestate obtained from a mesophilic anaerobic digester at Minhe Co., Ltd. after ultra-filtration treatment in a 400 L open raceway pond. The cultivation conditions were the same as that reported by Li et al. [25]. Then the mixture of *Chlorella* 1067 and digestate without separation was called as AW (TS: 0.21–0.47%, VS: 44.66–52.62%TS, TAN 268.24–485.50 mg/L) and directly used as the diluent in this study. AW also contained some metallic elements and trace elements, such as calcium (Ca), cobalt (Co), iron (Fe), magnesium (Mg), zinc (Zn), potassium (K), manganese (Mn), and others derived from the CM-based digestate. Anaerobically-digested sludge (ADS) was used as inoculum and was taken from a mesophilic anaerobic digester in the Little Red Door sewage treatment plant (Beijing, China). The ADS (TS: 1.47%, VS: 98.53%TS) was enriched with glucose in the lab under mesophilic conditions prior to use.

3.2. Experimental Design

The raw CM was diluted to 8%, 10%, and 12% (based on TS) from the original CM using AW and TW, respectively. Subsequently, CM was mixed with the ADS in a ratio of 2:5 (40 g:100 g) based on wet weight and placed in a set of 250 mL anaerobic reactors sealed with rubber stoppers. There were two tubes in the rubber stopper, one was above the liquid level to connect with the gas bag to accumulate the biogas produced, and the other was below the liquid level for taking the liquid sample. Two milliliters of liquid sample was taken once every 3–5 days during days 1–21, and once at the 35th day, respectively. The batch reactors were operated under mesophilic conditions (35 ± 1) °C. Seven trials were conducted, including three for diluted CM with AW at TS of 8, 10, and 12%; another three for the TW with the same TS range; and one only with inoculum was operated as a blank to correct the methane production of the AW group and TW group. Batch experiments were operated in a thermostat oscillator (SHA-B, Changzhou Guohua Electric Appliance Co., Ltd., Jintan, Jiangsu Province, China) at the target temperature. The batch experiments lasted 35 days. All trials were conducted in triplicate and all the results were reported as mean values with standard deviations.

3.3. Analytical Methods

The TS and VS were performed according to the APHA standard methods [36]. Element components (C, H, O, and N) of the CM were determined using an elemental analyzer (Vario MICRO Cube, Elementar Analysensysteme GmbH, Donaustraße, Germany).

The amount of CH_4 and CO_2 was determined by a gas chromatograph (GC 1490, Agilent Technologies, Santa Clara, CA, USA) equipped with a thermal conductivity detector and nitrogen gas as the carrier gas at a flow rate of 50 mL/min. The injector, oven, and detector temperatures were 150, 120, and 150 °C, respectively. The pH value was measured using a digital pH meter (FE20, Mettler Toledo Co., Inc., Shanghai, China). The total nitrogen (TN) was measured by the method of potassium persulfate oxidation using an UV–VIS spectrophotometer (UV-1800, Meipuda Instruments Co., Ltd., Shanghai, China). The TAN was determined using salicylic acid-hypochlorite spectrophotometry. The total organic carbon (TOC) was tested using a TOC analysis meter (TOC-VCPN, Shimadzu Company, Tokyo, Japan). The soluble chemical oxygen demand (SCOD) was measured using a potassium permanganate oxidation method. The free ammonia (FA) concentration was calculated based on TAN concentration and experimental conditions including temperature and pH using Equation (1) [37]:

$$\frac{NH_3}{TAN} = (1 + \frac{10^{-pH}}{10^{-(0.09018 + \frac{2729.92}{T(k)})}})^{-1} \tag{1}$$

where NH_3 is the concentration of FA in mg/L, TAN is the total ammonia nitrogen concentration in mg/L, pH is the pH value determined in the reactor, and $T(k)$ is the temperature (Kelvin).

3.4. Kinetics Analysis

Methane production and lag phase are both important factors in determining the efficiency of the AD process, thus, the modified Gompertz model (Equation (2)) [38] and logistic model (Equation (3)) [38] were applied to fit the lag time, methane production rate, and predict the methane production potential as a function of time:

$$P(t) = P_{max}exp\{-exp[R_me(\lambda - t)/P_{max} + 1]\} \tag{2}$$

$$P(t) = P_{max}/\{1 + exp[4R_m(\lambda - t)/P_{max} + 2]\} \tag{3}$$

where $P(t)$ is the cumulative methane production (mL/(gVS)) at a given time t(d). P_{max} is the maximum accumulative methane potential (mL/(gVS)). R_m is the maximum methane production rate (mL/(gVS.d)). λ is the lag phase (d) and e is the base of the natural logarithm (2.71828).

Data analysis for the two models was carried out with the Solver function of Microsoft Excel 2010 (Microsoft Corporation, Redmond, WA, USA) and the correlation coefficient (R^2) was used to evaluate the accuracy of the fitting results.

4. Conclusions

The novel idea of replacing TW with AW, to dilute CM for anaerobic digestion is feasible. The highest methane production of diluted CM (104.39 mL/gVS) was achieved with an AW of 8%TS. The result was markedly higher than that of the TW group (79.54–93.82 mL/gVS). A closed loop process can, thus, be built including digestate reuse and diluting CM with algae digestate water for biogas production. Thereby two benefits will be obtained. One is to enhance methane yield, recycle AW, and reduce water consumption; another is to decrease the energy consumption for separating microalgae and digestate. However, the AD process was susceptible to substrate concentration, the inoculum to substrate ratio, as well as TAN and FA concentrations. In consideration of the low C/N ratio of CM and AW, additional carbon sources can be added to improve the methane production, achieve nutrient balance, and reduce inhibition risk in future studies.

Author Contributions: The paper was a collaborative effort among the authors. Conceptualization: N.D. and Y.Z.; formal analysis: P.G.K.; funding acquisition: H.L.; investigation: X.R. and R.L.; writing—original draft: N.D.; and writing—review and editing: Y.Z., C.L., and H.L.

Funding: This research was funded by the National Natural Science Foundation of China (51506217) and the open fund of the Key Laboratory of Nonpoint Source Pollution Control, Ministry of Agriculture, P. R. China.

Acknowledgments: The authors also acknowledged the support by Shandong Minhe Biotech Limited Company.

Conflicts of Interest: The authors declare no conflict of interest.

References

1. Sakar, S.; Yetilmezsoy, K.; Kocak, E. Anaerobic digestion technology in poultry and livestock waste treatment—A literature review. *Waste Manag. Res.* **2009**, *27*, 3–18. [CrossRef] [PubMed]
2. Nasir, I.M.; Ghazi, T.I.M.; Omar, R. Anaerobic digestion technology in livestock manure treatment for biogas production: A review. *Eng. Life Sci.* **2012**, *12*, 258–269. [CrossRef]
3. Abouelenien, F.; Nakashimada, Y.; Nishio, N. Dry mesophilic fermentation of chicken manure for production of methane by repeated batch culture. *J. Biosci. Bioeng.* **2009**, *107*, 293–295. [CrossRef] [PubMed]
4. Bujoczek, G.; Oleszkiewicz, J.; Sparling, R.; Cenkowski, S. High solid anaerobic digestion of chicken manure. *J. Agric. Eng. Res.* **2000**, *76*, 51–60. [CrossRef]
5. Abouelenien, F.; Fujiwara, W.; Namba, Y.; Kosseva, M.; Nishio, N.; Nakashimada, Y. Improved methane fermentation of chicken manure via ammonia removal by biogas recycle. *Bioresour. Technol.* **2010**, *101*, 6368–6373. [CrossRef] [PubMed]
6. Magbanua, B.S., Jr.; Adams, T.T.; Johnston, P. Anaerobic co-digestion of hog and poultry waste. *Bioresour. Technol.* **2001**, *76*, 165–168. [CrossRef]
7. Boe, K.; Angelidaki, I. Serial CSTR digester configuration for improving biogas production from manure. *Water Res.* **2009**, *43*, 166–172. [CrossRef] [PubMed]

8. Niu, Q.; Takemura, Y.; Kubota, K.; Li, Y.Y. Comparing mesophilic and thermophilic anaerobic digestion of chicken manure: Microbial community dynamics and process resilience. *Waste Manag.* **2015**, *43*, 114–122. [CrossRef] [PubMed]

9. Kadam, P.C.; Boone, D.R. Influence of pH on ammonia accumulation and toxicity in halophilic, methylotrophic methanogens. *Appl. Environ. Microbiol.* **1996**, *62*, 4486–4492. [PubMed]

10. Sun, C.; Cao, W.X.; Banks, C.J.; Heaven, S.; Liu, R.H. Biogas production from undiluted chicken manure and maize silage: A study of ammonia inhibition in high solids anaerobic digestion. *Bioresour. Technol.* **2016**, *218*, 1215–1223. [CrossRef] [PubMed]

11. Yangin-Gomec, C.; Ozturk, I. Effect of maize silage addition on biomethane recovery from mesophilic co-digestion of chicken and cattle manure to suppress ammonia inhibition. *Energ. Convers. Manag.* **2013**, *71*, 92–100. [CrossRef]

12. Yenigün, O.; Demirel, B. Ammonia inhibition in anaerobic digestion: A review. *Process Biochem.* **2013**, *48*, 901–911. [CrossRef]

13. Calli, B.; Mertoglu, B.; Inanc, B.; Yenigun, O. Effect of high free ammonia concentrations on the performances of anaerobic bioreactors. *Process Biochem.* **2005**, *40*, 1285–1292. [CrossRef]

14. Angelidaki, I.; Ellegaard, L.; Ahring, B.K. Applications of the anaerobic digestion process. In *Biomethanation II. Advances in Biochemical Engineering/Biotechnology*; Ahring, B.K., Ed.; Springer: Berlin/Heidelberg, Germany, 2003; pp. 1–33.

15. Tian, H.L.; Fotidis, I.A.; Mancini, E.; Treu, L.; Mahdy, A.; Ballesteros, M.; González-Fernández, C.; Angelidaki, I. Acclimation to extremely high ammonia levels in continuous biomethanation process and the associated microbial community dynamics. *Bioresour. Technol.* **2018**, *247*, 616–623. [CrossRef] [PubMed]

16. Sung, S.; Liu, T. Ammonia inhibition on thermophilic anaerobic digestion. *Chemosphere* **2003**, *53*, 43–52. [CrossRef]

17. Matheri, A.N.; Ndiweni, S.N.; Belaid, M.; Muzenda, E.; Hubert, R. Optimising biogas production from anaerobic co-digestion of chicken manure and organic fraction of municipal solid waste. *Renew. Sustain. Energy Rev.* **2017**, *80*, 756–764. [CrossRef]

18. Li, C.; Strömberg, S.; Liu, G.J.; Negs, I.V.; Liu, J. Assessment of regional biomass as co-substrate in the anaerobic digestion of chicken manure: Impact of co-digestion with chicken processing waste, seagrass and Miscanthus. *Biochem. Eng. J.* **2017**, *118*, 1–10. [CrossRef]

19. Goel, A.; Gupta, S. Biogas from organic waste diluted with sugar mill waste water. *Asian J. Chem.* **2007**, *19*, 3435–3439.

20. Niu, Q.; Qiao, W.; Qiang, H.; Hojo, T.; Li, Y.Y. Mesophilic methane fermentation of chicken manure at a wide range of ammonia concentration: Stability, inhibition and recovery. *Bioresour. Technol.* **2013**, *137*, 358–367. [CrossRef] [PubMed]

21. Nie, H.; Jacobi, H.F.; Strach, K.; Xu, C.; Zhou, H.; Liebetrau, J. Mono-fermentation of chicken manure: Ammonia inhibition and recirculation of the digestate. *Bioresour. Technol.* **2015**, *178*, 238–246. [CrossRef] [PubMed]

22. Gamal-El-Din, H. Biogas from organic waste diluted with seawater. In *Biogas Technology, Transfer and Diffusion*; El-Halwagi, M.M., Ed.; Springer: Dordrecht, The Netherlands, 1986; pp. 417–423.

23. Xia, A.; Murphy, J.D. Microalgal cultivation in treating liquid digestate from biogas systems. *Trends Biotechnol.* **2016**, *34*, 264–275. [CrossRef] [PubMed]

24. Uggetti, E.; Sialve, B.; Latrille, E.; Steyer, J.P. Anaerobic digestate as substrate for microalgae culture: The role of ammonium concentration on the microalgae productivity. *Bioresour. Technol.* **2014**, *152*, 437–443. [CrossRef] [PubMed]

25. Li, R.R.; Duan, N.; Zhang, Y.H.; Liu, Z.D.; Li, B.M.; Zhang, D.M.; Lu, H.F.; Dong, T.L. Co-digestion of chicken manure and microalgae *Chlorella* 1067 grown in the recycled digestate: Nutrients reuse and biogas enhancement. *Waste Manag.* **2017**, *70*, 247–254. [CrossRef] [PubMed]

26. Li, J.; Wei, L.; Duan, Q.; Hu, G.; Zhang, G. Semi-continuous anaerobic co-digestion of dairy manure with three crop residues for biogas production. *Bioresour. Technol.* **2014**, *156*, 307–313. [CrossRef] [PubMed]

27. Karakashev, D.; Batstone, D.J.; Trably, E.; Angelidaki, I. Acetate oxidation is the dominant methanogenic pathway from acetate in the absence of Methanosaetaceae. *Appl. Environ. Microbiol.* **2006**, *72*, 5138–5141. [CrossRef] [PubMed]

28. Li, Y.Q.; Zhang, R.H.; Liu, X.Y.; Chen, C.; Xiao, X.; Feng, L.; He, Y.F.; Liu, G.Q. Evaluating methane production from anaerobic mono- and co-digestion of kitchen waste, corn stover, and chicken manure. *Energy Fuels* **2013**, *27*, 2085–2091. [CrossRef]

29. Moset, V.; Al-zohairi, N.; Møller, H.B. The impact of inoculum source, inoculum to substrate ratio and sample preservation on methane potential from different substrates. *Biomass Bioenergy* **2015**, *83*, 474–482. [CrossRef]

30. Raposo, F.; Banks, C.J.; Siegert, I.; Heaven, S.; Borja, R. Influence of inoculum to substrate ratio on the biochemical methane potential of maize in batch tests. *Process Biochem.* **2006**, *41*, 1444–1450. [CrossRef]

31. Hassan, M.; Ding, W.M.; Shi, Z.D.; Zhao, S.Q. Methane enhancement through co-digestion of chicken manure and thermo-oxidative cleaved wheat straw with waste activated sludge: A C/N optimization case. *Bioresour. Technol.* **2016**, *211*, 534–541. [CrossRef] [PubMed]

32. Ahring, B.K.; Angelidaki, I.; Johansen, K. Anaerobic treatment of manure together with industrial-waste. *Water Sci. Technol.* **1992**, *25*, 311–318. [CrossRef]

33. Wang, X.; Yang, G.; Feng, Y.; Ren, G.; Han, X. Optimizing feeding composition and carbon-nitrogen ratios for improved methane yield during anaerobic co-digestion of dairy, chicken manure and wheat straw. *Bioresour. Technol.* **2012**, *120*, 78–83. [CrossRef] [PubMed]

34. Koster, I.W.; Lettinga, G. Anaerobic digestion at extreme ammonia concentrations. *Biol. Wastes* **1988**, *25*, 51–59. [CrossRef]

35. Kayhanian, M. Ammonia inhibition in high-solids biogasification: An overview and practical Solutions. *Environ. Technol.* **1999**, *20*, 355–365. [CrossRef]

36. Eaton, A.D.; Clesceri, L.S.; Greenberg, A.E. *Standard Methods for the Examination of Water and Wastewater*, 21th ed.; American Public Health Association/American Water Works Association/Water Environment Federation: Washington DC, USA, 2005.

37. Hansen, K.H.; Angelidaki, I.; Ahring, B.K. Anaerobic digestion of swine manure: Inhibition by ammonia. *Water Res.* **1998**, *32*, 5–12. [CrossRef]

38. Tsapekos, P.; Kougias, P.G.; Egelund, H.; Larsen, U.; Pedersen, J.; Trénel, P. Mechanical pretreatment at harvesting increases the bioenergy output from marginal land grasses. *Renew. Energy* **2017**, *111*, 914–921. [CrossRef]

MDPI

St. Alban-Anlage 66

4052 Basel

Switzerland

Tel. +41 61 683 77 34

Fax +41 61 302 89 18

www.mdpi.com

Energies Editorial Office

E-mail: energies@mdpi.com

www.mdpi.com/journal/energies

MDPI
St. Alban-Anlage 66
4052 Basel
Switzerland

Tel: +41 61 683 77 34
Fax: +41 61 302 89 18

www.mdpi.com

MDPI

ISBN 978-3-03897-215-0

www.ingramcontent.com/pod-product-compliance
Lightning Source LLC
Chambersburg PA
CBHW051708210326
41597CB00032B/5410